Methods in Molecular Biology

Series Editor
John M. Walker
School of Life and Medical Sciences
University of Hertfordshire
Hatfield, Hertfordshire, UK

For further volumes:
http://www.springer.com/series/7651

For over 35 years, biological scientists have come to rely on the research protocols and methodologies in the critically acclaimed Methods in Molecular Biology series. The series was the first to introduce the step-by-step protocols approach that has become the standard in all biomedical protocol publishing. Each protocol is provided in readily-reproducible step-by step fashion, opening with an introductory overview, a list of the materials and reagents needed to complete the experiment, and followed by a detailed procedure that is supported with a helpful notes section offering tips and tricks of the trade as well as troubleshooting advice. These hallmark features were introduced by series editor Dr. John Walker and constitute the key ingredient in each and every volume of the Methods in Molecular Biology series. Tested and trusted, comprehensive and reliable, all protocols from the series are indexed in PubMed.

Manuscript Instructions https://drive.google.com/file/d/1K4Az5vpAFVzYadJ3jKx2-QSo7Dqa_SNA/view?usp=sharing

Cancer Epigenetics

Methods and Protocols

Edited by

Archana Dhasarathy

Department of Biomedical Sciences, University of North Dakota, Grand Forks, ND, USA

Motoki Takaku

Department of Biomedical Sciences, University of North Dakota, Grand Forks, ND, USA

Humana Press

Editors
Archana Dhasarathy
Department of Biomedical Sciences
University of North Dakota
Grand Forks, ND, USA

Motoki Takaku
Department of Biomedical Sciences
University of North Dakota
Grand Forks, ND, USA

ISSN 1064-3745 ISSN 1940-6029 (electronic)
Methods in Molecular Biology
ISBN 978-1-0716-5071-4 ISBN 978-1-0716-5072-1 (eBook)
https://doi.org/10.1007/978-1-0716-5072-1

This Humana imprint is published by the registered company Springer Science+Business Media, LLC, part of Springer Nature.
The registered company address is: 1 New York Plaza, New York, NY 10004, U.S.A.

Preface

Since the seminal discovery by Feinberg and Vogelstein that cancer cells are hypomethylated relative to their normal counterparts [1], the field of cancer epigenetics has proven to be transformative for modern oncology. Epigenetic mechanisms have recently been listed as one of the "hallmarks of cancer" [2]. Epigenetic changes, including DNA methylation, histone modifications, chromatin accessibility, and non-coding RNA-driven mechanisms, often precede genetic mutations and are detectable at the earliest stages of tumor development. In fact, a recent study by Parreno, Cavalli, and colleagues suggests that epigenetic changes, in the form of transcriptional gene silencing mediated by Polycomb group proteins, are enough to initiate cancer in the absence of genetic drivers [3]. Taken together, epigenetic mechanisms can offer profound insights into our understanding of how gene regulation, in addition to or by itself, can drive the initiation and progression of cancer. Therefore, epigenetic proteins are powerful biomarkers for the detection of cancer and are also strong candidates for developing therapeutic interventions.

The tools we use to study epigenetics in cancer, and manipulate the proteins involved, remain critical to our understanding of mechanisms in cancer initiation and progression. In this volume of *Methods in Molecular Biology*, we recognize that the tools we use shape the questions we are able to ask and answer. Since the dawn of massively parallel sequencing technologies, our approaches to studying cancer have dramatically increased, from analyzing single genes to spanning entire cancer genomes. Now, with single-cell, single-molecule, cell expansion, and spatial technologies at our fingertips, we can dissect even epigenomic heterogeneity of tumors with unprecedented resolution.

This book compiles protocols for detecting core epigenetic modifications, including DNA methylation and histone modifications, and highlights chromatin immunoprecipitation (ChIP), CUT&RUN, and advanced microscopy for visualizing chromatin architecture in situ. We include detailed protocols for exploring chromatin accessibility, such as ATAC-seq, single-molecule tracking, and specialized approaches, for studying transposable elements and single-cell multiomic landscapes. Finally, we present methods to detect long non-coding RNAs and microRNAs, and isolate them from challenging clinical samples like FFPE tissues. Other protocols include CRISPR-based tools, methods for translating epigenetic signatures into diagnostic and prognostic tools, such as using cell-free DNA and exosomal microRNAs, and high-throughput assays for identifying modulators of telomere maintenance pathways.

While we acknowledge that several other methods and tools exist, we hope this collection will be useful to researchers to uncover novel biomarkers and therapeutic targets, and will empower them to advance our understanding of cancer biology. We are deeply grateful to our contributors, many of whom have pioneered or refined the techniques described in this book. Their willingness to share detailed methodologies with their colleagues is a testament to the collaborative spirit that permeates science and helps advance its mission of exploration. We hope this volume will serve as a practical resource for researchers at all stages and prove to be a valuable guide on their journey to unlock the secrets of cancer epigenomes, and, ultimately, to improve patient outcomes.

Grand Forks, ND, USA — *Archana Dhasarathy*
Grand Forks, ND, USA — *Motoki Takaku*

References

1. Feinberg AP, Vogelstein B (1983) Hypomethylation distinguishes genes of some human cancers from their normal counterparts. Nature 301(5895):89–92. https://doi.org/10.1038/301089a0
2. Hanahan D (2022) Hallmarks of cancer: new dimensions. Cancer Discov 12(1):31–46. https://doi.org/10.1158/2159-8290.Cd-21-1059
3. Parreno V et al (2024) Transient loss of Polycomb components induces an epigenetic cancer fate. Nature 629(8012):688–696. https://doi.org/10.1038/s41586-024-07328-w

Contents

Contributors

TAIGA AJIRI • *Nagoya University, Nagoya, Aichi, Japan*

LIZ MARIE ALBERTORIO-SÁEZ • *EpiCypher Inc., Durham, NC, USA*

ELENI ANASTASIADOU • *Department of Clinical and Molecular Medicine, Sapienza University, Rome, Italy*

RHONDA BACHER • *Genetics Institute, UF Health Cancer Center, and Department of Biostatistics, College of Public Health & Health Professions, University of Florida, Gainesville, FL, USA*

COURTNEY A. BARNES • *EpiCypher Inc., Durham, NC, USA*

ZACHARY BIGELOW • *Department of Biomedical Sciences, School of Medicine and Health Sciences, University of North Dakota, Grand Forks, ND, USA*

JASON O. BRANT • *UF Health Cancer Center, and Department of Biostatistics, College of Public Health & Health Professions, University of Florida, Gainesville, FL, USA*

KUNANON CHATTRAIRAT • *Nagoya University, Nagoya, Aichi, Japan*

YU-LIN CHEN • *McArdle Laboratory for Cancer Research, Department of Oncology, University of Wisconsin-Madison, Madison, WI, USA*

THOMAS L. CLANTON • *Department of Applied Physiology and Kinesiology, College of Health and Human Performance, University of Florida, Gainesville, FL, USA*

PATRICK CONCANNON • *Genetics Institute, and Department of Pathology, Immunology, and Laboratory Medicine, College of Medicine, University of Florida, Gainesville, FL, USA*

MARTIS W. COWLES • *EpiCypher Inc., Durham, NC, USA*

CHRISTOPHER R. DAY • *Epigenetics and RNA Biology Laboratory, National Institute of Environmental Health Sciences (NIEHS), Research Triangle Park, Durham, NC, USA*

ARCHANA DHASARATHY • *Department of Biomedical Sciences, School of Medicine and Health Sciences, University of North Dakota, Grand Forks, ND, USA*

PHILIP A. EFRON • *Department of Surgery, College of Medicine and Sepsis and Critical Illness Research Center, University of Florida, Gainesville, FL, USA*

STEPHANIE ERAMO • *Structural Biology, Biochemistry, and Biophysics Graduate Program, University of Colorado Anschutz Medical Campus, Aurora, CO, USA; Department of Biochemistry and Molecular Genetics, University of Colorado Anschutz Medical Campus, Aurora, CO, USA*

KARLIE N. FEDDER-SEMMES • *EpiCypher Inc., Durham, NC, USA*

TESSA M. FIRESTONE • *EpiCypher Inc., Durham, NC, USA*

BROOKE FROELICH • *Department of Biomedical Sciences, School of Medicine and Health Sciences, University of North Dakota, Grand Forks, ND, USA*

MERRILL M. FRONEY • *Division of Chemical Biology and Medicinal Chemistry, Eshelman School of Pharmacy, University of North Carolina at Chapel Hill, Chapel Hill, NC, USA*

MARIE-PIERRE L. GAUTHIER • *Department of Biochemistry and Molecular Biology, College of Medicine, and UF Health Cancer Center, University of Florida, Gainesville, FL, USA*

CHITTIBABU GUDA • *Department of Genetics, Cell Biology, and Anatomy, University of Nebraska Medical Center, Omaha, NE, USA; Center for Biomedical Informatics Research and Innovation, University of Nebraska Medical Center, Omaha, NE, USA*

KAITLYN GURA • *Department of Biomedical Sciences, School of Medicine and Health Sciences, University of North Dakota, Grand Forks, ND, USA*
NATHAN W. HALL • *EpiCypher Inc., Durham, NC, USA*
ALLISON R. HICKMAN • *EpiCypher Inc., Durham, NC, USA*
MIKE HILL • *Department of Biomedical Sciences, School of Medicine and Health Sciences, University of North Dakota, Grand Forks, ND, USA*
SANKARASUBRAMANIAN JAGADESAN • *Department of Genetics, Cell Biology, and Anatomy, University of Nebraska Medical Center, Omaha, NE, USA*
YAGNA P. R. JARAJAPU • *School of Pharmacy, North Dakota State University, Fargo, ND, USA*
MICHAEL B. JARSTFER • *Division of Chemical Biology and Medicinal Chemistry, Eshelman School of Pharmacy, University of North Carolina at Chapel Hill, Chapel Hill, NC, USA*
ANDREA L. JOHNSTONE • *EpiCypher Inc., Durham, NC, USA*
MARK KADERLI • *EpiCypher Inc., Durham, NC, USA*
SRIKRISHNA KANDOORU • *Department of Biomedical Sciences, School of Medicine and Health Sciences, University of North Dakota, Grand Forks, ND, USA*
MICHAEL-CHRISTOPHER KEOGH • *EpiCypher Inc., Durham, NC, USA*
MICHAEL P. KLADDE • *Department of Biochemistry and Molecular Biology, College of Medicine, Genetics Institute, UF Health Cancer Center, and Sepsis and Critical Illness Research Center, University of Florida, Gainesville, FL, USA*
RAHUL M. KOHLI • *Department of Medicine, University of Pennsylvania, Philadelphia, PA, USA; Penn Epigenetics Institute, University of Pennsylvania, Philadelphia, PA, USA; Department of Biochemistry and Biophysics, University of Pennsylvania, Philadelphia, PA, USA*
SARA T. LEDEBOER • *Division of Chemical Biology and Medicinal Chemistry, Eshelman School of Pharmacy, University of North Carolina at Chapel Hill, Chapel Hill, NC, USA*
LESLIE LEWIS • *EpiCypher Inc., Durham, NC, USA*
JIAN LIU • *Edinburgh Medical School—Biomedical Sciences, College of Medicine and Veterinary Medicine, The University of Edinburgh, Edinburgh, UK; Zhejiang University—University of Edinburgh Institute (ZJU-UoE Institute), International Campus (Haining), Zhejiang University School of Medicine, Zhejiang University, Hangzhou, China; Department of Respiratory and Critical Care Medicine, The Second Affiliated Hospital, Zhejiang University School of Medicine, Zhejiang University, Hangzhou, China*
YANG LIU • *Epigenetics & RNA Biology Laboratory, Eukaryotic Transcriptional Regulation Group, National Institutes of Environmental Health Sciences (NIEHS), Research Triangle Park, NC, USA*
TREVOR LONG • *Department of Biology, School of Arts and Sciences, University of North Dakota, Grand Forks, ND, USA*
CHRISTIAN E. LOO • *Graduate Group in Biochemistry and Molecular Biophysics, University of Pennsylvania, Philadelphia, PA, USA*
RUI LU • *Department of Medicine, Division of Hematology/Oncology, University of Alabama at Birmingham, Heersink School of Medicine, Birmingham, AL, USA; O'Neal Comprehensive Cancer Center, University of Alabama at Birmingham, Heersink School of Medicine, Birmingham, AL, USA*
MANU • *Department of Biology, School of Arts and Sciences, University of North Dakota, Grand Forks, ND, USA*
SOFIA MAROTTA • *Department of Clinical and Molecular Medicine, Sapienza University, Rome, Italy*

MATTHEW R. MARUNDE • *EpiCypher Inc., Durham, NC, USA*
DANIELLE N. MARYANSKI • *EpiCypher Inc., Durham, NC, USA*
DEBORAH A. MORSE • *Department of Applied Physiology and Kinesiology, College of Health and Human Performance, University of Florida, Gainesville, FL, USA*
STELLA MRMIĆ • *Department of Clinical and Molecular Medicine, Sapienza University, Rome, Italy*
LUCIA MUNDO • *Department of Clinical and Molecular Medicine, Sapienza University, Rome, Italy*
KEVIN O. MURRAY • *Department of Applied Physiology and Kinesiology, College of Health and Human Performance, University of Florida, Gainesville, FL, USA*
NANCY H. NABILSI • *Department of Biochemistry and Molecular Biology, College of Medicine, and UF Health Cancer Center, University of Florida, Gainesville, FL, USA*
SERGEI NECHAEV • *Department of Biomedical Sciences, School of Medicine and Health Sciences, University of North Dakota, Grand Forks, ND, USA*
KELSEY NOLL • *EpiCypher Inc., Durham, NC, USA*
KATHERINE NOVITZKY • *EpiCypher Inc., Durham, NC, USA*
DAMIEN PARRELLO • *Department of Biomedical Sciences, School of Medicine and Health Sciences, University of North Dakota, Grand Forks, ND, USA*
SAMANTHA G. PATTENDEN • *Division of Chemical Biology and Medicinal Chemistry, Eshelman School of Pharmacy, University of North Carolina at Chapel Hill, Chapel Hill, NC, USA*
SRINIVAS RAMACHANDRAN • *Department of Biochemistry and Molecular Genetics, University of Colorado Anschutz Medical Campus, Aurora, CO, USA; RNA Bioscience Initiative, University of Colorado Anschutz Medical Campus, Aurora, CO, USA*
ALBERTO RIVA • *Bioinformatics Core, Interdisciplinary Center for Biotechnology Research, University of Florida, Gainesville, FL, USA*
JOSEPH RODRIGUEZ • *Epigenetics and RNA Biology Laboratory, National Institute of Environmental Health Sciences (NIEHS), Research Triangle Park, Durham, NC, USA*
TOMMASO SALAMONE • *Department of Clinical and Molecular Medicine, Sapienza University, Rome, Italy*
EUGENIO SANTACROCE • *Department of Clinical and Molecular Medicine, Sapienza University, Rome, Italy*
VICTORIA SILVA • *Department of Biomedical Sciences, School of Medicine and Health Sciences, University of North Dakota, Grand Forks, ND, USA*
JENNIFER SPENGLER • *EpiCypher Inc., Durham, NC, USA*
LU SUN • *EpiCypher Inc., Durham, NC, USA*
ZU-WEN SUN • *EpiCypher Inc., Durham, NC, USA*
AUSSIE SUZUKI • *McArdle Laboratory for Cancer Research, Department of Oncology, University of Wisconsin-Madison, Madison, WI, USA; Carbone Comprehensive Cancer Center, University of Wisconsin-Madison, Madison, WI, USA*
MOTOKI TAKAKU • *Department of Biomedical Sciences, School of Medicine and Health Sciences, University of North Dakota, Grand Forks, ND, USA*
QIAN TANG • *Zhejiang University—University of Edinburgh Institute (ZJU-UoE Institute), International Campus (Haining), Zhejiang University School of Medicine, Zhejiang University, Haining, China; Edinburgh Medical School—Biomedical Sciences, College of Medicine and Veterinary Medicine, The University of Edinburgh, Edinburgh, UK*
PANKAJ TRIVEDI • *Department of Experimental Medicine, Sapienza University, Rome, Italy*

AVINASH VEERAPPA • *Department of Genetics, Cell Biology, and Anatomy, University of Nebraska Medical Center, Omaha, NE, USA*
BRYAN J. VENTERS • *EpiCypher Inc., Durham, NC, USA*
PAUL A. WADE • *Epigenetics & RNA Biology Laboratory, Eukaryotic Transcriptional Regulation Group, National Institutes of Environmental Health Sciences (NIEHS), Research Triangle Park, NC, USA*
TONG WANG • *Graduate Group in Biochemistry and Molecular Biophysics, University of Pennsylvania, Philadelphia, PA, USA*
SAJAD A. WANI • *Department of Cancer Biology, University of Cincinnati College of Medicine, Cincinnati, OH, USA*
AMANDA R. WASYLISHEN • *Department of Cancer Biology, University of Cincinnati College of Medicine, Cincinnati, OH, USA*
ELLEN N. WEINZAPFEL • *EpiCypher Inc., Durham, NC, USA*
ABIGAIL WEIRICH • *Department of Biochemistry and Molecular Genetics, University of Colorado Anschutz Medical Campus, Aurora, CO, USA*
CAROLINA LIN WINDHAM • *EpiCypher Inc., Durham, NC, USA*
YAPING XU • *Zhejiang University—University of Edinburgh Institute (ZJU-UoE Institute), International Campus (Haining), Zhejiang University School of Medicine, Zhejiang University, Hangzhou, China; Edinburgh Medical School—Biomedical Sciences, College of Medicine and Veterinary Medicine, The University of Edinburgh, Edinburgh, UK*
PELIN YAŞAR • *Epigenetics and RNA Biology Laboratory, National Institute of Environmental Health Sciences (NIEHS), Research Triangle Park, Durham, NC, USA*
TAKAO YASUI • *Nagoya University, Nagoya, Aichi, Japan*
ZHENTING ZHANG • *Zhejiang University—University of Edinburgh Institute (ZJU-UoE Institute), International Campus (Haining), Zhejiang University School of Medicine, Zhejiang University, Hangzhou, China; Edinburgh Medical School—Biomedical Sciences, College of Medicine and Veterinary Medicine, The University of Edinburgh, Edinburgh, UK*
MINGQI ZHOU • *Department of Biochemistry and Molecular Biology, College of Medicine, Genetics Institute, and UF Health Cancer Center, University of Florida, Gainesville, FL, USA*
XINYUE ZHOU • *Department of Medicine, Division of Hematology/Oncology, University of Alabama at Birmingham, Heersink School of Medicine, Birmingham, AL, USA; O'Neal Comprehensive Cancer Center, University of Alabama at Birmingham, Heersink School of Medicine, Birmingham, AL, USA*
RUIYAO ZHU • *Department of Medicine, University of Pennsylvania, Philadelphia, PA, USA*

Part I

DNA Methylation and Its Detection Methods

Chapter 1

Enzymatic, Nondestructive Sequencing of 5-Methylcytosine with Direct Methylation Sequencing (DM-Seq)

Ruiyao Zhu, Christian E. Loo, Tong Wang and Rahul M. Kohli

Abstract

Here, we provide a detailed protocol for Direct Methylation Sequencing (DM-Seq), a method for localizing 5-methylcytosine (5mC) at single-base resolution within CpG contexts. DM-Seq relies solely on enzymatic transformation of DNA, thereby circumventing the generation of destructive or biased intermediates commonly seen with chemical sequencing methods, and provides results with high library yield, sample diversity, and fidelity. DM-Seq is particularly well-suited for low-input samples and applications where the detection of 5mC is desired without being confounded by 5-hydroxymethylcytosine. In this chapter, we present a comprehensive workflow for the entire method, from the synthesis and purification of the requisite substrates and enzymes to the construction and analysis of the sequencing library.

Keywords DNA methylation, DM-Seq, Epigenetic sequencing, Bisulfite, TAPS, Unnatural cytosine analogs, Methyltransferase enzymes, SAM analogs

1 Introduction

Dynamic modifications to cytosine bases play key epigenetic regulatory roles in mammals, both in normal development and in disease states, including cancer [1, 2]. Regarded as the fifth base of DNA, 5-methylcytosine (5mC) is generated by DNA methyltransferase enzymes (MTases) that employ the co-substrate, *S*-adenosyl-L-methionine (SAM), as a methyl donor [3]. In mammalian genomes, 5mC is most commonly found within cytosine-guanine dinucleotide (CpG) contexts, where the modifications play an essential role in a myriad of gene regulatory processes, such as transcriptional silencing, genomic imprinting, and cellular differentiation. DNA methylation can also be dynamically altered through oxidation by ten-eleven translocation (TET) enzymes [4]. Oxidation of 5mC can lead to the accumulation of 5-hydroxymethylcytosine (5hmC), a modification

These authors contributed equally: Ruiyao Zhu, Christian E. Loo.

Archana Dhasarathy and Motoki Takaku (eds.), *Cancer Epigenetics*, Methods in Molecular Biology, vol. 3005, https://doi.org/10.1007/978-1-0716-5072-1_1,

that can have independent epigenetic roles as a transcriptional activating mark or serve as an intermediate of the DNA demethylation pathways where 5mC is reverted to an unmodified C.

As normal modification patterns define cell lineage and aberrant DNA modifications are a frequent hallmark of many cancers, numerous methods have been developed to localize these modified cytosine bases in genomic DNA [5–7]. Most commonly, these methods rely on either chemical or enzymatic transformation of DNA, where some states of cytosine are converted while others are preserved, typically leading to selective C-to-T transitions in sequencing. A longstanding limitation of most of these methods has been the conflation of 5mC and 5hmC, an issue that exists with bisulfite-based sequencing (BS-Seq), the traditional method for studying DNA modification [8]. Another limitation specific to BS-Seq is its relatively poor preservation of DNA, arising from the introduction of non-amplifiable abasic sites due to depyrimidination of sulfonated intermediates in the process of chemical deamination [9]. To overcome these two major limitations of BS-Seq, we pioneered the use of DNA deaminase enzymes as a replacement for chemical deamination [10]. APOBEC-Coupled Epigenetic Sequencing (ACE-Seq) relies on the ability of the DNA deaminase APOBEC3A (A3A) to potently deaminate C and 5mC, but not 5hmC, when protected by β-glucosyltransferase (βGT). Building on this precedent and the established selectivity of A3A [11], TET enzymes were later incorporated into enzymatic methylation sequencing (EM-Seq), whereby TET oxidation of 5mC renders it protected from A3A, offering a non-destructive strategy for mapping DNA modifications, albeit one that still conflates 5mC and 5hmC [12]. Separately, TET enzymes were also incorporated into hybrid enzymatic-chemical methods, including TET-assisted pyridine borane sequencing (TAPS) and its variants for mapping 5mC and 5hmC together or 5mC alone [13]. While TAPS-based methods offer some advantages, they are also associated with limitations. First, TAPS relies upon TET oxidation of 5mC and 5hmC to 5-carboxycytosine (5caC), which can be difficult to achieve to completion. Second, TAPS results in the significant underdetection of methylation in regions that are highly modified due to amplification bias against the non-aromatic dihydrouridine (DHU) intermediate that is generated by pyridine-borane-mediated deamination [14].

Given that 5mC in CpG contexts is the most important modified base in mammalian genomes, we recognized the need for a non-destructive method that could faithfully localize 5mC without confounding from 5hmC [14]. In considering permutations of enzymes that might achieve this goal, we envisioned that if unmodified CpGs could be protected from enzymatic deamination by A3A, then we could directly detect 5mC alone (Fig. 1). We termed

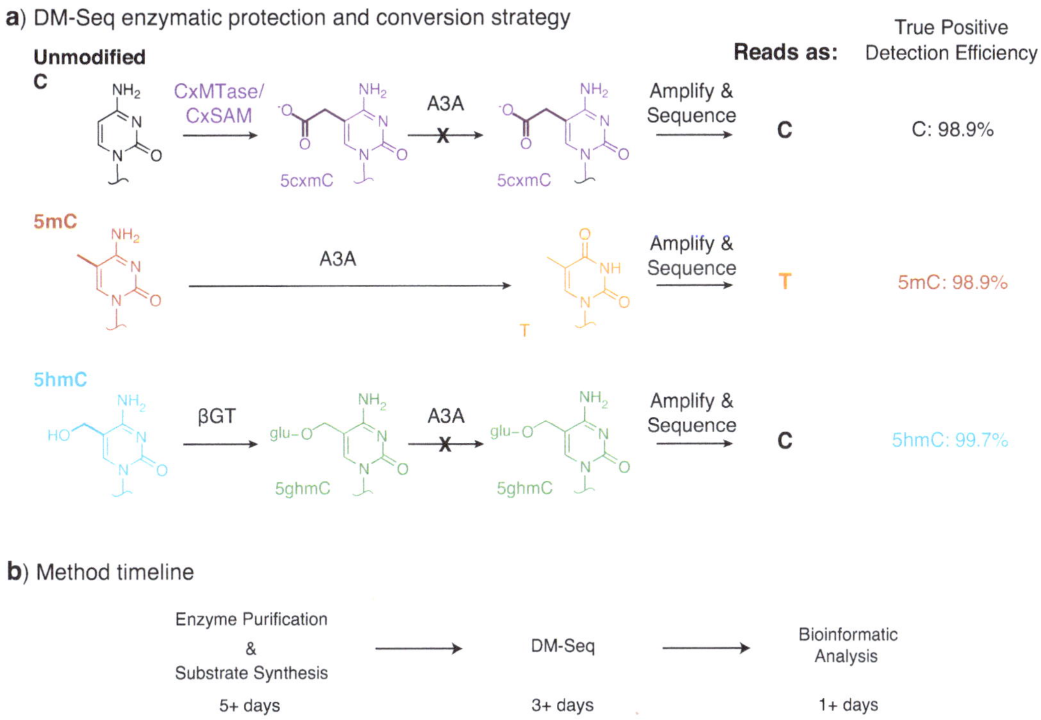

Fig. 1 DM-Seq overview. **a** DM-Seq employs a carboxymethyltransferase (CxMTase) and carboxy-SAM (CxSAM) to protect unmodified CpGs and βGT to protect 5hmC. Enzymatic protection of these two cytosines leaves only 5mC susceptible to deamination by A3A. **b** Overall timeline for the protocol from protein purification to analysis. Purification of the CxMTase and CxSAM is described (see Sect. 3.1 and 3.2), along with control DNAs (see Sect. 3.3) and the complete DM-Seq pipeline (see Sect. 3.4)

this method Direct Methylation Sequencing (DM-Seq). DM-Seq was enabled by several key innovations, including an engineered DNA MTase that can transfer a protecting group onto unmodified cytosines in CpG contexts [15], modified sequencing adapters containing 5-propynylcytosine to resist enzymatic deamination, and a copy-strand strategy that enhances unmodified CpG protection efficiency. In DM-Seq, the engineered MTase (M.MpeI N374K) is a variant that can accept carboxy-*S*-adenosylmethionine (CxSAM) instead of SAM as a co-substrate. This neomorphic carboxymethyltransferase (CxMTase) catalyzes the conversion of cytosines in unmodified CpGs into 5-carboxymethylcytosine (5cxmC), a modification that resists deamination by A3A. Here, to complement our prior report [14], we provide a detailed protocol that describes the preparation of key reagents and the complete workflow for DM-Seq.

2 Materials

2.1 CxMTase Preparation

2.1.1 CxMTase Expression

1. *E. coli* strain ER1821 (NEB)
2. CxMTase expression plasmid (Addgene, 197985)
3. LB Broth (1% [w/v] tryptone, 0.5% [w/v] yeast extract, and 0.5% [w/v] sodium chloride)
4. Carbenicillin (Carb)
5. LB agar plates with Carb
6. Anhydrotetracycline (Atc)
7. Refrigerated shaking incubator

2.1.2 Cell Lysis

1. Tabletop centrifuge, large
2. Lysis buffer: 50 mM Tris Cl, pH 7.5 at 25°C, 150 mM NaCl, 25 mM imidazole, 10% glycerol (v/v), 0.22 μm filtered
3. cOmplete EDTA-free Protease Inhibitor Cocktail (Sigma)
4. RNase A
5. Sonicator, microfluidizer, or French press for cell lysis
6. High-speed centrifuge
7. 0.22 μm syringe filter

2.1.3 CxMTase Purification

1. HisPur Cobalt Resin (Thermo)
2. Gravity column
3. High-salt wash buffer: 50 mM Tris Cl, pH 7.5 at 25°C, 1 M NaCl, 25 mM imidazole, 10% glycerol (v/v), 0.22 μm filtered
4. Elution buffer: 50 mM Tris Cl, pH 7.5 at 25°C, 150 mM NaCl, 150 mM imidazole, 10% glycerol (v/v), 0.22 μm filtered
5. Dialysis buffer: 20 mM Tris Cl, pH 7.5 at 25°C, 0.2 mM EDTA, 2 mM DTT, 150 mM NaCl, 10% glycerol (v/v), 0.22 μm filtered
6. 8% SDS-PAGE gel and Coomassie Blue or Qubit Protein Assay Kit (Thermo)
7. SnakeSkin Dialysis Tubing, 10,000 Da MWCO (Thermo)

2.1.4 Concentration and Storage

1. Amicon Ultra-4 centrifugal filter, 10,000 Da MWCO (EMD Millipore)
2. 0.2 mL 8-strip PCR tubes
3. 40% (v/v) Glycerol
4. Liquid nitrogen
5. −80°C Freezer

2.2 CxSAM Preparation and Characterization

1. *S*-adenosyl-L-homocysteine (SAH) (Sigma)
2. 150 mM ammonium bicarbonate solution
3. Solution of 15 mg of *S*-adenosyl-L-homocysteine (SAH) in 2.5 mL of 150 mM aqueous ammonium bicarbonate, prepare fresh and swirl to dissolve
4. Oven-dry a 50 mL round-bottom flask with a stir bar
5. Stir plate
6. Mineral oil bath
7. 2-iodoacetic acid (IAA) (Sigma)
8. Methanol
9. Swinging Bucket centrifuge
10. TLC (SiO_2) with SAH and IAA as standards, and a developing solution of 10:1 MeOH:1.5N ammonium bicarbonate
11. Spectrophotometer
12. UV lamp

2.3 CpG-methylated Lambda Spike-In Preparation

1. Lambda phage gDNA (300 ng/μL) (Thermo)
2. 10× NEBuffer 2 (NEB)
3. *S*-adenosylmethionine (SAM) (NEB)
4. M.SssI CpG Methyltransferase (NEB)
5. Nuclease-free H_2O (nfH_2O)
6. Zymo Genomic DNA Clean and Concentrator kit
7. Premium Bisulfite kit (Diagenode)

2.4 DM-Seq

2.4.1 Materials Used Throughout

1. Thermocycler
2. nfH_2O
3. SPRIselect Beads (Beckman Coulter)
4. Low EDTA TE buffer: 10 mM Tris Cl, pH 8.0 at 25°C, 0.1 mM EDTA, 0.22 μm filtered
5. Magnetic rack

2.4.2 Custom Synthesized Oligos (IDT)

1. Methylated copy primer:
   5′-GAMTGGAGTTMAGAMGTGTGMTMTTMGATM-3′
   M = 5-methylcytosine
2. Custom xGen stubby top adapter (IDT custom)
   5′-AXAXTXTTTXXXTAXAXGAXGXTXTTXXGATX*T-3′
   * = phosphorothioate; X = 5-propynylcytosine (5pyC)
3. Custom xGen stubby bottom adapter (IDT custom)
   5′-(P) GATXGGAAGAGXAXAXGTXTGAAXTXXAGTX-3′
   (P) = 5′ phosphate; X = 5-propynylcytosine (5pyC)

2.4.3 Spike-In Controls

1. Unmethylated pUC19 plasmid DNA (NEB)
2. CpG-methylated M.SssI-treated lambda phage gDNA
3. Hydroxymethylated T4-hmC DNA (contact authors)

2.4.4 Sample Shearing

1. Samples of interest
2. Sonicator for shearing DNA
3. Corresponding shearing tubes

2.4.5 DNA End Repair and Ligation

1. NEBNext Ultra II End Repair/dA-Tailing Module (NEB)
2. NEBNext Ultra II Ligation Module (NEB)
3. Duplexed Y-shaped adapters (15 μM) containing 5pyC

2.4.6 Copy Strand Synthesis

1. Methylated copy primer (IDT)
2. *Bst* DNA Polymerase, Large Fragment (NEB)
3. 10× NEB CutSmart Buffer (NEB)
4. Nucleotides: dATP, dGTP, dTTP (10 mM each)
5. 5-methyl-dCTP (10 mM)
6. Proteinase K

2.4.7 βGT Protection

1. UDP-glucose (NEB)
2. T4-βGT (NEB)
3. 10× NEB CutSmart Buffer (NEB)

2.4.8 Optional Degradation of Uncopied Strands

1. Mung Bean Nuclease (NEB)

2.4.9 CxMTase Protection

1. 5× CxMTase buffer: 50 mM Tris Cl, pH 7.9 at 25°C, 250 mM NaCl, 50 mM EDTA, 5 mM DTT, 0.22 μm filtered
2. CxMTase: purified MBP-M.MpeI-N374K enzyme (2.5 μM)
3. CxSAM (diluted to 3 mM)

2.4.10 A3A Deamination

1. Formamide
2. APOBEC3A (A3A) (NEB)
3. Bovine Serum Albumin (BSA) (NEB)
4. 10× APOBEC Reaction Buffer (NEB)

2.4.11 Optional qPCR

1. NEBNext Multiplex Oligos for Illumina-Dual Indices (NEB)
2. KAPA SYBR FAST qPCR Master Mix (2×) (Roche)
3. qPCR 384-well plate (Thermo)
4. Applied Biosystems Real-Time PCR Instrument

2.4.12 Indexing

1. NEBNext Multiplex Oligos for Illumina-Dual Indices (NEB)
2. KAPA HiFi HotStart Uracil+ ReadyMix (2×) (Roche)

2.4.13 Library Quantification and Quality Control

1. Agilent 2100 Bioanalyzer Instrument
2. Agilent Bioanalyzer High Sensitivity DNA kit
3. Qubit 3.0 Fluorometer (Thermo)
4. Qubit dsDNA BR Assay Kit (Thermo)
5. Qubit dsDNA HS Assay Kit (Thermo)

2.4.14 Bioinformatics

1. Trim Galore
https://www.bioinformatics.babraham.ac.uk/projects/trim_galore/
2. Bismark (v 0.14.3, Samtools and Bowtie2 required)
https://www.bioinformatics.babraham.ac.uk/projects/bismark/
3. Picard
https://broadinstitute.github.io/picard/

3 Methods

3.1 CxMTase Expression and Purification

The purpose of this step is to express and purify the CxMTase (Fig. 2a). The CxMTase is an N374K mutant of the M.MpeI CpG-specific MTase, expressed as an N-terminal MBP fusion with a C-terminal His tag that is used for purification.

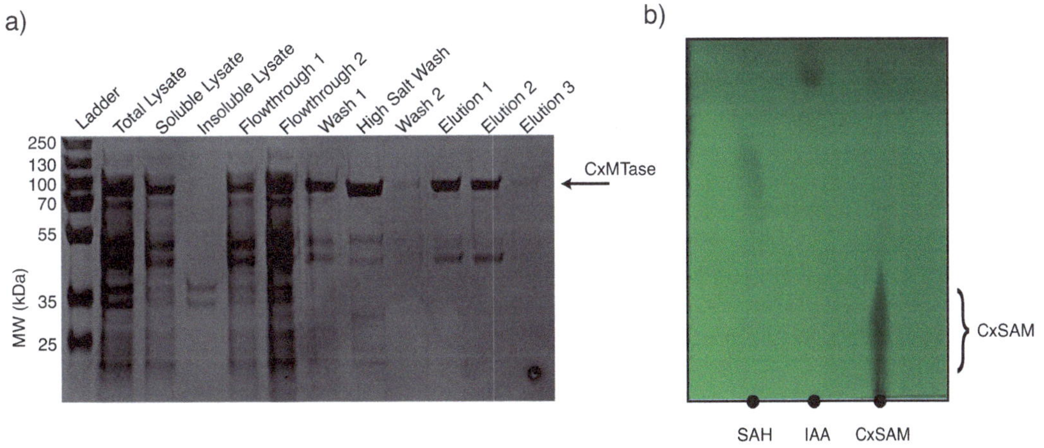

Fig. 2 Preparation of CxMTase and CxSAM for DM-Seq. **a** Representative gel for purification of CxMTase following expression. Note that the expected MW of the CxMTase is 91 kDa and the non-MBP-tagged CxMTase is 47 kDa (*see* **Note 4**). **b** Representative TLC monitoring for the synthesis of CxSAM from starting materials SAH and IAA

3.1.1 CxMTase Expression

1. Transform CxMTase expression plasmid encoding MBP-M. MpeI-N374K-His into ER1821 (*see* **Note 1**) and plate on an LB + Carbenicillin (Carb) (100 μg/mL) plate. Incubate overnight at 37°C.
2. Inoculate a single colony into 10 mL of LB + Carb (100 μg/mL) in a culture tube and incubate overnight at 37°C with shaking at 250 rpm.
3. Using a saturated overnight culture, inoculate a large culture (e.g., 1 L) of LB + Carb (100 μg/mL) with 10 mL of the overnight culture.
4. In a shaking incubator, grow at 37°C until the culture reaches an OD of 0.4.
5. Change incubator temperature to 16°C and allow culture to equilibrate (~1 h).
6. Induce expression by adding Anhydrotetracycline (Atc) to a final concentration of 20 ng/mL.
7. Grow culture at 16°C for an additional 12–16 h (overnight).

3.1.2 Cell Lysis

1. Pellet cells at 7000 rcf for 30 min at 4°C.
2. Decant supernatant and resuspend cells in 25 mL lysis buffer with one cOmplete EDTA-free Protease Inhibitor Cocktail tablet and 10 μL of RNase A per 1 L of bacterial culture.
3. *Optional:* Store resuspended cells at −80°C until ready to proceed.
4. Lyse cells with a sonicator, microfluidizer, or French press per the manufacturer's protocols.
5. Centrifuge lysed cells at 27,000 rcf for 30 min at 4°C.
6. Collect the supernatant, filter it through a 0.22 μm syringe filter, and place it on ice.

3.1.3 CxMTase Purification (See Note 2)

1. Use 4 mL of HisPur Cobalt Resin slurry per 1 L culture.
2. Load the slurry into a gravity column and let the resin storage solution drain.
3. Equilibrate the column by passing 25 column volumes of lysis buffer through the column.
4. Add filtered lysate to column, collecting flow-through (*see* **Note 3**).
5. After loading, wash with 25 column volumes of high salt wash buffer.
6. Add 1 column volume fractions of elution buffer sequentially and collect each elution fraction.

7. Analyze flow-through, wash, and elution fractions on an 8% SDS-PAGE gel and visualize with Coomassie Blue for 91 kDa MBP-M.MpeI-N374K-His (Fig. 2a, *see* **Note 4**).
8. Pool desired fractions and dialyze overnight at 4°C in 2 L of prechilled dialysis buffer.

3.1.4 Concentration and Storage

1. Quantify protein by either comparison to a BSA standard curve after running SDS-PAGE and visualizing with Coomassie Blue or using the Qubit Protein Assay Kit.
2. Concentrate protein, if necessary, to a final concentration of >2.5 μM using Amicon Ultra-4 centrifugal filters as per the manufacturer's instructions.
3. Add cold 40% (v/v) glycerol to dilute the dialyzed protein 2-fold.
4. Aliquot enzyme in 10 μL fractions in 8-strip PCR tubes.
5. Flash freeze with liquid nitrogen.
6. Store protein aliquots long-term at −80°C.

3.2 CxSAM Preparation and Characterization

The purpose of this step is to synthesize and purify CxSAM (Fig. 2b).

1. Pour a solution of SAH and ammonium bicarbonate into an oven-dried 50 mL round-bottom flask with a stir bar.
2. Weigh out 500 mg of 2-iodoacetic acid and add it to the above round-bottom flask.
3. Cover the reaction and incubate overnight at 37°C in a mineral oil bath, stirring.
4. Slowly add 20 mL of methanol to quench the reaction and transfer to a 125 mL glass bottle. Rinse the 50 mL round-bottom flask with 40 mL of methanol and transfer to the 125 mL glass bottle. Incubate at 4°C overnight, stirring.
5. Pour half of the resulting mixture into a 50 mL Falcon tube and harvest the precipitate by centrifugation at 4000 rcf for 30 min at 4°C. Repeat to harvest the remaining half.
6. Pour or pipette off supernatant.
7. Wash twice with 5 mL cold methanol, spin at 4000 rcf for 30 min at 4°C between washes to pellet.
8. Remove as much methanol as possible without disturbing the pellet, and leave the collection tube open to dry.
9. Dissolve CxSAM in minimal nfH_2O and place on ice.
10. Run a TLC (SiO_2), with SAH and IAA as standards, and develop with a solution of 10:1 MeOH:1.5N ammonium bicarbonate to confirm the generation of CxSAM. Visualize results with a UV lamp (Fig. 2b, *see* **Note 5**).

11. Quantify CxSAM using absorbance measured at 260 nm (extinction coefficient: 15,400 L mol^{-1} cm^{-1}).
12. Anticipated concentration is > 3 mM. Dilute to 3 mM final concentration.
13. Store at −80°C.

3.3 CpG-Methylated Lambda Spike-in Preparation (See Note 6)

The purpose of this step is to generate control DNA from the lambda phage that is enzymatically methylated at all CpG sites (mλ). The presence of both unmodified cytosines (CpH context) and 5mCpGs within this mλ sample provides controls to evaluate the efficiency of A3A deamination in DM-Seq, with other spike-in controls offering metrics for the efficiency of other enzymatic steps.

1. Set up the M.SssI methylation reaction as below:

Component	Volume (μL)
Lambda phage gDNA (300 ng/μL)	3
10× NEBuffer 2	5
1.6 μM SAM	1
nfH_2O	36
M.SssI CpG Methyltransferase	2

2. Incubate at 37°C for 4 h, hold at 4°C.
3. Spike in more enzyme and substrate after 4 h as below:

Component	Volume (μL)
1.6 μM SAM	1
M.SssI CpG Methyltransferase	1

4. Incubate at 37°C for 4 h, hold at 4°C.
5. Purify DNA with the Zymo Genomic DNA Clean and Concentrator kit, eluting in 50 μL EB.
6. *Quality Control:* Use the Qubit BR Assay Kit to quantify.
7. *Quality Control:* Prepare a BS-Seq library to confirm nearly complete methylation (>95% methylation at CpGs).

3.4 DM-Seq (See Note 7)

The protocol below allows for the generation of DM-Seq libraries using the reagents prepared above (Fig. 3). Sheared DNA is ligated to custom deaminase-resistant adapters, followed by strand copying with 5mC, which drives efficient protection of unmodified CpGs using the CxMTase and CxSAM in the following step. After enzymatic deamination, the modified DNA is amplified and sequenced.

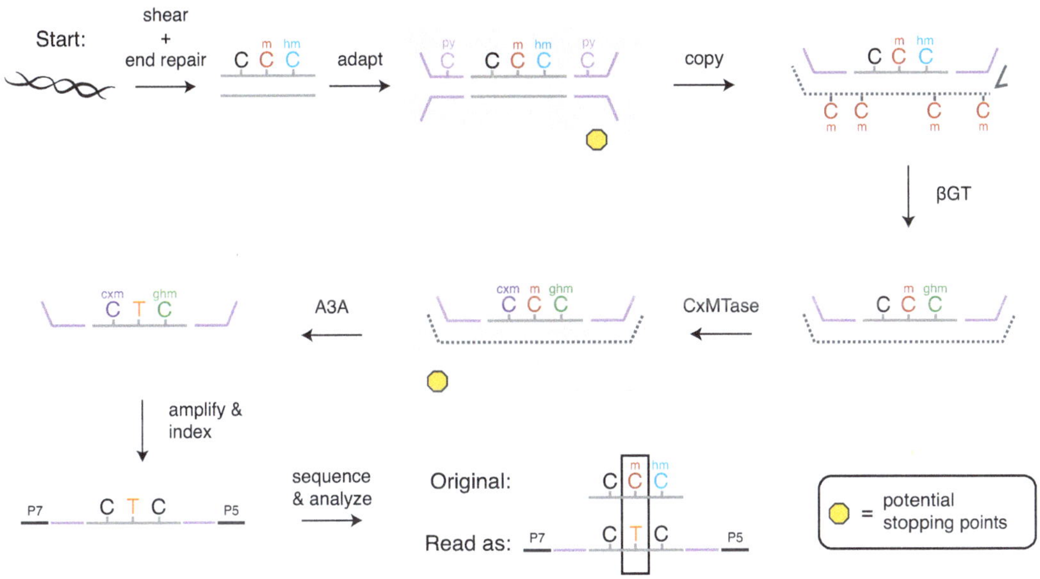

Fig. 3 Schematic overview of DM-Seq workflow. DM-Seq involves DNA shearing, end repair (see Sect. 3.4.1 and 3.4.2), the addition of DNA deaminase resistant 5pyC-containing adapters, and copying of the parent strand with 5mC in place of unmodified C (see Sect. 3.4.3). The 5hmC in the parent DNA strand are then protected by glucosylation, while unmodified CpGs are protected by DNA carboxymethylation (see Sect. 3.4.4–3.4.6). The enzymatically-protected DNA is then deaminated with A3A, which results in selective conversion of 5mC in CpGs to T (see Sect. 3.4.7). The enzymatically-modified DNA can then be amplified, sequenced (see Sect. 3.4.8–3.4.11), and analyzed to specifically localize 5mCpGs by the C-to-T conversion in sequencing reads

3.4.1 Sample Shearing (See Note 8)

1. Add between 10 and 200 ng of DNA of interest to a PCR tube (in a volume <47 μL).
2. Add spike-in control DNAs at a ratio of 1:100 total spike-ins: gDNA (m:m, 1:300 each). Note that more dilute concentrations can be used if not utilizing shallow (e.g., MiSeq) sequencing as a control.
   a. Unmethylated pUC19.
   b. CpG-methylated M.SssI-treated lambda phage gDNA (mλ).
   c. Hydroxymethylated T4-hmC phage DNA.
3. Bring DNA up to a final volume of 50 μL in low-EDTA TE buffer.
4. Use the Covaris sonicator as in a typical sonication protocol for an appropriate volume (50 μL).
   a. ~550 bp: 45 s duration, 500 cycles per burst, 25 watts, 10% duty factor.

3.4.2 DNA End Repair

1. Set up the end repair reaction as below, per sample:

Component	Volume (μL)
NEBNext Ultra II End Prep Enzyme Mix	3
NEBNext Ultra II End Prep Reaction Buffer	7
Fragmented sheared DNA (10–200 ng total)	50

2. Incubate samples using the following thermocycler method: 20°C for 30 min, 65°C for 30 min, 4°C hold. Set the heated lid to 75°C.
3. During the End Repair, prepare 15 μM stubby adapters by mixing equimolar amounts of Custom xGen stubby top and bottom adapters in low EDTA TE buffer. Anneal using the following thermocycler method: 95°C 5 min, cool down ramp at a rate of −0.1°C/s, 4°C hold (*see* **Note 9**).
4. Once the End Repair reaction is complete, proceed immediately to the Ligation Reaction as below, per sample:

Component	Volume (μL)
End Repair Reaction mixture (unpurified)	60
Annealed stubby adapters (15 μM, in low EDTA TE buffer)	2.5
NEBNext Ultra II Ligation Master Mix	30
NEBNext Ligation Enhancer	1

5. Incubate the samples using the thermocycler method: 20°C for 15 min, 4°C hold. Turn the heated lid off.
6. Perform a left-sided 1.2× SPRI purification, as per the manufacturer's instructions, and elute with 6 μL nfH_2O, yielding 5 μL adapter-ligated sample (*see* **Note 10**).

3.4.3 Copy Strand Synthesis and Degradation (See Note 11)

1. Set up the following annealing reaction as below, per sample (*see* **Note 12**):

Component	Volume
Adapter-ligated sample	Up to 5 μL in nfH_2O
10× NEB CutSmart Buffer	1 μL
Methylated copy primer (10 μM)	1 μL
10 mM dATP, dGTP, dTTP	1 μL
10 mM 5-methyl-dCTP	1 μL

2. Incubate samples using the thermocycler method: 95°C for 5 min, decreasing from 70°C to 40°C in steps of −5°C per min, 37°C for 5 min, 4°C hold.

3. Add 1 μL *Bst* DNA polymerase, Large Fragment to each sample (*see* **Note 13**).
4. Incubate samples using the thermocycler method: 65°C for 30 min, 4°C hold.

3.4.4 βGT Protection

1. Set up the following glycosylation reaction as below, per sample:

Component	Volume (μL)
Unpurified and copied sample	10
UDP-glucose	0.4
T4-βGT	1
10× NEB CutSmart Buffer	2
nfH_2O	6.6

2. Incubate the samples using the thermocycler method: 37°C for 1 h, 4°C hold.

3.4.5 Optional Degradation of Uncopied Strands

1. Add 1 μL of Mung Bean Nuclease to each sample and incubate samples using the thermocycler method: 30°C for 30 min, 4°C hold (*see* **Note 14**).
2. Perform a left-sided 1.2× SPRI purification and elute in 6 μL low EDTA TE buffer.

3.4.6 CxMTase Protection

1. Set up the following carboxymethylation reaction as below, per sample:

Component	Volume (μL)
Purified sample DNA	5
5× CxMTase buffer	2
MBP-M.MpeI N374K-His enzyme (2.5 μM)	2
CxSAM (3 mM)	1

2. Incubate the samples using the thermocycler method: 37°C for 16 h, 95°C for 5 min, 4°C hold (*see* **Note 15**).
3. Treat the post-carboxymethylation DNA with 0.5 μL Proteinase K and incubate at 37°C for 15 min.
4. Perform a left-sided 1.2× SPRI purification and elute in 16 μL nfH_2O.

3.4.7 A3A Deamination (See Note 16)

1. Preheat one thermocycler, and pre-program the thermocycler so that it is prepared to run the "ramp method" as specified below:

Temperature	Time
4°C	10 min
4°C–50°C	2:15 per degree of ramp (~2 h total)
50°C	10 min
4°C	Hold

2. Preheat a second thermocycler (with heated lid) to 85°C.
3. Set up the following denaturation reaction as below, per sample:

Component	Volume (μL)
Purified enzymatically-protected DNA	16
Formamide	4

4. Incubate the samples at 85°C for 10 min in the preheated thermocycler.
5. During this time, prepare the master mix for deamination reaction as below, per sample:

Component	Volume (μL)
nfH_2O	68
10× APOBEC reaction buffer	10
BSA	1
A3A	1

6. After DNA is fully denatured for 10 min in formamide, swiftly transfer samples to the PCR sample cooling rack pre-chilled to −20°C (*see* **Note 17**).
7. Add the master mix containing A3A (80 μL/reaction) to each denatured sample (20 μL) and mix, while still on the cooling rack.
8. Briefly spin down on a tabletop microfuge.
9. Rapidly transfer to the thermocycler prepared to run the "ramp method" and initiate the incubation.
10. Following deamination, perform a left-sided 1.2× SPRI purification and elute in 21 μL nfH_2O.

3.4.8 Optional qPCR (See Note 18)

1. Prepare the qPCR reaction as below, per sample:

Component	Volume (μL)
nfH_2O	3
KAPA SYBR FAST qPCR Master Mix (2×)	5
Purified deaminated DNA	1
2 μM indexing primers	1

2. Carefully pipette each sample into a 384-well plate and start the qPCR method (Applied Biosystems Real-Time PCR instrument): 95°C for 3 min, 25× (95°C for 15 s, 63°C for 30 s, 72°C for 30 s), 4°C hold.
3. For each sample, record the number of cycles required to reach the half-maximum value. The maximum value can be found by approximating where each sample curve plateaus. Use these values to assign cycles for the indexing PCR.

3.4.9 Indexing

1. Assign unique indices to each sample.
2. Set up the indexing reaction as below, per reaction:

Component	Volume (μL)
nfH_2O	5
KAPA HiFi HotStart Uracil+ ReadyMix (2×)	25
Purified, deaminated DNA	19
Combined i5/i7 indexing primers	1

3. Incubate samples with the following thermocycler method: 98°C 30 s, variable number of amplification cycles (98°C for 10 s, 65°C for 30 s, 68°C for 1 min), 4°C hold.
4. Perform a left-sided 0.8× SPRI purification and elute in 20 μL low EDTA TE buffer.

3.4.10 Library Quantification and Quality Control

1. Quantify all samples with the Qubit HS Assay Kit.
2. Ensure 100 pg/μL–10 ng/μL concentrations before loading on BioAnalyzer.
3. Use BioAnalyzer to estimate average library size, as well as assess the need for an additional SPRI purification due to adapter dimers.

3.4.11 Illumina Sequencing

1. Load sequencer per manufacturer instructions, utilizing library size and DNA concentration.

3.5 Bioinformatics

Bioinformatic analysis for DM-Seq is analogous to previously published pipelines for analysis of ACE-Seq data. The spike-in samples can generate metrics for DM-Seq, with pUC19 reflecting the efficiency of unmodified C protection, mλ reflecting the efficiency of 5mC deamination by A3A, and T4 hmC phage reflecting the efficiency of 5hmC protection, all in CpG contexts.

3.5.1 Trim Reads

1. Trim low-quality bases, adapters, and 5′ end.
trim_galore --fastqc --gzip --clip_R1 15 --length 20 SAMPLE.fastq.gz

3.5.2 Align Reads

1. Align reads to appropriate genome(s).
bismark --genome <path to SAMPLE_GENOME_FOLDER> --bowtie2 --fastq SAMPLE_trimmed.fq

3.5.3 Deduplicate Reads

1. Sort and index files for PCR deduplication.
   java -jar /path/to/picard.jar SortSam
   INPUT=SAMPLE_trimmed_bismark_bt2.bam
   OUTPUT=SAMPLE_sorted_trimmed_bismark_bt2.bam
   SORT_ORDER=coordinate CREATE_INDEX=true
2. Mark and remove PCR duplicates.
   java -jar /path/to/picard.jar MarkDuplicates
   ASSUME_SORTED=true
   REMOVE_DUPLICATES=true
   DUPLICATE_SCORING_STRATEGY=RANDOM
   INPUT=SAMPLE_sorted_trimmed_bismark_bt2.bam
   OUTPUT=SAMPLE_dedup_trimmed_bismark_bt2.bam
   METRICS_FILE=SAMPLE_dedup_trimmed_bismark_bt2.
   bam.metrics.
   txt VALIDATION_STRINGENCY=SILENT CREATE_IN-
   DEX=true

3.5.4 Filter Reads

A3A only acts on single-stranded DNA. As we have previously reported, we have found that ≥3 consecutive CpHs (unmethylated and significantly more favorable A3A substrates than 5mCs) that fail to convert are mechanistically suggestive of reannealing of double-stranded DNA after heat denaturation. Filtering of reads showing evidence of non-conversions due to strand reannealing can improve 5mC sequencing accuracy. Bismark has a stand-alone program, filter_non_conversion, that has the ability to filter based on consecutive CpHs that were not converted by the A3A enzyme.

1. Filter reads with three consecutive CpHs.
   filter_non_conversion --single --threshold 3 --consecutive
   SAMPLE_dedup_trimmed_bismark_bt2.bam

3.5.5 Extract and Analyze Reads

1. Extract reads with a MAPQ of ≥30 and obtain bedgraph files.
   samtools view -bq 30
   SAMPLE_filtered_dedup_trimmed_bismark_bt2.bam >
   SAMPLE_MAPQ30_filtered_dedup_trimmed_bismark_bt2.
   bam
2. Identify CpH and CpG C-to-T conversion rates for each genome in the sample.
   bismark_methylation_extractor --single-end --buffer_size 10G
   --gzip --bedGraph --counts --report
   SAMPLE_MAPQ30_filtered_dedup_trimmed_bismark_bt2.
   bam.

3.5.6 Extract Methylation Information

As the N374K mutant MTase acts in CpG contexts, DM-Seq specifically detects modifications in CpG contexts. Cytosines in CpH contexts are expected to be deaminated and read as T. The percentage read as T in the CpH context can thereby reflect A3A deamination efficiency. The above output report file can be used to quantify the accuracy of C, 5mC, and 5hmC detection. For unmethylated pUC19 plasmid DNA, CpHs should read as T, reflecting unmodified C conversion by A3A, while CpGs, protected by the CxMTase, should read as C. For the CpG-methylated M.SssI-treated lambda phage gDNA, CpHs should read as T, reflecting unmodified C conversion by A3A, while CpGs should also read as T, reflecting 5mC conversion by A3A. For hydroxymethylated T4-hmC DNA, all C bases should read as C, reflecting the protection by βGT and lack of conversion by A3A.

4 Notes

1. The CxMTase can modify the plasmid and genomic DNA; therefore, using a strain deficient in restriction of methylated DNA, such as ER1821 or T7 express, is necessary to avoid toxicity.
2. Since the CxMTase was expressed overnight at 16°C, precautions should be taken to keep everything cold during handling. Ideally, the purification should take place using only pre-chilled instruments when called for or carried out in a 4°C cold room.
3. The lysate should flow through the column at a rate of about 1 mL/min. At this rate, sufficient time is allotted for the protein to bind to the resin. If it flows through too fast, multiple passes of the lysate through the equilibrated HisPur Cobalt Resin may be required.
4. In the gel, fractions flanking a high salt wash are shown for washes that were performed with lysis buffer A. In the final protocol in this chapter, these washes were deemed unnecessary.
5. CxSAM will run with a lower retention factor (R_f) than SAH and IAA.
6. While easy to produce, other methylated CpG DNA spike-ins are also available for commercial purchase and are appropriate for use.
7. A master mix can be prepared for each of the reactions, with the exception that the annealed adapter should be added separately for the ligation reaction. Also, as these are enzymatic reactions, thorough mixing (without vortexing) is essential.
8. One strength of the enzymatic methods is that, as they are non-destructive, they can support the generation of libraries from longer DNA sequences relative to methods like BS-Seq. Shearing to longer lengths can potentially be partnered with longer read-length platforms.
9. Concentration of the adapter can be lowered to 1.5 μM instead of 15 μM if input DNA < 100 ng.
10. This is an optional stoppage point. If stopping here, store at −20°C.

11. We found that the copy strand does not readily sequence based on the spike-in controls. The fact that the 5mC bases are deaminated explains why the copy strands are not amplified during library construction.
12. We found that a methylated copy primer increases copy strand efficiency, possibly due to a higher annealing temperature than an unmethylated version of the primer.
13. *Bst* DNA polymerase, Large Fragment was chosen since it lacks 5′–3′ exonuclease activity but has excellent strand displacement activity.
14. Make sure the sample cools to 4°C before adding Mung Bean Nuclease. This step is an added precaution in the protocol and is optional.
15. This is the critical enzymatic reaction that will convert cytosine in unmodified CpGs to 5cxmC, which is resistant to A3A deamination. 16 h was selected for a convenient overnight protocol; however, shorter incubations may be possible. Omission of carboxymethylation and copy strand steps sequences 5hmC at base resolution, analogous to ACE-Seq.
16. Expression and purification of A3A can be performed based on the protocol provided in our previous publication [16]. Commercial A3A is also available.
17. As A3A acts only on single-stranded DNA, it is crucial to transfer the samples swiftly from high to low temperature to prevent reannealing.
18. The purpose of the qPCR is to determine an approximate number of indexing cycles for each sample, as over-indexing leads to bias in the final sequencing libraries. If you choose to skip this step, provided is a table of amplification cycles that have consistently worked for us.

Input (ng)	Cycle count
10	11 cycles
50	9 cycles
200	7 cycles

Acknowledgments

The authors would like to thank Christina M. Hurley for her critical reading of this book chapter. This work was supported by the National Institutes of Health through grants no. R01-HG010646 (to R.M.K.) and F31-HG012892 (to C.E.L.).

References

1. Schubeler D (2015) Function and information content of DNA methylation. Nature 517:321–326
2. Luo C, Hajkova P, Ecker JR (2018) Dynamic DNA methylation: in the right place at the right time. Science 361:1336–1340
3. Jurkowska RZ, Jeltsch A (2022) Enzymology of mammalian DNA methyltransferases. Adv Exp Med Biol 1389:69–110
4. Kohli RM, Zhang Y (2013) TET enzymes, TDG and the dynamics of DNA demethylation. Nature 502:472–479
5. Booth MJ, Raiber EA, Balasubramanian S (2015) Chemical methods for decoding cytosine modifications in DNA. Chem Rev 115:2240–2254
6. Wang T, Loo CE, Kohli RM (2021) Enzymatic approaches for profiling cytosine methylation and hydroxymethylation. Mol Metab 57:101314
7. Shen H, Laird PW (2013) Interplay between the cancer genome and epigenome. Cell 153:38–55
8. Huang Y, Pastor WA, Shen Y et al (2010) The behaviour of 5-hydroxymethylcytosine in bisulfite sequencing. PLoS One 5:e8888
9. Tanaka K, Okamoto A (2007) Degradation of DNA by bisulfite treatment. Bioorg Med Chem Lett 17:1912–1915
10. Schutsky EK, DeNizio JE, Hu P et al (2018) Nondestructive, base-resolution sequencing of 5-hydroxymethylcytosine using a DNA deaminase. Nat Biotech 36:1083–1090
11. Schutsky EK, Nabel CS, Davis AKF et al (2017) APOBEC3A efficiently deaminates methylated, but not TET-oxidized, cytosine bases in DNA. Nucleic Acids Res 45:7655–7665
12. Vaisvila R, Ponnaluri VKC, Sun Z et al (2021) Enzymatic methyl sequencing detects DNA methylation at single-base resolution from picograms of DNA. Genome Res 31:1280–1289
13. Liu Y, Siejka-Zielinska P, Velikova G et al (2019) Bisulfite-free direct detection of 5-methylcytosine and 5-hydroxymethylcytosine at base resolution. Nat Biotechnol 37:424–429
14. Wang T, Fowler JM, Liu L et al (2023) Direct enzymatic sequencing of 5-methylcytosine at single-base resolution. Nat Chem Bio 19:1004–1012
15. Wang T, Kohli RM (2021) Discovery of an unnatural DNA modification derived from a natural secondary metabolite. Cell Chem Biol 28:97–104
16. Wang T, Luo M, Berrios KN et al (2021) Bisulfite-free sequencing of 5-hydroxymethylcytosine with APOBEC-coupled epigenetic sequencing (ACE-Seq). Methods Mol Biol 2198:349–367

Check for updates

Chapter 2

FENGC: Targeted, Multiplexed, and Integrated Single-Molecule Profiling of Chromatin Accessibility and DNA Methylation

Mingqi Zhou, Marie-Pierre L. Gauthier, Nancy H. Nabilsi, Kevin O. Murray, Deborah A. Morse, Philip A. Efron, Alberto Riva, Patrick Concannon, Thomas L. Clanton, Jason O. Brant, Rhonda Bacher and Michael P. Kladde

Abstract

DNA methyltransferase accessibility protocol for individual templates (MAPit) enables localization of DNA-protein interactions by probing the native chromatin architecture in nuclei or permeabilized cells using exogenously applied DNA methyltransferases. By preserving the integrity of DNA in chromatin, MAPit enables simultaneous mapping of nucleosome positioning, transcription factor binding, and endogenous CpG methylation in a single assay—while maintaining linkage between regulatory modules along chromatin fibers. Although MAPit and other sequencing-based approaches can be applied genome-wide, there is a growing demand for orthogonal approaches that validate target discovery and offer deeper molecular resolution at defined loci. Here, we describe a detailed protocol for flap-enabled next-generation capture (FENGC), a cost-effective method for targeted, multiplexed enrichment of DNA sequences for epigenetic and genetic analysis. FENGC precisely excises and enriches tens to hundreds of user-defined targets, generating sequencing-ready libraries with superior on-target rates compared to existing methods, all at a fraction of the cost. When paired with MAPit, the MAPit-FENGC platform enables single-molecule resolution of CpG methylation patterns in the context of chromatin architecture and transcription factor occupancy.

Keywords Targeted sequencing, Flap endonuclease, Epigenetics, Chromatin accessibility, DNA methylation, Single molecule footprinting

1 Introduction

Chromatin consists of DNA and its full repertoire of associated proteins, forming chromosomes within eukaryotic nuclei [1]. Fractional occupancy—the quantitative degree to which nuclear proteins directly contact DNA—is governed by binding affinity, usually defined by the dissociation rate constant of macromolecular complexes such as nucleosomes and DNA-bound transcription factors (TFs). Regulated

Archana Dhasarathy and Motoki Takaku (eds.), *Cancer Epigenetics*, Methods in Molecular Biology, vol. 3005, https://doi.org/10.1007/978-1-0716-5072-1_2,

access to specific DNA sequences comprising *cis*-acting elements within enhancers and promoters is a central mechanism of transcriptional control [2]. Dynamic remodeling of chromatin structure, which modulates chromatin accessibility across the genome, reflects the convergence of gene regulatory networks, covalent modification of chromatinized DNA, and recruitment of chromatin remodeling complexes [3, 4]. Chromatin accessibility is also closely tied to cell state specification, and its measurement has been shown to more accurately define cell identity than transcriptomic assays alone [5].

In addition to proper modulation of accessibility at *cis*-acting elements, cells must tightly regulate C-5 DNA methylation (m^5C) at CpG motifs (hereafter, CG). Aberrant hypermethylation of CG islands (CGIs), particularly within GC-rich promoters of tumor suppressor genes, is a hallmark of cancer and often coincides with increased nucleosome occupancy [6]. A frequent driver of such epigenetic dysregulation is constitutive activation of the RAS/RAF/MEK/ERK signaling cascade [7]. In colorectal cancer (CRC), for instance, nearly half of tumors harbor mutually exclusive activating mutations in *KRAS* or *BRAF* [8]. These oncomutations are closely linked to widespread promoter silencing via DNA hypermethylation, a phenomenon known as the CpG island methylator phenotype (CIMP) [9]. Together, these epigenetic alterations contribute to transcriptional repression and cellular transformation, underscoring the need for techniques to precisely interrogate chromatin structure.

Therefore, a number of genome-wide methods for the global interrogation of chromatin accessibility have been developed. These include the widely used genomic nuclease-based methods, DNase-seq, MNase-seq, and ATAC-seq, which employ DNase I, micrococcal nuclease, and Tn5 transpososome, respectively, to preferentially cleave accessible or protein-free DNA sequences within chromatin [10–13]. Following library preparation and short-read sequencing, the ends of reads, which correspond to chromatin cleavage sites, are aligned to the reference genome. These nuclease-based approaches yield highly useful, albeit qualitative, assessments of chromatin structure, producing population-averaged, relative peaks of accessibility that distinguish "open" from "closed" regions.

Methyltransferase accessibility protocol for individual templates (MAPit) provides a robust alternative to nuclease-based assays for probing chromatin architecture [14, 15]. This approach utilizes an exogenously supplied C-5 DNA methyltransferase (DNMT) to selectively methylate accessible DNA sequences in cellular chromatin [14–20], biochemically reconstituted chromatin [21, 22], or in vitro DNA-TF binding reactions [23, 24]. DNMT-based probes preserve DNA integrity, thereby maintaining the spatial organization of individual gene regulatory modules. Furthermore, C-5 DNMTs enable nucleotide-resolution detection of 5-methylcytosine (m^5C) via bisulfite-mediated conversion of unmethylated cytosine (C) to uracil

(U), followed by PCR amplification of targeted genomic regions [14–20, 25, 26]. Bisulfite sequencing further determines the methylation status of all cytosines across every sequenced read or molecule [25]. Bisulfite sequencing of C-5 DNMT-probed chromatin thus offers a unique opportunity not only to localize TF binding or nucleosome positioning along DNA, but also to quantify their fractional occupancies with single-molecule precision [14, 15, 19, 20, 27, 28].

MAPit analysis of mammalian chromatin, which contains substantial levels of endogenous m^5CpG (hereafter, m^5CG), was enabled by the discovery and commercialization of M.CviPI, the recombinant chloroviral C-5 DNMT [29]. Due to its GC specificity, M.CviPI only partially overlaps with CG sites, specifically at GCG. Therefore, to avoid ambiguity with endogenous m^5CG, GCG sites are masked during sequencing analysis. This allows MAPit with M.CviPI to jointly survey native m^5CG at HCG (where $H \neq G$) and chromatin accessibility at GCH sites [15, 27, 30–32]. This approach is applicable to genome-wide studies using short-read sequencing, referred to as nucleosome occupancy and methylation sequencing (NOMe-seq), which yields population-averaged maps of Hm^5CG and chromatin accessibility [33].

By contrast, long-read sequencing further takes advantage of the molecular continuity of DNA and enables broader epigenetic phasing along individual chromatin fibers, offering single-molecule insights into epigenetic heterogeneity across cell populations [32]. Longer reads therefore support detailed, quantitative analysis of the interplay between TF binding, nucleosome positioning, and endogenous m^5CG.

Targeted sequencing has become indispensable for both foundational research and clinical diagnostics. Targeted detection of m^5C and genetic mutations reduces sequencing costs and enhances coverage depth, thereby improving metrics such as base-calling accuracy. Increased cost-effectiveness is also realized by maximizing the proportion of on-target sequences. Nevertheless, existing methods for multiplexed targeted sequencing remain limited. Hybridization-based enrichment using tiled biotinylated oligonucleotide baits is widely adopted but prohibitively expensive for many laboratories [34, 35]. More recently, CRISPR/Cas systems have shown promise for sequence enrichment [36–40], including MAPit/NOMe-seq [41]. However, the expense of recombinant Cas enzymes and the requirement for up to four guide RNAs per target constitute substantial challenges to scalability [41]. Both approaches also suffer from high off-target sequencing rates, low-efficiency recovery of target sequences, or both, necessitating large input amounts of DNA that constrain clinical applicability [34, 42–44].

Flap-enabled next-generation capture (FENGC) is a novel, cost-effective sequence enrichment alternative that supports profiling of both genetic and epigenetic heterogeneity [45] (Fig. 1a). The FENGC target-specific cleavage mechanism represents an advanced

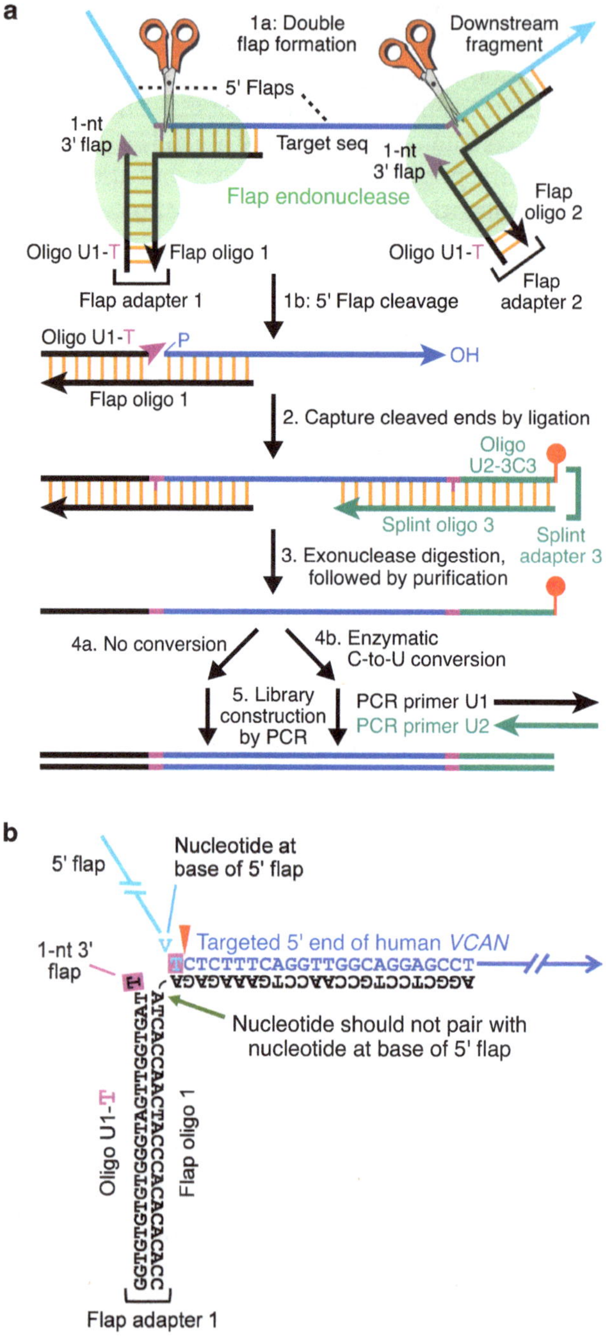

Fig. 1 FENGC target sequence enrichment. (**a**) FENGC workflow. Genomic DNA is denatured and hybridized with locus-specific flap adapters 1 and 2, forming double flaps (5′ flap and 1-nt 3′ flap) at both ends of each target sequence (Step 1a). The 1-nt 3′ flap arises from a one-nucleotide overlap (highlighted in purple) that displaces the 3′-terminal nucleotide of universal oligo U1-T. FEN1 cleaves at the scissile phosphodiester bond (scissors), releasing single-stranded target strands with ligatable ends: a 5′ phosphate and a 3′ hydroxyl (Step 1b). Thermostable ligase ligates the T of U1-T, which base pairs in the gap, to the 5′ phosphate of the target strand (Step 2). Concurrently, oligo U2-3C3 and splint oligos 3 form splint adapters that ligate to the 3′ hydroxyl ends.

adaptation of bisulfite patch PCR and MAPit-patch [32, 46]. Both patch methods rely on methylation-insensitive restriction endonucleases to cleave target sequences that may contain m^5CG and/or Gm^5C; only two four-base-pair-recognizing enzymes are currently commercially available for MAPit-patch [32]. In contrast, flap endonucleases, such as FEN1, recognize and cleave structured 5′ flaps with high efficiency and precision, regardless of DNA methylation status. To exploit this structural specificity, we devised a strategy in FENGC that assembles user-defined target sequences into double flaps, consisting of a 5′ flap and a one-nucleotide 3′ flap. Although FEN1 catalysis is sequence-independent, our preferred configuration requires a minimal two-base motif of T following a non-T to generate the double flap structure (Fig. 1b). This motif occurs frequently in genomic DNA (gDNA), providing an effective resolution of approximately one site per every five base pairs. These innovations make FENGC a robust, high-throughput, and scalable enrichment method, ideally suited for both discovery and targeted orthogonal validation in genome-wide and epigenome-wide association studies.

MAPit paired with FENGC (MAPit-FENGC), followed by non-destructive enzymatic-methyl sequencing (EM-seq) [47] and long-read Pacific Biosciences (PacBio) high-fidelity (HiFi) sequencing, achieves consistently high on-target rates (80–97%) from sub-microgram amounts of input DNA [45]. The performance and simplicity of MAPit-FENGC surpass hybridization- and CRISPR/Cas-based strategies, delivering at least a 16-fold reduction in cost per sample. In this chapter, we present a detailed MAPit-FENGC protocol, complete with best practices, technical notes, troubleshooting guidance, and updates of our previous MAPit protocol published in this series [15].

Fig. 1 (continued) Covalent modification of the U2-3C3 3′ hydroxyl (red lollipop) protects ligated products from exonuclease I and III digestion, which degrades unprotected DNA in the 3′ to 5′ direction, enriching target sequences (Step 3). All steps are performed sequentially in a single tube to minimize DNA loss. Enriched targets are then purified and amplified either by standard PCR (no conversion) or methyl-PCR (with enzymatic C-to-U conversion) to construct libraries for high-depth sequencing (Step 4). (**b**) Flap adapter-directed 5′ flap scission at the *VCAN* promoter (*see* Fig. 4). Shown is the structure and sequences of flap adapter 1 hybridized to the 5′ end of the upper strand of the human *VCAN* promoter, a known target of hypermethylation in CIMP. In multiplexed FENGC, flap adapter 2 targets the 3′ end (not shown). In both adapters, immediately 5′ of the FEN1 cleavage site (red arrowhead) a T base pairs with A in the flap oligo, displacing the 3′-terminal T of U1-T to form a 1-nt 3′ flap. Both T residues are highlighted in purple to emphasize their competitive overlap. To restrict displacement to one nucleotide, a V (A, C, or G; cyan color) is placed at the base of the 5′ flap to avoid pairing with the A in oligo U1-T (green arrow). The C immediately 3′ of the pictured cleavage site defines the target strand 5′ end, whereas the T immediately 5′ of the cleavage site defines the target strand 3′ end (not shown). Sequence coloration matches that in panel (**a**)

2 Materials

Unless indicated otherwise, prepare all solutions and reagents using molecular-grade water and store at room temperature.

Caution: Several hazardous compounds are used in these protocols, including but not limited to ethanol, digitonin, and sodium hydroxide used to pH solutions. Train personnel in the use of the materials listed herein and take appropriate precautions for use and waste disposal as set forth by manufacturers and institutional policies.

2.1 MAPit Reagents and Solutions

1. Optional: Bambanker serum-free cell freezing media (Bulldog Bio BB05).
2. Cell resuspension buffer 2×: 40 mM HEPES, pH 7.5; 140 mM NaCl; 0.5 mM ethylenediaminetetraacetic acid (EDTA), pH 8.0; 1 mM ethylene glycol tetraacetic acid (EGTA), pH 8.0; 1% (w/v) glycerol (*see* **Note 1**). Sterilize the solution by filtration through a 0.2 μm filter to avoid oxidation of the glycerol that would otherwise occur with autoclaving. Store at 4°C.
3. 1 M dithiothreitol (DTT): Dissolve crystals in molecular-grade water. Aliquot and store at −20°C.
4. 100 mM phenylmethylsulfonylfluoride (PMSF): Dissolve crystals in absolute isopropanol and store at −20°C.
5. Cell resuspension buffer 1×: Dilute 2× cell resuspension buffer to 1× with molecular-grade water and, after chilling on ice, add 20 mM DTT and 0.5 mM PMSF immediately before use (*see* **Note 1**).
6. 2.5% (w/v) Digitonin: Freshly dissolve digitonin powder in molecular-grade water by heating for 10 min at 95°C in a heat block. *Caution*: Digitonin has an acute oral toxicity. We recommend taring a 1.7-mL microcentrifuge tube, dispensing some powder to the tube in a chemical fume hood, and reweighing the tube to determine the amount of digitonin.
7. Cell permeabilization buffer: 1× Cell resuspension buffer containing 0.5% (w/v) digitonin (*see* **Note 2**).
8. Trypan blue, 0.4% (w/v).
9. 50 U/μL GpC methyltransferase (M.CviPI). Aliquot the enzyme to avoid contaminating the entire stock. To buffer against temperature fluctuations, store M.CviPI (and other enzymes) at −20°C in a benchtop cooler placed in a non-frost-free freezer. Always keep enzymes in a −20°C benchtop cooler when transporting or using them at the bench. Using these best practices, M.CviPI activity is stable for at least one year (*see* **Note 3**).
10. GpC methyltransferase (M.CviPI) storage buffer: 15 mM Tris–HCl, 0.2 mM NaCl, 1 mM DTT, 0.1 mM EDTA,

200 μg/mL acetylated bovine serum albumin (BSA), 50% (v/v) glycerol, pH 7.4 at 25°C. Not supplied with purchased M.CviPI. Use of a molecular-grade BSA is essential. Store at −20°C in a non-frost-free freezer.

11. 32 mM *S*-adenosyl-*L*-methionine (SAM): Supplied in 5 mM sulfuric acid, 10% (v/v) ethanol with purchased M.CviPI. Store at −80°C in 5 μL aliquots, each sufficient for two independent chromatin-probing reactions; adjust the aliquot volume as needed.
12. Methylation stop buffer 2×: 100 mM NaCl, 10 mM EDTA, pH 8.0, 1% (w/v) sodium dodecyl sulfate. Store at ambient temperature.
13. 20 mg/mL Proteinase K. Store up to 4 months at −20°C in a non-frost-free freezer.
14. Suitable gDNA purification kit or equivalent protocol that completely removes RNA.
15. Enzymatic conversion of cytosine (C) to uracil (U) (e.g., enzymatic methyl seq (EM-seq) conversion module, New England Labs) (*see* **Note 4**).
16. Formamide (≥99.5%; e.g., Sigma–Aldrich).
17. AMPure XP Beads or equivalent (e.g., Beckman Coulter).
18. 80% (v/v) Ethanol: 100% absolute ethanol diluted with molecular-grade water (*see* **Note 5**). *Caution*: Ethanol is flammable; do not autoclave or use ethanol-containing solutions near a heat source or open flame.
19. Optional: A mix of hot-start Platinum™ Taq DNA polymerase (ThermoFisher 10966026) and PfuTurbo Cx Hotstart DNA polymerase (Agilent 600410) in a 7 U:1 U ratio. Store at −20°C in a non-frost-free freezer in a benchtop cooler (*see* **Note 6**).
20. Optional: Deoxynucleotide triphosphates (dNTPs) for PCR.
21. Optional: 50 mM $MgCl_2$.
22. Optional: CoralLoad PCR buffer. Supplied with HotStarTaq Plus DNA Polymerase kit (e.g., Qiagen).
23. Optional: PCR primers for amplification of promoters from C-to-U deaminated gDNA, including human *HSPA5*, divergently transcribed with *HSPA5-DT* (AAACACCCCAATAaaTCAATC, melting temperature (T_m) = 53.4°C; GAGAAGAAAAAGTTTtAGATtttAtAG, T_m = 57.6°C, where lowercase "t" and "a" represent C and G originally present in the genome, respectively); human *COX8A* (tTGGTGTtAAAtTttTGAttTtAGA, T_m = 48.2°C; CATTCTCAaaACCACCTaAa, T_m = 49.1°C); and mouse *Mlh1* (GyTtTtAGAAATGAGytAATAGGAAGAG, T_m =

52.3-55.6°C, range due to the presence of degenerate bases; ACACTCTaAAaACTaTTTCTCTTCCAC, T_m = 55.0°C; where y represents GC or Gm^{5}C in GC motifs originally not methylated and methylated by M.CviPI, respectively, in chromatin) [27]) (*see* **Note 7**).

2.2 FENGC Reagents and Solutions

1. Thermostable FEN1 (flap endonuclease 1, e.g., recombinant enzyme purified from *E. coli* expressing the gene for thermostable *Thermococcus* 9° N, New England Biolabs).
2. Universal oligonucleotide 1 with 3′-terminal T (oligo U1-T): GGTGTGTGTGGGTAGTTGGTGATT; 10 μM stock; T_m = 60.9°C (Fig. 1) (*see* **Notes 8 and 9**).
3. Universal oligo 2 (oligo U2-3C3): /5Phos/ AGAGAATGAGGAAGGTGG*G*G*T*A*G/3SpC3/; 10 μM stock; T_m = 61.0°C, where * are phosphorothioate bonds and 3SpC3 is a three-carbon spacer modifying the 3′ -hydroxyl group; both covalent modifications are resistant to 3′ to 5′ exonuclease activity (*see* **Notes 8 and 9**).
4. Panels of target-specific oligos: These include flap oligos 1 and 2 (N_n**AATCACCAACTACCCACACACACC**) and splint oligos 3 (CTACCCCACCTTCCTCATTCTCT**A**N_n). The variable regions N_n, together with adjacent constant **A**, are designed by program FOLD to hybridize specifically to the 5′ end (flap oligo 1) and the 3′ end (flap oligo 2 and splint oligo 3) of each target sequence (Fig. 1; *see* **Notes 9 and 10**). In flap oligos 1 and 2, N_n**A** precedes a constant region complementary to oligo U1-T, whereas in splint oligos 3, **A**N_n follows a different constant region complementary to oligo U2-3C3 (*see* **Notes 9 and 10**). Prepare **separate** working stock solutions of all three oligos, each containing 0.5 μM of each constituent oligo (*see* **Note 11**).
5. PCR amplification primers for constructing sequencing libraries: These include universal primer 1 (primer U1; GGTGTGTGTGGGTAGTTGGTGAT; T_m = 63°C, 10 μM stock) and
   primer U2 (CTACCCCACCTTCCTCATTCTCT; T_m = 61°C, 10 μM stock) (*see* **Note 12**).
6. Thermostable DNA Ligase Kit (e.g., Ampligase, LGC Biosearch Technologies).
7. 20,000 U/mL Exonuclease I (e.g., New England Labs): Catalyzes 3′ to 5′ removal of nucleotides from linear single-stranded DNA (ssDNA; *see* **Note 13**).
8. 100,000 U/mL Exonuclease III (e.g., New England Labs): Catalyzes 3′ to 5′ removal of nucleotides from linear double-stranded DNA (*see* **Note 13**).

9. Enzymatic C-to-U conversion (e.g., EM-seq conversion module, New England Labs) (*see* **Note 4**). Do not substitute with bisulfite conversion.
10. Formamide (≥99.5%; e.g., Sigma–Aldrich).
11. 80% (v/v) Ethanol: 100% absolute ethanol diluted with molecular-grade water (*see* **Note 5**). *Caution*: Ethanol is flammable; do not autoclave or use ethanol-containing solutions near a heat source or open flame.
12. AMPure XP Beads or equivalent (e.g., Beckman Coulter).
13. KAPA HiFi HotStart Uracil+ DNA polymerase ReadyMix Kit (Roche 07959052001). Do not substitute with Platinum™ Taq DNA polymerase, PfuTurbo Cx Hotstart DNA polymerase, or HotStarTaq Plus DNA polymerase. In our experience with FENGC, none amplifies as well as the KAPA enzyme.

2.3 Computational Resources

All programs are publicly available at no charge. Refer to the accompanying README files for detailed instructions on installation and usage.

For determination of M.CviPI probing efficiency on single amplicons with open chromatin:

1. Optional: bisulfitePCRSeqAnalysis.m (https://github.com/jobrant/methylsanger-matlab).
   A MATLAB program that processes aligned C-to-U-converted Sanger sequencing data to generate plots showing the total fraction of each base at all cytosine positions.
   For FENGC sequence enrichment:
2. FENGC oligo designer (https://github.com/uf-icbr-bioinformatics/FOLD). Used for designing target sequences and their corresponding oligos 1–3. 2. A plasmid editor (ApE; https://jorgensen.biology.utah.edu/wayned/ape [48] or equivalent). Use to convert Sanger sequencing .ab1 files to .scf, one of two input files required by bisulfitePCRSeqAnalysis.m. Also, use Ape or equivalent in the few cases where FOLD cannot identify a matched target and oligos 1–3 that satisfy the condition requirements, necessitating manual design.
3. Primer3web (https://primer3.ut.ee; [49]) or equivalent. Primer3 is used by FOLD for T_m calculations, as well as stability of primer dimers and secondary structures (if installed locally). If not installed, FOLD uses a built-in, Primer3-based approach for T_m calculations. Primer3 (or equivalent) is also used directly when manual design of target sequences and oligos 1–3 is necessary.
4. ReAminator pipeline (https://github.com/jobrant/MAPit-Pipeline; [27]). Includes programs bsAlign, bsDraw, and all

necessary helper functions. bsAlign aligns C-to-U deaminated sequences to reference sequences, bsDraw extracts aligned reads from bsAlign that meet user-defined parameters (e.g., alignment length or bisulfite conversion efficiency). Extracted reads are saved to a locus-specific alignment FASTA file.

5. MethylMapper (https://github.com/uf-icbr-bioinformatics/methylmapper). Converts aligned FASTA files from bsAlign and bsDraw into numerical code used to generate color-coded plots for visualizing DNA methylation patterns and generates nucleotide frequency tables.
6. Methylscaper (https://rbacher.rc.ufl.edu/methylscaper; [50]). A Shiny R package for alignment and visualization of MAPit-FENGC single-molecule data. The program accepts FASTA files generated by Bismark or similar alignment tools and offers significant improvements over methylmapper. Notably, methylscaper enhances initial clustering of reads and provides functionality for manual weighting and refinement based on specific regions of an amplicon and/or clusters, enabling more accurate and customizable data interpretation.

3 Methods

All steps are performed on ice unless specified otherwise.

3.1 MAPit Probing of Accessible DNA in Native Chromatin

1. Pre-chill a centrifuge to 4°C.
2. Harvest tissue or cells at room temperature (*see* **Note 14**). If needed, process them to obtain a single-cell suspension, and chill the cells on ice. Thaw cells cryopreserved in Bambanker on ice. To decrease viscosity of Bambanker when thawed, add 800 μL of ice-cold 1× cell resuspension buffer (without DTT and PMSF) and gently tap each tube to mix.
3. Collect cell pellets by centrifugation for 5 min at 1000 × *g* at 4°C.
4. To wash the cells, resuspend the cell pellet in 500 μL of ice-cold 1× cell resuspension buffer, containing freshly added DTT and PMSF (*see* **Note 15**).
5. Count the washed single-cell suspension using a hemocytometer or automated cell counter. If available, pipette 10^6 cells to a new 1.7-mL microcentrifuge tube (*see* **Note 14**).
6. Collect cell pellets by centrifugation for 5 min at 1000 × *g* at 4°C.
7. Resuspend each cell pellet in 92 μL of ice-cold cell permeabilization buffer, i.e., 1× cell resuspension buffer containing freshly made and added 0.5% (w/v) digitonin (*see* **Note 16**).

Gently tap the tubes until a homogeneous single-cell suspension is observed.

8. Incubate the single-cell suspension for 10 min on ice. Proceed to Step 9 right away and to Step 10 after 8 min of permeabilization.
9. While permeabilization is in progress, make the M.CviPI working buffer solution (without enzyme) on ice in a chilled 1.7-mL microcentrifuge tube by adding the following components in the listed order. Scale the volumes as needed based on the number of experimental chromatin-probing reactions (*see* **Note 15**):
   (i) 2× Cell resuspension buffer, with 20 mM DTT and 0.5 mM PMSF 5.25 μL/reaction
   (ii) Molecular-grade water 0.85 μL/reaction
   Briefly vortex
   (iii) 32 mM SAM (*see* **Note 17**) 2.2 μL/reaction

   Briefly vortex (always mix contents before adding enzymes) This totals to 8.3 μL per reaction, which already includes 5% extra to account for pipetting error and leaves room for M.CviPI addition in Step 15. Postponing the addition of concentrated M.CviPI maintains as much activity as possible for the 15-min chromatin structure probing reaction in Step 16.
10. To determine if essentially 100% cell permeabilization has been achieved, mix 2 μL of the single-cell suspension with 2 μL of 0.4% (w/v) trypan blue staining solution (*see* **Note 16**).
11. Examine the cells by light microscopy at 100–400x magnification. Permeabilized, stained cells should appear blue and opaque, as they remain intact, round, and slightly swollen. In contrast, non-permeabilized cells appear clear and reflective (producing glare under incident light), and are noticeably smaller.
12. Transfer 90 μL of each permeabilized single-cell suspension to a 1.7-mL microcentrifuge tube pre-chilled on ice.
13. Pre-incubate the permeabilized single-cell suspensions for 2 min at 37°C.
14. Pre-incubate the methylation stop buffer for at least 10 min at 50°C.
15. Scaling for the total number of reactions as in Step 7, add 2.2 μL of 50 U/μL M.CviPI per reaction to the M.CviPI working buffer solution to make a M.CviPI working stock of 10.5 μL/reaction (*see* **Note 18**). Briefly vortex to mix.
16. For chromatin probing reactions containing 100 U M.CviPI:
   (a) Add 10 μL M.CviPI working stock solution to each 90 μL of pre-incubated single-cell suspension. Stagger the

addition of M.CviPI to each tube by 30 s to ensure all reactions are incubated the same length of time.

(b) Mix by gently pipetting up and down five times. Incubate for 15 min at 37°C. To minimize sedimentation of cells, gently tap each tube every ~5 min.

17. At the end of the 15-min incubation period, and in accordance with each sample's 30-s staggered start time, add 100 μL of 2× methylation stop buffer preheated to 50°C. Immediately vortex each tube briefly and vigorously to ensure complete cell lysis.
18. Add 2 μL of 20 mg/mL Proteinase K, achieving a final concentration of 200 μg/mL. Briefly vortex and incubate overnight at 50°C.
19. Purify gDNA according to the manufacturer's instructions using any suitable commercially available kit that completely removes RNA.
20. Quantify DNA concentration using a Nanodrop or equivalent spectrophotometer that measures microvolumes. Store the DNA samples at −20°C.

3.2 Assess M.CviPI Probing Efficiency

This section is optional but recommended for new users of MAPit and when purchasing new lots of M.CviPI. In such cases, validate MAPit efficacy before proceeding to the FENGC reaction and next-generation sequencing. The same single amplicon approach can be applied to individual regions to provide orthogonal corroboration of MAPit-FENGC results (see Sect. 3.3).

1. Select a region known to have highly accessible chromatin and non-detectable or background levels of m^{5}CG. Any such region can serve to initially determine the extent to which accessible GCH motifs were modified by M.CviPI. The best-suited regions are usually promoters of housekeeping genes that are transcriptionally active in most cell types. For human cells, we assay the promoters of *HSPA5* (*GRP78* [19]) or *COX8A*, whereas for cells isolated from mice, we assay the *Mlh1* promoter [27] (see Step 3).
2. Deaminate purified gDNA. Use the maximum permitted amount of purified gDNA input (e.g., 200 ng with the EM-seq conversion module; *see* **Note 19**) according to the manufacturer's instructions, with the following stipulations:

   (a) Use the formamide method to denature the gDNA and maintain single-strandedness, which is required for conversion of C to U, as denaturation with sodium hydroxide did not yield satisfactory results in our hands.

   (b) At the end of the conversion protocol, elute the gDNA with 20 μL of molecular-grade water, transfer the supernatant to a new low-bind PCR tube, repeat the elution with another 20 μL of molecular-grade water, and combine the supernatants for a final volume of 40 μL (*see* **Note 20**).

3. PCR amplify sequences from the known open region(s) selected in Step 1. Use of the primers indicated in Sect. 2.1 generates amplicons for human *HSPA5* (*GRP78*; *see* Step b), human *COX8A* (*see* Step b), and mouse *Mlh1* (*see* Step e) of 454, 727, and 724 bp, respectively. Two independent PCRs are performed for each DNA eluate to mitigate potential stochastic amplification bias [14].

   (a) Pipette 2 μL of each purified, deaminated gDNA sample (100 ng) into a PCR tube on ice.

   (b) For amplification of promoters of human *HSPA5* or *COX8A*: Make a working reaction mix by combining the following in order in a 1.7-mL microcentrifuge tube on ice. Scale the volumes as needed, making 5–10% extra to account for pipetting error:

      (i) Molecular-grade water 8.9 μL/reaction

      (ii) CoralLoad PCR buffer 1.5 μL/reaction

      (iii) dNTPs (2.5 mM in each) 1.2 μL/reaction

      (iv) *HSPA5* or *COX8A* primers (10 μM) 1.3 μL/reaction

   Briefly vortex

      (v) Platinum™ Taq : PfuTurbo Cx mix (7:1) 0.1 μL/reaction

   Briefly vortex to mix and tap spin if necessary.

   (c) Pipette 13 μL of the working reaction mix into each tube containing 2 μL of the gDNA sample.

   (d) To amplify *HSPA5* or *COX8A* promoters, incubate in a thermocycler with a heated lid as follows: Begin with an incubation of 3 min at 95°C; followed by 37 cycles of 30 s at 95°C, 30 s at 56°C, and 90 s at 72°C; a final extension of 10 min at 72°C; and end with a hold at 4°C.

   (e) For amplification of the promoter of mouse *Mlh1*: Scaling the volumes as needed, including 5–10% extra, make a working reaction solution by combining the following in order in a 1.7-mL microcentrifuge tube on ice:

      (i) Molecular-grade water 16.6 μL/reaction

      (ii) CoralLoad PCR buffer 2.5 μL/reaction

      (iii) 50 mM $MgCl_2$ 0.5 μL/reaction

      (iv) dNTPs (2.5 mM in each) 2.0 μL/reaction

      (v) *Mlh1* primers (10 μM in each) 1.3 μL/reaction

   Briefly vortex

      (vi) Platinum™ Taq : PfuTurbo Cx mix (7:1) 0.1 μL/reaction.

Briefly vortex to mix and tap spin if necessary.

(f) Pipette 23 μL of *Mlh1* working reaction mix to each tube containing 2 μL of purified, deaminated gDNA sample.

(g) To amplify mouse *Mlh1* promoter, incubate in a thermocycler with a heated lid as follows: Begin with an incubation of 3 min at 95°C; followed by 37 cycles of 30 s at 95°C, 30 s at 50°C, and 90 s at 72°C; a final extension of 10 min at 72°C; and end with a hold at 4°C.

4. Purify the PCR products with AMPure XP beads (or equivalent). Perform all steps at room temperature according to the manufacturer's instructions (https://www.beckman.com/reagents/genomic/cleanup-and-size-selection/pcr/bead-ratio), with minor modifications.

(a) Per the instructions, for PCR products of the MAPit test region size range (454–727 bp), add 0.8× the volume of AMPure XP beads relative to the reaction volume.

(b) Briefly vortex, then incubate for 5 min on the benchtop to bind DNA to the beads.

(c) Wash the beads with bound DNA twice with 200 μL of 80% (v/v) ethanol each time.

(d) Air-dry the beads until all visible liquid evaporates and they remain dark brown.

(e) Elute the DNA with 11.5 μL of molecular-grade water (half the volume of the 23 μL PCR) and transfer to a microcentrifuge tube.

5. Quantify the PCR product. Use a Nanodrop or equivalent spectrophotometer that measures microvolumes.

6. Submit the purified PCR product for Sanger sequencing to your preferred vendor or core facility (*see* **Note 21**).

7. Analyze the PCR product sequences for DNA methylation. It is unnecessary to mask GCG sites if appropriate regions with non-detectable endogenous m^5CG were chosen. Consult the README file that accompanies bisulfitePCRSeqAnalysis.m for detailed instructions on the analysis of Sanger sequences of EM-seq-converted amplicons. The analysis generates stacked bar charts of the base frequencies at the C in all GC sites and CG sites in the amplicon (Fig. 2).

3.3 FENGC Enrichment of Target Sequences

FENGC is a streamlined, "one-pot" method for sequence enrichment, involving the sequential addition of reagents (Steps 1–6) to low-bind PCR tubes. This process enables double flap formation, 5′ flap cleavage, ligation of universal primers to the liberated target DNA strands, and 3′ to 5′ exonuclease digestion (Fig. 1). Because DNA recovery during purification is inherently inefficient, requiring only a single purification step significantly minimizes input DNA loss.

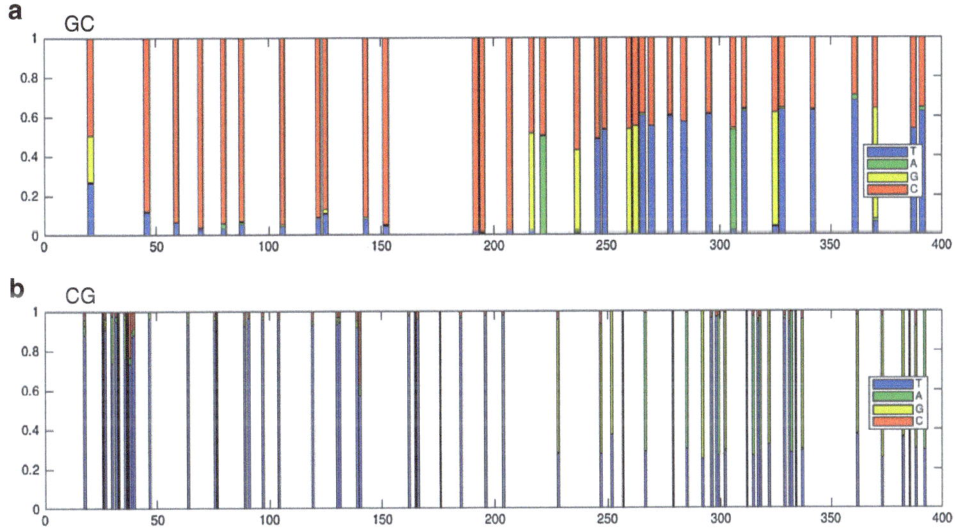

Fig. 2 Assessment of M.CviPI probing efficiency at the *HSPA5* promoter. Chromatin accessibility in digitonin-permeabilized HCT116 colorectal cancer cells was probed using 100 U of M.CviPI. Genomic DNA was extracted via phenol:chloroform, bisulfite-converted to induce C-to-U transitions, and PCR-amplified using primers described in Sect. 2.1, Step 23. The resulting amplicon was sequenced with primer GAGAAGAAAAAGTTTtAGATtttAtAG, and the Sanger electropherogram was analyzed using bisulfitePCRSeqAnalysis.m to estimate base proportions at cytosine positions within **a** GC and **b** CG motifs. Stacked bar charts display the relative abundance of each base (color-coded per key) across the sequenced region. High proportions of C in GC sites (*upper* panel, red), especially within the first ~ 200 nt, reflect efficient methylation by M.CviPI and correspondingly poor bisulfite conversion—indicative of accessible chromatin. In contrast, high proportions of T in CG sites (*lower* panel, blue) suggest minimal endogenous m^{5}CG methylation and thus efficient C-to-U conversion. Mixed base calls beyond position 210 likely result from declining sequencing quality at increasing distance from the primer binding site

1. Choose at least 10 target regions of interest, e.g., promoters. Depending on the number of PCR cycles used during library preparation (see Step 10, below), more targets may increase the amplification yield. To date, we have multiplexed 453 target regions with no reduction in assay performance (*see* **Note 10**).
2. Design targets and panels of target-specific oligos, flap oligos 1 and 2, as well as splint oligos 3, using program FOLD (*see* **Notes 9–11**).
3. Double flap formation and 5′ flap cleavage (Fig. 1, steps 1a and 1b; Fig. 1b; *see* **Note 9**). The time required for this reaction is 15 min of hands-on time and approximately 3 h of incubation.
   (a) Add 20 μL containing 150–800 ng gDNA to low-bind PCR tubes (*see* **Note 22**).
   (b) Combine the following in a 1.7-mL microcentrifuge tube on ice to make a working reaction mix. Scale the volumes

as needed based on the number of FENGC reactions, making 5–10% extra to account for pipetting error:

(i) Molecular-grade water 5.5 μL

(ii) 10× ThermoPol® reaction buffer 3.5 μL

(iii) 10 μM U1-T oligo (for additional options, *see* **Note 9**) 1.0 μL

(iv) 0.5 μM Flap oligos 1 (working stock solution) 2.0 μL

(v) 0.5 μM Flap oligos 2 (working stock solution) 2.0 μL

Briefly vortex

(vi) Thermostable FEN1 (32 U/μL) 1.0 μL

Briefly vortex, then tap spin the tube if needed to collect all contents at the bottom.

(c) Add 15 μL of this working reaction mix to each low-bind PCR tube containing 20 μL gDNA.

(d) Thermocycle as follows: Begin with incubations of 3 min at 95°C, then 20 min at 65°C; followed by 14 cycles of 30 s at 95°C and 10 min at 65°C; conclude with a hold at 20°C. Proceed to Step 4 (recommended) or store overnight at 4°C.

4. Target sequence capture by hybridization and ligation (Fig. 1a, step 2; *see* **Note 9**). The time required for this reaction is approximately 5 min of hands-on time followed by a 14 h incubation.

(a) Make a PCR working reaction solution in a 1.7-mL microcentrifuge tube on ice by combining the following per reaction reagent volumes:

(i) Molecular-grade water 0.5 μL

(ii) 10 μM U2-3C3 oligo 1.0 μL

(iii) 0.5 μM Oligos 3 mixture 2.0 μL

(iv) 10× Thermostable ligase buffer 4.5 μL

Vortex.

(v) Thermostable ligase (e.g., 5 U/μL Ampligase) 2.0 μL

Briefly vortex, then tap spin the tube if needed.

(b) Add 10 μL of this working reaction mix to each low-bind PCR tube containing 35 μL.

(c) Incubate in a thermocycler with a heated lid as follows: Begin with an incubation of 3 min at 95°C; followed by 100 cycles of 30 s at 94°C and 8 min at 65°C; conclude with a hold at 10°C. Depending on the thermocycler, this step requires 14.5–16 h to conclude, making it a good step to run overnight. After overnight ligation, the volume may decrease by 2–5 μL, leaving ~45 μL, which does not affect the exonuclease treatment.

5. Enrichment of captured target strands via exonucleolytic degradation in the 3′ to 5′ direction (Fig. 1a, step 3; *see* **Note 13**).
   (a) Tap spin the low-bind PCR tubes or strips.
   (b) Make an exonuclease mix consisting of 1 μL exonuclease I (20 U) and 1 μL exonuclease III (100 U).
   (c) Briefly vortex, then tap spin the tube if needed.
   (d) Add 2 μL of the mix to each PCR tube, vortex, and tap spin.
6. Incubate in a thermocycler with a heated lid for 60 min at 37°C for the exonuclease treatment, followed by 20 min at 80°C to heat denature the exonucleases.
7. Purify the enriched target strands with AMPure XP beads (or equivalent). Perform all steps at room temperature, with these modifications:
   (a) To bind the enriched ssDNA, add 2× volumes (90 μL) of AMPure XP beads to each tube. More AMPure beads are required to bind ssDNA.
   (b) Briefly vortex to incubate for 5 min on the benchtop to bind DNA to the beads.
   (c) Wash the beads with bound DNA twice with 200 μL of 80% (v/v) ethanol each time.
   (d) Air-dry the beads until all visible liquid evaporates and they remain dark brown.
   (e) Elute the DNA with 28 μL of molecular-grade water and transfer to a new low-bind PCR tube.
8. At this point, the enriched, ssDNA sequences can be amplified directly for genotyping, deaminated for epigenetic analysis, or both (Fig. 1a, steps 4a, 4b, and 5).
   (a) For genotyping, proceed directly to PCR amplification (Step 10).
   (b) For epigenetic analysis of endogenous CG methylation alone or paired with MAPit to determine chromatin accessibility, convert DNA using EM-seq or equivalent (Step 9).
   (c) To perform parallel genetic and epigenetic analyses, double the input DNA amount and divide the eluate of enriched sequences, using half for direct PCR and the remainder for EM-seq. Keep these reactions separate, and do not pool them until they have been differentially barcoded or indexed.

9. Enzymatic conversion of unmethylated C-to-U (e.g., EM-seq conversion module). Follow the manufacturer's instructions (*see* **Note 4**), with the following minor modifications:
   (a) Use the formamide method to denature the DNA and maintain single-strandedness.
   (b) At the end of the protocol, elute the converted fragments with 20 μL of molecular-grade water, transfer the supernatant to a new low-bind PCR tube, and repeat the elution with another 20 μL of molecular-grade water.
   (c) Combine the supernatants for a total volume of 40 μL eluate.
10. PCR amplification of enriched target strands to prepare sequencing libraries (see **Note 23**). Two independent PCRs are performed for each combined FENGC eluate to mitigate potential stochastic amplification bias [14]. However, initially, conduct only one PCR to assess reaction efficacy.
    (a) Make a PCR working reaction solution in a 1.7-mL microcentrifuge tube on ice by combining the following per reaction reagent volumes:
        (i) KAPA HiFi HotStart Uracil+ Ready Mix 25 μL
        (ii) 10 μM U1 primer 2.5 μL
        (iii) 10 μM U2 primer 2.5 μL
    (b) Briefly vortex, and if needed, tap spin to collect all volume at the bottom.
    (c) Pipette 30 μL of PCR working reaction mix to each PCR tube on ice, containing 20 μL of purified target strands. Briefly vortex and tap spin the tubes if needed.
    (d) Incubate in a thermocycler with a heated lid as follows: Incubate 3 min at 95°C; 20–30 cycles of 20 s at 98°C, 15 s at 62°C, and 40 s at 72°C; incubate 5 min at 72°C; followed by a hold at 4°C.
11. Preliminary examination of PCR product yields. Examine 2 μL of each PCR by agarose gel electrophoresis (Fig. 3a).
12. Pending observation of a visible PCR product of the proper size, set up the second PCR using the remaining 20 μL of AMPure eluate. Examine 2 μL of the second reaction by agarose gel electrophoresis. If successful, pool both PCRs in a microcentrifuge tube for a total of 100 μL.
13. FENGC PCR product purification (see Sect. 3.2, Step 4).
    (a) Add 0.8× volumes (80 μL) of AMPure XP beads (or equivalent).

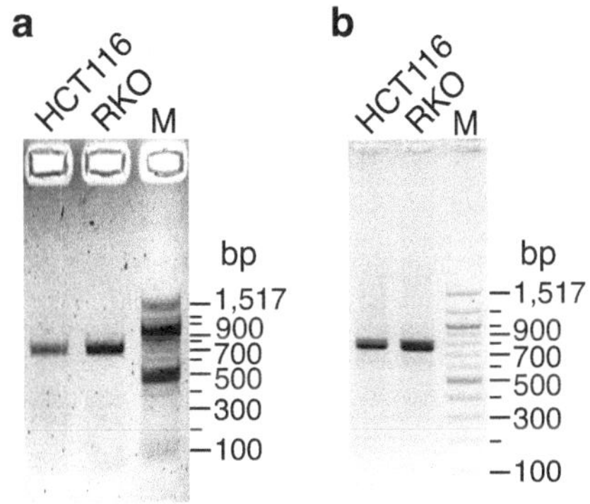

Fig. 3 MAPit-FENGC of a panel of target promoters prone to epigenetic silencing by CIMP. Cultured colorectal cancer lines HCT116 and RKO, harboring Ras pathway oncogenic driver mutations *KRAS*^{G12V} and *BRAF*^{V600E}, respectively, were harvested, permeabilized, and probed with M.CviPI. Purified gDNA (800 ng) was subjected to FENGC using a panel of oligos 1–3 designed by FOLD to enrich 121 CGI-containing promoters, 114 of which are known to undergo CG hypermethylation (CIMP) and transcriptional silencing in CRC and melanoma. The target strands had a mean length of 759 nt (median = 760 nt; range = 656–864 nt) and a mean GC content of 62.8% (median = 62.0%; range = 43–82%). Enriched sequences were C-to-U converted via EM-seq and PCR amplified using KAPA HiFi HotStart Uracil+ DNA polymerase ReadyMix Kit and primers U1 and U2. Shown is 1.5% agarose gel electrophoresis analysis of each crude amplification product (2 μL of 100 μL) (**a**), and 50 ng of each product after primer dimer removal with AMPure XP beads, quantified by Qubit (**b**). M, 2 μL (100 ng) of 100 bp DNA ladder (New England Biolabs). The expected PCR product size corresponds to the median target length of the multiplexed amplicons plus 47 bp added by the universal primers (760 bp + 47 bp = 807 bp)

(b) Wash the beads with bound DNA twice with 200 μL of 80% (v/v) ethanol each time.

(c) Elute the PCR fragments with 20 μL of molecular-grade water, transfer the supernatant to a new low-bind PCR tube, repeat the elution with another 20 μL of molecular-grade water, and pool the two eluates for a total of 40 μL.

14. PCR product quantification and final quality evaluation. Accurately measure the purified PCR product using a Qubit Fluorometer. Expected DNA concentrations are 120 ng/μL (range = 30 ng/μL–290 ng/μL).

15. Analyze 50 ng of each PCR alongside a molecular weight marker by agarose gel electrophoresis to verify efficient removal of primer dimer (Fig. 3b).

3.4 DNA Sequencing and Data Analysis

1. Optional: Short-read assessment. For users new to FENGC, initial short-read sequencing (e.g., 50,000–100,000 total Illumina reads) can be used to evaluate assay performance. Specifically, this step assesses the proportion of reads that align to the target reference sequences with ≥10× coverage. For high-complexity genomes such as human and mouse, these benchmarks should exceed 80% and 90%, respectively. However, due to the reproducibility and robustness of FENGC in our hands, and our preference for amplicons >600 bp, we routinely proceed directly to long-read sequencing (*see* **Note 23**).
2. FENGC amplicon barcoding and SMRTbell library preparation.
   (a) For multiplexed sequencing, purified amplicons are ligated to barcoded SMRTbell adapters with T-tailed overhangs (PacBio 101-791-700 version 07 and SMRTbell Barcoded Adapter Plate 3.0 102-009-200), following the manufacturer's protocol with one modification. Specifically, the input amount of each amplicon is increased 3–6× during DNA damage repair, end repair, A-tailing, and barcode ligation steps to ensure sufficient barcoded material for flexible pooling strategies should additional sequencing be required.
   (b) Equimolar amounts of amplicons ligated to barcoded SMRTbell adapters are pooled.
   (c) Circular, covalently closed molecules are enriched with exonuclease treatment using the SMRTbell Enzyme Clean-Up Mix (PacBio 101-932-600). Typical library yields are ~70 ng, adequate for 2–3 SMRT cells.
3. PacBio sequencing.
   (a) Sequencing is performed on the Sequel IIe platform using diffusion loading of libraries at 120 pM and 20–30 h movie times. Our core facility uses Sequencing Kit 2.0 (PacBio 101-389-001) and chemistry bundle version 11.0.
   (b) HiFi reads are demultiplexed using standard SMRT Link (PacBio) protocols, with a minimum of ≥5 passes per polymerase read. A single Sequel IIe SMRT cell typically yields 3–4 million polymerase reads.

   More recently, a library pool was diffusion loaded at 300 pM for 30-h movie collection on a Revio SMRT cell. The Revio SPRQ polymerase kit (PacBio 103-520-100) and instrument chemistry bundle version 25.2 were used. As with the Sequel IIe, other steps followed the recommended protocol using the PacBio loading calculator. This sequencing run produced 10.9 million HiFi reads, demonstrating scalability, with 99.28% of bases achieving Q30. PacBio now recommends using this Q30 metric for read filtering, rather than the number of sequencing passes.

4. Data processing analysis. In conventional amplicon bisulfite sequencing, conversion occurs prior to PCR amplification, necessitating the removal of primer binding sequences from reference templates. In contrast, MAPit-FENGC leverages C-to-U conversion after target enrichment, enabling

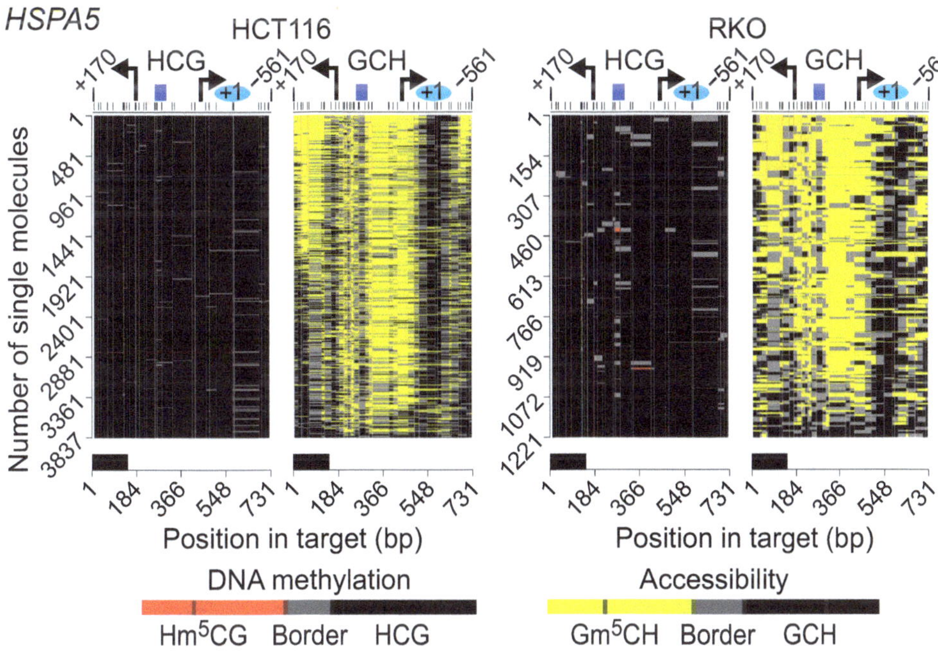

Fig. 4 MAPit-FENGC of the control promoter, *HSPA5*. The purified MAPit-FENGC PCR amplification products from HCT116 and RKO cells (Fig. 3b) were ligated to PacBio barcodes and sequenced on a PacBio HiFi Sequel IIe. Reads were filtered for ≥5 polymerase passes and ≥95% for reference coverage and HCH conversion, respectively. High-quality reads represented 82% (99/121) of targeted promoters with ≥20× coverage, including 77% of regions in the upper 50% GC content range (62%–82%)—compelling benchmarks for such regions. On the other hand, for eight promoters with the highest GC contents (75–82%), one locus yielded high-quality results, five loci exhibited low epiallelic diversity or duplicates, and two loci returned no on-target reads. Shown are methylscaper heatmaps of human *HSPA5* (–561 to +170 relative to the TSS, *left*-pointing arrow), which is divergently transcribed with *HSPA5-DT* (*right*-pointing TSS arrow), downloaded directly without clustering refinement. In HCT116 (*left* pair of panels) and RKO (*right* pair), endogenous DNA methylation (HCG; *left* panel of each pair) and chromatin accessibility (GCH sites methylated by M.CviPI; *right* panel) are displayed separately. Single molecules in each pair of panels are shown in the same top-to-bottom order. Consecutive Hm^{5}CG and Gm^{5}CH are connected by *yellow* and *red* lines, respectively; unmethylated HCG or GCH are connected by *black* lines; borders between methylated and unmethylated regions are *gray* (see keys at bottom). Similar accessibility patterns are observed in both cell types, including: (i) a short footprint at the *HSPA5* TSS, (ii) a sequence-specific TF footprinted upstream of *HSPA5* TSS (*blue rectangle*), and (iii) larger footprints corresponding to the +1 nucleosome (*blue oval*) located downstream of the *HSPA5-DT* TSS. This region represents an unmethylated, highly accessible promoter architecture consistent with active (or poised) transcription. Additionally, widespread accessibility across thousands of molecules rules out incomplete cell permeabilization or limiting M.CviPI activity as explanations for observed inter- and intra-target differences. A schematic of the locus is shown at the *top*, with *vertical ticks* marking HCG and GCH motifs. All features and distances are drawn to scale. *Black bar*, nucleosome core particle (147 bp); *pink bar*, CGI

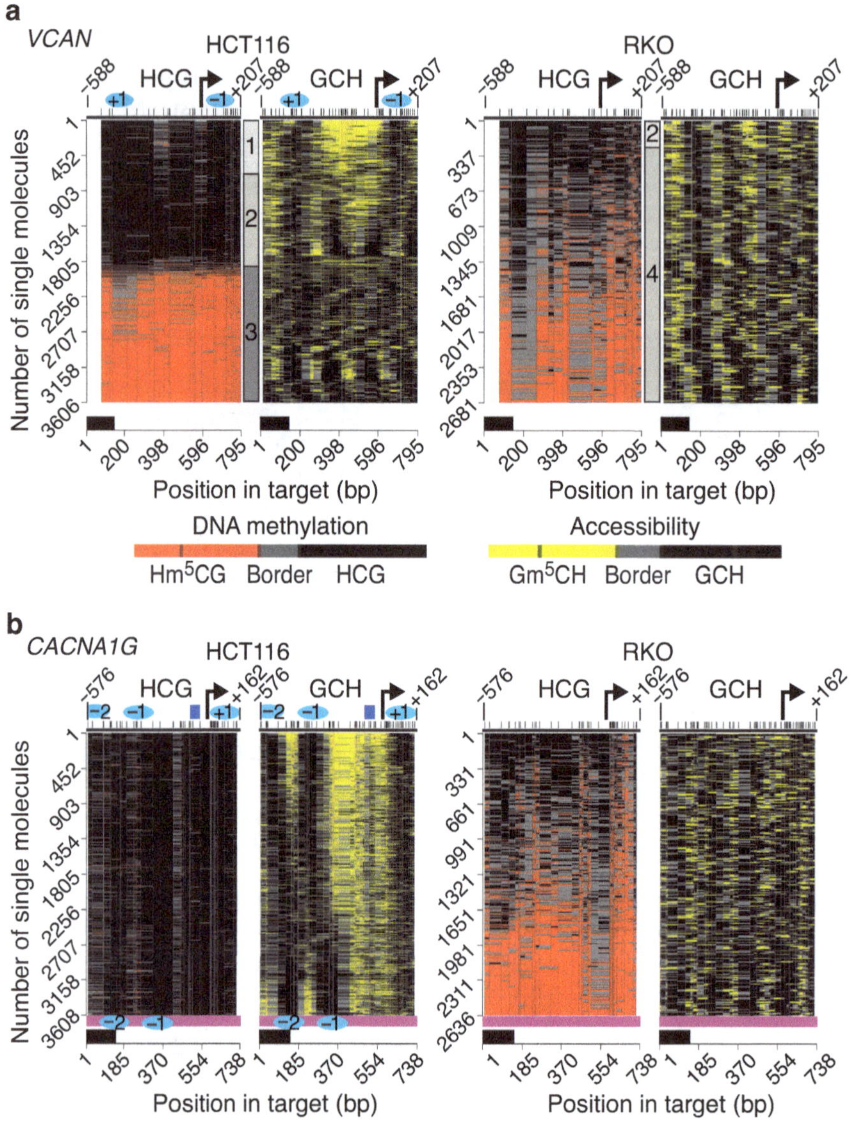

Fig. 5 MAPit-FENGC concurrently identifies tumor-specific differences in native DNA methylation and chromatin accessibility. Methylscaper heatmaps of (**a**) *VCAN* (–588 to +207) and (**b**) *CACNA1G* (–576 to +162) from the same experiment described in Figs. 3 and 4. These promoters are representative of 66 out of 121 CIMP targets that showed clear differences in Hm^{5}CG and chromatin accessibility between HCT116 and RKO cells. In HCT116, *VCAN* (**a**) exemplifies the power of assaying multiple epigenetic layers in a single assay. Three distinct clusters of molecules are evident (*bars* labeled 1–3 between the HCG and GCH panels): Cluster 1 consists of largely unmethylated molecules with variably sized nucleosome-free regions (NFRs) upstream of the transcription start site (TSS), potentially accommodating RNA polymerase II pre-initiation complex assembly and active transcription. Cluster 2 comprises largely unmethylated molecules containing arrays of two to three nucleosomes, connected by accessible linker DNA but lacking large NFRs, likely representing transcriptionally quiescent chromatin copies. To better distinguish between clusters 1 and 2, refinement of methylscaper clustering focused on the molecules with low Hm^{5}CG abundance. Cluster 3 molecules are hypermethylated with diffuse background accessibility, likely transcriptionally silenced. This group did not require further cluster refinement. Strong negative correlations between DNA methylation and chromatin accessibility are observed.

methylation profiling of all HCG and GCH sites within each amplicon, including those overlapping oligo binding regions.

(a) Reference sequences are created to span the region between the two FEN1 cleavage sites, beginning immediately 3′ of the 5′ T and ending at the 3′ T (see Fig. 1).

(b) Reference sequences and HiFi CCS reads exported by SMRT Link that pass quality metrics (after conversion of BAM to FASTA format) are used as inputs to the Python-based reAminator pipeline [27].

   (i) bsAlign aligns HiFi reads in FASTA format to the supplied reference sequences.

   (ii) bsDraw extracts aligned reads, filters out those not meeting user-defined thresholds for conversion efficiency at HCH sites and minimum coverage (recommended: ≥95% for both), and saves the filtered reads to locus-specific alignment FASTA files.

   (iii) MethylMapper converts these aligned FASTA files into numerical codes used to generate nucleotide frequency tables and color-coded methylation heatmaps. The clustering quality of these heatmaps is generally sufficient for loci with high levels of m^5CG; however, performance diminishes when epigenetic heterogeneity is high.

(c) For loci that cluster poorly in MethylMapper, methylscaper improves initial clustering by performing sequence-specific alignment and unsupervised hierarchical clustering based on the first principal component of methylation across both HCG and GCH sites [50]. This clustering can be further refined through user-guided selection of specific subregions or clusters of molecules (Figs. 4 and 5).

Fig. 5 (continued) The chromatin architecture at *VCAN* is consistent with monoallelic expression on chromosome 5, given the lack of evidence in the literature for parent-of-origin imprinting at this locus. In RKO, two clusters are apparent: one resembling cluster 2 in HCT116, and another (cluster 4) characterized by arrays of 4–5 randomly positioned nucleosomes with accessible linkers and variable extents of Hm^5CG. It is plausible that Hm^5CG has spread over the *VCAN* TSS from two downstream CGIs (not included in the assayed amplicon). The absence of large NFRs in RKO suggests transcriptional silencing of *VCAN*. In contrast, the *CACNA1G* promoter in HCT116 cells is largely unmethylated and contains prominent, variable-length NFRs, correlating with the positions of upstream nucleosomes −1 and −2. A sequence-specific TF is also footprinted in a substantial fraction of NFRs (*blue rectangle*), consistent with active transcription. In RKO, the promoter lacks large NFRs and individual promoter copies are variably methylated with randomly positioned nucleosomes, resembling the *VCAN* promoter in RKO. All features and distances are drawn to scale, and all symbols are as defined in the legend to Fig. 4

4 Notes

1. 1× Cell resuspension buffer serves as the base solution for the three major steps of the MAPit protocol: (1) washing cells, (2) cell permeabilization, and (3) probing accessible DNA with a suitable DNMT, such as the GC-methylating enzyme M.CviPI. M.CviPI does not require $MgCl_2$ as a cofactor; therefore, 0.25 mM EDTA and 0.5 mM EGTA are included in the cell resuspension buffer and the chromatin-probing reaction. DTT and PMSF must be added to ice-cold 1× cell resuspension buffer immediately prior to use. Freshly added DTT is critical to maintain full M.CviPI activity. The cell resuspension buffer composition has been optimized to preserve native chromatin structure within permeabilized cells and freshly isolated nuclei. Cells can also be cryopreserved in Bambanker, a serum-free cell freezing medium.
2. The final concentration of digitonin in the cell permeabilization buffer may require optimization depending on the cell type. We have determined that M.CviPI retains full activity at 0.5% (w/v) digitonin; higher concentrations have not been evaluated. In general, though, increasing digitonin concentration tends to improve methylation efficiency at accessible GC sites in chromatin, which is advantageous for single-molecule analysis.
3. Purchasing the higher concentration enzyme stock (New England Biolabs M0227B-HI) enables the addition of 100 U M.CviPI diluted in a 10 μL volume. This minimizes the final amount of M.CviPI storage buffer, including glycerol, introduced into the chromatin-probing reaction. As noted in the Introduction, any suitable C-5 DNMT may be used as the chromatin structural probe, either alone or in combination with additional C-5 DNMTs.
4. Enzymatic conversion is a non-degradative alternative to bisulfite treatment for C-to-U deamination or conversion of DNA [47]. Because neither EM-seq nor bisulfite deaminates m^5C or 5-hydroxymethyl cytosine, they cannot discriminate between these two modifications.
5. This recipe for 80% (v/v) ethanol compensates for the hygroscopic nature of absolute ethanol in an open bottle, which may decrease to ~95% due to water absorption.
6. Until all MAPit reagents have been validated for consistent efficacy, we recommend assessing M.CviPI probing efficiency by assaying one or more control, open chromatin regions (see Sect. 3.2). Hot-start Platinum™ Taq DNA polymerase is effective for amplifying single regions from converted gDNA, particularly when used in a 7 U : 1 U ratio with PfuTurbo

Cx Hotstart DNA polymerase. This Pfu enzyme has been engineered for both improved proofreading activity and read-through on uracil-containing template strands in converted DNA. This polymerase mix provides a practical balance between cost and amplification performance on deaminated gDNA. However, if yields of individual amplicons are insufficient, KAPA HiFi HotStart Uracil+ DNA polymerase may be used as an alternative, although optimization of PCR conditions may be required. Single-amplicon PCR using either the Platinum™ Taq:PfuTurbo Cx mix or the KAPA HiFi enzyme also serves as a useful orthogonal validation method for MAPit-FENGC results (see Sect. 3.2). The Platinum™ Taq : PfuTurbo Cx mix and deamination sequencing primers are listed as optional reagents for those opting to forego such validations.

7. Consult our previous *Methods in Molecular Biology* chapter for recommendations and further description of deamination sequencing primer design [15].

8. Both universal oligos are C-less, bypassing C-to-U conversion during enzymatic deamination.

9. After DNA denaturation, FENGC enables assembly of multiple, user-defined target sequences into pairs of double flaps, each consisting of a 5′ flap and a one-nucleotide (1 nt) 3′ flap (see Fig. 1, step 1a). Based on molecular kinetic considerations, oligo U1-T likely first anneals to the constant 3′ tails of flap oligos 1 and 2, respectively, forming flap adapters 1 and 2. These adapters then hybridize to their respective 5′ and 3′ target ends, resulting in tripartite complexes with two double flaps, each ultimately defining one end of the target sequence. The 5′ flaps formed by target-annealed flap adapters 1 vary in length, whereas 5′ flaps formed by hybridization of flap adapters 2 constitute the target strands themselves, which have a fixed target-specific length. The 1 nt 3′ flap is generated when the constant **A** in each flap oligo hybridizes to the "overlapping" T at the target ends, displacing the 3′-terminal T of oligo U1-T. To limit the length of the 3′ flap to 1 nt, the overlapping T is designed to follow a non-T base (i.e., 5′-VT-3′, where V ≠ T, an A, C, or G).
Importantly, altering the order of oligo hybridization—specifically, annealing flap oligos to target-specific sequences prior to their annealing to oligo U1-T—does not disrupt double flap formation. Following flap assembly, FEN1 precisely and uniformly cleaves the 5′ flaps at the phosphodiester bond immediately 3′ of the T preceded by a non-T (5′-VT-3′), liberating target strands with a ligatable 5′ phosphate and 3′ hydroxyl [51, 52] (Fig. 1, step 1b).

After 5′ flap cleavage, the liberated single-stranded DNA (ssDNA) targets are captured by thermostable ligase-mediated ligation to universal oligos (Fig. 1, step 2). Because ssDNA targets are captured, their sizes are indicated in nucleotides (nt) and polarity as + or – for the upper and lower genome strands. At the 5′ termini of cleaved target strands, flap oligo 1 acts as a splint, enabling the initially unpaired T of oligo U1-T to pair with the constant **A** of flap oligo 1, followed by ligation to the target 5′ phosphate.

Concurrently, oligos 3 (nested relative to oligos 2) and U2-3C3 hybridize to form adapters 3. Their addition is required because the annealing sites for flap adapters 2 are severed from the target sequences. Oligos 3 in adapters 3 serve as splints, annealing to the 3′ terminus of each target and facilitating ligation of the 5′ phosphate of oligo U2-3C3 to the 3′ hydroxyl of each target strand. Parenthetically, oligo U1-T also ligates to the 5′ phosphate of the released downstream fragments; additional flap and splint adapters can be included in FENGC reactions to conduct serial "walks" of sequences downstream and/or upstream.

The main text highlights a minimal motif of VT at both target strand ends. Given the 59.1% AT base composition of the human genome, this VT motif occurs approximately once every ~4.8 bp at a single end.

This enables high-resolution targeting—a substantial improvement over the ~200–300 bp average spacing of 4-bp restriction enzyme recognition sites—allowing precise enrichment of short genomic regions, even in densely methylated contexts. Moreover, while a 4.8 bp resolution at each end is more than sufficient for flexible target design, FENGC is not limited to the VT configuration. Similar minimal motifs, specifically, BA (where B ≠ A) and DC (B ≠ A), can also be used, enabled by flap adapters containing oligos U1-A and U1-C, respectively, which assemble and cleave 5′ flaps with comparable efficiency. Interestingly, although U1-G-containing flap adapters efficiently cleave double flaps at HG sites, PCR with the current primers U1 and U2 yields high-molecular-weight smears for reasons that remain unclear. Overall, the motif flexibility significantly enhances the density and precision of targetable sites for enrichment, independent of DNA methylation status.

10. FENGC oligonucleotide designer is a publicly available program (https://github.com/uf-icbr-bioinformatics/FOLD) for rapid design of hundreds of targets and target-specific oligos that meet specified parameters (*see* README file for details). FOLD accepts input cohorts of candidate regions up to 3 kb. However, in our hands, MAPit-FENGC has been

successfully multiplexed for 10–453 targets up to 1 kb without any detectable loss in performance. Based on this experience, we see no technical barrier to scaling the method to many hundreds or even thousands of targets, budget permitting.

To begin designing target sequences and their corresponding oligos 1–3, upload an organism-specific gene annotation file, such as a GTF or GFF file, that lists genomic features, including genes, transcripts, and exons, for the organism, to FOLD. Then, upload a file listing the HUGO gene names of all targets of interest. It is advisable to include control promoters with open chromatin, such as human *COX8A* and *HSPA5*. After indicating a preferred target sequence length (up to 1 kb), FOLD defaults to allow a maximum of 15% total increase and/or 5% decrease in target length. In practice, we limit the increase in length to 5–10% as a command-line option, maintaining the default 5% decrease. This strikes a good balance between (i) maintaining similar target sequence lengths to minimize PCR bias toward short sequences during library construction and (ii) maximizing the likelihood of obtaining a single optimal target sequence for all genes with a single exon 1 or TSS, while limiting the maximum allowed number of repeat bases (default value of six). The user must provide a repeat-masked genome reference file, i.e., repeats indicated in lower-case letters, when executing FOLD.

For genes transcribing multiple isoforms, the default setting, all isoforms, can be changed to only the longest isoform. Additional command line options include (i) preferred distance upstream and downstream of transcription start sites (TSSs); (ii) specifying regions of interest based in reference genome coordinates; (iii) limiting T_m range (we initially enter 55–65°C, and increase to 66°C if FOLD is unable to find a target and oligos meeting specified conditions for too many genes or transcript isoforms; (iv) checking for formation of primer dimers and secondary structures; and (v) other options as listed in the program's README file. For each HUGO gene and TSS for which FOLD computes a solution, the program exports a table listing the HUGO gene name, genome coordinates, base-pair length, and % GC for each target sequence. In addition, for each target, the table lists features of corresponding oligos 1–3, including their sequence, length, T_m (calculated by built-in method based on Primer3; [49]), and % GC. For the few targets FOLD cannot find oligos that meet the condition requirements, targets and oligos are designed manually (*see* **Note 9**).

11. Once optimal targets and oligos have been selected, we order oligos 1–3 in 96-well plates (Eurofins; 4 nmole scale/oligo, standard desalting, with each quantified and resuspended in

molecular-grade water to 100 μM). Shipping these plates at ambient temperature has not posed any problems. Depending on the synthesis yield, the total received volume of 100 μM solution for each oligo typically ranges from 30 to 160 μL. These oligo amounts are sufficient for 3000–16,000 FENGC reactions! Therefore, for the average lab, each panel of oligos most likely constitutes a one-time cost. For these reasons, we usually choose at least 32 targets, which fills one 96-well plate with oligos 1–3. Three separate master mixes of oligos 1–3 are made, usually by mixing 2 μL of each oligo, and if needed, diluting the solution to 0.5 μM final concentration in each constituent oligo. Mixing 2 μL of each oligo for 200 targets creates 400 μL of 0.5 μM master mix, enough for 200 FENGC reactions. Mixing only 2 μL of each oligo retains a 100 μM stock solution in case the concentrations of specific oligos need to be rebalanced or analysis of different combinations of targets is desired.

12. Primer U1 has the same sequence as oligos U1-T, U1-A, and U1-C, except that it does not have their 3′-terminal nucleotide. Primer U2 is complementary to oligo U2-C3C and has the same constant 5′ sequence as splint oligos 3, except that it lacks the constant **A** (**T** or **G**) that anneals to each target sequence.

13. The three-carbon-spacer covalent modification of the 3′-terminal nucleotide of oligo U2-C3C protects it against 3′ to 5′ exonuclease degradation. Therefore, only oligo U2-C3C and target sequences to which it has ligated are resistant to exonucleases I and III—non-targeted genomic sequences and flap and splint adapters are degraded to undetectable levels. After overnight ligation with thermostable ligase, the volume may decrease by 2–5 μL; however, this reduction does not impact the subsequent exonuclease treatment.

14. MAPit with M.CviPI yields reproducible, high-quality results across many human and mouse cultured cell lines tested to date with isolated nuclei [15, 27, 28, 30, 33], and more recently, following permeabilization [45]. The protocol has also been applied successfully to primary cells isolated by negative or positive selection, as well as to dispersed single-cell suspensions from neurospheres and primary rat colonic epithelium [15, 45]. When using trypsin–EDTA to detach adherent cells or dissociate tissue, add at least two volumes of growth medium containing 10% (v/v) fetal bovine serum, which contains natural trypsin inhibitors. For best results, use 10^6 cells/chromatin-probing reaction, with a minimum input of 200,000 cells per reaction. However, high-quality

MAPit-FENGC data can be obtained for a lower percentage of targets with as few as 25,000 cells per reaction.

15. Probing chromatin accessibility of 10^6 cells with 100 U M.CviPI under the conditions employed in the updated protocol presented herein achieves saturating levels of methylation at accessible GC motifs. Nevertheless, if the amount of M.CviPI activity is to be varied to perform a dose response or include a control reaction with 0 U M.CviPI, increase the total number of cells accordingly, but do not exceed 10^6 cells/reaction. In such cases, all of the cells should be washed together in a single 1.7-mL microcentrifuge tube using up to 1.5 mL of ice-cold 1× cell resuspension buffer, containing freshly added DTT and PMSF (i.e., it is not necessary to wash cells in multiple tubes if more than three M.CviPI concentrations are used).

16. Resuspend the washed cell pellets in 92 μL of ice-cold cell permeabilization buffer, multiplied by the number of data points (e.g., number of total M.CviPI doses). After permeabilization, the shade of blue resulting from trypan stain entry is cell-type specific. However, under the employed conditions—using 0.5% (w/v) digitonin, 10× more digitonin than is customarily used—fields with non-permeabilized cells are rarely, if ever, observed.

17. SAM is the universal methyl donor and a cofactor for methyltransferases. To ensure that SAM is not limiting in the chromatin-probing/methylation reaction, we increased its final concentration to 670 μM, compared with the manufacturer-recommended 160 μM. In addition, *S*-adenosyl-*L*-homocysteine (SAH), the byproduct formed after methyl group transfer from SAM to DNA, has a binding affinity to DNA methyltransferase (DNMT) catalytic sites comparable to that of SAM. Therefore, using a higher SAM concentration is expected to enhance SAH replacement by SAM and maintain sufficient levels of active methyl donor throughout the 15-min methylation reaction.

18. If 0 U M.CviPI reactions are performed, add 2.2 μL of M.CviPI storage buffer instead of enzyme. To perform a dose response, use M.CviPI storage buffer as the diluent to make serial dilutions of the 50 U/μL M.CviPI stock. The goal in both cases is to keep the reaction constituents consistent between all methylation reactions.

19. An alternative conversion method is bisulfite treatment, which requires a minimum input of 0.5 μg gDNA, and preferably 1–2 μg, especially for amplicons >700 bp in length. The increased DNA requirement is due to its chemical degradation during hydrolytic deamination. For this reason, we do

not recommend the use of bisulfite conversion for deamination of enriched target strands in FENGC.

20. DNA purification of converted fragments in the EM-seq conversion module protocol requires that the AMPure XL beads not be dried as thoroughly as in other protocols. To address the resulting residual ethanol, we found that performing two sequential elutions with 20 μL of molecular-grade water effectively dilutes the ethanol and improves PCR yield during library preparation.
21. To assess M.CviPI probing efficiency, Sanger sequencing offers a relatively rapid and cost-effective means to survey multiple GC and CG sites within regions of open chromatin. As an alternative, a quantitative methylation-sensitive restriction endonuclease assay may be used as previously described [15]; however, this method interrogates only a single GC site and only provides an approximation of GC methylation levels.
22. FENGC, whether used for DNA methylation detection alone or in combination with MAPit, reliably enriches 10–453 targets ranging from 282 to 959 nucleotides in length, using 150–800 ng of input genomic DNA, and obtaining 80,000–100,000 total unfiltered PacBio HiFi reads. These metrics span several oligo panels, each designed to target sequences of similar lengths to minimize PCR amplification bias. To ensure adequate representation of epigenetic heterogeneity across the population of sequenced molecules, it is advisable to aim for approximately 1000 total HiFi reads per target per sample. This number can be adjusted as needed, given that the total number of targets in a panel determines the overall read depth attained per SMRT cell.

    The updated enrichment protocol described herein consistently achieves 80–97% on-target alignment and ≥10× coverage for approximately 95% of targets in complex genomes such as mouse and human. This excludes sequences with extreme GC content, such as CGI-containing promoters, which yield usable data for 82% of panel targets (*see* Figs. 3–5). When input DNA is reduced to 75 ng—while maintaining oligo concentrations and sequencing depth—primer dimer formation increases during PCR amplification, and on-target alignment decreases to 64%. Nonetheless, ≥10× coverage is still achieved for ~95% of targets, and the resulting GCH chromatin accessibility and HCG methylation profiles remain consistent with those obtained using higher input amounts.

    Under previous, lower-performance conditions [45], genotyping of 119 targets (mean of 450 nt in length), using 500 ng of non-converted human gDNA and 27,000 total

PacBio HiFi reads, achieved 82% on-target alignment and ≥10× coverage for 83% of targets.

23. Longer amplicons and sequencing reads enable phasing of epigenetic features, such as co-occupancy of transcription factors and nucleosomes, at single-molecule resolution [15, 27, 32]. To leverage this capability, we sequence FENGC amplicons using the PacBio single-molecule real-time (SMRT) platform, which employs circular sequencing. For amplicons up to one kilobase, SMRT circular sequencing allows multiple passes over both the upper and lower template strands of the same circular DNA molecule, generating high-fidelity (HiFi) reads derived from circular consensus sequences (CCS). In a recent run on the PacBio Revio system, 11 FENGC libraries targeting regions of 720–750 bp achieved an average of 49 passes per molecule, with 99.2% of total bases reaching Q30 quality (99.9% accuracy). The use of Oxford Nanopore sequencing has also been described [41].

Acknowledgments

This work was supported by grants 1RMGM139690 from the NIH to P.A.E. and M.P.K., HT94252410275 from the DoD, US Army Medical Research Acquisition to T.L.C., HDTRA1-16-1-0048 from the DoD, Defense Threat Reduction Agency to P.C. and M.P.K., and R01CA155390 from the National Cancer Institute to M.P.K. We express our gratitude to the University of Florida Interdisciplinary Center for Biotechnology Research (UF-ICBR, RRID:SCR_019152) for assistance and troubleshooting during the development of FENGC. We thank Hengye Chen and Lu (Lucy) Bai for writing and providing bisulfitePCRSeqAnalysis.m, and for granting permission to share it publicly.

References

1. Kornberg RD, Lorch Y (1999) Twenty-five years of the nucleosome, fundamental particle of the eukaryote chromosome. Cell 98(3):285–294. https://doi.org/10.1016/s0092-8674(00)81958-3
2. Klemm SL, Shipony Z, Greenleaf WJ (2019) Chromatin accessibility and the regulatory epigenome. Nat Rev Genet 20(4):207–220. https://doi.org/10.1038/s41576-018-0089-8
3. Buenrostro JD, Wu B, Litzenburger UM, Ruff D, Gonzales ML, Snyder MP, Chang HY, Greenleaf WJ (2015) Single-cell chromatin accessibility reveals principles of regulatory variation. Nature 523(7561):486–490. https://doi.org/10.1038/nature14590
4. Corces MR, Buenrostro JD, Wu B, Greenside PG, Chan SM, Koenig JL, Snyder MP, Pritchard JK, Kundaje A, Greenleaf WJ, Majeti R, Chang HY (2016) Lineage-specific and single-cell chromatin accessibility charts human hematopoiesis and leukemia evolution. Nat Genet 48(10):1193–1203. https://doi.org/10.1038/ng.3646
5. Li H, Sharma A, Luo K, Qin ZS, Sun X, Liu H (2020) DeconPeaker, a deconvolution model to identify cell types based on chromatin accessibility in ATAC-seq data of mixture

samples. Front Genet 11:392. https://doi.org/10.3389/fgene.2020.00392
6. Esteller M, Dawson MA, Kadoch C, Rassool FV, Jones PA, Baylin SB (2024) The epigenetic hallmarks of cancer. Cancer Discov 14(10):1783–1809. https://doi.org/10.1158/2159-8290.CD-24-0296
7. Ryan MB, Corcoran RB (2018) Therapeutic strategies to target RAS-mutant cancers. Nat Rev Clin Oncol 15(11):709–720. https://doi.org/10.1038/s41571-018-0105-0
8. Roth AD, Tejpar S, Delorenzi M, Yan P, Fiocca R, Klingbiel D, Dietrich D, Biesmans B, Bodoky G, Barone C, Aranda E, Nordlinger B, Cisar L, Labianca R, Cunningham D, Van Cutsem E, Bosman F (2010) Prognostic role of *KRAS* and *BRAF* in stage II and III resected colon cancer: Results of the translational study on the PETACC-3, EORTC 40993, SAKK 60-00 trial. J Clin Oncol 28(3):466–474. https://doi.org/10.1200/JCO.2009.23.3452
9. Toyota M, Ahuja N, Ohe-Toyota M, Herman JG, Baylin SB, Issa JP (1999) CpG island methylator phenotype in colorectal cancer. Proc Natl Acad Sci USA 96(15):8681–8686. https://doi.org/10.1073/pnas.96.15.8681
10. Boyle AP, Davis S, Shulha HP, Meltzer P, Margulies EH, Weng Z, Furey TS, Crawford GE (2008) High-resolution mapping and characterization of open chromatin across the genome. Cell 132(2):311–322. https://doi.org/10.1016/j.cell.2007.12.014
11. Hesselberth JR, Chen X, Zhang Z, Sabo PJ, Sandstrom R, Reynolds AP, Thurman RE, Neph S, Kuehn MS, Noble WS, Fields S, Stamatoyannopoulos JA (2009) Global mapping of protein-DNA interactions *in vivo* by digital genomic footprinting. Nat Methods 6(4):283–289. https://doi.org/10.1038/nmeth.1313
12. Allan J, Fraser RM, Owen-Hughes T, Keszenman-Pereyra D (2012) Micrococcal nuclease does not substantially bias nucleosome mapping. J Mol Biol 417(3):152–164. https://doi.org/10.1016/j.jmb.2012.01.043
13. Buenrostro JD, Giresi PG, Zaba LC, Chang HY, Greenleaf WJ (2013) Transposition of native chromatin for fast and sensitive epigenomic profiling of open chromatin, DNA-binding proteins and nucleosome position. Nat Methods 10(12):1213–1218. https://doi.org/10.1038/nmeth.2688
14. Jessen WJ, Hoose SA, Kilgore JA, Kladde MP (2006) Active *PHO5* chromatin encompasses variable numbers of nucleosomes at individual promoters. Nat Struct Mol Biol 13(3):256–263. https://doi.org/10.1038/nsmb1062
15. Pardo CE, Nabilsi NH, Darst RP, Kladde MP (2015) Integrated DNA methylation and chromatin structural analysis at single-molecule resolution. Methods Mol Biol 1288:123–141. https://doi.org/10.1007/978-1-4939-2474-5_9
16. Kladde MP, Xu M, Simpson RT (1996) Direct study of DNA-protein interactions in repressed and active chromatin in living cells. EMBO J 15(22):6290–6300
17. Kladde MP, Xu M, Simpson RT (1999) DNA methyltransferases as probes of chromatin structure vivo. Methods Enzymol 304:431–447. https://doi.org/10.1016/s0076-6879(99)04026-4
18. Fatemi M, Pao MM, Jeong S, Gal-Yam EN, Egger G, Weisenberger DJ, Jones PA (2005) Footprinting of mammalian promoters: use of a CpG DNA methyltransferase revealing nucleosome positions at a single molecule level. Nucleic Acids Res 33(20):e176. https://doi.org/10.1093/nar/gni180
19. Gal-Yam EN, Jeong S, Tanay A, Egger G, Lee AS, Jones PA (2006) Constitutive nucleosome depletion and ordered factor assembly at the *GRP78* promoter revealed by single molecule footprinting. PLoS Genet 2(9):e160. https://doi.org/10.1371/journal.pgen.0020160
20. Lin JC, Jeong S, Liang G, Takai D, Fatemi M, Tsai YC, Egger G, Gal-Yam EN, Jones PA (2007) Role of nucleosomal occupancy in the epigenetic silencing of the *MLH1* CpG island. Cancer Cell 12(5):432–444. https://doi.org/10.1016/j.ccr.2007.10.014
21. Bouazoune K, Miranda TB, Jones PA, Kingston RE (2009) Analysis of individual remodeled nucleosomes reveals decreased histone-DNA contacts created by hSWI/SNF. Nucleic Acids Res 37(16):5279–5294. https://doi.org/10.1093/nar/gkp524
22. Dechassa ML, Sabri A, Pondugula S, Kassabov SR, Chatterjee N, Kladde MP, Bartholomew B (2010) SWI/SNF has intrinsic nucleosome disassembly activity that is dependent on adjacent nucleosomes. Mol Cell 38(4):590–602. https://doi.org/10.1016/j.molcel.2010.02.040
23. Duan R, Porter W, Samudio I, Vyhlidal C, Kladde M, Safe S (1999) Transcriptional activation of *c-fos* protooncogene by 17β-estradiol: mechanism of aryl hydrocarbon receptor-mediated inhibition. Mol Endocrinol 13(9):1511–1521. https://doi.org/10.1210/mend.13.9.0338
24. Dong L, Wang W, Wang F, Stoner M, Reed JC, Harigai M, Samudio I, Kladde MP, Vyhlidal C, Safe S (1999) Mechanisms of transcriptional activation of *bcl-2* gene expression by 17β-estradiol in breast cancer cells. J Biol Chem 274(45):32099–32107. https://doi.org/10.1074/jbc.274.45.32099

25. Frommer M, McDonald LE, Millar DS, Collis CM, Watt F, Grigg GW, Molloy PL, Paul CL (1992) A genomic sequencing protocol that yields a positive display of 5-methylcytosine residues in individual DNA strands. Proc Natl Acad Sci USA 89(5):1827–1831. https://doi.org/10.1073/pnas.89.5.1827
26. Darst RP, Pardo CE, Ai L, Brown KD, Kladde MP (2010) Bisulfite sequencing of DNA. Curr Protoc Mol Biol 91:7.9.1–7.9.17. https://doi.org/10.1002/0471142727.mb0709s91
27. Stees JR, Hossain MA, Sunose T, Kudo Y, Pardo CE, Nabilsi NH, Darst RP, Poudyal R, Igarashi K, Huang S, Kladde MP, Bungert J (2016) High fractional occupancy of a tandem Maf recognition element and its role in long-range β-globin gene regulation. Mol Cell Biol 36(2):238–250. https://doi.org/10.1128/MCB.00723-15
28. Krebs AR, Imanci D, Hoerner L, Gaidatzis D, Burger L, Schübeler D (2017) Genome-wide single-molecule footprinting reveals high RNA polymerase II turnover at paused promoters. Mol Cell 67(3):411–422. https://doi.org/10.1016/j.molcel.2017.06.027 e414
29. Xu M, Kladde MP, Van Etten JL, Simpson RT (1998) Cloning, characterization and expression of the gene coding for a cytosine-5-DNA methyltransferase recognizing GpC. Nucleic Acids Res 26(17):3961–3966. https://doi.org/10.1093/nar/26.17.3961
30. Kilgore JA, Hoose SA, Gustafson TL, Porter W, Kladde MP (2007) Single-molecule and population probing of chromatin structure using DNA methyltransferases. Methods 41(3):320–332. https://doi.org/10.1016/j.ymeth.2006.08.008
31. Pondugula S, Kladde MP (2008) Single-molecule analysis of chromatin: changing the view of genomes one molecule at a time. J Cell Biochem 105(2):330–337. https://doi.org/10.1002/jcb.21849
32. Nabilsi NH, Deleyrolle LP, Darst RP, Riva A, Reynolds BA, Kladde MP (2014) Multiplex mapping of chromatin accessibility and DNA methylation within targeted single molecules identifies epigenetic heterogeneity in neural stem cells and glioblastoma. Genome Res 24(2):329–339. https://doi.org/10.1101/gr.161737.113
33. Kelly TK, Liu Y, Lay FD, Liang G, Berman BP, Jones PA (2012) Genome-wide mapping of nucleosome positioning and DNA methylation within individual DNA molecules. Genome Res 22(12):2497–2506. https://doi.org/10.1101/gr.143008.112
34. Samorodnitsky E, Datta J, Jewell BM, Hagopian R, Miya J, Wing MR, Damodaran S, Lippus JM, Reeser JW, Bhatt D, Timmers CD, Roychowdhury S (2015) Comparison of custom capture for targeted next-generation DNA sequencing. J Mol Diagn 17(1):64–75. https://doi.org/10.1016/j.jmoldx.2014.09.009
35. Kozarewa I, Armisen J, Gardner AF, Slatko BE, Hendrickson CL (2015) Overview of target enrichment strategies. Curr Protoc Mol Biol 112:7.21.1–7.21.23. https://doi.org/10.1002/0471142727.mb0721s112
36. Shin G, Grimes SM, Lee H, Lau BT, Xia LC, Ji HP (2017) CRISPR-Cas9-targeted fragmentation and selective sequencing enable massively parallel microsatellite analysis. Nat Commun 8: 14291. https://doi.org/10.1038/ncomms14291
37. Bennett-Baker PE, Mueller JL (2017) CRISPR-mediated isolation of specific megabase segments of genomic DNA. Nucleic Acids Res 45(19):e165. https://doi.org/10.1093/nar/gkx749
38. Nachmanson D, Lian S, Schmidt EK, Hipp MJ, Baker KT, Zhang Y, Tretiakova M, Loubet-Senear K, Kohrn BF, Salk JJ, Kennedy SR, Risques RA (2018) Targeted genome fragmentation with CRISPR/Cas9 enables fast and efficient enrichment of small genomic regions and ultra-accurate sequencing with low DNA input (CRISPR-DS). Genome Res 28(10):1589–1599. https://doi.org/10.1101/gr.235291.118
39. Lee J, Lim H, Jang H, Hwang B, Lee JH, Cho J, Lee JH, Bang D (2019) CRISPR-Cap: multiplexed double-stranded DNA enrichment based on the CRISPR system. Nucleic Acids Res 47(1):e1. https://doi.org/10.1093/nar/gky820
40. Quan J, Langelier C, Kuchta A, Batson J, Teyssier N, Lyden A, Caldera S, McGeever A, Dimitrov B, King R, Wilheim J, Murphy M, Ares LP, Travisano KA, Sit R, Amato R, Mumbengegwi DR, Smith JL, Bennett A, Gosling R, Mourani PM, Calfee CS, Neff NF, Chow ED, Kim PS, Greenhouse B, DeRisi JL, Crawford ED (2019) FLASH: a next-generation CRISPR diagnostic for multiplexed detection of antimicrobial resistance sequences. Nucleic Acids Res 47(14):e83. https://doi.org/10.1093/nar/gkz418
41. Battaglia S, Dong K, Wu J, Chen Z, Najm FJ, Zhang Y, Moore MM, Hecht V, Shoresh N, Bernstein BE (2022) Long-range phasing of dynamic, tissue-specific and allele-specific regulatory elements. Nat Genet 54(10):1504–1513. https://doi.org/10.1038/s41588-022-01188-8
42. Lin Y, Cradick TJ, Brown MT, Deshmukh H, Ranjan P, Sarode N, Wile BM, Vertino PM, Stewart FJ, Bao G (2014) CRISPR/Cas9 systems have off-target activity with insertions or

deletions between target DNA and guide RNA sequences. Nucleic Acids Res 42(11):7473–7485. https://doi.org/10.1093/nar/gku402

43. Zhang XH, Tee LY, Wang XG, Huang QS, Yang SH (2015) Off-target effects in CRISPR/Cas9-mediated genome engineering. Mol Ther Nucleic Acids 4:e264. https://doi.org/10.1038/mtna.2015.37
44. Guo C, Ma X, Gao F, Guo Y (2023) Off-target effects in CRISPR/Cas9 gene editing. Front Bioeng Biotechnol 11:1143157. https://doi.org/10.3389/fbioe.2023.1143157
45. Zhou M, Gauthier M-PL, Nabilsi NH, Wang A, Murray KO, Azari H, Owens WS, Newman JRB, Pardo-Palacios FJ, Conesa A, Riva A, Clanton TL, Reynolds BA, Concannon P, Brant JO, Bacher R, Kladde MP (2022) Flap-enabled next-generation capture (FENGC): precision targeted single-molecule profiling of epigenetic heterogeneity, chromatin dynamics, and genetic variation. bioRxiv 1–44. https://doi.org/10.1101/2022.11.08.515732
46. Varley KE, Mitra RD (2010) Bisulfite patch PCR enables multiplexed sequencing of promoter methylation across cancer samples. Genome Res 20(9):1279–1287. https://doi.org/10.1101/gr.101212.109
47. Schutsky EK, DeNizio JE, Hu P, Liu MY, Nabel CS, Fabyanic EB, Hwang Y, Bushman FD, Wu H, Kohli RM (2018) Nondestructive, base-resolution sequencing of 5-hydroxymethylcytosine using a DNA deaminase. Nat Biotechnol 1083–1090. https://doi.org/10.1038/nbt.4204
48. Davis MW, Jorgensen EM (2022) ApE, a plasmid editor: a freely available DNA manipulation and visualization program. Front Bioinform 2:818619. https://doi.org/10.3389/fbinf.2022.818619
49. Koressaar T, Remm M (2007) Enhancements and modifications of primer design program Primer3. Bioinformatics 23(10):1289–1291. https://doi.org/10.1093/bioinformatics/btm091
50. Knight P, Gauthier M-PL, Pardo CE, Darst RP, Kapadia K, Browder H, Morton E, Riva A, Kladde MP, Bacher R (2021) Methylscaper: an R/Shiny app for joint visualization of DNA methylation and nucleosome occupancy in single-molecule and single-cell data. Bioinformatics 37(24):4857–4859. https://doi.org/10.1093/bioinformatics/btab438
51. Friedrich-Heineken E, Henneke G, Ferrari E, Hubscher U (2003) The acetylatable lysines of human Fen1 are important for endo- and exonuclease activities. J Mol Biol 328(1):73–84. https://doi.org/10.1016/s0022-2836(03)00270-5
52. Liu Y, Kao HI, Bambara RA (2004) Flap endonuclease 1: a central component of DNA metabolism. Annu Rev Biochem 73:589–615. https://doi.org/10.1146/annurev.biochem.73.012803.092453

Part II

Genome-Wide Methods for Mapping Chromatin-Associated Proteins

Chapter 3

Chromatin Immunoprecipitation Sequencing in Cancer Cells

Zachary Bigelow, Sergei Nechaev and Archana Dhasarathy

Abstract

Analysis of DNA co-immunoprecipitated with proteins of interest has been foundational to our understanding of chromatin and its regulation. Despite the emergence of crosslink-free techniques such as CUT&RUN and CUT&TAG, Chromatin Immunoprecipitation followed by sequencing (ChIP) remains a method of choice for profiling of protein binding at individual DNA sites and, when combined with sequencing (ChIP-seq), across the genome. ChIP-seq can provide insights into transcription factor binding sites, histone modification patterns and chromatin structure, and has broad applications in the study of cancer and other diseases. Here, we outline the protocols for ChIP-seq, including immunoprecipitation, library preparation for high-throughput sequencing, and bioinformatics analysis of sequencing data that we routinely perform in our lab groups. We also include troubleshooting tips for each step. We hope this guide will be helpful to laboratories that wish to adopt this method for their cancer biology studies.

Keywords Chromatin, Immunoprecipitation, ChIP-seq, Genomics, Cancer, Epigenetics

1 Introduction

Chromatin Immunoprecipitation (ChIP) has long been used to detect protein binding to DNA. The technique was first developed by Gilmour and Lis, using UV light to crosslink proteins to DNA [1]. Shortly thereafter, Solomon and Varshavsky adopted the use of formaldehyde, which is a more efficient crosslinker than UV [2, 3]. Formaldehyde reacts with nucleophilic groups (e.g., amino acid) to form a Schiff's base, which is then stabilized by reacting with another functional group (e.g., nucleotide on DNA or another amino acid), forming a methylene bridge [4]. Importantly, this is a reversible reaction, which allows for the downstream study of immunoprecipitated DNA by PCR or sequencing. Subsequently, other reagents have been described to facilitate crosslinking of proteins that are in close proximity to, but are not in direct contact with, the DNA. For

These authors contributed equally: Sergei Nechaev, Archana Dhasarathy

Archana Dhasarathy and Motoki Takaku (eds.), *Cancer Epigenetics*, Methods in Molecular Biology, vol. 3005, https://doi.org/10.1007/978-1-0716-5072-1_3,

instance, bi-functional NHS-esters like Disuccinimidyl glutarate (DSG) and Dithiobis-succinimidyl propionate (DSP), or dimethyl 3,3′-dithiobispropionimidate (DTBP), an imidoester crosslinker, have been used with some success by research groups [5, 6]. However, to date, formaldehyde remains a mainstay of the ChIP technique.

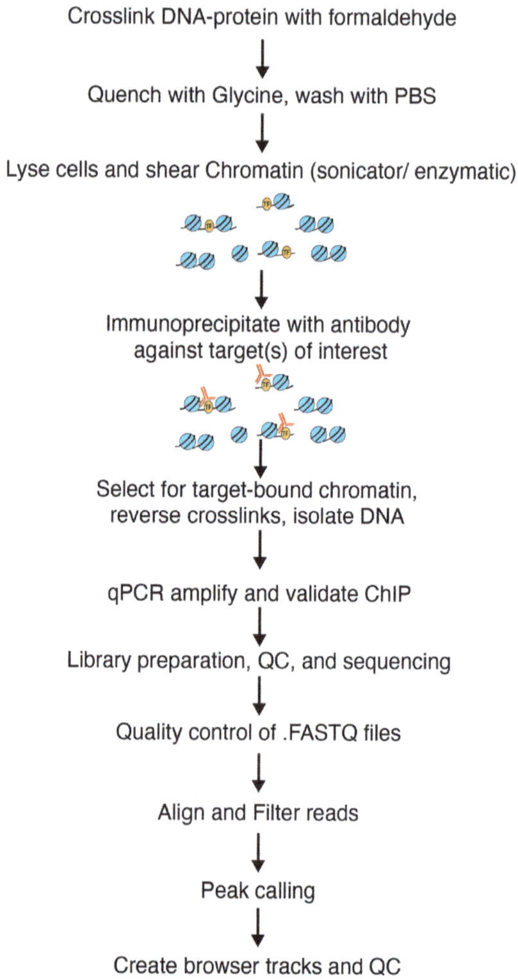

Fig. 1 Overview of ChIP-sequencing. The ChIP-seq workflow begins with the crosslinking of proteins to DNA in cells or tissue, quenching with glycine, and washing to remove formaldehyde. The cells are lysed and crosslinked chromatin is sheared (e.g., by sonication or enzymatic digestion) to fragment the DNA. Immunoprecipitation is then performed using an antibody specific to the protein of interest, capturing DNA fragments bound by the protein. After reversal of crosslinking, the purified DNA is amplified using primers designed across regions of interest to validate the ChIP experiment. Libraries are prepared and sequenced using high-throughput sequencing platforms. Key data analysis steps include quality control (e.g., FastQC), alignment to a reference genome (e.g., using Hisat2 or Bowtie2), peak calling (e.g., with MACS2 or HOMER) to identify regions of significant enrichment, and visualization (e.g., generating bigWig tracks for IGV or UCSC Genome Browser). Further downstream analyses may involve signal counting, differential binding analysis, and motif discovery

Following cross-linking of proteins to DNA, ChIP involves shearing of the DNA into smaller fragments, immunoprecipitation of the protein–DNA complexes using specific antibodies, and analysis of the associated DNA fragments (Fig. 1). When combined with next-generation sequencing (ChIP-seq), this technique can aid in the identification of binding sites for DNA-associated proteins genome-wide. One of the earliest techniques to benefit from massively parallel sequencing [7–9], ChIP-seq has been heavily utilized in cancer biology to map transcription factor binding sites and histone modifications that drive oncogenesis in cell lines, human tissue samples, and mouse models [4, 10], leading to collections such as the NIH Epigenomics Roadmap project [11]. Although a much-utilized technology, ChIP-seq is being gradually replaced with newer technologies such as CUT&RUN and CUT&Tag. However, ChIP-seq is still a mainstay of cancer epigenetics research, due to its ability to detect the binding of transcription factors, chromatin proteins, and histone modifications, and to reveal transient interactions that may be missed by the newer technologies. ChIP also remains a method of choice to analyze individual loci using qPCR without sequencing. We outline below a comprehensive ChIP-seq protocol and basic bioinformatics pipeline utilized by our laboratories, along with some troubleshooting steps at each point.

2 Materials

2.1 Cell Culture Reagents

1. Growth media for cells of interest (e.g., DMEM/F-12 with 10% Fetal Bovine Serum (FBS)).
2. 5% CO_2 incubator.
3. 0.25% Trypsin-EDTA.
4. Phosphate-Buffered Saline (PBS; calcium-magnesium free, sterile).
5. Additional supplements/antibiotics as needed.
6. Additional media as needed for spike-in controls, if using.

2.2 Crosslinking and Quenching Reagents

1. Methanol-free formaldehyde, 1% final in FBS-free cell culture media (*see* **Notes 1–3**).
2. Glycine, 2.5M solution in nuclease-free water. Filter-sterilize or autoclave if long-term storage is planned.

2.3 Cell lysis and Sonication

1. Sonication buffer (0.5 or 1% SDS (*see* **Note 4**), 20 mM Tris, pH 8.0, 2 mM EDTA, 0.5 mM EGTA, q.s. to 10 mL with nuclease-free water). Immediately before use, add protease inhibitors (*see* **Note 5**) and PMSF (0.5 mM).

2. A sonicator instrument, such as Covaris S220, or alternatives. Some sonicators, including Covaris, require special tubes to be purchased, while others are compatible with standard 1.5 or 15 mL tubes.

2.4 Pre-clearing and Immunoprecipitation

1. Immunoprecipitation (IP) buffer: (0.5% Triton X-100, 2 mM EDTA, 20 mM Tris-HCl, pH 8.0, 150 mM NaCl, 10% Glycerol, q.s. to 50 mL with nuclease-free water).
2. A test tube rotator or nutator is placed in a cold room or deli fridge.
3. Benchtop microcentrifuge chilled to 4°C.

2.5 Wash Buffers

1. Protein A/G beads (can be magnetic or Agarose-based).
   If purchasing Protein A and Protein G beads separately, mix equal volumes of each to make a Protein A/G bead mix, for instance, 500 μL each.
2. Low salt wash (0.1% SDS, 1% Triton X-100, 2 mM EDTA, 20 mM Tris-HCl, pH 8.0, 150 mM NaCl).
3. High salt wash (0.1% SDS, 1% Triton X-100, 2 mM EDTA, 20 mM Tris-HCl, pH 8.0, 500 mM NaCl).
4. Lithium chloride wash (20 mM Tris-HCl, pH 8.0, 2 mM EDTA, 250 mM LiCl, 1% IGEPAL CA-630 detergent, 1% Sodium deoxycholate).
5. Tris-EDTA (TE) wash (10 mM Tris-HCl, pH 8.0, 1 mM EDTA, pH 8.0).

Prepare at least 50–100 mL of each wash buffer.

2.6 Elution Buffer

1. Elution buffer, 10 mL (1% SDS and 0.1 M Sodium bicarbonate in nuclease-free water). Prepare just before use.

2.7 Crosslink Reversal and DNA Isolation

1. Five-molar (5M) NaCl.
2. 1M Tris pH 7.8.
3. Phenol: Chloroform: Isoamyl alcohol (25:24:1) (pH 8.0) (not required if a commercial DNA preparation kit is used).
4. Chloroform (not required if a commercial DNA preparation kit is used).
5. Co-precipitant (e.g., GlycoBlue; Invitrogen). Not required if a commercial DNA preparation kit is used.
6. Sodium acetate (3M) or Ammonium acetate (7.5M). Not required if a commercial DNA preparation kit is used.
7. Ethyl alcohol (100% or 96% and 70% (v/v)).
8. Tris-EDTA (TE) buffer pH 8.0.
9. RNase A/T1 (such as RNase Cocktail enzyme mix, ThermoFisher).
10. Proteinase K (such as 20 mg/mL solution, ThermoFisher).

2.8 Materials for DNA Size Estimation and Protein Immunoblotting

1. A horizontal electrophoresis apparatus, such as the ThermoFisher Owl B1A Mini Gel Electrophoresis System.
2. Agarose for gel electrophoresis.
3. Tris-Borate EDTA (TBE) buffer.
4. Ethidium Bromide or SYBR Gold DNA stain.
5. A vertical electrophoresis system, such as the Bio-Rad Mini-Protean 3 Cell system.
6. Commercial pre-cast or self-poured polyacrylamide gels.
7. 2×Laemmli sample loading buffer containing Dithiothreitol (DTT) or Beta-mercaptoethanol added just before use.
8. Tris-Glycine SDS gel running buffer and the same with 20% methanol for gel transfer.
9. A Western Blot setup including a Transfer System, Western blot supplies including PVDF transfer membrane, pre-stained protein standards, western blot detection reagents, an imager suitable for western blot detection, and primary antibodies for proteins of interest, and appropriate secondary antibodies.

2.9 qRT-PCR Reagents

1. Validated primers for positive and negative control loci (*see* **Note 6**).
2. SYBR Green PCR master mix (e.g., SsoAdvanced mix, Bio-Rad). Design primers such that amplicons are between ~80 and 250 bp.
3. A real-time PCR instrument with compatible plasticware.

2.10 ChIP-seq Library Preparation

1. A DNA library preparation kit based on ligation of linkers (not tagmentation), such as NEBNext or NEXTFLEX® Rapid DNA-seq 2.0.
2. Fluorescent dye for real-time PCR (we use EvaGreen Dye, 20× in water, at 1:20 final dilution (*see* **Note 7**)).
3. Qubit or similar instrument based on fluorescence-based quantification of DNA, including Qubit dsDNA detection reagents.
4. A capillary electrophoresis system, such as Tapestation or Bioanalyzer. If one is not available in the lab or core facility, libraries can be evaluated using agarose gel electrophoresis. This uses more material and may increase the potential for contaminating the lab space with post-PCR library DNA.

2.11 Bioinformatics Analyses

1. A UNIX-based workstation with at least 64 GB RAM and at least 16 available CPU cores (*see* **Note 8**).
2. Software for data analysis. For our ChIP-seq analysis example, the following software is used: quality control (FastQC [12] or MultiQC [13]); read trimmer (Trimmomatic [14] or Cutadapt [15]), aligner (for example, HISAT2 [16] or Bowtie

2 [17]), SAMtools[18], Picard Tools (MarkDuplicates) [19], BEDTools (bedtools genomecov) [20], deepTools (Bamcoverage, MultiBamSummary, MultiBigWigSummary) [21], peak caller (MACS2 [22] and/or HOMER software [23]). Additional software may be needed for graphing or advanced analyses.

3. In addition to the disk space for data analysis, access to at least 1–2 TB of frequently backed up data storage space.
4. On user's personal computers, install Integrative Genomics Viewer (IGV) browser [24] for data visualization, and set up an account on the UCSC genome browser [25, 26].

3 Methods

3.1 Before You Begin: Determining Optimal Conditions for ChIP

1. Grow cancer cells of choice in 15 cm dishes or suspension cells in T75 or T175 flasks to near confluence (roughly 1–2 × 10^7 cells required). We have tested multiple types of cancer cell lines, including breast, ovary, bladder, and synovial cancer lines.
2. Before starting ChIP, pre-chill a centrifuge that can hold 15 or 50 mL conical tubes and prepare 1% formaldehyde in media. If using suspension cells, a room-temperature centrifuge should be available for collecting cells prior to crosslinking. Initialize the sonication instrument if not stopping immediately after crosslinking. This is particularly important if the sonicator is connected to a chiller, as it takes time to cool down.
3. Begin crosslinking by replacing growth media with 1% formaldehyde in room-temperature FBS-free media. Prepare enough for the size and number of culture dishes used immediately before cross linking. For example, for 15 cm culture dishes, prepare 15 mL of media with formaldehyde per dish. For adherent cells, pour the media out of the dish and add the same volume of media with formaldehyde. For suspension cells, transfer the suspension into a 15 mL or 50 mL conical tube and spin for 3 min at 300g in a room temperature centrifuge, withdraw media and resuspend the cell pellet in media with formaldehyde. Continue crosslinking at room temperature with gentle agitation for a total of 10 min (*see* **Notes 1–3**).
4. Quench crosslinking by adding 2.5M glycine to a final concentration of 125 mM and incubating for 5 min with gentle agitation.
5. Wash twice with ice-cold PBS on plates (for adherent cells) or by centrifugation (for suspension cells). For adherent cells,

add ~10 mL of ice-cold PBS and scrape with a cell scraper to dislodge the cells, then collect them by pipetting into a Falcon tube. Collection may need to be repeated with extra PBS, so a larger size (50 mL) conical tube should be used for adherent cells (*see* **Note 9**).

6. Centrifuge cells at 300 × *g* for 3 min at 4°C and remove all traces of PBS.

   ***Stopping point**: If desired, cell pellets can be flash-frozen in liquid nitrogen and stored at −80°C for future use.*

7. Add protease inhibitors to the Sonication buffer (*see* **Note 5**). Resuspend the pellet in ~900–950 μL ice-cold Sonication buffer so that the final volume of the suspension is as close to 1 mL as possible, incubate on ice for 10 min to allow for lysis, and transfer into a 1 mL Covaris tube (*see* **Note 10**).

8. Perform a sonication time course to determine optimal sonication conditions. This needs to be done at least once for each cell line. A typical sonication time course would be 0, 4, 6, 8, 10, and 12 min using the Covaris High Cell Program (Output = 140, Duty cycle = 5), and adjusting accordingly as needed (*see* **Note 10** and Fig. 2).

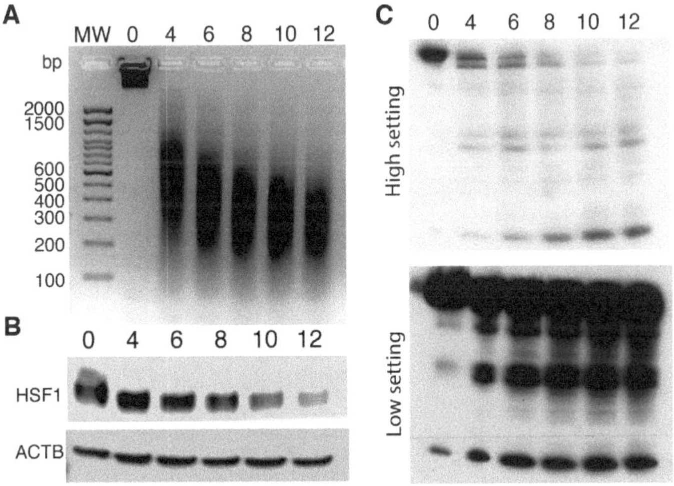

Fig. 2 Optimization of sonication conditions. Effect of the sonication time in MCF7 cells on (**A**) Genomic DNA size, samples loaded on a 1% agarose gel, and (**B**) Immunoblot for protein levels of transcription factor HSF1 with Beta-Actin as loading control. (**C**) Optimization of sonication times: effect of Covaris high (Covaris Medium Cell program (Output = 140, Duty cycle = 5) vs. low (Output = 120, Duty cycle = 3) sonication settings on protein levels of RNA Pol II in a mouse mammary gland epithelial cell line, NMuMG

9. After each sonication time point, remove 25 μL of lysate for DNA and another 25 μL for protein analysis. Add back 50 μL of lysis buffer to the remaining sample to maintain the volume at 1 mL during sonication.
10. For DNA preparation, add 224 μL of TE buffer and 1 μL of RNase cocktail to 25 μL of each sonication time point sample and incubate at 37°C for 30 min to remove RNA. Next, add 1 μL of Proteinase K and incubate at 65°C for at least 2 h or overnight to remove proteins and reverse cross-links. Then, phenol-extract DNA (*see* **Note 9**), resuspend in 25 μL of TE, and load ~3–5 μL on a 1% Agarose gel to observe shearing (typically 200–500 bp on average) (*see* **Note 11**). Stain with ethidium bromide, SYBR Gold, or similar DNA intercalating dye to visualize and determine the best shearing condition (usually ~200–500 bp; *see* **Note 12**).
11. For protein, to 25 μL of the sonicated lysate, add 25 μL of 2× Laemlli loading buffer containing DTT or Beta-mercaptoethanol. Heat at 95°C for 5 min, load and run 10–20 μL on an SDS-PAGE gel. Transfer proteins onto a PVDF or nitrocellulose membrane and immunoblot with antibodies of interest. Save the remaining sample at −80°C in case you need to rerun it.
12. Once you determine the right shearing conditions (DNA sheared to the right sizes, and protein still intact), you can proceed with ChIP. *This is an important step that is often overlooked, especially for transcription factors, which are less abundant targets with less robust antibodies than histone marks.* Fig. 2A, B show the effect of sonication conditions on DNA and protein levels of the transcription factor HSF1 in MCF7 cells. Note the decrease in protein after the 10-min sonication point. Sonication strength may also need optimization, as seen in Fig. 2C. For example, in another cell line, NMuMG, we observe loss of Pol II signal at higher vs. lower sonication settings.

3.2 ChIP-seq. Formaldehyde Crosslinking

1. Grow adherent cells in 15 cm dishes or suspension cells in T75 or T175 flasks to near confluence (~$1–2 \times 10^7$ cells per condition). If using spike-in controls such as *Drosophila* S2 cells, grow these separately.
2. Remove media from the 15cm dishes. Add spike in control cells resuspended in serum-free DMEM to the dishes to a final concentration of 5-10% of the total cell number (ignore if not using spike-in controls). Replace media on dishes with FBS-free media containing 1% final formaldehyde for 10 min at room temperature (*see* **Note 13**).
3. Quench crosslinking with glycine at the final concentration of 200 mM (add 1.5 mL of 2.5M stock to 15 mL of media per dish).

4. Wash cells 2× with ice-cold PBS and collect as described above (see Sect. 3.1 and **Note 1**) into a 15 or 50 mL conical tube.
5. Centrifuge cells at 300 × *g* for 3 min at 4°C and remove PBS.

Stopping point: Pellets can be flash-frozen and stored at −80°C.

6. Resuspend the pellet in 1 mL ice-cold Sonication Buffer, incubate on ice for 10 min to allow cell lysis. Take an aliquot from samples (25 μL each) before sonication for DNA and western blot analysis. Transfer 1 mL of the sample into a Covaris tube (*see* **Note 10**).
7. Sonicate using appropriate conditions as determined in Sect. 3.1 above.
8. After sonication, transfer contents to a 1.5 mL Eppendorf tube and centrifuge at full speed in a chilled tabletop centrifuge. This precipitates the insoluble chromatin, seen as a small pellet, while leaving a mostly clear supernatant (*see* **Note 14**). Save sample aliquots (25 μL each) after sonication for DNA extraction and for western blot analysis.

Stopping point: Instead of stopping after crosslinking (step 5 above), you can stop here by flash-freezing the sonicated material in liquid nitrogen and storing it at −80°C until use. This way, ChIP material can be stored until sonication quality has been verified.

3.3 Immunoprecipitation

1. Prepare the beads. Add enough Protein A+G beads for all samples and for pre-clearing to a 1.5 mL or 2 mL centrifuge tube (see **Note 15**).
2. Centrifuge at 400 × *g* to precipitate beads if using agarose/sepharose beads, or place on a magnet to pellet if using magnetic beads and remove supernatant (*see* **Note 15**).
3. Wash beads twice with IP buffer and resuspend in a volume of IP buffer equal to the original bead volume to create a 50% suspension.
4. Pre-clear ChIP sample with protein A/G beads equilibrated in IP buffer as follows: Dilute 200–400 μL of sonicated cell lysate from the previous step with IP buffer to 2 mL in a 2 mL conical tube for each IP. IPs can be scaled, for instance, 1 mL of sonicated lysate can be diluted with 9 mL of IP buffer in the 15 mL tube. In this case, adjust the volume of beads proportionally (*see* **Note 16**).
5. Add 25 μL of 50% bead slurry per 2 mL of diluted ChIP material and preclear for 1–2 h at 4°C on rotator/nutator in the cold room.
6. Spin samples for 1 min at 1000 × *g* in a centrifuge to pellet the beads if using agarose beads. Magnetic beads could also be briefly centrifuged at a higher speed to help dislodge beads

from the tube cap and then placed on a magnetic stand to separate the beads.

7. Transfer the supernatant containing the precleared IP material into a new 2 mL Eppendorf tube.
8. Save two separate aliquots each containing 1% of the IP material (20 μL) as inputs (*see* **Note 17**). For ChIP qPCR without sequencing, take 10% (200 μL of IP material) as input.
9. Add immunoprecipitating antibody to each tube (*see* **Note 18**).
10. Place IP material at 4°C on a nutator/rotator overnight.
11. The next morning, set the input samples aside at 4°C. Add equilibrated Protein A/G bead slurry, 100 μL per Sample. A 2 mL Eppendorf tube should fit this volume with some room left for mixing. The sample can also be divided into two 1 mL aliquots prior to adding beads: in this case, add 50 μL bead slurry to each tube.
12. Incubate at 4°C on a nutator/rotator for 1–2 h.

3.4 Washes and Elution

1. If using agarose-based beads, centrifuge samples in a pre-chilled (4°C) microcentrifuge at 1000 × *g* for 1 min to pellet beads and remove supernatant by pipetting. Place magnetic beads on a magnetic rack and remove supernatant.
2. Perform 1× Low salt wash, 2× High salt washes, 1× LiCl wash, and 2× washes with TE. For each wash, add 1 mL of ice-cold wash solution to beads, and leave for 5 min on a rotator at room temperature, then briefly centrifuge for 15 s at 1000 × *g* at 4°C (if using agarose) or place on a magnetic rack (if using magnetic beads) to separate beads from supernatant (*see* **Note 19**).
3. After the last TE wash, spin for 20 s or place on a magnetic rack to remove remaining TE buffer.
4. Add 270 μL room temperature Elution buffer and mix by gently flicking the tube.
5. Incubate beads with the elution buffer at room temperature on a rotator/nutator for 15 min, then centrifuge and save 250 μL of the supernatant in a fresh tube. The ~20 μL left above the beads ensures no bead carryover.
6. Add 250 μL Elution buffer and repeat incubation for 15 min, spin and combine the second eluate with the first for a total volume of 500 μL.
7. Add 480 μL Elution buffer to 20 μL Input samples to make a final volume of 500 μL. If using 200 μL input, add 300 μL of Elution buffer.

3.5 Crosslink Reversal and DNA Isolation

1. Add 20 μL of 5 M NaCl to 500 μL of each eluted sample and input, and place in a water bath at 65°C for 6 h to overnight to reverse cross-links.
2. Add 15 μL of 1M Tris, pH 7.8, and 2 μL Proteinase K to each sample, incubate for an additional 1 h at 65°C.
3. Add 500 μL Phenol-Chloroform-Isoamyl alcohol, mix by vortexing for ~10 s and spin for 5 min in a room-temperature microcentrifuge at 16,000 x *g* or maximum speed. If using a column-based DNA isolation kit, follow the kit instructions instead, and continue with the next section (3.6).
4. Transfer the aqueous (upper) phase into a new tube without taking any phenol phase (*see* **Note 20**).
5. Add equal volume of chloroform per initial sample volume (e.g., for a 500 μL starting sample volume, add 500 μL chloroform) and vortex.
6. Centrifuge 5 min at 4°C or room temperature in a microcentrifuge at maximum speed.
7. Transfer the supernatant to a new tube and precipitate DNA with 2 μL co-precipitant (such as GlycoBlue). Add 1/10th volume of 3M Sodium Acetate or 1/3rd volume of 7.5 M Ammonium Acetate, mix by flicking the tube several times, then add 2.5× volume ethanol.
8. Mix the tube by inverting several times and place at −20°C for at least 10 min or overnight.
9. Centrifuge for 10 min at maximum speed and carefully remove supernatant without disturbing the pellet.
10. Wash pellets with 70% EtOH, centrifuge at full speed for 5 min.
11. Pipette out supernatant carefully without disturbing the pellet, centrifuge empty tubes briefly (10 s) and remove all liquid with a P20 or P10 pipette. Air-dry briefly (~30 s).
12. Resuspend each ChIP and input sample in 50 μL of nuclease-free water (25 μL per tube if using two tubes per sample). The undiluted input sample concentration from 20 μL of IP material corresponds to 1% of the total material (10% if 200 μL input was taken). After checking with qPCR, the rest of the ChIP sample will be used for library preparation. Therefore, when resuspending, do not exceed the sample volume allowed by the library kit manufacturer.
13. Make serial dilutions of both input samples: 1:10, 1:100, and 1:1000, which correspond, respectively, to 0.1%, 0.01%, and 0.001% input. If 10% input was taken, which is recommended if only qPCR and no sequencing is desired, the dilutions will correspond to 1%, 0.1%, and 0.01% of input. Leave ChIP samples undiluted.

Stopping point. Samples can be stored for several days at −20°C.

3.6 ChIP-seq Sample Quality Assessment

1. ChIP-sequencing is more sensitive to high background and has more stringent requirements for DNA quantity and ChIP quality than ChIP-qPCR. The signal enrichment at positive over negative control region amplicons should be calculated to avoid sequencing samples with low ChIP efficiency or high background contamination. The exact quality metrics are specific to each antibody and cell type and should be determined empirically by relating sequencing outcomes to qPCR quality control metrics for ChIP material and for libraries.
2. Dilute 5 μL of each ChIP sample and all input dilutions to 50 μL with nuclease-free water to make working solutions for qPCR. Perform real-time PCR for at least one positive and one negative control amplicon. The working solutions are sufficient for 24 qPCR reactions, allowing for testing 6 amplicons in duplicate and for one repeat. Set up the reaction as follows:

Working solution (template)	2 μL
Forward and reverse primers	(0.5 μL each of a 10 mM stock)
2× SYBR green master mix	5 μL
Nuclease-free water	2 μL
Total = 10 μL	

Amplify using a real-time PCR machine under the following conditions:

Step 1: 98°C for 30 s
Step 2: 98°C for 10 s,
Step 3: 60°C for 30 s,
Step 4: 72°C for 1 min.
Plate read.
Repeat Steps 2–4 39 times (40 total cycles)
Hold at 4°C.

Set each qPCR reaction in duplicate (or triplicate if new to qPCR). If using ChIP-qPCR without sequencing, samples and input dilutions from step 3.5 can be used for qPCR.

3. After qPCR, use the Ct values for input and its serial dilutions to generate a standard curve (Fig. 3A, B).
4. Use the standard curve to determine the % Input of your immunoprecipitated samples (Fig. 3C, D). Calculate the ratio between percent inputs at each amplicon and the negative control amplicon for your protein of interest (*see* **Note 21** and Fig. 4).

3.7 Library Preparation

1. Preparation of ChIP-seq libraries follows ligation-based protocols for whole-genome DNA amplification. Any ligation-based genomic DNA library preparation kit is, in principle, suitable for ChIP-seq. Because DNA in ChIP material is almost always

A

Input samples	% Dilution	Log10	Avg Ct value
Input_10%	10	1	20.23
Input_1%	1	0	23.31
Input_0.1%	0.1	-1	26.51
Input_0.01%	0.01	-2	30.37

B

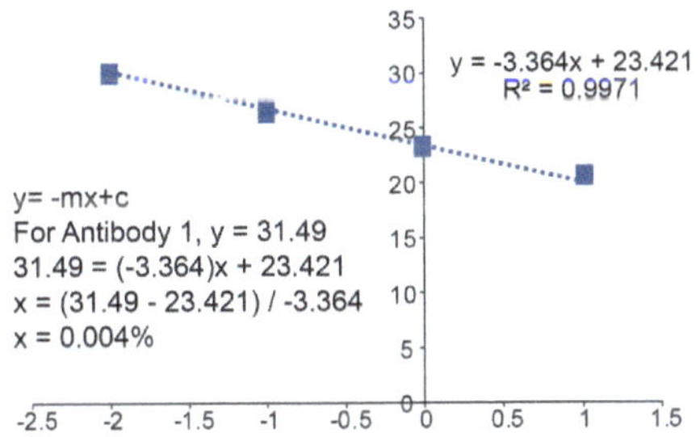

C

IPs	y (Avg Ct)	x	% Input
IgG	31.49	-2.4	0.0040
H3K4Me3	24.13	-0.2	0.6134
Pol II	30.16	-2.0	0.0099
TF	29.73	-1.9	0.0133

D

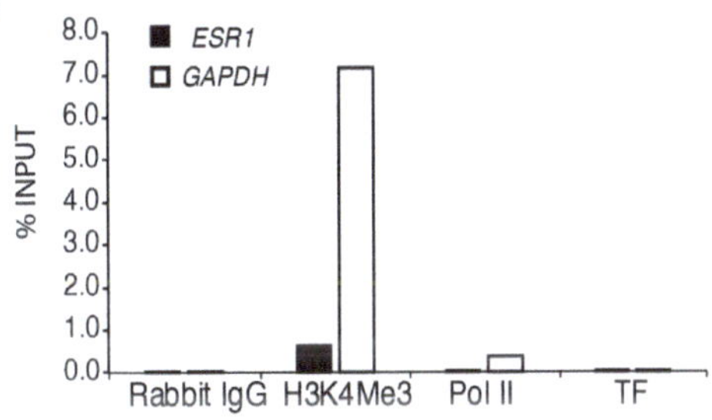

E

	% INPUT ESR1	% INPUT GAPDH
Rabbit IgG	0.004	0.0062
H3K4Me3	0.613	7.1547
Pol II	0.010	0.3591
TF	0.013	0.014

Fig. 3 Calculation of percent input example. (**A**) The percent input values (10, 1, 0.1, and 0.01) are converted to their corresponding $\log_{10}$ values (1, 0, −1, and −2). (**B**) A scatter plot is then generated using the $\log_{10}$ values and their respective Ct values, and linear regression is performed to determine the slope and intercept ($y = mx + c$). (**C**) This equation can be used to interpolate Ct values obtained for immunoprecipitated samples (IPs) and to calculate their corresponding percent input. (**D**) Example graph generated after calculation of percent input values for IgG, H3K4Me3, Pol II, and a transcription factor ChIP. (**E**) Table demonstrating differences in percent input for the same factors at different loci

in limiting amounts, choose a kit that works with the lowest amounts of starting material. The NEBNext DNA library kit is a good starting point. Do not use Tn5 (tagmentation)-based DNA library kits, as these are designed to work with the specific cuts left by the Tn5 enzyme and will not work well for ChIP.

2. Prepare a library from the rest of the ChIP material using a DNA library kit of choice. Complete steps before PCR, including End Repair, A-tailing, and Adapter ligation per kit protocol. It is recommended to also prepare a library from at least one input sample.
3. Test-amplify the libraries. Because of low amounts of ChIP DNA, the number of PCR cycles needs to be adjusted for each library to generate a sufficient amount of DNA without over-amplification. Instructions included in library kits may not be sufficient. Instead, use a defined portion of the ChIP material in test amplification using real-time PCR. For example, for the NEBNext kit, the reaction volume at this point is 15 μL. To

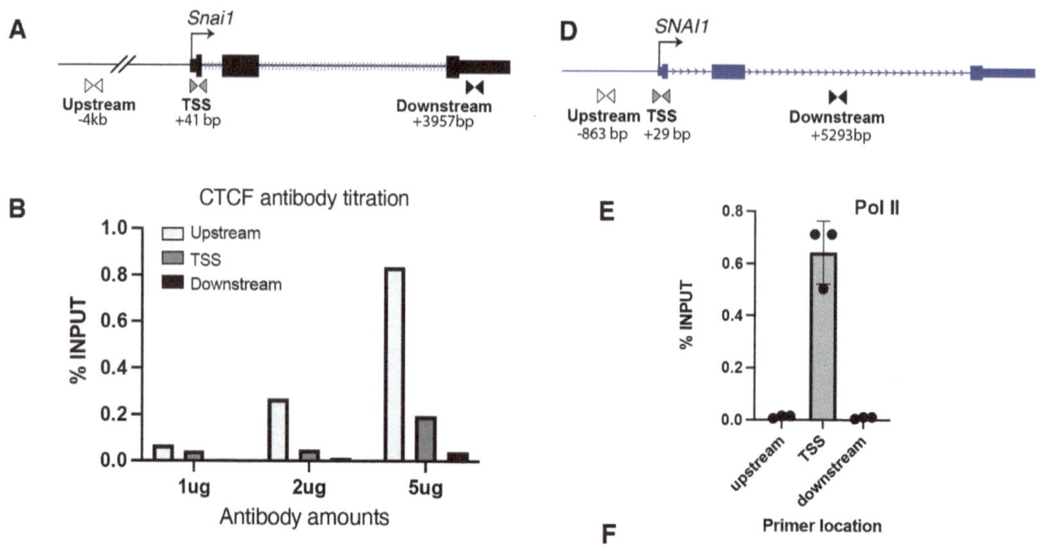

C

	% INPUT			
Antibody amt	4Kb up	TSS	Down	Signal to noise
CTCF 1ug	0.067	0.042	0.006	10.72 ; 6.71
CTCF 2ug	0.269	0.048	0.013	20.39 ; 3.67
CTCF 5ug	0.836	0.196	0.039	21.19 ; 4.98

F

	% INPUT		
Primer location	R1	R2	R3
Upstream	0.015	0.015	0.006
TSS	0.712	0.712	0.504
Downstream	0.011	0.011	0.004

Fig. 4 Primer locations and antibody titration examples. (**A**) Schematic depicting locations of primers across the mouse *Snai1* gene. (**B**) Percent input calculated at each of the three primer locations using ChIP material with three different antibody concentrations of CTCF (1, 2, 5 μg). Two positive control regions for CTCF show different signal-to-noise ratios relative to a downstream region. In this experiment, 5 μg of antibody was chosen as it had a better signal-to-noise ratio for both regions. (**C**) Table with percent input calculations. (**D**) Schematic depicting locations of primers across the human *SNAI1* gene. (**E**) Percent input calculated at each of the three primer locations using anti Pol II antibody. (**F**) Table with percent input calculations. The signal-to-noise for RNA Pol II is clearly evident when comparing the TSS region with two regions upstream and downstream of the TSS

withdraw 1/16th of this reaction, add 1 μL of water, mix and take 1 μL of the mix for test amplification. To 1 μL of this material, add PCR reagents from the kit and PCR primers compatible with the kit, 5% by volume (1.25 μL) water-dissolved Eva Green dye, and q.s. to 25 μL with nuclease-free water. Indexing is not important at this point—we dedicate one index primer for test amplification of all samples. It is also possible to use generic qPCR reagents and primers. Amplify in a real-time instrument using PCR cycling conditions suggested by the kit. Determine the number of PCR cycles corresponding to ~1/4 the maximum saturated fluorescence intensity [27] (*see* **Note 22**).

4. Amplify the rest of the ChIP sample (15 μL) without the fluorescent dye using PCR reagents and conditions from the kit, with the appropriate index primers. Use the number of PCR cycles based on the test amplification: In the above example, the number of cycles for library amplification is the number

determined in test amplification minus 4. ChIP-seq libraries for abundant targets (histone modifications) should normally take no more than 15 PCR cycles.

5. Following amplification, follow PCR cleanup instructions from the kit to remove primers, with no size selection (*see* **Note 23**).

6. Assess libraries for quantity using Qubit, and quality using capillary electrophoresis (Agilent Bioanalyzer of TapeStation). The yield of primer-free library DNA normally falls between 100 and 1000 ng. The fragment size distribution in a library should mirror that of the ChIP material used to prepare the library, plus the size contributed by the linkers; an extra ~125 nt for the NEBNext kit. There should be no empty linker dimers and little high molecular weight bands present (Fig. 5A) (*see* **Note 24**).

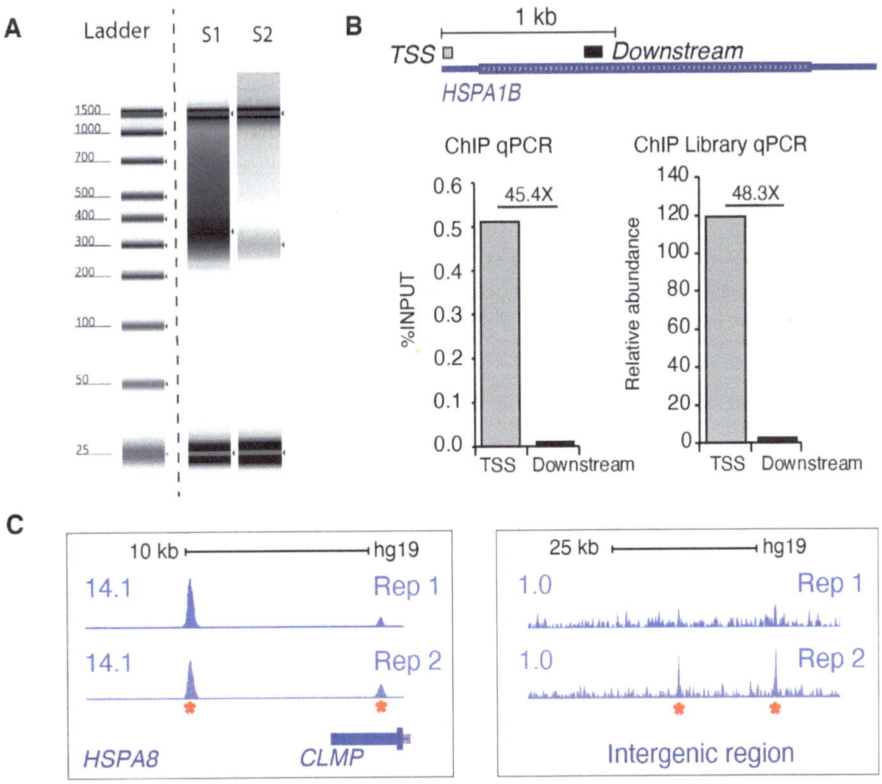

Fig. 5 ChIP library qc and quantification. (**A**) Bioanalyzer gel image of two library samples representing good (S1) and bad (S2) preparations. (**B**) Quantification of signal to noise using qPCR of pre (left; ChIP qPCR) and post (right; ChIP library qPCR) library ChIP samples. The location of the primers are indicated in the cartoon above. (**C**) Example genome browser tracks for ChIP-seq library preps. HSF1 ChIP-seq tracks from two replicates of MCF7 cells treated with heat shock for 60 min to induce HSF1 binding are shown [34]. The left box indicates a known heat-shock responsive region bound strongly by HSP1 (*HSP8*) that shows enrichment in both replicates, and the right track is an intergenic region that shows HSF1 enrichment only in replicate 2

7. Check the library for the signal-to-noise ratio between known positive and negative control genomic regions. Libraries prepared from input samples can be used to normalize the signal. In the absence of a library prepared from an input sample, it is possible to use Ct values obtained on genomic DNA. To perform qPCR, dilute 1 μL of each ChIP-seq and at least one input library to 100 μL with nuclease-free water and use 2 μL of these diluted templates for real-time PCR as above (Fig. 5B).
8. Calculate positive-to-negative control signal fold-enrichment. First, calculate Delta Ct values for the input library between each amplicon and a negative control amplicon. Second, use the Delta Cts obtained from the input library to adjust Ct values in each ChIP-seq library for the same amplicon. Lastly, from the adjusted Ct values, for each ChIP-seq library, calculate the fold-differences (ratios) between each amplicon and the negative control amplicon (Fig. 5B; right). The fold differences for libraries should be as high or higher as those for ChIP material.

3.8 Sequencing

ChIP-seq libraries are often sequenced on high-throughput instruments such as Illumina NovaSeq along with many other samples. Individual libraries must therefore contain unique indexes that are usually introduced with primers during PCR. Library preparation kits are compatible with multiple indexing strategies, including single, dual, and UMI indexing. Some considerations are listed below:

1. It is highly recommended to verify indexing requirements with the sequencing facility before preparing libraries. For example, some facilities no longer allow for the use of single indexes.
2. Libraries can be stored at −20°C in water or TE until use, but −80°C storage is preferred to minimize evaporation long-term.
3. Frozen libraries can be courier-shipped to a facility of choice. Commercial facilities frequently run paired-end (2×150 nt) sequencing as the default or the only choice. Single-end sequencing and/or shorter reads may be adequate for ChIP-seq, but are rarer today. Given the plummeting costs of sequencing, the depth of sequencing should be at least 40 million reads per sample for a point-source target such as a transcription factor or H3K4Me3 modification, and 80M reads per sample for broader targets such as H3K27Me3.

3.9 Data Analysis

Primary processing of ChIP-sequencing data involves technical quality control, alignment to a reference genome, visualization and identification of enrichment regions (peak calling), and quantifying signal at specified features such as genes, promoters, or peak regions, etc. Advanced analyses include overlapping with and quantitation of signal around known genomic features, comparison between samples to identify changed regions, and comparing different datasets, such as RNA, etc. These analyses are highly dependent on a specific project and are not discussed here.

3.9.1 Quality Control

1. **Run FastQC on your fastq files**. The files can be gzipped. You can run FastQC with multithreading enabled to speed up the analysis. Below is an example of how to execute FastQC for paired-end reads:

   For Read 1:

   ```
   fastqc -t 10 -o output_dir file_R1.fastq.gz
   ```

   For Read 2:

   ```
   fastqc -t 10 -o output_dir file_R2.fastq.gz
   ```

   - The "-t" option specifies the number of threads to use. Adjust the number of threads (10 in this case) based on your system's resources.
   - The "-o" option specifies the output directory to save FastQC reports.

2. **Review the results**. FastQC generates an HTML report that includes basic metrics such as the number of sequencing reads in the file, read sequence length, and potential technical issues like poor basecall quality or unexpected adapter contamination. Open the resulting ".html" file in your browser to review the quality metrics.
3. If your samples are part of a series and will be compared side-by-side, ensure that they have similar sequencing depths. Significant differences in sequencing depth may affect the quality and comparability of the data. If base sequencing quality, sequencing length or the depth of sequencing are not as expected, contact the sequencing facility.

3.9.2 Groom Fastq Files

1. **Adapter Trimming**. Up to 10% of ChIP-sequencing reads at the most commonly sequenced 150 nt length may reach into adapter sequences, which should be trimmed before alignment. We use Trimmomatic for this task, but other trimming tools are also suitable.
2. **Run Trimmomatic**. Below is an example command to trim paired-end ChIP-seq reads using Trimmomatic:

   ```
   java -jar trimmomatic.jar PE -threads 14 \
   file_R1.fastq.gz file_R2.fastq.gz \
   file_1_paired_at.fq.gz file_1_Unpaired_at.fq.gz \
   file_2_paired_at.fq.gz file_2_Unpaired_at.fq.gz \
   ILLUMINACLIP:TruSeq3-PE-2.fa:2:30:10:8:TRUE MIN-
   LEN:30
   ```

Trimmomatic parameters:

- -threads 14: Adjusts the number of threads based on available resources.
- ILLUMINACLIP: Specifies the path to the TruSeq3-PE-2.fa file containing adapters to be trimmed.
- :2:30:10:8—technical parameters that have to do with the detection of partial adapter sequence and mismatches. We did not optimize these parameters. Refer to the Trimmomatic manual for further information.
- TRUE. This **keepBothReads** option is important for keeping both reads of the pair if the insert (ChIP fragment) is shorter than the sequence read (usually 150 nt). This TRUE option must be set to preserve shorter inserts for downstream paired-end analyses.
- MINLEN:30: Drops reads that are shorter than 30 nt after adapter trimming.

3. **Run FastQC on trimmed reads**. After trimming, use FastQC to verify proper adapter trimming. The command below will run FastQC on all fastq files containing paired trimmed reads. We usually discard unpaired files generated by Trimmomatic:

```
fastqc -t 10 -o output_dir file*paired*fq.gz
```

4. **Review trimming results**. FastQC reports for samples before and after adapter trimming will provide insights into the quality of the trimmed reads, including:

- **Retention rate:** The number of reads retained after trimming. This will help estimate the abundance of empty reads without inserts (if any) and reads containing fragments that are too short.
- **Adapter content:** Confirm that adapter sequences have been successfully removed.
- **Read size distribution:** Compare the distribution of read lengths before and after trimming for both R1 and R2 reads (see Fig. 6 and **Note 25**).

3.9.3 Alignment to a Reference Genome

1. **Create a genome index**. If the appropriate genome index for HISAT2 does not exist, generate it using the following command:

```
hisat2-build -p 14 genome.fa genome
```

Hisat2-build parameters:

- -p 14: Specifies the number of processors to use (adjust based on available resources).

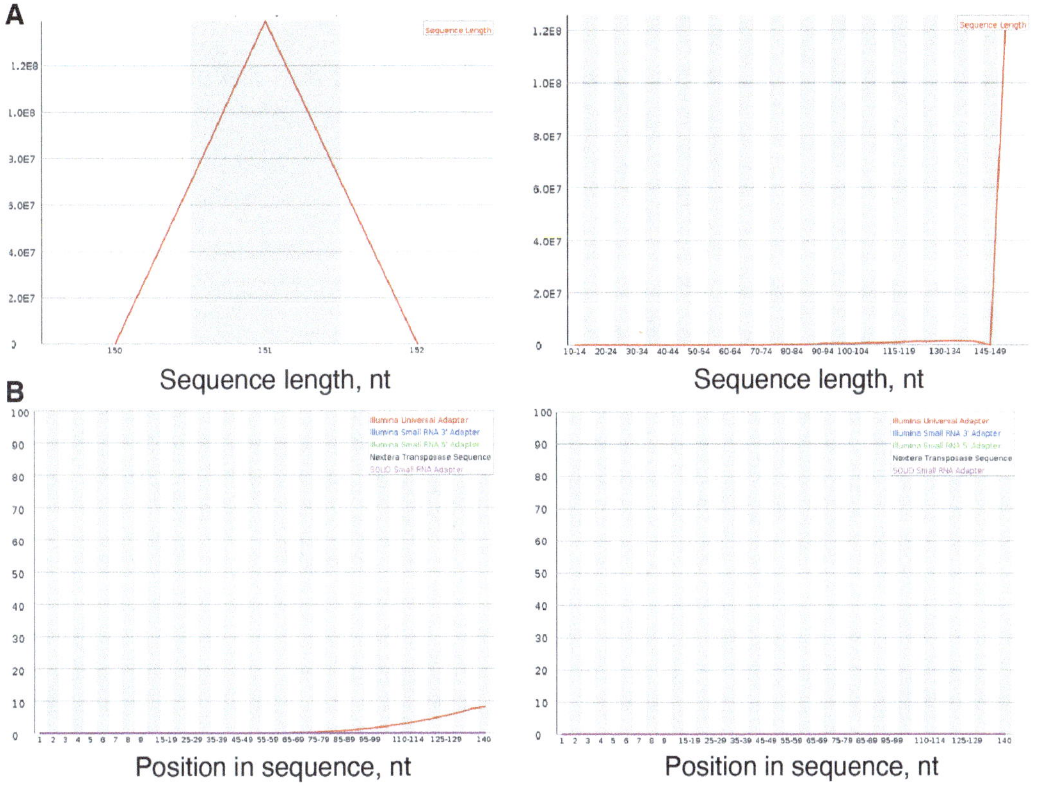

Fig. 6 Adapter trimming from a ChIP-seq library. (**A**) FastQC report of sequencing read lengths distribution for untrimmed (left) and Trimmomatic-trimmed (right) fastq files from a ChIP-seq fastq file showing the differences in size distribution. Trimming results in up to 10% of ChIP-seq reads being shorter than the untrimmed uniform 150 nt sequence length (*see* **Note 23**). (**B**) FastQC report of adapter content in untrimmed (*left*) and trimmed (*right*) fastq files from a ChIP-seq experiment showing successful removal of adapters after trimming

- genome.fa: The path to the unzipped genome text file containing one chromosome per line. The above example refers to the genome.fa file that should be located in the current directory. The file can be downloaded from a genome database. Pay attention to the names of chromosomes in the genome.fa file as they must match any database files (such as genomic features) that these data will be compared with. Chromosome names are case sensitive.
- genome: A user-specified prefix for output index files.
- Add any additional sequences (e.g., virus chromosomes) to the .fa file as needed.
- Genome indexes without customization can be placed into a commonly accessible directory and shared among multiple users.

2. **Align trimmed reads to the genome**. Once the genome index is created, align the reads. We give an example of using HISAT2 for aligning paired-end read files after adapter trimming, but other aligners of choice can be used.

```
hisat2 -x genome -1 file_1_paired_at.fq.gz -2
file_2_paired_at.fq.gz \
-p 14 --no-spliced-alignment -I 0 -X 2000 | sam-
tools view -f2 -b - > \
file.bam 2>file_hisat2_stderr.txt
```

Hisat2 parameters:

- -x genome: Specifies the genome index created in the previous step.
- -1 and –2: Paths to the paired-end trimmed fastq files.
- --no-spliced-alignment: Disables spliced alignment, which is typically used for RNA-seq data.
- -I 0 and -X 2000: Optional parameters that define the minimum and maximum distance between mapped positions of reads within a pair to keep.
- - > The standard input from hisat2 ("-" in sam file format) is converted by samtools view into a BAM file (file.bam) without retaining the sam file.
- The 2> operator redirects the standard error output from hisat2 (including alignment statistics) to a text file (file_hisat2_stderr.txt).
- It is important to record alignment statistics for reporting and troubleshooting.

3. **Sort and index the BAM file**. After alignment, the BAM file should be sorted and indexed. Indexing generates a .bai file that should be present in the same directory as the sorted bam file for downstream applications.

```
samtools sort -@ 12 -o file_sorted.bam file.bam
samtools index file_sorted.bam
```

Samtools parameters:

- -@ 12 the number of parallel threads for sorting.
- -o output file name.

4. **Mark duplicates**. Because ChIP involves random DNA fragmentation, molecules that have both identical ends should be rare. Therefore, high levels of molecules with identical ends (duplicates) are likely due to technical reasons at the usual depths of sequencing <<100M reads per sample. Duplicates should be excluded from calculations. To identify duplicate reads, which may be used Picard's MarkDuplicates tool:

```
java -jar picard.jar MarkDuplicates I=file_sort-
ed.bam O=file_marked.bam M=file_dupInfo.txt
```

This command marks duplicate reads in the BAM file, and the output is saved as file_marked.bam. Retain the duplicate information in file_dupInfo.txt for downstream analysis. To remove duplicate reads from the BAM file rather than flagging them, use --REMOVE_DUPLICATES option.

5. **Cleanup**. After marking duplicates, the original BAM file should be deleted to save server space:

```
rm file.bam
```

6. **Alignment and troubleshooting**. For ChIP-seq experiments, concordant (paired) alignment fraction against human or mouse genomes should typically be above 75%. All samples in the same series should have similar alignment percentages. If the alignment percentage is significantly lower, consider checking for incorrect trimming, wrong reference genome, sample contamination, or presence of heterologous DNA such as spike-ins.

3.9.4 Track Visualization

1. **Generate a BigWig file**. To visualize sequencing data, a BigWig (.bw) file can be generated from the BAM file using, for example, bamCoverage:

```
bamCoverage -b file_marked.bam -bs 10 -o file.bw -p
max --normalizeUsing CPM --ignoreDuplicates
```

Bamcoverage parameters:

- -b file_marked.bam: Input BAM file (typically after duplicates marked), sorted and indexed.
- -bs 10: Specifies the bin size (e.g., 10 bp).
- -o file.bw: Output file in BigWig format.
- -p max: Utilizes the maximum available number of processors. -p max/2 is an option for shared workstations.
- --normalizeUsing CPM: Normalizes the output to counts per million (CPM), which helps when comparing samples.
- --ignoreDuplicates: discards duplicate reads that were marked as duplicates (sam flag 0×400).

2. **Alternatively, generate a BedGraph file**. Generating BedGraph files is useful when additional manipulation, such as filtering by chromosome, regions, etc is needed. BedGraph files, but not bw files, can also be uploaded to a UCSC Genome Browser to create custom tracks. If you prefer to create a BedGraph, use bedtools:

```
bedtools genomecov -bga -i file_marked.bam -g
chromosome_sizes > file.bdg
```

Bedtools genomecov parameters:

- -bga: Generates a BedGraph with all positions in the genome, including an option of including regions of zero coverage.
- -i file_marked.bam: Input BAM file.
- -g chromosome_sizes: The path to the file (in this example, the file is expected to be in the current directory. The file provides chromosome size information for the genome assembly used for alignment. The file for each genome assembly can be downloaded from the UCSC genome browser website. For example, the file for hg38 is available at https://hgdownload.soe.ucsc.edu/goldenPath/hg38/bigZips/hg38.chrom.sizes
- The output is written to file.bdg in BedGraph format.

3. **Convert BedGraph to BigWig**. If a .bw file is needed from the BedGraph file, the BedGraph file can be converted to BigWig. If using bedGraphToBigWig, normalization would need to be done manually using -scale option.

```
bedGraphToBigWig file.bdg chromosome_sizes file.bw
```

4. **Visualize data**. The .bw file can be used for visualization in genome browsers:
   - **IGV Browser**: Drag-and-drop the .bw file onto an IGV browser window with an appropriate preloaded genome. IGV is a cross-platform, highly configurable browser that is useful for quick visualization, quick troubleshooting, and the generation of publication-quality data. Autoscale or Group Autoscale options may need to be checked to show tracks properly. To save an image, use the SVG (not PNG) format for editing in vector graphing software.
   - **UCSC Genome Browser**: Offers wider access to pre-annotated tracks, making it easier to share visualizations with external collaborators or anonymous reviewers.
   - .bw files need to be hosted on an FTP server and referred to when setting up custom tracks on the UCSC browser.
   - Bed or BedGraph files can be uploaded directly to the UCSC browser without hosting.

5. **Normalize data**. When comparing multiple samples, normalization based on sequencing depth (e.g., Counts Per Million, CPM) is recommended, as shown above in this section. Using spike-in samples for ChIP-seq samples is possible, but it is crucial to count cells properly between samples and add spike-in cells before crosslinking the samples. The same antibody must also recognize its target in both species. If using spike-ins, samples should be scaled using a scale factor option (--scaleFactor x for bamCoverage, where x is calculated as a ratio of the

number of reads mapping to the spike-in species genome in a reference sample (arbitrarily chosen sample, usually the sample with the lowest sequencing coverage) and a sample). After normalization, examine all tracks in a genome browser.

3.9.5 Peak Calling

Identifying enrichment regions is a common step in ChIP-seq data analysis. Numerous peak callers are suitable for ChIP-seq data, so the choice depends on preferences and the type of peaks expected from a particular protein. For proteins known to generate broad enrichment regions (e.g., H3K27Me3), HOMER is effective, while for point-source peaks, MACS2 is commonly used.

1. **Run MACS2 peak calling**. Use the following command to call peaks with MACS2 for paired-end BAM files:

```
macs2 callpeak -t $file -f BAMPE --outdir $macs/${sam-
ple}_macs -n $sample -B --SPMR -g hs -q 0.01
```

   Macs2 parameters:

   -t $file: Input BAM file (e.g., sorted and marked BAM).

   -f BAMPE: Specifies the format (paired-end BAM).

   --outdir $macs/${sample}_macs: Output directory for MACS2 results.

   -n $sample: Prefix for output files, typically the sample name.

   -B --SPMR: Generates signal pileup track normalized per million reads, which can be used to visualize signal intensities.

   -g hs: Effective genome size (e.g., human, hs). The size can also be specified by a number, e.g., 2.7e9 for the human genome.

   -q 0.01: False discovery rate (FDR) threshold for significant peaks.

2. **Interpret MACS2 output**: The output will include several files, such as:

   - .narrowPeak: Peak intervals and scores. The file can be directly visualized in genomic browsers.
   - .bedGraph: Signal tracks that can be uploaded to the UCSC browser for visualization.
   - .xls: Summary of peak statistics.

3. **Run HOMER peak calling.**

   1. **Create a tag directory**. HOMER requires a tag directory to store information about aligned reads:

```
makeTagDirectory directory file_sorted.bam
```

   MakeTagDirectory parameters:

   - directory: The output directory where the tag directory will be created.
   - file_sorted.bam: Input BAM file.

2. **Find peaks**. Identify peaks using HOMER, specifying the style of peaks (e.g., factor for transcription factors):

```
findPeaks directory -style factor -o output
```

FindPeaks parameters:

- directory: Tag directory created in the previous step.
- -style factor: Specifies peak calling style (e.g., factor for point-source peaks). Additional options are available to call peaks as broad regions.
- -o output: Specifies the output file for the identified peaks.
- Peak calling is available with and without control (input) libraries. Refer to the manual to change parameters.

3. **Convert HOMER output to BED format**. Convert HOMER's peak output into a BED file for visualization in a genome browser and compatibility with downstream applications:

```
pos2bed.pl output > output.bed
```

4. **Visualize peaks in a genome browser**. After calling peaks, it's important to visually inspect them in a genome browser alongside your .bw track. Supported formats for visualization: .bed, .narrowPeak, .broadPeak, etc. Genome browsers such as **IGV** or **UCSC** can directly display peak files.
5. **Best practices**.
   - **Double-check peak calls**: Do not rely solely on automated peak calls. Visual inspection of tracks (bw or BedGraph files) next to peak call intervals is crucial to ensure accuracy and consistency.
   - **Compare replicates**: When analyzing replicates, ensure consistent peak calling strategies and verify results visually to detect any discrepancies between replicates.

3.9.6 Signal Counting

1. **Run MultiBamSummary**. To quantify signal over specified genomic intervals, use the following command:

```
multiBamSummary BED-file -b file_1.bam file_2.bam
file_3.bam -BED feature_list.bed -p max --outRaw-
Counts multibamsummary_counts.txt
```

MultiBamSummary Parameters:

- -b: Specifies the BAM files to include in the analysis. In this example, file_1.bam, file_2.bam, and file_3.bam are used.
- --BED feature_list.bed: A BED file containing the genomic intervals for which the signal will be counted. For example, this could a list of coordinates for genomic features such as genes, gene promoter regions, or a list of peaks.

- -p max: Specifies the number of processors to use (maximum available in this case). Adjust as necessary based on system resources and dataset size.
- --outRawCounts multibamsummary_counts.txt: The output file that will contain the raw counts for each interval and BAM file.

2. **Interpret output**. The output will be a table (multibamsummary_counts.txt) with rows corresponding to the intervals from the BED file and columns representing the counts from each BAM file. This provides a straightforward way to compare signals across multiple BAM files for specific genomic regions that can be used for calculations in Excel, R, etc.

3.9.7 Dealing with Replicates

There is no single right way to treat replicates. The options must be selected based on the data, sequencing coverage and question at hand:

1. **Process individual replicates independently**. One common approach for handling biological replicates is to process each replicate independently, including creating separate peak lists for each replicate. This allows for downstream comparisons using methods such as correlation analysis or principal component analysis (PCA). **Best for:** Quantifying signal in fixed genomic intervals (e.g., around genes).
2. **Independent peak calling**. When working with peak calling, individual replicates normally yield different peak sets. To combine peaks between replicates:
    - **Call peaks:** Run peak calling on each individual replicate.
    - **Merge peak lists:** Combine peak lists from individual replicates using logical operations AND or OR. Using AND merging, wherein a peak must be present in both replicates, yields fewer, but more robust peaks. Using OR merging, when a peak can be found in one or both replicates, results in a larger number of less robust peaks, but allows for higher sensitivity. The strategy for peak merging should be established for each project.
    - **Quantify signal:** Quantify the signal for each replicate using the combined peak list.

3. **Combined peak calling**. Peaks can be called on all replicates combined. However, consider the following:
    - **Normalization:** Ensure that one replicate doesn't dominate due to technical variations (e.g., different sequencing depths). It is recommended to downsample all replicates to the sample with the lowest coverage. Raw (unsorted) fastq files can be downsampled by writing the first X sequences (4X lines) into a new file: head -n <4X> input_file.fastq > downsampled.fastq OR for gzipped files: zcat input_file.fastq.gz | head -n <4X> | gzip > downsampled.fastq.gz

- **Drawbacks:** This approach can potentially obscure differences between replicates, so use with caution depending on the specific research question.

4. **Peak calling considerations**

   - **Depth of coverage:** The number of called peaks is highly dependent on sequencing depth. Peak saturation may not occur even with very high read counts (e.g., over 150 million reads per sample).
   - **Peak calling parameters:** Adjust peak calling parameters based on your dataset and research objectives. Verify peaks visually and based on orthogonal approaches, including previous datasets, known sequence motif enrichment, etc.

4 Notes

1. Methanol-free formaldehyde solutions are often prepared from paraformaldehyde to ensure purity and stability. We use the one from Pierce (Cat# 28908), which is sold as a 16% solution packaged in ampoules under inert gas to prevent oxidation and maintain freshness. Methanol present in standard 37% formaldehyde solutions can lead to increased cell permeability and cause over-fixation of chromatin. If methanol-free formaldehyde is used, use a new ampoule for each experiment to be opened immediately before crosslinking.
2. The length of fixation time is critical and may need to be determined empirically for proteins that either do not contact DNA directly (e.g., chromatin remodeler subunits) or have low residence time on DNA (e.g., transcription factors).
3. FBS contains primary amines, which will hinder crosslinking. PBS can be used instead of DMEM media. However, when studying histone variants such as H2A.Z or H3.3, these might be fragile and not detectable when cells are resuspended in PBS [28]. To avoid these issues, our laboratories have been using FBS-free DMEM for crosslinking.
4. The SDS concentration will need to be optimized for your protein of interest. One percent is optimal for lysis but can be problematic for the antibody immunoprecipitation step. We have obtained better results with 0.5% SDS for certain antibodies.
5. We have stored IP and wash buffers without protease inhibitors at 4°C for up to 2 weeks. You can use either Complete Protease Inhibitor tablet (e.g., Roche/SIGMA) or add protease inhibitors individually (Leupeptin, Leustatin, and Aprotinin). It has been suggested that adding individual

inhibitors increases the efficiency of ChIP, relative to the inhibitor cocktail [29].

6. Primers should be designed both at positive control locations, such as a known protein of interest binding site, as well as at negative control locations where you expect no binding of the protein. We use manual design, but any primer design tool, such as Primer3 or similar, can be used. For manual design, we download the DNA sequence from the UCSC genome browser [30] into a program such as ApE [31]. We design primers around the TSS, 1 kb upstream and 1 kb downstream to start with. Additional regions can be designed based on the transcription factor or histone modification of interest. The distance between primer amplicons should be larger than the sonicated DNA fragment length, i.e., if the fragment sizes are ~500 bp, then amplicons should be at least 600 bp apart (Fig. 4). Designing both positive and negative control primers within the same locus helps estimate the quality of DNA fragmentation. Validate the specificity of each primer pair by performing a test PCR on genomic DNA, followed by agarose gel electrophoresis and, preferably, by Sanger sequencing of PCR products. Discard primers unless they yield the expected PCR product with no additional products.
7. SYBR green dye can be used at a 1:20,000–40,000 dilution. We have also used SYBR Gold at 1:10,000 dilution.
8. To maintain workstations, some familiarity with UNIX administration is required, so for smaller groups and those new to bioinformatics, shared remote-access workstation(s) should be considered. For large-scale analysis, access to a high-performance cluster (HPC) at your institution is strongly advised. Online cloud-based data analysis platforms such as Galaxy [32] are becoming broadly available at a low cost or for free. These are excellent for analysis without access to high-performance computing. We find that it is still beneficial to do some data analysis locally. Users tend to choose to work on local machines upon gaining more familiarity with data analysis. We recommend that every new user gain some familiarity with bioinformatics to improve communication with bioinformaticians and to handle publicly available data more readily.
9. Sometimes adherent cells start lysing at this point. If so, make the best effort to collect the material with 1–2 extra PBS washes. In our experience, lysis leads to decreased yield but does not negatively affect ChIP quality.
10. We use the Covaris S220 sonicator and describe the protocol using this sonicator. Adjust settings on Covaris to the High cell (Output = 140, Duty cycle = 5) or Low cell (Output = 105, Duty cycle = 2) as needed. For other types of sonicators

(e.g., Bioruptor, etc.), please follow the manufacturer's recommended steps. Other chromatin fragmentation methods, such as MNase digestion, can also be employed, but all of these need to be standardized per cell line.

11. You can also use column-based DNA isolation (such as the DNA purification columns from Cell Signaling) instead of phenol-chloroform extraction.
12. The smaller the DNA fragments, the better the resolution. For histone marks, the ideal fragment size would be around 200 bp (slightly larger than a mononucleosome). Chromatin remodeling complexes might occupy somewhat larger fragments. However, longer shearing times might result in protein degradation, so we recommend testing to determine the smallest fragment size distribution that still retains intact protein.
13. The amount of formaldehyde and/or crosslinking timing can be varied, as well as additional crosslinkers used, depending on the target protein. For instance, mild or no crosslinking (0–0.5%) can be used for histone proteins that are tightly bound to DNA, while up to 2% formaldehyde has been used for chromatin complexes that do not directly touch DNA. Time can be varied too, for instance, shorter crosslinking times of 5 min for histones, or longer times (up to 30 min) for proteins that are not in direct contact with DNA can be tried. Bifunctional crosslinkers can be used for protein-protein crosslinking before the addition of formaldehyde [5, 6]. Temperature is another variable that can be changed for optimal crosslinking, for instance, 4°C, room temperature, or 37°C.
14. If you see a large pellet, this means that you either over-crosslinked your cells or shearing failed. Insufficient maintenance of sonicator instruments, such as the Covaris, can result in non-shearing or poor shearing.
15. Do not vortex or centrifuge beads above ~2000 × *g*. Beads can be pre-blocked as follows: wash 20 μL Protein A+G Dynabeads with 300 μL of IP buffer. Add BSA solution [466 μL of IP buffer and 34 μL of BSA Fraction V (from 75 mg/mL, Final conc 5 mg/mL)] and rotate overnight at 4°C. Note that the capacity of Dynabeads, e.g., protein A, is 0.3 μg/μL of human IgG. Therefore, 1mg Dynabeads can capture 8 μg human IgG. By extrapolation, 10 μL Dynabeads contain around 0.3 mg beads.
16. Dilution with IP buffer helps to lower the final concentration of SDS, which can hinder the antigen-antibody reaction at high concentrations. In this example, we use a five-fold dilution, e.g., 0.4 mL sonicated lysate plus 1.6 mL of IP buffer for a total of 2 mL. If the antibody is known to be

especially sensitive to SDS, further dilution may be needed (e.g., 0.2 mL sonicated lysate plus 1.8 mL of IP buffer).

17. We recommend incubating the input along with the IP tubes to ensure it undergoes similar incubations as the rest of the samples.
18. To determine how much antibody to add, start with the datasheets from the company. You can increase or decrease the recommended amount; usually, it is good to titrate a new antibody (e.g., 1, 2, 5, 10 μg) to determine the optimum concentration. More is not necessarily better. Antibody titration has been directly linked to improved consistency between replicates [33].
19. Additional high salt washes can be included for stringency.
20. Phase-lock tubes can be used for easier separation.
21. Histone marks, which are generally ubiquitous, show between 0.5% to about 25% of the input at known enrichment regions (Fig. 3D). RNA polymerase II and some highly bound proteins show between 0.1% and 5% enrichment at active promoters, while transcription factor ChIPs typically show <1% enrichment (Figs. 3D, 4B and 4E). Accordingly, it generally takes more material for successful ChIP-seq of transcription factors.
22. We do not quantify DNA after ChIP because amounts of DNA below detection limits are expected, especially important for low-abundance targets such as transcription factors. Instead, we estimate the amount indirectly, by extracting and measuring genomic DNA from a 1% aliquot, and then extrapolating the amount of DNA in chromatin. The amount of chromatin for ChIP-seq in human cancer cells [34]. Given a typical yield of 0.3–0.6 μg/μL of DNA in sonicated ChIP material, 250–500 μL of ChIP material is typically used to dilute to 2 mL for HSF1 ChIP-seq.
23. Alternatively, we have done full-scale amplification in the presence of SYBR dye without test amplification. In this case, set up a PCR reaction per kit recommendation, except also add 5% (v/v) of Eva Green dye and set the real-time PCR to run for 40 cycles. Monitor amplification after completing the cycle and pause the run to remove the sample(s) once the desired fluorescence is achieved. After removing the sample, the rest of the samples can continue amplification as needed. Stopping and resuming reactions is not supported by every PCR instrument. We use the Bio-Rad CFX instrument that can be stopped and resumed manually. The Applied Biosystems (Thermofisher) QuantStudio should also allow one to stop, retrieve samples and resume reactions.

24. It is in principle possible to have the library made from the ChIP material by a commercial or core facility. However, because the ChIP DNA is at a low concentration and thus likely to fail DNA quantity requirements, the facility must be willing to prepare libraries under these nonstandard circumstances, including adjusting PCR cycles. It is therefore recommended to prepare libraries by the same laboratory that did the ChIP.
25. ChIP-seq data used in this work are under the Gene Expression Omnibus accession GSE209687.

Acknowledgments

This work was supported by the NSF award 1750379 to S.N., the National Institutes of Health grant P20GM104360 to S.N. and A.D., and the IDeA Networks of Biomedical Research Excellence INBRE grant UND0026601 to A.D. We thank the UND Genomics core for assistance with sequencing and analysis of the ChIP-seq samples. The Genomics core is supported by the National Institute of General Medical Sciences of the National Institutes of Health under Award Number U54GM128729 and Award number 2P20GM104360-06A1.

References

1. Gilmour DS, Lis JT (1984) Detecting protein-DNA interactions in vivo: distribution of RNA polymerase on specific bacterial genes. Proc Natl Acad Sci USA 81(14):4275–4279. https://doi.org/10.1073/pnas.81.14.4275
2. Solomon MJ, Larsen PL, Varshavsky A (1988) Mapping protein–DNA interactions in vivo with formaldehyde: evidence that histone H4 is retained on a highly transcribed gene. Cell 53(6):937–947. https://doi.org/10.1016/S0092-8674(88)90469-2
3. Solomon MJ, Varshavsky A (1985) Formaldehyde-mediated DNA-protein crosslinking: a probe for in vivo chromatin structures. Proc Natl Acad Sci USA 82(19):6470–6474. https://doi.org/10.1073/pnas.82.19.6470
4. Hoffman EA, Frey BL, Smith LM et al (2015) Formaldehyde crosslinking: a tool for the study of chromatin complexes. J Biol Chem 290(44):26404–26411. https://doi.org/10.1074/jbc.R115.651679
5. Aoki T, Wolle D, Preger-Ben Noon E et al (2014) Bi-functional cross-linking reagents efficiently capture protein-DNA complexes in Drosophila embryos. Fly (Austin) 8(1):43–51. https://doi.org/10.4161/fly.26805
6. Fujita N, Wade PA (2004) Use of bifunctional cross-linking reagents in mapping genomic distribution of chromatin remodeling complexes. Methods 33(1):81–85. https://doi.org/10.1016/j.ymeth.2003.10.022
7. Barski A, Cuddapah S, Cui K et al (2007) High-resolution profiling of histone methylations in the human genome. Cell 129(4):823–837. https://doi.org/10.1016/j.cell.2007.05.009
8. Johnson DS, Mortazavi A, Myers RM et al (2007) Genome-wide mapping of in vivo protein-DNA interactions. Science 316(5830):1497–1502. https://doi.org/10.1126/science.1141319
9. Robertson G, Hirst M, Bainbridge M et al (2007) Genome-wide profiles of STAT1 DNA association using chromatin immunoprecipitation and massively parallel sequencing. Nat Methods 4(8):651–657. https://doi.org/10.1038/nmeth1068
10. Lloyd SM, Bao X (2019) Pinpointing the genomic localizations of chromatin-associated

proteins: the yesterday, today, and tomorrow of ChIP-seq. Curr Protoc Cell Biol 84(1):e89. https://doi.org/10.1002/cpcb.89
11. Bernstein BE, Stamatoyannopoulos JA, Costello JF et al (2010) The NIH roadmap epigenomics mapping consortium. Nat Biotechnol 28(10):1045–1048. https://doi.org/10.1038/nbt1010-1045
12. Andrews S (2010) FastQC: a quality control tool for high throughput sequence data [online]. http://www.bioinformatics.babraham.ac.uk/projects/fastqc/
13. Ewels P, Magnusson M, Lundin S et al (2016) MultiQC: summarize analysis results for multiple tools and samples in a single report. Bioinformatics 32(19):3047–3048. https://doi.org/10.1093/bioinformatics/btw354
14. Bolger AM, Lohse M, Usadel B (2014) Trimmomatic: a flexible trimmer for Illumina sequencedata. Bioinformatics30(15):2114–2120. https://doi.org/10.1093/bioinformatics/btu170
15. Martin M (2011) Cutadapt removes adapter sequences from high-throughput sequencing reads. EMBnet J 17(1):10–12. https://doi.org/10.14806/ej.17.1.200
16. Kim D, Paggi JM, Park C et al (2019) Graph-based genome alignment and genotyping with HISAT2 and HISAT-genotype. Nat Biotechnol 37(8):907–915. https://doi.org/10.1038/s41587-019-0201-4
17. Langmead B, Salzberg SL (2012) Fast gapped-read alignment with Bowtie 2. Nat Methods 9(4):357–359. https://doi.org/10.1038/nmeth.1923
18. Li H, Handsaker B, Wysoker A et al (2009) The sequence alignment/map format and SAMtools. Bioinformatics 25(16):2078–2079. https://doi.org/10.1093/bioinformatics/btp352
19. Picard tools. https://broadinstitute.github.io/picard/
20. Quinlan AR, Hall IM (2010) BEDTools: a flexible suite of utilities for comparing genomic features. Bioinformatics 26(6):841–842. https://doi.org/10.1093/bioinformatics/btq033
21. Ramirez F, Ryan DP, Gruning B et al (2016) deepTools2: a next generation web server for deep-sequencing data analysis. Nucleic Acids Res 44(W1):W160–165. https://doi.org/10.1093/nar/gkw257
22. Zhang Y, Liu T, Meyer CA et al (2008) Model-based analysis of ChIP-Seq (MACS). Genome Biol 9(9):R137. https://doi.org/10.1186/gb-2008-9-9-r137
23. Heinz S, Benner C, Spann N et al (2010) Simple combinations of lineage-determining transcription factors prime cis-regulatory elements required for macrophage and B cell identities. Mol Cell 38(4):576–589. https://doi.org/10.1016/j.molcel.2010.05.004
24. Robinson JT, Thorvaldsdottir H, Winckler W et al (2011) Integrative genomics viewer. Nat Biotechnol 29(1):24–26. https://doi.org/10.1038/nbt.1754
25. Raney BJ, Barber GP, Benet-Pages A et al (2024) The UCSC Genome Browser database: 2024 update. Nucleic Acids Res 52(D1):D1082–D1088. https://doi.org/10.1093/nar/gkad987
26. UCSC Genome Browser. http://genome.ucsc.edu
27. Buenrostro JD, Giresi PG, Zaba LC et al (2013) Transposition of native chromatin for fast and sensitive epigenomic profiling of open chromatin, DNA-binding proteins and nucleosome position. Nat Methods 10(12):1213–1218. https://doi.org/10.1038/nmeth.2688
28. Jin C, Felsenfeld G (2007) Nucleosome stability mediated by histone variants H3.3 and H2A.Z. Genes Dev 21(12):1519–1529. https://doi.org/10.1101/gad.1547707
29. de Jonge WJ, Brok M, Kemmeren P et al (2020) An optimized chromatin immunoprecipitation protocol for quantification of protein-DNA interactions. STAR Protoc 1(1):100020. https://doi.org/10.1016/j.xpro.2020.100020
30. Perez G, Barber GP, Benet-Pages A et al (2024) The UCSC Genome Browser database: 2025 update. Nucleic Acids Res. https://doi.org/10.1093/nar/gkae974
31. Davis MW, Jorgensen EM (2022) ApE, a plasmid editor: a freely available DNA manipulation and visualization program. Front Bioinform 2:818619. https://doi.org/10.3389/fbinf.2022.818619
32. Galaxy C (2024) The Galaxy platform for accessible, reproducible, and collaborative data analyses: 2024 update. Nucleic Acids Res 52(W1):W83–W94. https://doi.org/10.1093/nar/gkae410
33. Caride A, Jang JS, Shi GX et al (2023) Titration-based normalization of antibody amount improves consistency of ChIP-seq experiments. BMC Genomics 24(1):171. https://doi.org/10.1186/s12864-023-09253-0
34. Dastidar SG, De Kumar B, Lauckner B et al (2023) Transcriptional responses of cancer cells to heat shock-inducing stimuli involve amplification of robust HSF1 binding. Nat Commun 14(1):7420. https://doi.org/10.1038/s41467-023-43157-7

Chapter 4

Chromatin Immunoprecipitation from Mouse Lung Tissue

Zhenting Zhang, Yaping Xu and Jian Liu

Abstract

Chromatin immunoprecipitation (ChIP) can selectively enrich chromatin associated with certain proteins, thus enabling observation of epigenetic alterations. With ChIP methods, various epigenetic signatures can be identified, allowing for the inference of transcriptional characteristics of chromatin regions. These biological features may lead to activation or inhibition of cancer-associated genes, thus underscoring the importance of ChIP studies in cancer research. ChIP from tissues can be more complicated than using cells grown in culture, and needs some additional optimization. We describe here the process of ChIP using mouse lung tissues (including tumors), and the DNA thus isolated can be further analysed using sequencing methods.

Keywords Chromatin immunoprecipitation (ChIP), ChIP-Seq, Cancer research, Epigenetic research, Transcription regulation

1 Introduction

Chromatin immunoprecipitation (ChIP) is a key step to selectively enrich chromatin associated with specific proteins of interest, and observe the dynamics in their interactions, which is often followed by sequencing [1]. Chromatin structure misregulation plays a major role in cancer development and progression. As the protein–DNA interactions epigenetically alter gene expression, ChIP is an important tool in cancer research, useful for studying transcription factor binding to genes and histone modification changes during tumor progression. During the ChIP procedure, a 1% paraformaldehyde (PFA) solution is often used to cross-link proteins to DNA, preserving interactions and effectively freezing the chromatin structure in its current state. This step captures a moment-in-time image of protein–DNA interactions within the cell. The crosslinked chromatin is released by nuclear lysis and then fragmented by sonication or enzymatic digestion to break the chromatin into smaller, more manageable pieces, thereby increasing resolution by

These authors contributed equally: Zhenting Zhang, Yaping Xu.

Archana Dhasarathy and Motoki Takaku (eds.), *Cancer Epigenetics*, Methods in Molecular Biology, vol. 3005,
https://doi.org/10.1007/978-1-0716-5072-1_4,

creating shorter DNA fragments associated with the protein(s) of interest. Next, specific antibodies are added to target the protein or histone modification being studied. Protein A/G beads are then introduced to bind these antibodies, allowing for the separation and purification of the protein–DNA complexes of interest from the rest of the chromatin.

ChIP techniques enable researchers to identify and map specific histone modifications (epigenetic markers) across the genome, each associated with distinct characteristic features and biological functions [1]. Particularly, various lysine methylation states on the histone H3 subunit are regarded as five "core histone marks" and associated with different transcription regulatory regions like enhancers and promoters. This information can be integrated to divide chromatin into regions with diverse "chromatin states," which further represent their epigenetic characteristics and transcriptional status.

In cancer research, ChIP plays an important role by helping to identify epigenetic mechanisms that amplify oncogenes or inhibit tumor-suppressor genes, which can lead to cancer initiation and progression. To name but a few, ChIP-Seq was used to map the genome-wide binding sites of nuclear factor erythroid-derived-2-like 2 (NRF2), a transcription factor whose overactivation contributed to cancer initiation and drug resistance [2]. In combination with other techniques, ChIP-Seq was used to profile active enhancers at the genome-wide level and enabled identification of enhancers with oncogenic properties [3]. ChIP-Seq helped reveal the oncogenic epigenetic effects resulting from the loss of the SETD2 gene in clear cell renal cell carcinoma (ccRCC) [4].

Mice are widely accepted animal models in cancer research, due to their genetic and physiological similarities to humans. Thus, studying the epigenetic regulatory patterns in mice can provide valuable insight into the complexity of tumors and offer the possibility of developing new therapeutic strategies to treat cancer. By understanding gene expression and epigenetic regulation patterns in these mouse models, researchers can gain critical insights to combat cancer more effectively. This article focuses on ChIP from mouse tissues, and specifically on the process of extraction of ChIP DNA to further investigations, including sequencing.

2 Materials

2.1 Reagents

1. DPBS
2. Paraformaldehyde (PFA)
3. Glycine
4. Tris-HCl (pH 8.0)
5. Sodium chloride
6. Magnesium chloride (1M)
7. NP-40 (or IGEPAL CA-630)

8. EDTA (0.5M)
9. EGTA (0.5M)
10. Sodium deoxycholate
11. Dithiothreitol (DTT)
12. Sodium dodecyl sulfate (SDS)
13. Phenylmethanesulphonyl fluoride (PMSF)
14. Triton X-100
15. Proteinase inhibitors (100 ×)
16. Sodium bicarbonate
17. Proteinase K (20 mg/mL)
18. Phenol-chloroform isoamyl alcohol
19. Chloroform
20. Anhydrous ethanol
21. Sodium acetate
22. GenElute™-LPA (from Sigma)

2.2 Commercial Kits

Qubit™ dsDNA HS Assay Kit (from Invitrogen)

2.3 Equipment

1. Dounce tissue grinders
2. 70 μm nylon mesh strainer
3. Centrifuge
4. Ultra-sound machine (e.g., S220, from Covaris)
5. Protein A/G magnetic beads
6. Magnetic stand
7. Rotator
8. Block heater
9. Qubit 4 (from Invitrogen)

2.4 Buffer Formula

1. **Hypotonic buffer** [5]

Tris-HCl (pH 7.5)	10 mM
NaCl	10 mM
$MgCl_2$	3 mM
NP-40 or IGEPAL CA-630	0.1%
ddH_2O	N/A

Add protease inhibitors just before use

2. **Lysis buffer** [6]

Tris-HCl (pH 8.0)	20 mM
NaCl	150 mM
EDTA	1 mM
EGTA	0.5 mM
Sodium deoxycholate	0.4%
SDS	0.1%
NP-40 (or IGEPAL CA-630)	1%
DNase/RNase-free distilled water	N/A
Add fresh before use:	
PMSF	1 mM
Proteinase inhibitors	

3. **2× IP buffer**

Tris-HCl (pH 7.5)	20 mM
NaCl	150 mM
EDTA	1 mM
NP-40 or Triton X-100	1%
Sodium deoxycholate	0.5%
Add fresh before use:	
PMSF	1 mM
Proteinase inhibitors	

4. **Low-salt wash buffer** [7]

Tris-HCl (pH 8.0)	20 mM
NaCl	150 mM
EDTA	2 mM
SDS	0.1% w/v
Triton X-100	1% v/v

5. **High-salt wash buffer** [7]

Tris-HCl (pH 8.0)	20 mM
NaCl	500 mM
EDTA	2 mM
SDS	0.1% w/v
Triton X-100	1% v/v

6. **LiCl wash buffer** [7]

EDTA	1 mM
Sodium deoxycholate	1% w/v
NP-40 (or IGEPAL CA-630)	1% v/v

7. **Elution buffer** [7]

$NaHCO_3$	0.1 M
SDS	1% w/v

3 Methods

3.1 Lung Tissue Cells Fixation

This step aims to preserve protein–DNA interactions in the lung tissue cells.

1. Before starting: Pre-cool DPBS, put tissue grinders on ice. Add 1 mL DPBS to the tubes.
2. Weigh 300–500 mg of the lung tissue samples, wash them with chilled DPBS twice, and put it into the narrow part of the grinder tube.
   Note: 40–50 mg lung tissue is enough for per IP sample; these mouse tissues should be collected and cut into small pieces, then frozen by liquid nitrogen or dry ice immediately, preserved in the −80°C refrigerator.
3. On the ice, grind 10 times with the "Loose" pestle and another 20 times with the "Tight" pestle until there is no obvious resistance (no obvious tissue chunks should be seen after the grinding process). The tissue pestles should not be removed from the liquid surface during the process to avoid making bubbles.
4. Place a 70 μm nylon mesh strainer over a 2 mL microcentrifuge tube. Pre-wet the center of the strainer with DPBS and then filter the lung tissue mixture with the strainer.
5. Centrifuge at 500 × *g* at 4°C for 5 min and discard the supernatant.
6. Add 1 mL of 1% PFA. After mixing, incubate the samples at room temperature for 10 min to fix the cells.
7. Add 50 μL of 2.5 M Glycine to every 1 mL of 1% PFA. Incubate at room temperature for 5 min to quench the crosslinking reaction.
8. Centrifuge at 2500 × *g* 4°C for 5 min and discard the supernatant.
9. Wash the cells with 1 mL DPBS. Centrifuge at 2500 × *g* 4°C for 5 min and discard the supernatant.

3.2 Cell Permeabilization

1. Resuspend cells with 750 μL hypotonic buffer and incubate on ice for 5 min.
2. Centrifuge at 800 × *g* at 4°C for 5 min and discard the supernatant.
3. Repeat steps 3.2.1 and 3.2.2 with 500 μL hypotonic buffer.
4. Resuspend cells with an appropriate amount of lysis buffer (~2 million cells/100 μL) and incubate on ice for 5 min.

3.3 Sonication

1. Pre-cool the Covaris ultra-sound machine S220.
2. Transfer the sample mix to the appropriate Covaris tubes. Ensure no bubbles are present in the sample mix.
3. Run the sonication program to fragment the chromatin regarding the substances of interest (Table 1).
4. Transfer the sample to a fresh 1.5 mL microcentrifuge tube.
5. Centrifuge at 14,000 rpm 4°C for 10 min.
6. Transfer the supernatant to a fresh 1.5 mL microcentrifuge tube.

3.4 Immune Precipitation

1. Take 5–10 percent of the input chromatin. Dilute the chromatin solution with 2× IP buffer. This buffer can be prepared as follows.
2. For each sample, add 20–30 μL Protein A/G magnetic beads to each of the fresh 1.5 mL microcentrifuge tubes. Resuspend beads with 400 μL PBST (PBS + 0.1% Tween 20).
3. Place the tube on the magnet to separate the beads from the solution, and remove the supernatant.
4. Add your antibody (typically 1–10 μg) diluted in 400 μL PBST, then transfer to the 1.5 mL microcentrifuge tube with activated protein A/G beads. Incubate with rotation for 30 min at room temperature (**Note 1**).
   Note: For the usage of the antibody, please refer to the antibody protocol, or you can refer to the published article.
5. Place the tube on the magnet and remove the supernatant. Wash the beads with 400 μL PBST twice.

Table 1
Sonication programs depend on the substance of interest

	Transcription factors	Histones
Peak power	175	175
Duty factor	20	10
Cycle burst	200	200
Duration	20 min	10 min

6. Add the IP samples to the tube, then mix the beads with the IP sample, and then incubate them overnight at 4°C.

3.5 Wash Antibody–Beads Complex

1. Place the tube on the magnetic stand, then remove the supernatant.
2. Add 1 mL low-salt wash buffer, mix well, and rotate 4°C for 5 min.
3. Spin briefly and place on the magnetic stand. Carefully remove the supernatant.
4. Repeat Step 3.5.2 and Step 3.5.3 with 1 mL high salt wash buffer.
5. Then repeat with 1 mL LiCl wash buffer.
6. At room temperature (instead of 4°C), repeat Step 3.5.2 and Step 3.5.3 with 1 mL TE buffer, containing 10 mM Tris-HCl (pH 8.0) and 1 mM EDTA [7].
7. Add 1 mL TE buffer, transfer bead-containing TE solution to a new tube, and mix well.
8. Place the tube on the magnet, then remove the supernatant.
9. Elute DNA with 100 μL of freshly prepared elution buffer to each sample. Heat it at 65°C on a block heater for 1 h. Shake it every 20 min to ensure full mixing.

3.6 Reverse Crosslinking

1. Add 10 μL 5M NaCl, 10 μL 10% SDS, and 10 μL Proteinase K (20 mg/mL) to the samples.
2. Heat at 65°C in a block heater overnight for de-crosslinking.

3.7 Phenol-Chloroform Isoamyl Alcohol (PCI) DNA Extraction

1. Add the same volume of PCI to each sample. Invert or vortex the tubes to mix thoroughly and transfer the sample to a phase lock tube.
2. Centrifuge at 13,000 × *g* at room temperature for 5 min.
3. Add the same volume of chloroform. Invert the tubes to mix.
4. Centrifuge at 13,000 × *g* at room temperature for 5 min, then carefully transfer the colorless aqueous phase to a fresh 1.5 mL EP tube.
5. Add 2.5 volumes of pre-cooled 100% ethanol (750 μL), 23 μL of 3 M Sodium Acetate and 1 μL LPA. Invert or vortex the tubes.
6. Incubate at −20°C for at least 30 min.
7. Centrifuge at 13,000 × *g* (or maximum speed of the machine) 4°C for 10 min and pipette out the supernatant.
8. Add 500 μL of pre-cooled ethanol to wash the pellet. Invert the tubes.
9. Centrifuge at 13,000 × *g* 4°C for 1 min and pipette out all the supernatant. Ensure no liquid drop is left in the tube. Air dry for 3–5 min.

10. Add 20–50 μL TE buffer (pH 8.0). Tap the tubes by hand to mix well. Take 1 μL as an aliquot to measure DNA concentration with Qubit (**Note 2**).
Note: For the histone, normally, the IP efficiency is around 5% of input DNA. However, for transcription factors, the IP efficiency will be quite low; normally, the efficiency is around 0.4% of input DNA.

4 Notes

1. Antibodies used in ChIP should be specific and have high affinity to the target protein. If no ChIP/ChIP-Seq antibodies are available, the ones for IP or IHC can probably work (not always). As crosslinking may alter the epitopes, a Western blot can be performed to test the antibodies.
2. ChIPed DNA can be cloned by PCR/qPCR and further investigated with sequencing technologies.

Acknowledgments

This work was supported by grants to Dr. Jian Liu from the Natural Science Foundation (NSF) of China (General Grant: 82172899 and 82472637), the NSF of Zhejiang Province (Distinguished Young Scholars: LR22H160002).

References

1. Nakato R, Sakata T (2021) Methods for ChIP-seq analysis: a practical workflow and advanced applications. Methods 187:44–53. https://doi.org/10.1016/j.ymeth.2020.03.005
2. Namani A, Liu K, Wang S, Zhou X, Liao Y, Wang H, Wang X, Tang X (2019) Genome-wide global identification of NRF2 binding sites in A549 nonsmall cell lung cancer cells by ChIP-Seq reveals NRF2 regulation of genes involved in focal adhesion pathways. Aging 11(24). https://doi.org/10.18632/aging.102590
3. Li QL, Lin X, Yu YL, Chen L, Hu QX, Chen M, Cao N, Zhao C, Wang CY, Huang CW, Li LY, Ye M, Wu M (2021) Genome-wide profiling in colorectal cancer identifies PHF19 and TBC1D16 as oncogenic super enhancers. Nat Commun 12(1):6407. https://doi.org/10.1038/s41467-021-26600-5
4. Xie Y, Sahin M, Sinha S, Wang Y, Nargund AM, Lyu Y, Han S, Dong Y, Hsieh JJ, Leslie CS, Cheng EH (2022) SETD2 loss perturbs the kidney cancer epigenetic landscape to promote metastasis and engenders actionable dependencies on histone chaperone complexes. Nat Cancer 3(2):188–202. https://doi.org/10.1038/s43018-021-00316-3
5. Hiraike Y (2023) Chromatin immunoprecipitation with mouse adipocytes using hypotonic buffer to enrich nuclear fraction before fixation. STAR Protoc 4(1):102093. https://doi.org/10.1016/j.xpro.2023.102093
6. Texari L, Spann NJ, Troutman TD, Sakai M, Seidman JS, Heinz S (2021) An optimized protocol for rapid, sensitive and robust on-bead ChIP-seq from primary cells. STAR Protoc 2(1):100358. https://doi.org/10.1016/j.xpro.2021.100358
7. Haring M, Offermann S, Danker T, Horst I, Peterhansel C, Stam M (2007) Chromatin immunoprecipitation: optimization, quantitative analysis and data normalization. Plant Methods 3:11. https://doi.org/10.1186/1746-4811-3-11

Chapter 5

High-Efficiency Genomic Mapping of Chromatin-Associated Targets with CUT&RUN

Tessa M. Firestone, Bryan J. Venters, Katherine Novitzky, Liz Marie Albertorio-Sáez, Courtney A. Barnes, Karlie N. Fedder-Semmes, Nathan W. Hall, Allison R. Hickman, Mark Kaderli, Carolina Lin Windham, Matthew R. Marunde, Danielle N. Maryanski, Kelsey Noll, Leslie Lewis, Jennifer Spengler, Martis W. Cowles, Zu-Wen Sun, Michael-Christopher Keogh, Andrea L. Johnstone, Ellen N. Weinzapfel and Lu Sun

Abstract

The precise regulation of chromatin composition is critical to gene expression and cellular identity, and thus a key component in development and disease. Robust assays to study chromatin features, including histone post-translational modifications (PTMs) and chromatin-associated proteins (e.g., transcription factors or PTM readers), are crucial for understanding their function and identifying novel therapeutic strategies. To this end, Cleavage Under Targets and Release Using Nuclease (CUT&RUN) has emerged as a powerful tool for high-resolution epigenomic profiling. The approach has been successfully applied to numerous cell and tissue types, providing insights into target genomic distribution with unprecedented sensitivity and throughput. Here, we provide a detailed CUT&RUN protocol from sample collection through data analysis, including best practices and defined controls to ensure specific, efficient, and robust target profiling.

Keywords Chromatin, Chromatin-associated protein(s), Nucleosome, CUT&RUN, ChIP-seq

---

Abbreviations

CAP	Chromatin-associated protein
ChIP	Chromatin ImmunoPrecipitation
CUT&RUN	Cleavage Under Targets and Release Using Nuclease
ConA	Concanavalin A

The authors Tessa M. Firestone and Bryan J. Venters contributed equally.

Archana Dhasarathy and Motoki Takaku (eds.), *Cancer Epigenetics*, Methods in Molecular Biology, vol. 3005,
https://doi.org/10.1007/978-1-0716-5072-1_5,

dNuc	designer nucleosome
mAb	monoclonal antibody
NE	Nuclei extraction
NGS	Next generation sequencing
pAb	polyclonal antibody
pAG-MNase	protein A-protein G-Micrococcal Nuclease
RT°C	Room temperature
PTM	Post-translational modification
QC	Quality control
S:N	Signal:Noise

1 Introduction

1. **Chromatin Mapping Approaches**

Chromatin refers to the structural organization of the genome in eukaryotic cells, and is central to gene expression, cell division, and DNA repair [1, 2]. This protein-nucleic acid assemblage is regulated by diverse modifications (including histone post-translational modifications (PTMs) [3, 4], DNA methylation [5]), chromatin-associated proteins (CAPS; including transcription factors [6], chromatin readers and editors [7], and remodeling enzymes [8]). The dysregulation of chromatin features can operate in concert with genetic mutations and environmental factors [9–12] to alter gene expression profiles and drive cancer development [13–19]. As such, mapping the genomic localization of histone PTMs and CAPs can reveal relationships between the chromatin landscape and disease phenotypes. To this end, researchers have historically relied on Chromatin Immuno Precipitation (ChIP)-based approaches, most notably *ChIP-seq*. This has provided fundamental insights into chromatin structure and function and advanced the application of epigenomics across many biomedical fields. However, the inherent limitations of *ChIP-seq*, which require high cell numbers and deep sequencing to overcome high background and technical fragility, have precluded its widespread application to drug research and clinical studies. Further, *ChIP-seq* can suffer from technical artifacts arising from experimental manipulations, including heavy crosslinking and its reversal. Assays with improved sensitivity, robustness, and scalability are needed to better inform (pre) clinical research and identify novel diagnostic and therapeutic opportunities.

To address these needs, a new epigenomic mapping method has recently emerged: **C**leavage **U**nder **T**argets **& R**elease **U**sing **N**uclease (CUT&RUN) [20, 21]. The revolutionary approach CUT&RUN grants the ability to map shifts in histone PTMs and CAPs within cancerous states and in response to treatments with ultra-sensitivity and low material consumption [22–24].

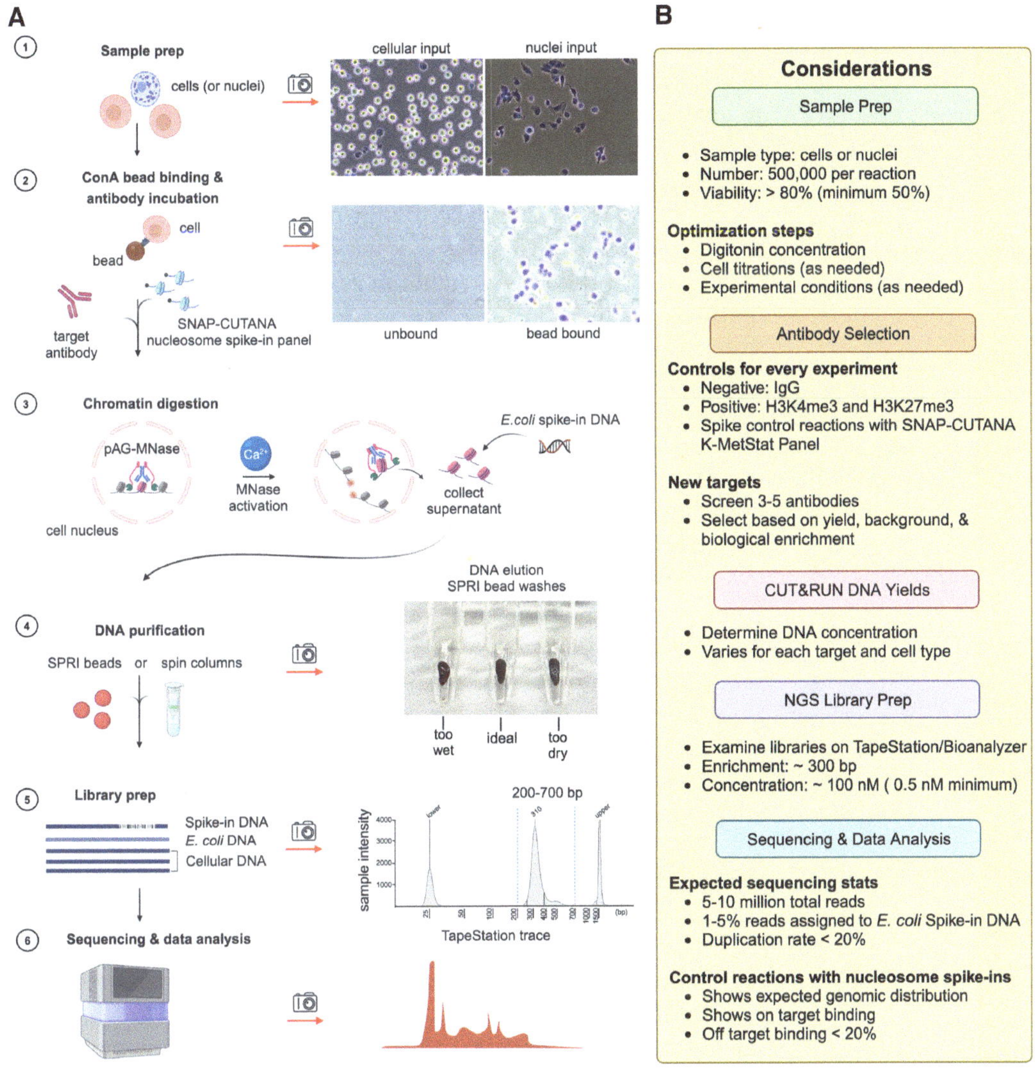

Fig. 1 The CUT&RUN workflow. **A** Schematic of a representative CUT&RUN experiment. Camera icons depict expected observations at key quality control (QC) steps. (1) The quality of monodispersed cells or nuclei is determined by Trypan Blue staining, where the dye is excluded by native "viable" cells (K562 shown), but enters permeabilized cells or isolated nuclei (MCF7 shown) (*see* **Notes 17–19**). (2) Cells or nuclei are immobilized to ConA magnetic beads. Efficient bead binding is confirmed by examination of bound and unbound fractions under a microscope (HEK293 shown). The bead slurry is then aliquoted to 8-strip tubes for CUT&RUN. If applicable for the reaction target, SNAP-CUTANA™ nucleosome spike-ins (EpiCypher) are added (*see* **Notes 1 and 8–10**). Target-specific antibody (*see* **Notes 2–10**) is added and reactions incubated overnight at 4°C. (3) pAG-MNase is added, and activated by further addition of Ca^{2+}, leading to selective cleavage and diffusion of antibody-labeled chromatin fragments to the supernatant. Reactions are supplemented with exogenous *E. coli* spike-in DNA to monitor, to control for input, and support normalization after NGS (*see* **Notes 40**). (4) Bead-bound cells are removed on a magnet, and target-enriched DNA is purified from the supernatant. If using SPRI beads, avoid over-drying (they should retain a damp dark brown color) before elution. (5) CUT&RUN DNA is quantified by fluorometric assays (e.g., Qubit™, Invitrogen™) and used to prepare sequencing libraries. Purified libraries are examined on a TapeStation or Bioanalyzer (Agilent Technologies)

Identifying distinct epigenomic features within the chromatin landscape (e.g., promoters and enhancers) expands opportunities for discovering new biomarkers, drug targets, and relationships between target genes and their regulatory pathways. In the approach (Fig. 1), cells or nuclei are immobilized to Concanavalin A (ConA) coated magnetic beads, permeabilized, and incubated with target-specific antibody. A pAG-MNase (protein A-protein G-Micrococcal Nuclease) fusion is added to selectively cleave and release antibody-bound chromatin to the supernatant. Bead-coupled cells remain on the magnet, while CUT&RUN enriched DNA is purified from the supernatant and prepared for multiplexed next-generation sequencing (NGS). Finally, CUT&RUN reaction libraries are sequenced to a depth of 5–10 million total reads (assuming a human genome), which provides sufficient coverage for high-resolution genomic mapping.

Importantly, CUT&RUN offers dramatic improvements in throughput, sensitivity and cost over *ChIP-seq* [25–27]. The high Signal-to-Noise (S:N) of the new approach is achieved by fewer, lower stringency washes and the direct partition of target-enriched chromatin fragments (signal) from bead-coupled cells (noise). This enables reduced cell numbers (5–500 thousand cells per reaction) and sequencing depth (5–10 million total reads per reaction, assuming a human sample), compared to the >1 million cells and >30 million reads needed for *ChIP-seq* [28, 29]. To further improve throughput and reproducibility at reduced cost, this CUT&RUN protocol is adapted for 8-strip tubes and multichannel pipettors. In our experience, the practiced user can process 24–48 reactions over a two- to three-day procedure. This throughput supports the inclusion of control reactions for each biological sample and encourages deeper experimental design, allowing the user to examine more replicates, time points, inhibitor doses, or additional targets.

Studies using CUT&RUN have delivered high-quality datasets for multiple histone PTMs and CAPs [30–34] from a diversity of sample inputs, including cell lines, primary tissues, and model organisms [35–51]. To illustrate its feasibility, we used CUT&RUN to study cellular responses after inhibition of the chromatin remodeler SMARCA2 (BRM) (Fig. 2). BRM014 allosterically inhibits SMARCA2 catalytic activity [52] without impacting global abundance [53, 54]; while the AU-15330 and ACBI1 proteolysis targeting chimeras (PROTACs) selectively engage and degrade the ATPase [54, 55], leading to signal loss on chromatin (Fig. 2b). This highlights a major advantage of CUT&RUN for parallelized drug

Fig. 1 (continued) to determine molarity and confirm enrichment of mononucleosome-sized fragments (~300 bp) (*see* **Notes 48–52**). (6) Pooled CUT&RUN libraries are sequenced (e.g., Illumina), aiming for 5–10 million total reads per reaction (assuming a human genome), and data analyzed (*see* **Notes 53–59**). **B** Important considerations at each step to ensure a successful CUT&RUN experiment.Created in BioRender. Firestone T (2024) https://BioRender.com/e48e774

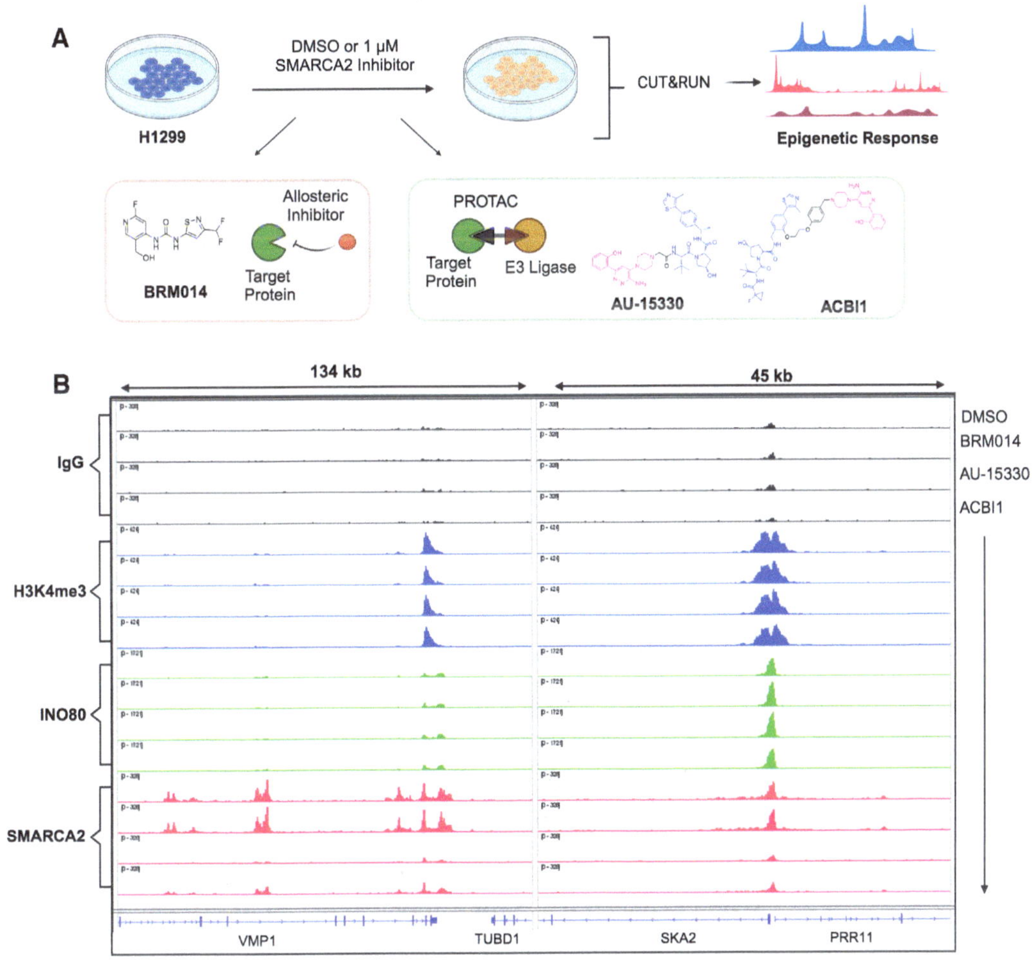

Fig. 2 Application of CUT&RUN to study different classes of SMARCA2/BRM inhibitors. **A** Design of a parallelized CUT&RUN experiment testing various inhibitors of the SMARCA2 (BRM) chromatin remodeling ATPase. H1299 cells were treated for 4 h with BRM014 [52], AU-15330 [54], or ACBI1[55] (all 1 μM); or DMSO vehicle control (0.01%). Cells were lightly cross-linked (0.1% formaldehyde for 1 min) (*see* **Note 16**) and processed by CUT&RUN. **B** Browser tracks (IGV) at representative genomic locations [81] after indicated treatment, showing signal for IgG (negative control), H3K4me3 (control; active promoters), INO80 (control; a non-targeted remodeler), and SMARCA2 (inhibitor target). BRM014 or DMSO had no impact on SMARCA2 or INO80 enrichment, while the PROTACs (AU-15330 and ACBI1) reduced SMARCA2 but not INO80 binding to chromatin, confirming their selective degrader mechanism of action. RPKM normalized reads are group-scaled by target and aligned to the human genome build hg38. Created in BioRender. Firestone T (2024) https://BioRender.com/i49s432

screening studies: the workflow can be easily scaled for additional compounds, doses or cell types.

This chapter provides a detailed CUT&RUN protocol optimized for high throughput and S:N at reduced sequence depth (Fig. 1). We discuss the criteria for identifying the best antibodies for a given target (e.g., histone PTM or CAP); methods to prepare CUT&RUN

workflow, including sample quality, cell permeabilization, antibody performance, pAG-MNase activity, and library preparation. This includes reactions controlled by DNA-barcoded nucleosomes (e.g., *EpiCypher* SNAP-CUTANA™ K-MetStat or epitope tag panels) (*see* **Note 1**) that can be seamlessly integrated into the workflow and greatly increase confidence in the resulting CUT&RUN datasets. Finally, we include sections on library preparation for *Illumina* NGS with considerations for sequencing depth, quality metrics, and data analysis.

2 Materials

2.1 Equipment

Aluminum Block compatible with strip tubes: (e.g., Thermo Fisher Scientific NC9820161) (optional)

Automated capillary electrophoresis instrument: (e.g., *Agilent* Bioanalyzer or *Agilent* TapeStation)

Cell Counter: (e.g., Invitrogen AMQAX2000)

8-channel multichannel pipettes: (e.g., VWR 76169-250)

Fluorometer: (e.g., Invitrogen Qubit 4 Q33238)

Hemocytometer slides: (e.g., Sigma-Aldrich Z359629)

Illumina sequencer: (e.g., NextSeq1000/2000)

Incubator: (e.g., Millipore Sigma Z763314)

Magnetic rack for 8-strips: (e.g., EpiCypher 10-0008)

Magnetic rack for 1.5 mL tubes: (e.g., EpiCypher 10-0012)

Microcentrifuge: (e.g., Thermo Fisher Scientific 75002435)

Minicentrifuge: (e.g., Thermo Fisher Scientific 75004086)

Microscope: (e.g., Nikon Eclipse Ts2)

Nutator: (e.g., VWR 82007-202)

Thermocycler with heated lid: (e.g., *BioRad*, *Applied Biosystems*, *Eppendorf* (or comparable).

Vortex Mixer: (e.g., Scientific Industries SI-0236)

2.2 Consumables and Reagents (*Optional)

Automated capillary electrophoresis consumables (e.g., *Agilent* High Sensitivity DNA Analysis Kit 5067-4626; or D1000 ScreenTape 5067-5582 and D1000 Reagents 5067-5583)

$CaCl_2$ (e.g., *Sigma-Aldrich* C1016)

Cell counter slides (e.g., *Invitrogen* C10228)

Cell culture and appropriate growth medium

ConA Beads (e.g., *EpiCypher* 21-1401)

Control Antibodies

- IgG Negative Control (e.g., *EpiCypher* 13-0042)
- H3K4me3 Positive Control (e.g., *EpiCypher* 13-0060)
- H3K27me3 Positive Control (e.g., *EpiCypher* 13-0055)
- H3K27ac Positive Control*, labile PTM (e.g., *EpiCypher* 13-0059)

CUTANA™ ChIC/CUT&RUN Kit* (*EpiCypher* 14-1048)

CUTANA™ Wash Buffer Enhancer Set (*EpiCypher* 14-1803)

Digitonin (5% stock in DMSO) (e.g., *EpiCypher 21-1004*)

DMSO (e.g., *Sigma-Aldrich* D8418)

E. coli Spike-in DNA (e.g., EpiCypher 18-1401)

EDTA (0.5 M pH 8) (e.g., *Sigma-Aldrich* E5134)

EGTA (0.5 M pH 8) (e.g., *Sigma-Aldrich* E3889)

Formaldehyde* (37% w/v)(e.g., *Sigma-Aldrich* 252549)

Fluorometer reagents: (e.g., *Invitrogen* 1X dsDNA HS Kit Q33230)

Glycerol (e.g., *Sigma-Aldrich* G5516)

Glycine* (2.5 M) (e.g., *Sigma-Aldrich* 50046)

Glycogen (20 mg/mL) (e.g., Sigma-Aldrich, Roche 10930193001)

HEPES (*e.g.*, *Sigma-Aldrich* H3375)

HEPES (0.5 M, pH 7.9) [autoclaved or filter sterilized] [stocks for Sect. 2.3]

HEPES (0.5 M, pH 7.5) [autoclaved or filter sterilized] [stocks for Sect. 2.3]

KCl (e.g., *Sigma-Aldrich* P3911)

Library prep kit(s) (e.g., *EpiCypher* 14-1001-48s1 or 14-1001-48s2 [each provides different 48X multiplex barcode sets] or NEB NEB-Next Ultra II DNA Library Kit E7645, E7600S, & E7780S)

Low-retention filter pipette tips: any vendor

$MnCl_2$ (e.g., *Sigma-Aldrich* 203734)

Molecular biology grade water (e.g., *Corning* 46-000-CM)

Multichannel reagent reservoirs: (e.g., Thermo Fisher Scientific 14-387-072)

NaCl (e.g., *Sigma-Aldrich* S5150-1L)

Nuclei Extraction Buffer* (e.g., *EpiCypher* 21-1026)

100% Ethanol (200 proof) (e.g., *Millipore Sigma* EX0276)

1.5, 15, 50 mL tubes: any vendor

1X PBS* (calcium and magnesium free) (e.g., *Cytiva* SH30256.01) pAG-MNase (e.g., *EpiCypher* 15-1016)

Propidium Iodide* (e.g., *Thermo Fisher Scientific* BMS500PI)

Protease Inhibitor (e.g., *Roche* 11873580001)

RNase A (10 mg/mL) (e.g., *Thermo Fisher Scientific* EN0531)

SDS (10% solution): any vendor

SNAP-CUTANA Nucleosome Panels* (*see* **Note 1**)

- e.g., K-MetStat (*EpiCypher* 19-1002)
- e.g., HA Tag (*EpiCypher* 19-5002)
- e.g., DYKDDDDK Tag (*EpiCypher* 19-5001)

Spermidine trihydrochloride (molecular weight 254.63) (e.g., *Sigma-Aldrich* S2501).

- Spermidine (1 M) [stocks for Sect. 2.3]

Solid-Phase Reversible Immobilization (SPRI) Bead-Based Reagent

- e.g., CUTANA™ DNA Purification Beads (*EpiCypher* 21-1407)
- e.g., SPRIselect (e.g., *Beckman Coulter* B23318)

Test antibodies (*see* **Notes 2–10**)

Tris Base (e.g., *Genesee Scientific* 18–145)

- Tris (1 M, pH 8.0) [autoclaved or filter sterilized] [stocks for Sect. 2.3]
- Tris (1 M, pH 8.5) [autoclaved or filter sterilized] [stocks for Sect. 2.3]

Triton X-100 (10% solution) (e.g., *Sigma-Aldrich* T8787)

Trypan Blue (0.4%) (e.g., *Invitrogen* T10282)

Trypsin* (0.05%) (e.g., *Cytiva* SH30236.01)

0.2 mL 8-strip tubes with attached flat caps: (e.g., EpiCypher 10-0009)

2.3 Buffer Recipes (* Optional)

All buffers are made with molecular biology-grade water.

Wash, Antibody, and **Nuclei Extraction Buffers** are made fresh (same day) for each experiment by adding spermidine, digitonin or protease inhibitor (as individually required) to respective precursor buffers.

Cell Permeabilization Buffer can be made on Day 1 and stored at 4°C for use on Day 2. **Stop** and **Bead Activation Buffers** can be filter sterilized and stored at 4°C for ~6 months.

5% Digitonin: Prepare 5% (w/v) digitonin in DMSO. Stable for 6 months at −20°C.

2.5 M Glycine*: Dissolve 9.4g in 50 mL molecular grade water. Filter sterilize. Stable at room temperature (RT°C) for 6 months.

25X Protease Inhibitor: Dissolve 1 protease inhibitor tablet in 2 mL of water. Stable for 12 weeks at −20°C.

1M Spermidine: Dissolve 1 g Spermidine trihydrochloride (MW 254.63) in 3.93 mL water. Stable at −20°C for 6 months, store in single-use aliquots (*see* **Note 11**).

Bead Activation Buffer: 20 mM HEPES pH 7.9, 10 mM KCl, 1 mM $CaCl_2$, 1 mM $MnCl_2$. Filter sterilize. Stable at 4°C for 6 months.

Nuclei Extraction Buffer: 20 mM HEPES pH 7.9 (pH with KOH), 10 mM KCl, 0.10% Triton X-100, 20% glycerol. Filter sterilize. Stable at 4°C for 6 months. On **Day 1** of each experiment, supplement with 1X protease inhibitor and 0.5 mM spermidine and store at 4°C. Discard after **1 day**.

Pre-wash Buffer: 20 mM HEPES pH 7.5, 150 mM NaCl. Filter sterilize. Stable at 4°C for 6 months.

Wash Buffer: **Pre-wash Buffer** freshly supplemented with 1X protease inhibitor, 0.5 mM spermidine and 1X Wash Buffer Enhancer Set. Use within **Day 1** of each experiment (*see* **Notes 12 and 29**).

Cell Permeabilization Buffer: **Wash Buffer** freshly supplemented with digitonin (to 0.01%) on **Day 1** of each experiment. Stable at 4°C overnight. *See* Sect. 4.2 for digitonin optimization.

Antibody Buffer: **Cell Permeabilization Buffer** freshly supplemented with EDTA, pH 8.0 (to 2 mM). Use within **Day 1** of each experiment (Sect. 4.3.3).

Stop Buffer: 340 mM NaCl, 20 mM EDTA, pH 8.0, 4 mM EGTA, pH 8.0, 50 μg/mL RNase A, 50 μg/mL glycogen. Filter sterilize. Stable at 4°C for 6 months.

0.1X TE Buffer: 1 mM Tris pH 8, 0.1 mM EDTA pH 8.0. Filter sterilize. Stable at RT°C for 6 months.

Sequencing Buffer: Illumina RSB with Tween 20. Stable at 4°C for 6 months.

3 Antibody Selection

Delivering the full high-sensitivity/low background potential of CUT&RUN requires antibodies of high specificity and efficiency. The presumptuous adoption of unsuitable antibodies is a major and widespread issue in biomedical research [56–58]. This problem is especially relevant in genomic mapping assays, where antibodies validated by ELISA or immunoblot are often assumed to perform similarly in the very different biophysical context of chromatin [59]. Such surrogate validations have caused great harm, with many of the resulting "approved" histone PTM antibodies showing high cross-reactivity and/or poor efficiency in ChIP with spike-in nucleosome standards [59–61]. Further, even appropriately ChIP-validated antibodies are not guaranteed to similarly perform in CUT&RUN (*see* **Note 2**), demonstrating the necessity for assay-specific validation. Finally, antibodies as biological products often display lot-to-lot variability: polyclonals (pAbs) >> monoclonals (mAbs), but still experimentally observed for the latter [58, 62].

Encompassing all the above: for accurate and robust CUT&RUN, antibodies need to be thoroughly vetted, directly validated for in-assay performance, and continuously monitored. Reagent efficiency and specificity metrics are particularly critical. **High efficiency** is the greatest possible capture of on-target events and can be explored by reducing cell numbers and/or using samples with reduced target abundance. **High specificity** refers to low cross-reactivity, and can be assayed by target deletion/depletion or DNA-barcoded nucleosome spike-ins. Here, we describe how to select histone PTM and CAP antibodies that generate high-quality CUT&RUN profiles (Fig. 3).

3.1 Obtain Antibodies

1. Source three to five antibodies to the PTM or CAP of interest. Each should be distinct (i.e., not the same reagent distributed by multiple entities), which can be achieved by ensuring different immunogens, host species or mAb clone identifiers (*see* **Notes 3–5**).

3.2 Plan Your Experiment

2. Start with 500 k cells per reaction (*see* **Note 6**) in a native or lightly fixed form (*see* **Notes 13–16**). When establishing the CUT&RUN approach or testing novel detection reagents, it is often most convenient to use an easily grown cell line wherein the target is known to be expressed and binds chromatin (*see* **Notes 6 and 7**).
3. Antibodies to histone PTMs or epitope tags can be validated and monitored in situ using SNAP-CUTANA DNA-barcoded nucleosome spike-ins (*see* **Notes 1 and 8**).
4. Include parallel reactions with negative (IgG) and positive (H3K4me3, H3K27me3) control antibodies and appropriate SNAP-CUTANA K-MetStat nucleosome spike-ins (*see* **Notes 9 and 10**).
5. Relative efficiency for each antibody can be determined by titrating cell numbers with capability for the intended study identified by bracketing 500 k (the initial recommendation) to the likely available input (if precious or rare) (e.g., see Fig. 3c and **Note 6**).

3.3 Antibody Performance: Analysis and Comparison

Confirm positive and negative controls performed as expected. Troubleshoot the workflow as needed (*see* **Notes 9 and 10**). For novel targets, select the antibody with best balance of CUT&RUN yields, S:N and distinct genomic enrichment (Fig. 3d).

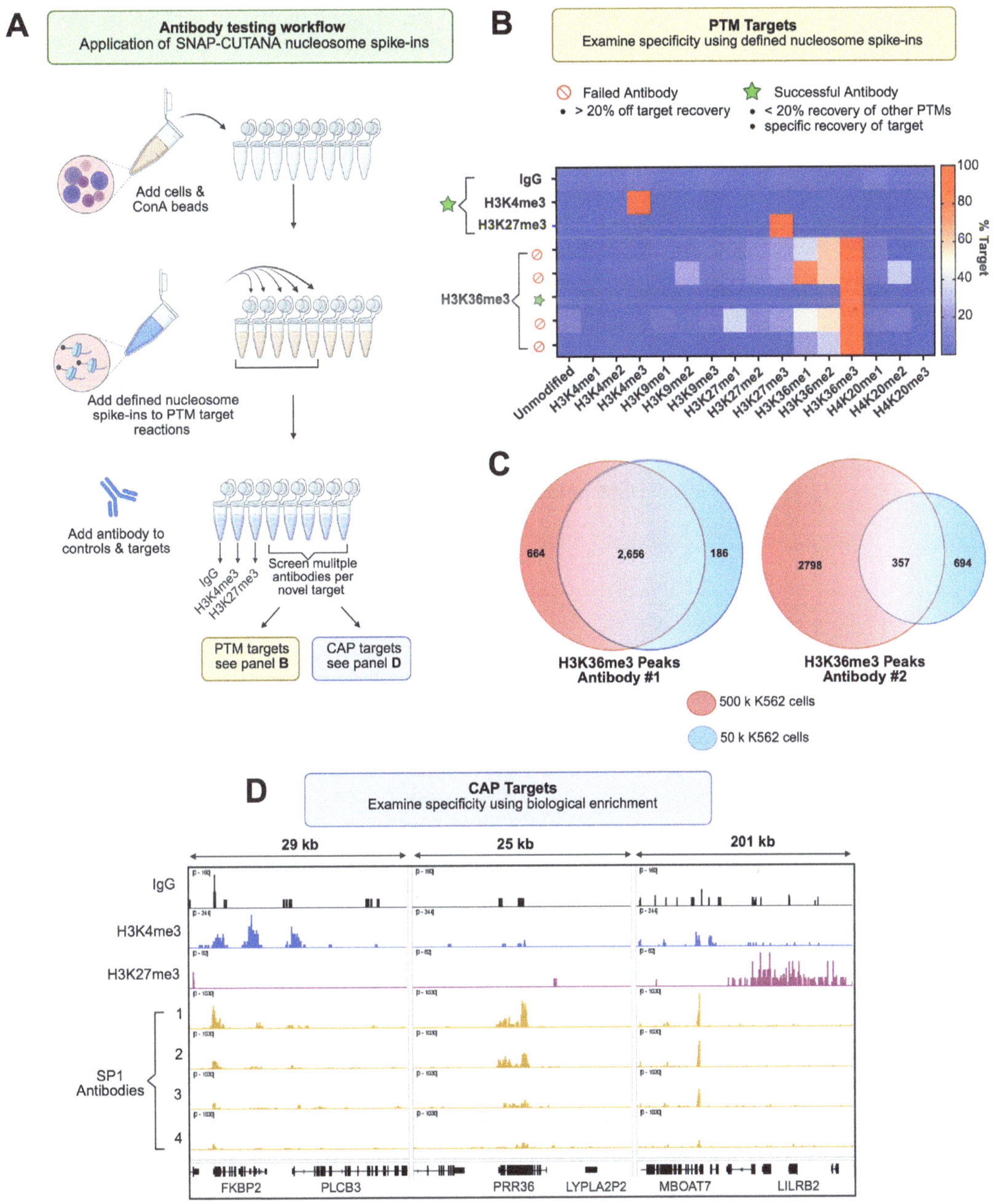

Fig. 3 Antibody selection guidelines for robust CUT&RUN profiling. Created in BioRender. Culture T (2025) https://BioRender.com/n17l732 **A** Schematic for antibody testing. Parallel reactions with cells/nuclei include controls (e.g., IgG [Negative], anti-H3K4me3 [Positive] and anti-H3K27me3 [Positive]; each with SNAP-CUTANA K-MetStat spike-in nucleosome panel) and the antibodies under test (with appropriate SNAP-CUTANA spike-in if available). **B** For PTMs and epitope-tagged targets, antibody specificity is determined by relative recovery of on- vs. off-target DNA-barcoded spike-in nucleosomes (see Sect. 3.3). Representative data show pass/fail metrics for five anti-H3K36me3 tested with the K-MetStat Panel (signal for each antibody by row; for each spike-in nucleosome by column). Heatmap data (key on right) reflects percentage cross-reactivity (>20% = fail) after normalization to on-target (set to 100). **C** Antibody efficiency in CUT&RUN is determined by peak recovery after cell dilution. When challenged by reduced cell numbers, an efficient antibody (#1) retains peaks lost by a mediocre reagent (#2) (500 –> 50 k; both anti-H3K36me3 in K562 cells)

6. Antibodies with acceptable specificity to PTMs or epitope-tags should show <20% cross-reactivity to SNAP-CUTANA spike-in panels (Fig. 3).
7. Antibodies with acceptable efficiency can be identified by maintaining peak enrichment at reduced sample input (Fig. 3c). Lower efficiency reagents, in contrast, will manifest decreased peak number, lower FRiP scores and degraded peak enrichment; eventually reaching IgG-like profiles.
8. Success with an "antibody screening to a novel target" campaign can be determined by various considerations (e.g., see Fig. 3d and **Note 10**)
   a. Peak enrichment metrics indicate high Signal:Noise (S:N).
   b. Two or more independent reagents yield highly similar peak locations.
   c. The genomic distribution of peaks is consistent with reported target function or correlated epigenetic marks (not being blind to a reported function being wrong).

4 Method

4.1 Sample Prep

Perhaps the major variables to CUT&RUN experimental success are the antibody (**3**) and the sample. The latter is impacted by a range of considerations:

1. Source and processing (e.g., tissues, primary cells or lines; cells or nuclei; fresh, frozen or lightly cross-linked) (*see* **Notes 13–16 and 21**).
2. Cell quality (e.g., material from a necrotic tissue, dying culture or after poor cryo-preservation will show increased background) (*see* **Notes 17–19**).
3. Cell numbers (starting recommendation is 500 k cells/reaction; though as above, high quality antibodies will often give success from fewer) (*see* **Notes 6, 17**, and **18**).
4. Effective cell permeabilization [e.g., by Digitonin (Sect. 4.2)] and/or nuclei isolation (*see* **Notes 13–15 and 23**).

Fig. 3 (continued) **D** For targets without an available spike-in, multiple antibodies are compared to identify the one with the best balance of high yields, robust S:N and rational enrichment. In the example of four anti-SP1 antibodies tested, #1 best generates strong peak structures associated with active transcription (i.e., high correlation with H3K4me3/low with H3K27me3). Each CUT&RUN reaction was performed with 500 k native HeLa cells; RPKM normalized reads were group-scaled and aligned to the genome build hg19. Created in BioRender. Culture T (2025) https://BioRender.com/n17l732

This section describes sample processing workflows to generate reliable CUT&RUN data from native or lightly-fixed cells or nuclei (Fig. 4). We have found that data quality for many targets using cryopreserved material is indistinguishable from fresh samples (and the former greatly increases convenience). Nuclei extraction is recommended when performing CUT&RUN on immune cells and often required when working with tissues (*see* **Notes 13–15**).

Of note, the heavy cross-linking strategies of *ChIP-seq* (e.g., 1% formaldehyde for 10 min) are not suitable for CUT&RUN since they dramatically reduce yields and S:N (*see* **Note 16**). However, CUT&RUN is compatible with frozen and/or lightly cross-linked

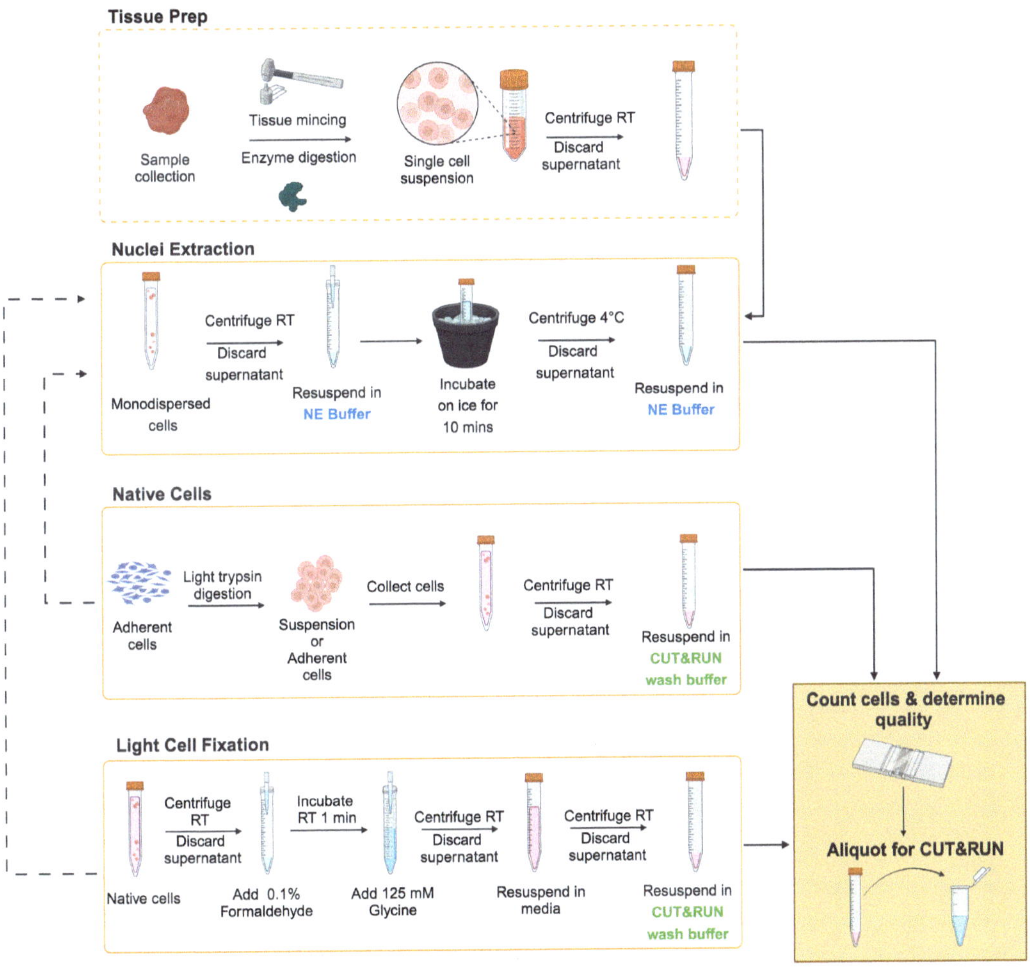

Fig. 4 Sample preparation for CUT&RUN. Schematic includes the preparation of tissues, native cells, lightly-fixed cells or nuclei for CUT&RUN (*see* **Notes 12–16**). For tissues, the extraction of the nuclei is strongly recommended. Samples are resuspended in CUT&RUN Wash Buffer (cells) or Nuclei Extraction Buffer (nuclei), counted, and checked for integrity, morphology and debris. Created in BioRender. Firestone T (2024) https://BioRender.com/j87y014

material (0.1% formaldehyde for 1 min). Instances where light cross-linking is particularly helpful (or even essential) include stabilizing labile PTMs (e.g., acetylation, ubiquitination or phosphorylation) and CAPs after disruptive treatment time courses.

Optimal samples are monodispersed cells (or nuclei) with normal morphology and minimal cell lysis debris. Low starting cell viability and/or poor morphology generally increases assay background. We recommend checking samples by microscopy at three points during early CUT&RUN steps: (i) initial cell harvest; (ii) after cells have been washed in CUT&RUN **Wash Buffer** (or after nuclei have been extracted), just prior to bead binding; and (iii) after binding to ConA beads. Performing these QC steps will identify problem samples and contribute to assay success.

The recommended starting point is 500 k cells (or nuclei) per CUT&RUN reaction. When collecting cells, process a minimum of 20% excess to account for QC checks and sample loss during wash steps. While high-performance antibodies will yield robust S:N from fewer cells (*see* **Note 6**), it is recommended to first validate reagents at standard cell numbers before titrating to lower input levels.

4.1.1 Counting Cells and Examining Sample Integrity

Cell quality/membrane integrity can be checked at CUT&RUN **Day 1** by Trypan Blue exclusion. Alternatively, the membrane impermeant fluorescent DNA stain propidium iodide (PI; used as per manufacturer's recommendations) improves accuracy with compatible automated cell counters, especially when using samples from trypsinized adherent cell lines or primary tissue that are often accompanied by higher debris levels. Either stain can be used to examine samples at initial processing (Sect. 4.1.2), to monitor effective permeabilization with Digitonin (Sect. 4.2), after resuspending in CUT&RUN **Wash Buffer** (Sect. 4.3.2), or to confirm nuclei extraction (*see* **Notes 17–19**).

1. Add 10 μL of Trypan Blue (0.4%) to 10 μL cells and gently pipette ten times to mix (*see* **Note 18**).
2. Immediately transfer 10 μL of the mix to a hemocytometer or automated cell counter.
3. View under a brightfield microscope or in a dedicated cell counter. Dead cells (and isolated nuclei) are permeable to Trypan Blue and will stain blue. Cells with an intact membrane will exclude the dye and thus not stain. For fresh material, aim for viability >90% (or as high as feasible given sample type) (*see* **Note 19**).
4. Examine cellular/nuclear morphology (and the amount of sub-cellular debris), remembering that a low-quality sample can compromise CUT&RUN.

4.1.2 Cell Processing

Fresh native cells

1. For cell cultures, aim for 70% confluency at harvest vs. overgrown cells.
2. Count starting material and confirm cellular integrity/morphology (Sect. 4.1.1). Collect and process ~20% excess cells to account for sample loss through handling and QC checks.
   a. Suspension cells can be transferred directly to 15–50 mL collection tubes.
   b. Adherent cells are recovered from growth vessels by gentle scraping or mild trypsin treatment.
      i. Scraping: Remove growth media and replace with fresh serum-free media sufficient to cover the growing surface. Using a cell scraper, lightly rub across the culture area and pipette the collected cells to a new tube. Add fresh media to rinse the plate/flask and collect any remaining cells. Visually inspect the plate/flask to ensure efficient cell harvest.
      ii. Trypsin treatment: Remove growth media and gently wash cells once with PBS. Add trypsin (0.05% in PBS) sufficient to cover the growing surface. Incubate at 37°C for 5–10 min, visually inspecting at regular intervals to ensure efficient cell dissociation. Inactivate trypsin with a 10X volume of complete media and pipette the cell suspension to a collection tube (*see* **Notes 20 and 21**).
3. Count cells (as per Sect. 4.1.1). Transfer the number needed for CUT&RUN (500 k <u>per reaction</u> plus 20% excess) to a collection tube and directly continue (Sect. 4.3.2).
4. If cells are to be used for CUT&RUN at a later time, transfer them to cryopreservation media and use a slow-freeze method (Sect. 4.1.3).

Cell fixation (optional)

Light cross-linking (0.1% formaldehyde for 1 min) can preserve labile targets (e.g., lysine acylation, phosphorylation or ubiquitination) and should be considered in such cases (*see* **Notes 16 and 22**). Depending on their fragility, cross-linking can result in 25–50% cell loss, so process more material accordingly. When testing new targets or antibodies with cross-linked material, always test native cells in parallel. Note that cross-linked cells can require an extra step to reverse cross-links prior to DNA purification on **Day 2** of CUT&RUN (Sect. 4.3.5).

1. Confirm sample integrity (Sect. 4.1.1). For cell cultures, aim for 70% confluency vs. overgrown. This protocol can process up

to 50 M cells. For >50 M, divide the material and process in parallel.

2. Suspension cells can be transferred to appropriately sized collection tubes and collected by centrifugation (600 x *g* for 3 min at RT°C). Resuspend in fresh serum-free medium.
3. Adherent cells can be cross-linked directly in the culture plate/flask. Remove growth media and replace with fresh serum-free medium.
4. Add 37% formaldehyde (2.7 μL per mL of serum-free media) directly to the culture to the desired final concentration (0.1%) and quickly swirl to mix.
5. Incubate as desired (generally RT°C 1 min) and quench cross-linking by adding 2.5 M glycine to a final concentration of 125 mM. Quickly swirl to mix.
6. Adherent cells can now be transferred to collection tubes (as in Sect. 4.1.2a).
7. Collect cells by centrifugation (600 x *g* for 3 min at RT°C) and resuspend in fresh serum-free media. Proceed directly to CUT&RUN (Sect. 4.3) unless:
   a. If cell cryopreservation is intended, proceed to Sect. 4.1.3.
   b. If nuclei extraction is intended, proceed to Sect. 4.1.4.

4.1.3 Cell Cryopreservation (Optional)

This section describes how to process cryopreserved cells while minimizing lysis (thus avoiding a source of CUT&RUN background). Freezing allows the user to bank samples for later study and can hugely increase convenience.

1. Prepare Cryopreservation media (90% complete growth media + 10% DMSO).
2. Harvest cells. Count and confirm high viability/expected morphology (Sect. 4.1.1).
3. Resuspend in Cryopreservation media (~5 M cells/mL).
4. Aliquot cells to 1.5 mL Eppendorf tubes.
5. Slow freeze (−1°C/min) aliquots in an isopropanol-filled chiller at −80°C. Keep at −80°C for long-term storage (duration will need to be empirically determined for each study system, though we have seen acceptable stability for a diversity of sample/target types over 12 months) [63].

   ---Thawing cells---
6. Immediately before use, place tubes at 37°C. Work quickly to avoid cell lysis.
7. When cells are almost thawed (i.e., a small amount of ice is still visible in the tube), remove from the heat block. Gently pipette to complete thawing/ensure a mono-suspension and remove a

small aliquot to count cells and examine their morphology (*see* **Notes 18 and 19**).

8. Collect cells by centrifugation (600 x *g* for 3 min at RT°C). Carefully remove and discard supernatant (while avoiding the cell pellet).
9. Resuspend cell pellet in **Wash Buffer** (105 μL per reaction) and proceed to CUT&RUN (Sect. 4.3.3).

4.1.4 Nuclei Extraction

(Required for tissues and many immune cells/Optional for cell lines)

Nuclei Extraction (NE) is well-suited for many sample types, but particularly tissues (*see* **Notes 13–15 and 23**) (Fig. 4). If cross-linking is intended, this should be performed before nuclei are isolated. Delicate samples should be processed with an additional 25–50% cells to account for material loss. **NE Buffer** is fully compatible with CUT&RUN, and nuclei in **NE Buffer** can be directly added to activated ConA beads.

1. Prepare **NE Buffer** (2 mL per 5 M cells) fresh on the day of sample collection.
2. Count fresh or frozen cells and confirm integrity by trypan blue (or PI) staining (Sect. 4.1.1).
3. Collect cells by centrifugation (600 x *g* for 3 min at RT°C). Carefully remove and discard supernatant (while avoiding the cell pellet). Harvest 500 k cells per reaction plus ~20% excess to account for sample loss.
4. Resuspend the cell pellet in cold **NE Buffer** (100 μL per reaction).
5. Incubate on ice for 10 min.
6. Spin cells by centrifugation (600 x *g* for 3 min at RT°C). Pellet appearance should change from sticky pale-yellow cells to white fluffy nuclei. Carefully remove and discard supernatant (while avoiding the pellet).
7. Gently resuspend nuclei in cold **NE Buffer** (105 μL per reaction). For samples prone to clumping, resuspend nuclei in cold Wash Buffer (105 μL per reaction) instead. If Wash Buffer was used, proceed straight to CUT&RUN
8. Remove 10 μL of sample and complete a post NE quality check by trypan blue (or PI) staining on a manual or automated counter (Sect. 4.1.1) (Fig. 5). 95%–99% trypan blue positive nuclei with minimal cellular debris indicates successful extraction.
9. If proceeding straight to CUT&RUN, move to Sects. 4.3.1 and 4.3.3 of the workflow. Alternatively, nuclei can be cryopreserved for later use (Sect. 4.1.5).

4.1.5 Cryopreservation of Nuclei for CUT&RUN

1. Dispense aliquots (~5 M nuclei/mL **NE Buffer**) sufficient for intended reactions (e.g., 1 mL = eight reactions plus 20% excess to account for material loss).

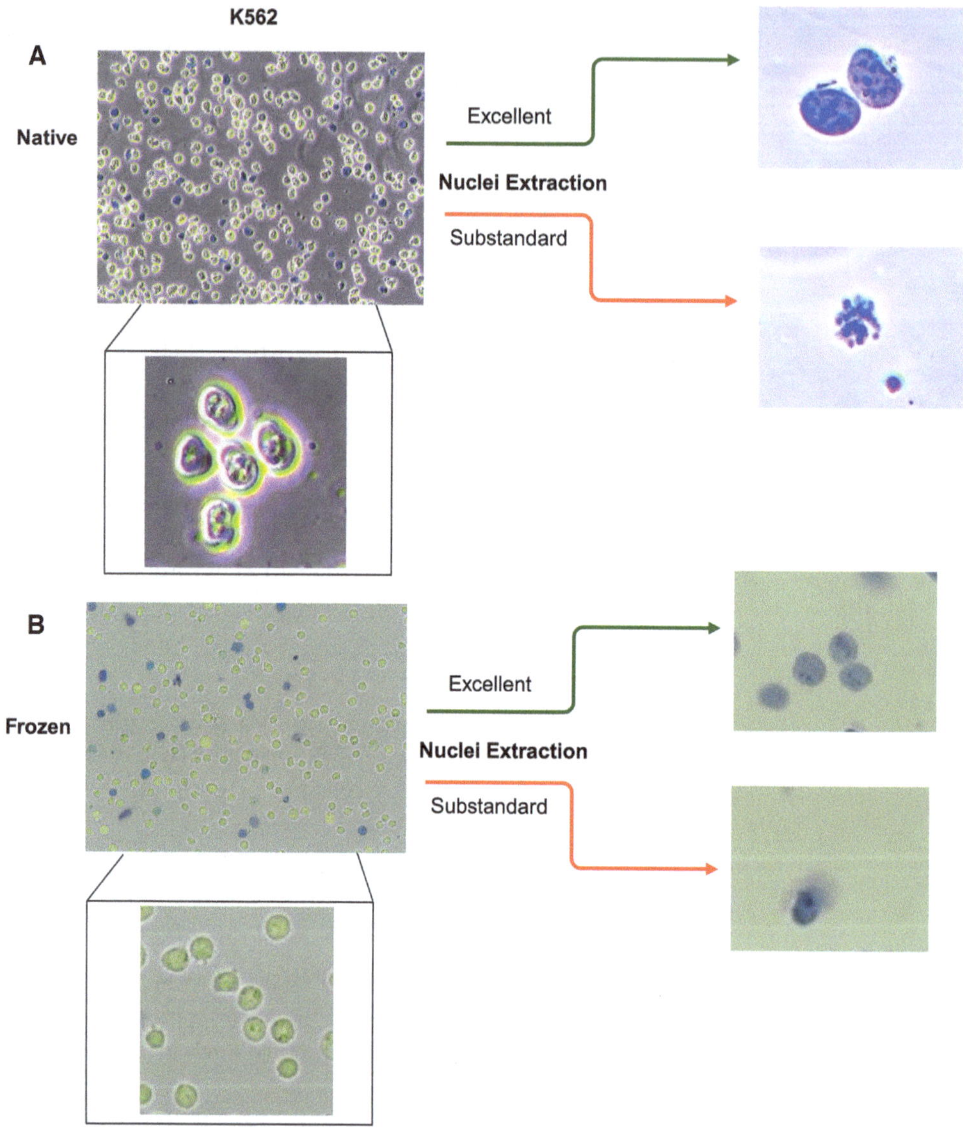

Fig. 5 Nuclei extraction quality control. Microscope images show representative morphology after nuclei extraction from **A** native or **B** Cryopreserved K562 cells. Trypan blue staining shows that starting cells have high potential viability (dye exclusion) and expected morphology. Successfully extracted nuclei are Trypan blue positive with intact nuclear membranes; critical for ConA bead binding. A disrupted nuclear membrane indicates poor quality material: if the predominant species, such a sample will yield poor quality CUT&RUN data- Created in BioRender. Firestone T (2024) https://BioRender.com/w21q580

2. Slow freeze (−1°C/min) aliquots in an isopropanol-filled chiller at −80°C. Store at −80°C for long term storage (Sect. 4.1.3).
3. When ready to use an aliquot, quick thaw at 37°C to minimize nuclei lysis and chromatin fragmentation.

4. Thawed nuclei in **NE Buffer** can be directly added to activated ConA beads (Sect. 4.3.3).

4.2 Digitonin Concentration Optimization (Optional) (Day 1)

Effective cell permeabilization with digitonin is essential for a successful CUT&RUN, and it is recommended to identify the optimal detergent concentration for a new sample type. Poor permeabilization prevents antibodies and pAG-MNase from entering cells, while excess can result in cell lysis. The default digitonin concentrations in **Cell Permeabilization** and **Antibody Buffers** (0.01%) are optimal for common cell lines (e.g., K562, MCF7, H1299) (*see* **Note 23**).

For new cell types, it is recommended to optimize digitonin concentrations as below. We recommend using the minimal amount that permeabilizes >95% of cells.

1. Prepare five **Digitonin Buffers**, titrating detergent concentration (0.05, 0.01, 0.001, 0.0001, 0.00001%) in **Wash Buffer**. Prepare 0.05% DMSO in **Wash Buffer** as a negative control.
2. Harvest enough cells for six CUT&RUN reactions (~3.1 M cells; 500 k cells per reaction plus excess).
3. Collect cells by centrifugation (600 x *g* for 3 min at RT°C). Carefully remove and discard supernatant. Resuspend pellet in 620 μL RT 1X PBS.
4. Aliquot 100 μL cells to six 1.5 mL tubes. Assign one of the buffers from Step 1 to each tube.
5. Collect cells by centrifugation (600 x *g* for 3 min at RT°C). Carefully remove and discard supernatant. Resuspend each cell pellet in 100 μL of the assigned buffer.
6. Incubate at RT°C for 10 min.
7. Remove an aliquot of each sample and determine % permeabilization (aiming for >95%) with Trypan blue (Sect. 4.1.1) or an alternate membrane impermeant dye.

4.3 CUT&RUN

Below supports the user in performing CUT&RUN via a component-built assay that provides ease of customization and experimental scaling. Consolidated kit solutions containing pre-validated reagents are also available (*e.g., EpiCypher* CUTANA™ ChIC/CUT&RUN Kit) to speed technique adoption.

4.3.1 ConA Bead Binding and Antibody Incubation (Day 1)

In this section of the workflow (see Sect. 4.3.1–4.3.3) ConA beads (*see* **Note 24**) are activated (*see* **Note 25**) and bound to prepared cells in the CUT&RUN **Wash Buffer**. SNAP-CUTANA nucleosome spike-in controls are added to appropriate reactions before the target-specific primary antibody is added and incubated overnight.

ConA Bead Activation

1. Transfer 11 μL per planned reaction of ConA beads to a 1.5 mL tube (*see* **Note 24**).
2. Place tube on a compatible magnetic rack until slurry clears (2–5 min). Pipette to remove and discard supernatant. Do not disrupt the pelleted beads while on the magnet.
3. Remove tube from magnet and add 100 μL per reaction Bead Activation Buffer (*see* **Notes 25 and 27**). Resuspend beads by pipette mixing or gentle vortexing (medium setting 5–7: used through this protocol). Do not scrape beads, as this can damage the surface coating. Flash spin to collect.
4. Return to magnet until slurry clears (2–5 min). Pipette to remove and discard supernatant.
5. Repeat Steps 3 and 4 for a total of two washes. Beads should tightly collect on the magnet (*see* **Note 26**).
6. Remove tube from magnet and resuspend beads in 11 μL per reaction **Bead Activation Buffer**.
7. Aliquot 10 μL activated beads into 8-strip PCR tubes. Keep tubes on ice (*see* **Note 28**).

Initial Cell Harvest

For efficient capture by ConA beads, cells/nuclei with intact membranes should maintain morphology and be monodispersed (Fig. 1a (Step 1) & Fig. 5) (*see* **Note 29**).

1. Count starting cells (as in Sect. 4.1.1) and harvest 500 k per reaction (plus 20% excess). Collect cells by centrifugation (600 x *g* for 3 min at RT°C). Pipette to remove supernatant, leaving enough liquid to keep the cell pellet submerged (~50 μL). The pellet should be visible, and the supernatant should be clear.
2. Resuspend cells in 100 μL per reaction of RT **Wash Buffer** and gently pipette mix (*see* **Note 29**).
3. Repeat the above step for a total of two washes (*see* **Notes 30 and 31**).
4. After the final wash, resuspend cells in 105 μL per reaction **Wash Buffer**.
5. Count cells and examine morphology by Trypan Blue staining (as in Sect. 4.1.1).
6. Capture representative sample images and proceed (to Sect. 4.3.3).

Sample Binding to Solid Support

1. For each reaction, add 100 μL cells or nuclei to 10 μL activated ConA beads in 8-strip tubes. When first using this protocol or working with a new sample type, process an extra reaction in parallel to directly evaluate cell integrity and effective bead-binding (i.e., QC reaction) (*see* **Note 32**). Gently vortex and flash spin in a tabletop mini centrifuge to collect liquid (where ConA beads should remain dispersed).

2. Incubate sample-bead slurry at RT°C for 10 min (where ConA beads should bind cells/nuclei).
3. Place tubes on an 8-strip tube compatible magnetic rack until slurry clears (2–5 min). Remove supernatant and save 10 μL (*aka. unbound fraction*) into a new 1.5 mL tube.
4. Remove from the magnet. Immediately add 50 μL cold **Antibody Buffer** to each reaction (creating the *bead fraction*). Gently pipette to resuspend. If working with > 8 reactions, work with one 8-strip at a time to avoid ConA bead dry-out.
5. Transfer 10 μL of the *cell:bead slurry* from the QC reaction to a new 1.5 mL tube.
6. Visually inspect the *unbound fraction* and *bead fraction* on a brightfield or phase contrast microscope following the Trypan Blue staining protocol (Sect. 4.1.1). This QC examines if the sample has successfully bound to ConA beads (Fig. 1a (Step 2)). Permeabilized cells/nuclei should appear blue (>95%) and be surrounded by beads (*see* **Note 33**) (Fig. 1a (Step 2)).
7. Add **SNAP-CUTANA Spike-in Panels** to designated reactions (as applicable). Do not vortex the spike-in panel stock tube. Instead, flash spin and pipette mix prior to dispensing. If using 500 k cells per reaction, add 2 μL SNAP-CUTANA Spike-in Panel; if using <500 k cells, dilute panel appropriately (*see* **Notes 34 and 35**).
8. Gently vortex 8-strip tubes and flash spin to collect liquid.
9. Add 0.5 μg target-specific antibody to each reaction. Pipette up and down 3–5 times to clear tip (*see* **Note 36**).
10. Gently vortex reactions to mix and flash spin to collect liquid. Incubate on a nutator overnight at 4°C with 8-strip caps elevated at a 45° angle. Do not put tubes on a rotator: when rotating end-over-end, ConA beads stick to the lids, leading to bead drying and sample loss.

--- OVERNIGHT 4 °C ---

4.3.2 Targeted Chromatin Release and DNA Purification (Day 2)

On protocol **Day 2**, reactions are washed to remove excess primary antibody, and pAG-MNase is used to cleave and release antibody-bound chromatin fragments into solution (*see* **Note 37**). Bulk chromatin and cellular components stay bound to ConA beads and are easily removed by magnetic separation. CUT&RUN enriched DNA is then purified and readied for library preparation. Analyzing this CUT&RUN-enriched DNA on a TapeStation or Bioanalyzer will generally be uninformative, as raw yields are typically below their limit of detection. Instead, CUT&RUN DNA yields can be quantified by fluorometric assays (e.g., Qubit) following the manufacturer's protocol.

Chromatin Digestion

ConA beads become sticky after overnight incubation at 4°C; pipette carefully to resuspend, and regularly vortex to keep beads in solution (*see* **Notes 38 and 39**). Ensure all liquid is ejected from pipette tips at each step to avoid bead loss. If concerned by sample loss, vortex mix reactions instead.

1. Remove reactions from 4°C and flash spin to collect liquid. Bead settling after overnight incubation is common and does not indicate experimental failure. Place on an 8-strip tube magnet until slurry clears and the beads form a tight pellet (2–5 min). Remove **Cell Permeabilization Buffer** from 4°C and place on ice.
2. Use a multichannel pipette to carefully remove and discard supernatant. If processing ≥8 reactions, complete buffer exchange one 8-strip at a time to avoid ConA bead dry-out.
3. Keeping tubes on the magnetic rack, immediately add 200 μL of cold **Cell Permeabilization Buffer**. Leave for ~10–30 s, and then pipette to remove and discard buffer.
4. Repeat for a total of two washes with **Cell Permeabilization Buffer**.
5. Prepare a master mix of pAG-MNase (2.5 μL per intended reaction) in cold **Cell Permeabilization Buffer** (50 μL per intended reaction), with extra volume to account for pipetting loss (*see* **Note 38**).
6. Remove tubes from magnet and resuspend beads in 52.5 μL cold pAG-MNase master mix per reaction. Gently vortex and/or pipette mix and flash spin to collect liquid (*see* **Note 39**).
7. Incubate reactions RT°C for 10 min on a nutator with caps elevated at 45° angle.
8. Gently vortex tubes to mix and flash spin. Place on magnetic rack until slurry clears and the beads form a tight pellet (2–5 min). Pipette to remove supernatant.
9. Keeping tubes on the magnet, add 200 μL cold **Cell Permeabilization Buffer** per reaction. Leave for ~10–30 s, and then pipette to remove and discard buffer.
10. Repeat for a total of two washes. Then place tubes on ice (using a pre-chilled aluminum block can greatly improve handling).
11. Prepare a master mix of $CaCl_2$ (1 μL of 100 mM stock per intended reaction) in cold **Cell Permeabilization Buffer** (50 μL per intended reaction, with extra volume to account for pipetting loss) (*see* **Note 38**).

12. Resuspend each reaction in 51 μL cold **$CaCl_2$ master mix**. Keep tubes on ice or a pre-chilled aluminum block throughout.
13. Gently vortex and/or pipette mix to ensure beads are resuspended and flash spin to collect liquid.
14. Incubate reactions 4°C for 2 h on a nutator with caps elevated at a 45° angle.

Chromatin Release

During the $CaCl_2$ incubation, prepare the **Stop Buffer** and DNA purification reagents. *E. coli* spike-in DNA can be added to **Stop Buffer** as a master mix and used to monitor CUT&RUN library prep and aid sequencing normalization (*see* **Note 40**). Add 0.5 ng *E. coli* DNA per 500 k cells (scaling linearly for fewer cells).

1. Prepare **Stop Buffer master mix** by combining 1 μL *E. coli* DNA (0.5 ng/μL) and 33 μL **Stop Buffer** per reaction. *E. coli* can be added to any target reaction. Gently vortex to mix, flash spin, and place on ice (*see* **Note 41**).
2. Set a thermocycler to 37°C.
3. After the 2 h incubation (Sect. 4.3.4 Step 14), vortex and flash spin the 8-strip tubes. Add 33 μL **Stop Buffer master mix** to each reaction. Gently vortex to mix, flash spin, and place in 37°C thermocycler for 10 min (*see* **Note 42**).
4. Vortex and flash spin the 8-strip tubes. Place on 8-strip tube compatible magnet until slurry clears (30 s–2 min).
5. Transfer supernatant to new 8-strip PCR tubes. **Do not discard supernatant: this is the target-enriched chromatin**!
6. For native materials, proceed to DNA purification (Sect. 4.3.6). To reverse cross-links, continue below.

4.3.3 Cross-linking reversal (Optional)

CUT&RUN cross-linking can cause cell lysis and/or reactions to become clumpy or sticky. If this occurs, do not complete cross-linking reversal; instead, move directly to DNA purification (Sect. 4.3.6).

7. Add 0.8 μL 10% SDS to each CUT&RUN-enriched chromatin sample. Pipette up and down 3–5 times to fully clear tips.
8. Add 1 μL of 20 μg/μL Proteinase K to each reaction. Pipette up and down 3–5 times to fully clear tips. Gently vortex to mix and flash spin.
9. Place tubes in a thermocycler at 55°C. Incubate 1 h to overnight.
   --- ***cross-link reversal* 1 H or OVERNIGHT at 55°C** ---
10. Flash spin tubes and proceed to DNA purification (Sect. 4.3.6).

DNA Purification

Many DNA purification approaches, including spin columns and magnetic beads, are compatible with the CUT&RUN workflow. However, this protocol uses SPRI bead-based reagents (*see* **Note 43**) to streamline DNA purification in 8-strip tubes [64].

1. Prepare 85% ethanol (EtOH), fresh 100% EtOH, and molecular-grade H_2O. Allow 500 μL per reaction (includes extra volume to account for pipetting loss).
2. Thoroughly vortex SPRI bead reagent in manufacturer's buffer to an even suspension (this will maximize DNA yields and reduce potential for sample loss).
3. Slowly add 151 μL SPRIselect or CUTANA DNA Purification Beads reagent at 1.8X ratio (*see* **Note 44**) to each sample. Ensure tips are free of bead droplets before addition (*see* **Note 45**).
4. Thoroughly pipette to mix and then vortex 8-strip on medium setting for 10–15 s. The solution is viscous, and thorough mixing is critical to good DNA yield recovery. Flash spin tubes to collect liquid and incubate at RT°C for 5 min.
5. Place tubes on a magnet until slurry clears (2–5 min). For washes in Steps 5–8, process one strip at a time to avoid beads drying out.
6. Carefully pipette to remove and discard supernatant, taking care not to disturb the beads. Keeping samples on the magnet, immediately add 180 μL 85% EtOH to each. Leave for ~30 s, and then pipette to remove and discard.
7. Repeat the previous step for a total of two washes.
8. Flash spin tubes to collect excess EtOH.
9. Place tubes back on the magnet and remove residual EtOH with a 10 μL multichannel pipette while taking care not to disturb the beads.
10. Remove tubes from magnetic rack and air dry 2–3 min at RT°C. Beads should remain dark brown and just barely damp (*see* **Note 46**) (Fig. 1a (Step 4)).
11. Add 17 μL of **0.1X TE Buffer** to each tube. Pipette to mix beads and vortex to fully resuspend. Flash spin tubes to collect liquid.
12. To elute DNA, incubate samples 2–5 min at RT°C.
13. Vortex tubes, flash spin, and return to the magnet. Incubate additional 2 min at RT°C.
14. Transfer 15 μL supernatant containing CUT&RUN-enriched DNA to new 8-strip PCR tubes.
15. Quantify eluted DNA with the Qubit Fluorometer and 1X dsDNA HS Assay Kit (*see* **Note 47**).

16. Safe point to pause: store DNA at −20°C or continue to library preparation (Sect. 4.4).

--- **Safe pause point. Store at −20°C** ---

4.4 Library Prep (Day 2 or 3)

Library prep is a multistep process whereby CUT&RUN DNA fragments are prepared for sequencing (most often paired-end on *Illumina* platforms). A range of commercial library prep kits are available (and should supply all needed reagents, including barcoded primers to distinguish each reaction library (*see* **Note 48**), but selecting one optimized for CUT&RUN (e.g., see Sect. 2.2) will improve data quality and consistency.

With our preferred library prep kit 5 ng CUT&RUN DNA per library (*see* **Note 49**) will provide the best balance of yield and high-quality sequencing data. If using a different kit follow its respective instructions but use the below CUT&RUN-specific PCR parameters optimized for low inputs and small DNA fragments (200–700 bp) (*see* **Notes 50–52**). Irrespective of the kit selected, do not shear or fragment DNA prior to library prep.

Temperature °C	Time (s)
98°C	45
98°C	15
60°C	10
72°C	60
4–12°C	∞

A major predictor of CUT&RUN success prior to sequencing is the fragment distribution of purified sequencing libraries, which should show a primary peak at ~300 bp (~170 bp mononucleosome + adaptors) on the Bioanalyzer or TapeStation (Fig. 1a (Step 5)). Note that CUT&RUN library yields are not the only success metric, as these are influenced by target abundance, antibody efficiency, cell type and/or number.

Good libraries provide sufficient DNA to sequence at standard concentrations (>2–100 nM), but quality data can be generated from as low as 0.5 nM (*see* **Note 52**). If present, adapter dimers will be between 100–175 bp. If they comprise >5% of the total DNA, we recommend their removal prior to sequencing by gel purification kits [e.g., Qiagen 28704] or right-shifted size selection via SPRI beads (Sect. 4.5.1).

--- **Safe pause point. Store at −20°C** ---

4.5 Sequencing and Data Analysis (Day 3+)

Select an appropriate sequencing platform based on multiplexing requirements and desired sequencing depth. Paired-end sequencing with at least 50 bp reads (*aka.* PE100) is recommended, as this improves genomic mapping accuracy and distinguishes the nucleosome spike-in panels. CUT&RUN requires only 5–10 million total reads for appropriate peak-generation/genome coverage compared to the ~30 million reads required by *ChIP-seq* (*see* **Note 53**). Below describes how to analyze and visualize sequencing data generated from CUT&RUN libraries on a Mac OS or Linux operating system. The self-assembled methods are approachable, but if expertise and timelines are limited, the user might consider an online data analysis platform (*e.g.*, *EpiCypher* CUTANA™ Cloud) for streamlined bioinformatics.

4.5.1 Pooling Libraries and Loading Sequencer

1. Pool equimolar libraries from individual reaction libraries for multiplex sequencing.
   a. Dilute each reaction library to the same concentration. 2–4 nM is recommended (if possible from the available library), but confirm appropriate for the intended sequencing platform (*see* **Note 54**).
   b. Pool equal volumes of each library in one tube. Ensure only unique indexes are pooled so samples can be effectively demultiplexed.
   c. Dilute pool to the appropriate concentration for the intended sequencing platform.
2. Load the correct volume onto a temperature equilibrated sequencing cartridge for the platform of choice (*see* **Note 55**).

   Adapter Dimer Removal (optional).

   Adapter dimers can be removed from pooled sequencing libraries (at recommended 2–4 nM) using a 1:1 ratio of [library volume: SPRIselect or CUTANA DNA Purification Beads]. Save ~10–20 μL of <u>pre</u>-cleaned library pool to confirm adapter dimer removal.

   a. Pool equal volumes of normalized libraries (e.g., 2–5 μL per) in a 1.5 mL tube.
   b. Slowly add SPRIselect or CUTANA DNA Purification beads to a final ratio of 1:1. Vortex to mix. Incubate RT°C for 5 min.
   c. Place a 1.5 mL tube onto a compatible magnet until the slurry clears (2– 5 min).
   d. Remove supernatant. Keeping tubes on the magnet, add 180 μL of 85% EtOH (freshly prepared) to the tube. Leave the sample in EtOH for ~30 s.
   e. Remove supernatant and repeat EtOH wash for a total of two washes.
   f. Place samples in a tabletop mini centrifuge and flash spin to collect excess EtOH.

g. Place samples back on the magnet and gently remove residual EtOH with a 10 μL volume multichannel pipette. Do not disturb the pellet.

h. Remove samples from magnetic rack and air dry 2–3 min RT°C. Beads should remain dark brown and damp.

i. Add 12–15 μL of **0.1X TE Buffer** or **Sequencing Buffer** to each reaction. Pipette mix while washing the pelleted side to resuspend beads. Follow pipette mixing with a gentle vortex and flash spin.

j. Incubate RT for 2–5 min to elute DNA.

k. Place a 1.5 mL tube onto a compatible magnet until the slurry clears (30 s–2 min).

l. Carefully transfer 10.5 μL supernatant to a clean new 1.5 mL tube as the post-cleaned library pool.

m. Run pre- and post-cleaned library pools on TapeStation to confirm successful adapter dimer removal.

4.5.2 SNAP-CUTANA Spike-In Analysis

Any reaction spiked with an *EpiCypher* SNAP-CUTANA nucleosome panel (e.g., K-MetStat, HA Tag, or DYKDDDDK Tag (*see* Note 1) can be analyzed using the pipeline below. The shell script searches for and counts DNA barcodes corresponding to specific dNucs. The relative on- vs. off- target recovery validates antibody performance and the experimental workflow, before diving into genomic alignments. Additionally, spike-in results guide any needed optimization/troubleshooting (e.g., Fig. 3b). Since the same amount of nucleosome panel is added to each reaction, they can also be utilized as a normalization tool.

1. Download R1 & R2 paired-end sequencing files (fastq.gz) for each spiked-in reaction. Unzip and save the fastq files in a new folder.
2. Download the Shell script (.sh) and corresponding analysis files (.xlsx) and save these to the folder created in **Step 1**. Applicable files are on the relevant product page at *epicypher.com* [e.g., SNAP-CUTANA -K-MetStat panel, #19-1002].
3. Open the shell script file in a plain text editing program (*Microsoft* Word or PDF editing programs are currently incompatible). Scroll past barcode sequences and find the analysis script.
4. The script is a loop that counts the reads aligned to each DNA barcode in a given reaction, with each nucleosome spike-in usually represented by two barcodes (A & B). Create one loop per reaction (below is customizable):
   a. Copy lines between # template loop begin ## and # template loop end ##.
   b. Paste the loop under the last done. Paste one copy per reaction.

c. In the first loop, replace Sample1_R1/2.fastq with the correct R1 & R2 fastq file names for one reaction. Repeat each loop and save.

5. In Terminal, set the directory to your folder. To change the directory location, type "cd" followed by a space and drag the folder into the terminal. Press return.
6. Run the script in the terminal. Type "sh" followed by a space and drag the .sh file from your files into Terminal to copy the location. Press return.
7. Open the panel analysis Excel file and fill in the reaction names and set the relevant **on-target** in column B. The first reaction is set to the negative control (IgG); all additional reactions select a target from the drop-down menu.
8. Copy the R1 barcode read counts from the first loop in the terminal and paste into the yellow cells in the Excel sheet, matching the corresponding sample in **Column C**. Repeat for R2 barcodes and paste into yellow cells in **Column D**. Repeat for each loop/reaction.
9. The Excel file automatically analyses the spike-in data.

   a. Calculates the total reads for each DNA barcode in **Column E** (R1 + R2).
   b. Calculates the total barcode read counts for each nucleosome in **Column F** (A + B).
   c. Expresses total read counts for each spike-in nucleosome as a percentage of on-target read counts found in **Column G & J**. This provides on-target recovery and specificity for each antibody.
   d. **Column J** displays an auto-populate table (**Output Table**), with the percent recovery of each nucleosome normalized to the on-target mark. A color gradient is used to visualize recovery of each nucleosome normalized to on-target (Fig. 3).
   e. For each reaction, the percent of unique sequencing reads assigned to each spike-in can be calculated. Type the total number of unique reads in the yellow cell titled **Uniq align reads** found in **Column B**. The percent total barcode reads are added to the **Output Table**.

4.5.3 *E. coli* Spike-In Analysis

In this CUT&RUN protocol, the same scaled amount of *E. coli* DNA is added to related reactions (Sect. 4.3.5) and (*see* **Note 40**). Therefore, the spike-in can be used as a normalization tool to scale signals across reactions. Aim for 1–5% *E. coli* reads in the total sequencing reads: >10% often indicates a poor CUT&RUN reaction (e.g., antibody inefficiency or poor sample permeabilization). Any reaction spiked with *E. coli* DNA can be analyzed using the pipeline below. Normalizing is valid for multiple CUT&RUN data sets generated from the same antibody target, or experimental vs. IgG negative control antibody. Applying a scaling factor across CUT&RUN data from different antibody targets (e.g., H3K4me3 vs. H3K27me3) is invalid because there is no expectation that the *E. coli* bandwidth should be the same for epitopes of differing genomic abundance.

1. Data from demultiplexed CUT&RUN sequencing libraries requires separate alignment to the *E. coli* genome (K-12 MG1655), while filtering duplicate reads.
2. Calculate spike-in bandwidth by dividing total uniquely aligned reads to the *E. coli* genome by total uniquely aligned reads to the sample reference genome (e.g., hg38). $\left(\frac{count\ of\ Ec\ reads\ in\ reaction\ A\ (e.g.\ H3K4me3)}{Total\ uniquely\ aliged\ reads\ in\ reaction\ A\ (e.g.\ H3K4me3}\right)$
3. The normalization scaling factor is calculated such that the *E. coli* spike-in signal is equal across paired reactions [65]. $\left(\frac{Normalization\ factor\ \#\ (e.g.\ 1)}{Spike\ in - bandwidth\ (e.g.\ reaction\ A}\right)$
4. Use normalization scaling factor with the --scaleFactor option enabled in deeptools bamCoverage tool to generate normalized bigwig files for visualization (e.g., on Integrative genome viewer (IGV)) (*see* **Note 56**).

4.5.4 Genomic Sequencing Data analysis

CUT&RUN produces clean genomic mapping profiles using many of the same data analysis pipelines and peak calling tools developed for *ChIP-seq* (albeit with some important exceptions). Various segments of the bioinformatics pipeline transform raw sequencing reads into data for visualization and interpretation (Fig. 6). The initial stages of analysis involve read pre-processing, demultiplexing, sample identification and processing, and genome alignment. Within this, sequence files from individual reactions are identified by their assigned barcode and filtered to remove duplicate reads, unaligned reads, and reads to repetitive DAC exclusion list regions. The filtered files are then processed by secondary analyses, including peak calling and motif enrichments. Below describes our recommended pipeline to analyze CUT&RUN sequencing data, using SNAP-CUTANA nucleosome and *E. coli* spike-in controls to identify assay success/quality, and troubleshooting (if needed).

Download CUT&RUN fastq files and align raw reads to the reference genome of choice (e.g., human hg38, mouse mm10, *Arabidopsis* TAIR10) with Bowtie2, generating a BAM file.

1. BAM files are filtered to remove duplicate reads, multi-aligned reads, and reads within DAC exclusion list regions, generating a set of **uniquely aligned reads**. Ideally,>70% of total reads in a sequencing library uniquely align to the genome.

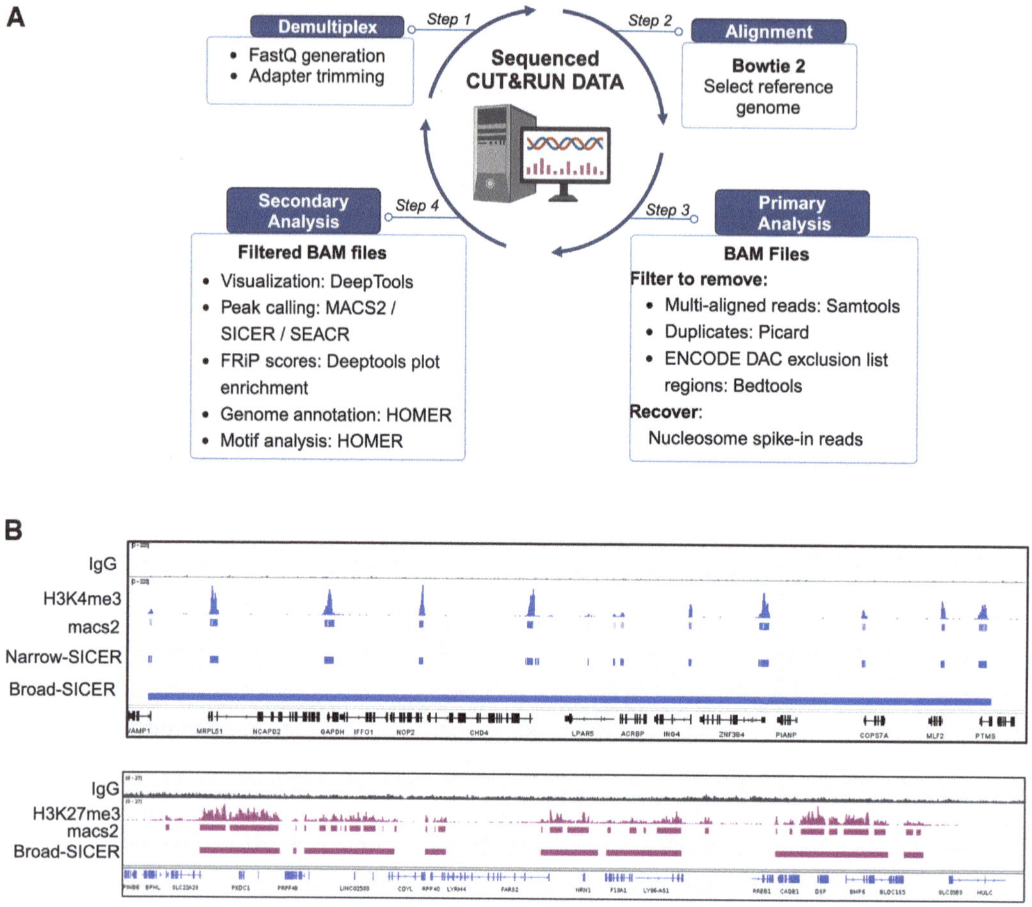

Fig. 6 CUT&RUN data analyses pipelines and peak calling examples. **A** Bioinformatic workflow and recommended toolkits to analyze raw fastq sequencing files from CUT&RUN experiments. **B** Comparison of peak-calling algorithms. MACS2 is preferred for calling targets with sharp, narrow peaks (1–2 kb; e.g., H3K4me3), while SICER is ideal for targets that enrich over broad regions (>10 s of kbs; e.g., H3K27me3). All experiments used 500 k K562 cells/reaction; RPKM reads aligned to the hg38 genome. H3K4me3 peaks were called by MACS2, H3K27me3 peaks were called by SICER. Created in BioRender. Firestone T (2024) https://BioRender.com/a79b961

2. Filtered BAM files are converted using deepTools2 to sequence depth (and optionally spike-in) normalized enrichment tracks (Bigwig files) that enable visualization (e.g., on IGV) (*see* **Note 57**).
3. Reads from SNAP-CUTANA barcoded nucleosomes or *E. coli* DNA spike-ins can be used to validate the workflow and identify/troubleshoot failed reactions (Sects. 4.5.2 and 4.5.3).
4. Different peak "shapes" (i.e., patterns of target localization) are called with varying efficiencies on narrow (suitable for H3K4me3) or broad (suitable for H3K27me3) settings using a

variety of programs, such as MACS2, SICER, SEACR [66], or CUT&RUNTools 2.0 [67, 68] (Fig. 6b). It is recommended the user consider (or at least understand the pros/cons of) several peak callers and select one that best represents the target of interest [68] (*see* **Note 58**).

5. For computing S:N, use software such as Deeptools plotEnrichment to calculate fractions of reads in peaks (FRiP), and compare FRiP scores from experimental vs. controls reactions [69] (*see* **Note 59**) (Fig. 6c).
6. Spike-in based normalization can be accomplished using reads from *E. coli* aligned reads (Sect. 4.5.3).
7. Comparative analyses between two or more data sets (antibody or cell titrations, drug treatments, different targets, etc.) can be done with Deeptools heat maps or Venn overlaps of called peaks (Fig. 3c).

5 Notes

5.1 Antibody Selection

1. SNAP-CUTANA™ Spike-in Controls (*EpiCypher*) are panels of semi-synthetic, DNA-barcoded nucleosomes carrying defined PTMs or common epitope tags. Panels are bound to streptavidin magnetic beads, and added to CUT&RUN reactions just prior to antibody. During incubation, antibodies bind their target epitope in the spike-in panel, allowing targeted cleavage and release of the relevant barcoded nucleosome from bead to supernatant alongside target-enriched sample chromatin. Off-target spike-in nucleosomes remain bead-bound and are magnetically removed alongside bead-bound cells or nuclei. After DNA purification, library prep, and NGS, the relative recovery of on- and off-target spike-in nucleosomes is determined from sequencing data by quantifying their unique DNA barcodes, offering a direct readout of antibody performance and assay success. Spike-in reads can also be used to derive a normalization standard, creating a quantifiable way to measure target-enrichment differences induced by disease or drug treatment. Available SNAP-CUTANA Spike-in Controls can be found at: https://www.epicypher.com/products/nucleosomes/snap-cutana-spike-in-controls.
2. Due to the distinct nature of immunotethering approaches vs. ChIP, antibodies must be empirically tested for potential adoption in CUT&RUN. In ChIP assays, bead-bound antibodies capture target fragments in the milieu of "sticky" and typically denatured chromatin which drives background. This is addressed by stringent washing that can also remove

on-target binding, leading to the low Signal:Noise (S:N) inherent to ChIP. Many frequently used "ChIP-grade" antibodies have been selected by the scientific community because of their high yield, which can be a product of high efficiency and/or off-target binding [70]. However, even the most specific antibodies in ChIP could require high-stringency washing to remove off-target binding, and this element is not present in the CUT&RUN workflow.

3. Testing five independent antibodies may seem excessive, but this initial investment can help to identify the best reagent, increasing confidence in all subsequent data. From our extensive testing (hundreds of targets to date), three to five antibodies are typically required to find one that works convincingly (absent a confirmed knockout). When selecting reagents, antibodies "validated" for ChIP, immunofluorescence, or FACS are a good place to start, but they do not guarantee success. Multiple companies often distribute the same reagent, making it important to select truly independent antibodies by looking for distinct host species, immunogens, or clone identifiers. The candidate pool can include monoclonals (mAbs), polyclonals (pAbs) or recombinants, with monoclonals and recombinants often preferred to reduce (but not completely avoid) potential lot-to-lot variability. However, pAbs provide a range of paratopes and may thus be more resistant to clonal liabilities, such as sensitivity of a PTM-specific mAb to nearby PTMs (e.g., anti-H3K9me3 cannot accommodate co-incident H3S10ph) [71].

4. After comparing data from all test antibodies, decide which (if any) are suitable for further experiments based on rational pass/fail metrics: e.g., % cross-reactivity to barcoded nucleosome standards (*see* **Note 1**), or independent reagents yield the same peak structure that makes biological sense (such as anti-TF enriches regions containing the cognate DNA-motif). To sidestep potential future lot variability, now is the time to secure sufficient material from the test lot to cover all intended studies.

5. For proteins that lack usable antibodies, epitope-tag fusions (e.g., FLAG, HA, GST, or 6HIS) and related anti-tag antibodies can be explored. Epitope tags can be truly enabling but come with caveats: potentially impacting protein folding, stability or cellular localization. Further, the tag might not be accessible when the protein is part of a complex [72].

6. It is recommended to initially test and validate antibodies at 500 k cells (the standard CUT&RUN reaction) and then downscale inputs to explore antibody efficiency: when do weaker peaks drop out? If the intended study requires rare or

precious cell populations, it may be useful to perform antibody validation using a relevant but more easily accessible cell line or type.

7. For CAPs, confirm the target is expressed and localized to chromatin (if feasible). Consider whether this might be impacted by conditions of cell growth and/or stimulation.
8. SNAP-CUTANA spike-in panels are added to the designated CUT&RUN reactions before the target antibody. They should only be added to reactions mapping a target within the panel; i.e., the K-MetStat Panel adds little discernible value to a reaction mapping lysine acetylation or a transcription factor. Reads assigned to spike-in barcodes should comprise ~1% of total (0.1–5% is acceptable). Results are used to gauge experimental success and guide troubleshooting.
9. The negative control antibody (IgG) should show no preference for any targets in a SNAP-CUTANA panel and low genomic enrichment (potential over-digestion can be identified by a characteristic signal at the open chromatin over active promoters [defined by H3K4me3]). Positive controls (H3K4me3 and H3K27me3) should display strong enrichment for the respective on-target spike-in nucleosome with <20% off-target binding. In genomic data, positive control reactions should also show high signal to noise, as measured by the Fraction of Reads in Peaks (i.e., FRiP score).
10. When control antibodies (*see* **Note 9**) generate the expected spike-in percentages and genomic feature enrichments, the CUT&RUN data is reliable and the workflow validated. Alternatively, >20% off-target binding to standards and high genomic background indicate experimental issues. Sequence data from the standards can inform on failed individual reactions of a particular target, sample, or the entire workflow.

5.2 Sample Preparation

11. Spermidine is required in buffers because it compensates for the absence of magnesium and acts to stabilize chromatin. Buffers containing magnesium can prematurely activate pAG-MNase, causing DNA degradation.
12. Although rarely observed, the addition of **CUT&RUN Wash Buffer** can lyse particularly fragile cell types or samples. If a potential concern, test compatibility on a small volume of cells prior to harvesting for CUT&RUN.
13. For immune cells (e.g., T, B, monocytes), there is a potential concern that binding to ConA (an immunostimulatory lectin) can cause inadvertent activation. To remove all doubt, the experimenter might want to consider using extracted nuclei instead of whole cells.

14. Tissue samples require mono-dispersion (i.e., to a single cell suspension) and nuclei extraction. Tissues can be processed by mechanical maceration, douncing and/or enzymatic digestion (trypsin, collagenase or dispase) [25, 73–77]. When dispersing clumps by pipetting, tips can be cut off to increase bore size and reduce the likelihood of clogging or shearing. If clumping is observed after nuclei extraction, resuspend in 1X PBS.
15. Using extracted nuclei can increase S:N for chromatin-associated protein targets that are also/primarily cytoplasmic (e.g., GATA3 [78]).
16. Light to moderate formaldehyde cross-linking (0.1% to 1% for 1 min) can help preserve labile targets (e.g., histone lysine acylations or monoubiquitylations); likely by inactivating the targeting enzymes (e.g., lysine deacetylases or deubiquitinases). However, cross-linking may also lead to cell/bead clumping and loss of material. Further, heavy cross-linking (1% formaldehyde for 10 min; as used for ChIP) is hugely detrimental to CUT&RUN, as the resulting large DNA:protein complexes are unable to diffuse across the nuclear membrane, compromising DNA yields. We have not tested other cross-linkers and so cannot comment on useful conditions. For a novel sample type/target, always test native and cross-linked conditions in parallel.
17. Resuspension in **CUT&RUN Wash Buffer** can increase permeability to Trypan Blue and reduce apparent cell "viability." Instead, focus on total cell counts and cellular morphology/integrity. If cell counts fall and morphology deteriorates, it may be better to restart with more cells, and be careful when resuspending in **CUT&RUN Wash Buffer** to avoid losing material during early wash steps.
18. Count cells immediately after adding Trypan Blue and avoid prolonged incubation. Thawed cells may have increased permeability and thus appear to have reduced viability. As above, it is recommended to instead focus on total cell counts, cellular morphology, and the amount of contaminating cell debris. If the majority of cells appear normal and there is minimal cell lysis and debris, move forward to CUT&RUN.
19. Viability >90% is optimal but consistently lower with some primary cells, cell lines, and/or after drug treatment. Thawing cryopreserved material will result in slightly reduced viability. Reduced viability is often accompanied by reduced DNA yields and a higher sequencing background, but can still yield good data.
20. Trypsin will be removed in subsequent washes.

21. If used effectively, trypsinization will disaggregate clumps to monodispersed cells without damage. Optimize cell detachment by adjusting enzyme concentration and duration. Avoid extensive digestion since it can damage cells and impede binding to ConA; if extensive trypsinization is required, consider nuclei extraction.
22. Labile PTMs can be preserved by crosslinking (Sect. 4.1.2.b) or adding appropriate inhibitors to incubation buffers: e.g., protein phosphatase or KDAC inhibitor cocktails.
23. **Antibody Buffer** and **Cell Permeabilization Buffer** contain **digitonin**, a detergent to permeabilize the cell membrane. Permeabilization is NOT required when using nuclei. However, adding the detergent to **Cell Permeabilization** and **Antibody Buffers** improves bead handling (absent digitonin, they can form a film on the sides of 8-strip tubes).

5.3 ConA Bead Binding and Antibody Incubation

24. ConA beads should never be frozen or dried out, or appear black, grainy or clumped. This will/indicate damaged beads, and the CUT&RUN will fail.
25. Bead activation occurs through a series of washes that remove ConA beads from their storage buffer to facilitate their cell-isolating properties based on interactions with glycoproteins on the cell / nuclear surface.
26. If working with 16 or more reactions, complete each wash with no less than 1 mL **Bead Activation Buffer**.
27. ConA bead activation requires coordination of metal ions (Calcium and magnesium) by the lectin tetramer to ensure the ConA beads can effectively bind glycoproteins on the cell/nuclear surface. ConA beads will become sticky (often manifesting as a brown residue in the pipette tip) after washing with **Bead Activation Buffer**.
28. Keep activated beads on ice until ready to combine with samples. It is advised that ConA beads be used within 4 h of activation.
29. **CUTANA Wash Buffer Enhancer Set** is added at a 1x final concentration to **Wash Buffer** to reduce sample clumping, a frequent cause of poor sequencing performance. It is recommended for minimizing technical variability and improving data quality.
30. Cell loss through centrifugation, supernatant aspiration, and resuspension steps can be a major concern. This can be minimized (but not completely avoided) by longer centrifugations and careful aspiration that does not disturb the cell pellet.
31. If sample cells are "fragile" and prone to lysis in **CUT&RUN Wash Buffer**, reduce to one wash step (*see also* **Note 17**). Cell lysis increases CUT&RUN background and is thus to be avoided.

32. If cell numbers are limiting, use a small aliquot (e.g., 1–5 μL) of the H3K27me3 reaction for QC checks: when using a good antibody, this high-abundance target can spare the material.
33. At recommended volumes, ConA beads are in excess of cells; thus, observing some free-floating beads does not indicate binding failure. However, if the unbound cell/nuclei fraction is >50% consider restarting the experiment. Ensure beads were undamaged before starting and properly activated (*see* **Notes 23 and 25**).
34. SNAP-CUTANA Spike-in Panels must be added before the primary antibody.
35. If using <500 k cells/reaction, dilute spike-ins with **Antibody Buffer** for an appropriate working stock (to maintain the preferred ratio of [sample:spike-in]). This is prepared on the day of the experiment and used immediately.

No. of cells	Panel dilution in antibody buffer
250 k	1:2
100 k	1:5
50 k	1:10
< 50 k	1:10

36. Take particular care when pipetting antibodies from high viscosity glycerol storage buffers or accuracy will suffer. In the standard reaction, the primary antibody (0.5 μg) is already in excess and adding more can increase the background signal.

5.4 Chromatin Digestion and DNA Purification

37. Protein A and Protein G both bind antibody Fc domains but exhibit different binding affinities to isotypes from different species (https://assets.thermofisher.com/TFS-Assets/LSG/Application-Notes/TR0034-Ab-binding-proteins.pdf). pAG-MNase (as used in the protocol) includes both Protein A and Protein G to ensure a broader tethering compatibility for a variety of anti-chromatin antibodies.
38. Adding pAG-MNase/$CaCl_2$ via a master mix minimizes experimental variability and reduces the possibility of forgetting to add a reagent to an individual reaction.
39. To allow effective reactant diffusion, ConA beads/cells should remain in suspension through pAG-MNase binding and cleavage. Beads/cells may become clumpy through incubation, but gentle pipette mixing (with tips cut off to increase bore size) can usually resolve this issue. Flash spin to settle liquid from lids before magnetic capture/opening tubes.

40. *E. coli* DNA is added to monitor library preparation/normalize sequencing data without committing excess reads, though the precise bandwidth consumed will depend on target abundance and antibody efficiency. Under optimized conditions, a negative control (IgG) or low abundance target (e.g., H3K4me3) may have 10%–20% *E. coli* reads, while high abundance targets (e.g., H3K27me3) have 0.1%–1%. *E. coli* percentages outside the usual ranges often indicate a poor CUT&RUN reaction.

41. **Stop Buffer** chelates free Ca^{2+} to quench pAG-MNase activity.

42. Samples are heated to promote chromatin release and RNase A digestion. Higher NaCl concentrations (up to 350 mM in the final eluate) may increase CUT&RUN yields.

43. SPRI reagents are paramagnetic beads coated with carboxy molecules designed to reversibly bind and size-select nucleic acids. Changing the ratio of SPRI reagent to sample alters the size range of DNA captured. Increasing SPRIselect or CUTANA DNA Purification Beads ratios captures smaller fragments, while decreasing ratios capture larger fragments [79, 80].

44. Multiply CUT&RUN reaction volume by the ratio number for the amount of SPRIselect or CUTANA DNA Purification Bead reagent (e.g., 84 μL reaction X 1.8 = 151 μL beads). This bead-to-sample ratio enables capture of both regular nucleosomal fragments and small TF-associated fragments.

45. Each lot of SPRI beads should be tested prior to use by mixing varying bead ratios (0.8–2X) with a low range DNA ladder [*ThermoFisher* SM1213]. A passing ratio shows >90% recovery of intended target-sized DNA fragments.

46. Prior to the elution step (Sect. 4.3.6), SPRI beads should appear slightly damp and matte brown. Glossy beads are too wet and need more time to air dry; cracked light brown beads are over-dried. Proceeding in either extreme may impact yields.

47. Low DNA yields and/or those similar to IgG (negative control) do not necessarily indicate assay failure and may simply reflect target abundance and/or antibody efficiency. For instance, with suggested antibodies (Sect. 2.2), low abundance H3K4me3 typically yields close to IgG (~2–5 ng from 500 k K562 cells) while high abundance H3K27me3 yields should be > than IgG under the same conditions.

5.5 Library Preparation

48. Individual CUT&RUN libraries are prepared with 8bp barcodes at their 5 prime (i5 index) and 3 prime (i7 index) ends. For multiplexing, each individual library must be assigned a unique i5 and i7 index pair. Library prep kits will impact how many libraries can be multiplexed in a single sequencing run

(e.g., NEB Multiplex Oligos allows up to 24 or 96 libraries). In a given sequencing run, we recommend multiplexing at least five CUT&RUN libraries to ensure index diversity. For low complexity libraries, consider including spike-in PhiX DNA based on *Illumina* recommendations.

49. For CUT&RUN reactions with yields below 5 ng DNA, use as much as possible for library preparation.

50. Making libraries from low DNA yield reactions increases susceptibility to adapter dimers. Maintain 4°C during adapter ligation setup to minimize their formation.

51. We recommend 14X PCR cycles for standard CUT&RUN reactions. If yields are low (e.g., due to reduced cell inputs) this can be increased to 16X or 18X PCR cycles. However, since starting from less complexity the read duplication rate in sequencing data will likely increase.

52. If individual libraries are <1 nM, it is recommended to repeat library prep with more starting material. Alternatively, add as much as you can of lower-concentration libraries to the sequencing pool; deeper sequencing (>10 million reads) may be necessary to fully capture read diversity.

5.6 Sequencing and Data Analysis

53. Suggested numbers for sequencing depth (5–10 M total reads/3–8 M mapped unique reads) assume human genome size. All things being equal, the suggested sequence depth for other species scales by genome size. As reference, some common genome builds: 3100 Mb *H. sapien* (hg19, hg38); 2500 Mb *M. musculus* (mm10); 140 Mb *D. melanogaster* (dm6); 135 Mb *A.thaliana* (TAIR10); 12 Mb *S.cerevisiae* (S288C).

54. Do not dilute libraries below 2 nM. Pooling should use 2–5 μL of normalized libraries.

55. Carefully pipette diluted libraries to avoid bubble formation that easily arises from the detergent in the **Sequencing Buffer**.

56. Reactions with the highest *E. coli* spike-in bandwidth will receive the largest reduction in signal after normalization.

57. Each sample is processed and normalized by Reads Per Kilobase per Million mapped reads (RPKM). This calculation is done by Deeptools with a recommended bin size set to 20 bp [59].
RKM (per bin) = number of reads per bin/(number of mapped reads (in millions)* bin length (kb))

58. Various secondary analysis methods exist to identify and visualize statistically enriched narrow and broad peaks. MACS2 is a widely used peak caller that excels at detecting narrow peaks: a signal profile localized to short genomic regions, usually several hundred base pairs (e.g., H3K4me3 located at promoters). SICER2 is recommended for calling broad peaks: a signal profile that covers genomic regions spanning several kilobases

(e.g., H3K27me3 created by the Polycomb complex or H3K36me3 located at gene bodies). Further, the peak calling tool and settings used should accommodate target dynamics within a given experiment (e.g., early time-course or low-dose treatment with an EZH2 inhibitor may reduce H3K27me3 levels but paradoxically appear to increase peak numbers if the caller was optimized to join short enrichment gaps that now increase in size). If uncertain of which kind of signal profile will be generated, we recommend visualizing the data with the generated bigwig files; this will give an indication of how the profiles of diverse targets differ.

59. FRiP (Fraction of Reads in Peaks) scores reflect a peak caller understanding of S:N, where low FRiP scores identify reactions with a small number of called peaks and/or most reads outside the same. This could be due to multiple technical reasons (related to antibody, sample or protocol) as described through this protocol. However, target distribution and peak caller appropriateness/settings can have a dramatic impact on FRiP scores (even in a successful experiment), so it is important to examine browser tracks and bioinformatic analyses together while interrogating the data.

Acknowledgments

EpiCypher is supported by the National Institutes of Health through grants R44GM123869, R44DE029633, R44HG010640 and R43AI165019.

Conflict of Interest

The authors declare competing financial interests. EpiCypher is a commercial developer and supplier of reagents (e.g., antibodies, pAG-MNase) and platforms (e.g., CUTANA® CUT&RUN, SNAP-CUTANA™ spike-in nucleosomes) used in this study. All authors own shares in EpiCypher, with M-CK also a board member of the same.

References

1. Allis CD, Jenuwein T (2016) The molecular hallmarks of epigenetic control. Nat Rev Genet 17(8):487–500. https://doi.org/10.1038/nrg.2016.59
2. Ma Y, Kanakousaki K, Buttitta L (2015) How the cell cycle impacts chromatin architecture and influences cell fate. Front Genet 6:19. https://doi.org/10.3389/fgene.2015.00019
3. Flebbe H, Hamdan FH, Kari V et al (2019) Epigenome mapping identifies tumor-specific gene expression in primary rectal cancer. Cancers 11(8). https://doi.org/10.3390/cancers11081142
4. Noberini R, Restellini C, Savoia EO et al (2019) Profiling of epigenetic features in clinical samples reveals novel widespread changes in cancer.

Cancers 11(5). https://doi.org/10.3390/cancers11050723
5. Gao F, Ji G, Gao Z et al (2014) Direct ChIP-bisulfite sequencing reveals a role of H3K27me3 mediating aberrant hypermethylation of promoter CpG islands in cancer cells. Genomics 103(2-3):204–210. https://doi.org/10.1016/j.ygeno.2013.12.006
6. Walker CJ, Miranda MA, O'Hern MJ et al (2015) Patterns of CTCF and ZFHX3 mutation and associated outcomes in endometrial cancer. J Natl Cancer Inst 107(11). https://doi.org/10.1093/jnci/djv249
7. Gough SM, Lee F, Yang F et al (2014) NUP98-PHF23 is a chromatin-modifying oncoprotein that causes a wide array of leukemias sensitive to inhibition of PHD histone reader function. Cancer Discov 4(5):564–577. https://doi.org/10.1158/2159-8290.CD-13-0419
8. Lobo J, Henrique R, Jerónimo C (2018) The role of DNA/histone modifying enzymes and chromatin remodeling complexes in testicular germ cell tumors. Cancers 11(1). https://doi.org/10.3390/cancers11010006
9. Bai ZT, Bai B, Zhu J et al (2018) Epigenetic actions of environmental factors and promising drugs for cancer therapy. Oncol Lett 15(2):2049–2056. https://doi.org/10.3892/ol.2017.7597
10. Tracey R, Manikkam M, Guerrero-Bosagna C et al (2013) Hydrocarbons (jet fuel JP-8) induce epigenetic transgenerational inheritance of obesity, reproductive disease and sperm epimutations. Reprod Toxicol 36:104–116. https://doi.org/10.1016/j.reprotox.2012.11.011
11. Jimenez-Chillaron JC, Nijland MJ, Ascensao AA et al (2015) Back to the future: transgenerational transmission of xenobiotic-induced epigenetic remodeling. Epigenetics 10(4):259–273. https://doi.org/10.1080/15592294.2015.1020267
12. Zhang Y (2015) Detection of epigenetic aberrations in the development of hepatocellul. Methods Mol Biol. https://doi.org/10.1007/978-1-4939-1804-1_37
13. Lowe BR, Maxham LA, Hamey JJ et al (2019) Histone H3 mutations: an updated view of their role in chromatin deregulation and cancer. Cancers 11(5). https://doi.org/10.3390/cancers11050660
14. Sandoval J, Esteller M (2012) Cancer epigenomics: beyond genomics. Curr Opin Genet Dev 22(1):50–55. https://doi.org/10.1016/j.gde.2012.02.008
15. Weinstein JN, Collisson EA, Mills GB et al (2013) The Cancer Genome Atlas Pan-Cancer analysis project. Nat Genet 45(10):1113–1120. https://doi.org/10.1038/ng.2764
16. Zhao S, Allis CD, Wang GG (2021) The language of chromatin modification in human cancers. Nat Rev Cancer 21(7):413–430. https://doi.org/10.1038/s41568-021-00357-x
17. Terekhanova NV, Karpova A, Liang WW et al (2023) Epigenetic regulation during cancer transitions across 11 tumour types. Nature 623(7986):432–441. https://doi.org/10.1038/s41586-023-06682-5
18. Chan KM, Fang D, Gan H et al (2013) The histone H3.3K27M mutation in pediatric glioma reprograms H3K27 methylation and gene expression. Genes Dev 27(9):985–990. https://doi.org/10.1101/gad.217778.113
19. Corujo D, Buschbeck M (2018) Post-translational modifications of H2A histone variants and their role in cancer. Cancers 10(3). https://doi.org/10.3390/cancers10030059
20. Schmid M, Durussel T, Laemmli UK (2004) ChIC and ChEC; genomic mapping of chromatin proteins. Mol Cell 16(1):147–157. https://doi.org/10.1016/j.molcel.2004.09.007
21. Skene PJ, Henikoff S (2017) An efficient targeted nuclease strategy for high-resolution mapping of DNA binding sites. eLife 6:e21856. https://doi.org/10.7554/eLife.21856
22. Gillis NE, Boyd JR, Tomczak JA et al (2022) Thyroid hormone dependent transcriptional programming by TRbeta requires SWI/SNF chromatin remodelers. Nucleic Acids Res 50(3):1382–1395. https://doi.org/10.1093/nar/gkab1287
23. Asberry AM, Liu S, Nam HS et al (2022) Reprogramming landscape highlighted by dynamic transcriptomes in therapy-induced neuroendocrine differentiation. Comput Struct Biotechnol J 20:5873–5885. https://doi.org/10.1016/j.csbj.2022.10.031
24. Bagheri M, Mohamed GA, Mohamed Saleem MA et al (2024) Pharmacological induction of chromatin remodeling drives chemosensitization in triple-negative breast cancer. Cell Rep Med 5(4):101504. https://doi.org/10.1016/j.xcrm.2024.101504
25. Janssens DH, Wu SJ, Sarthy JF et al (2018) Automated in situ chromatin profiling efficiently resolves cell types and gene regulatory programs. Epigenet Chromatin 11(1):74. https://doi.org/10.1186/s13072-018-0243-8
26. Skene PJ, Henikoff JG, Henikoff S (2018) Targeted in situ genome-wide profiling with high efficiency for low cell numbers. Nat Protoc 13(5):1006–1019. https://doi.org/10.1038/nprot.2018.015
27. Meers MP, Bryson TD, Henikoff JG et al (2019) Improved CUT&RUN chromatin profiling tools. Elife 8. https://doi.org/10.7554/eLife.46314
28. Jung YL, Luquette LJ, Ho JW et al (2014) Impact of sequencing depth in ChIP-seq experiments. Nucleic Acids Res 42(9):e74. https://doi.org/10.1093/nar/gku178

29. Brand M, Rampalli S, Chaturvedi CP et al (2008) Analysis of epigenetic modifications of chromatin at specific gene loci by native chromatin immunoprecipitation of nucleosomes isolated using hydroxyapatite chromatography. Nat Protoc 3(3):398–409. https://doi.org/10.1038/nprot.2008.8
30. Lazar JE, Stehling-Sun S, Nandakumar V et al (2020) Global regulatory DNA potentiation by SMARCA4 propagates to selective gene expression programs via domain-level remodeling. Cell Rep 31(8):107676. https://doi.org/10.1016/j.celrep.2020.107676
31. Coelho DR, Palma FR, Paviani V et al (2023) SOCS1 regulates a subset of NFkappaB-target genes through direct chromatin binding and defines macrophage functional phenotypes. iScience 26(4):106442. https://doi.org/10.1016/j.isci.2023.106442
32. Musa RE, Lester KL, Quickstad G et al (2024) BRD4 binds to active cranial neural crest enhancers to regulate RUNX2 activity during osteoblast differentiation. Development 151(2). https://doi.org/10.1242/dev.202110
33. Thakur J, Henikoff S (2018) Unexpected conformational variations of the human centromeric chromatin complex. Genes Dev 32(1):20–25. https://doi.org/10.1101/gad.307736.117
34. Wolf BK, Zhao Y, McCray A et al (2023) Cooperation of chromatin remodeling SWI/SNF complex and pioneer factor AP-1 shapes 3D enhancer landscapes. Nat Struct Mol Biol 30(1):10–21. https://doi.org/10.1038/s41594-022-00880-x
35. Hainer SJ, Bošković A, McCannell KN et al (2019) Profiling of pluripotency factors in single cells and early embryos. Cell 177(5):1319–1329.e1311. https://doi.org/10.1016/j.cell.2019.03.014
36. Zheng XY, Gehring M (2019) Low-input chromatin profiling in Arabidopsis endosperm using CUT&RUN. Plant Reprod 32(1):63–75. https://doi.org/10.1007/s00497-018-00358-1
37. Patty BJ, Hainer SJ (2021) Transcription factor chromatin profiling genome-wide using uliCUT&RUN in single cells and individual blastocysts. Nat Protoc 16(5):2633–2666. https://doi.org/10.1038/s41596-021-00516-2
38. Brahma S, Henikoff S (2022) CUT&RUN profiling of the budding yeast epigenome. Methods Mol Biol 2477:129–147. https://doi.org/10.1007/978-1-0716-2257-5_9
39. Carpenter MD, Fischer DK, Zhang S et al (2022) Cell-type specific profiling of histone post-translational modifications in the adult mouse striatum. Nat Commun 13(1):7720. https://doi.org/10.1038/s41467-022-35384-1
40. Emerson FJ, Lee SS (2022) CUT&RUN for chromatin profiling in *Caenorhabditis elegans*. Curr Protoc 2(6):e445. https://doi.org/10.1002/cpz1.445
41. Qasim MN, Valle Arevalo A, Paropkari AD et al (2022) Genome-wide profiling of transcription factor-DNA binding interactions in *Candida albicans*: a comprehensive CUT&RUN method and data analysis workflow. J Vis Exp (182). https://doi.org/10.3791/63655
42. Buchert EM, Fogarty EA, Uyehara CM et al (2023) A tissue dissociation method for ATAC-seq and CUT&RUN in Drosophila pupal tissues. Fly 17(1):2209481. https://doi.org/10.1080/19336934.2023.2209481
43. Hartana CA, Lancien M, Gao C et al (2023) IL-15-dependent immune cross-talk between natural killer cells and dendritic cells in HIV-1 elite controllers. Cell Rep 42(12):113530. https://doi.org/10.1016/j.celrep.2023.113530
44. Morillo RC, Harris CT, Kennedy K et al (2023) Genome-wide profiling of histone modifications in *Plasmodium falciparum* using CUT&RUN. Life Sci Alliance 6(1). https://doi.org/10.26508/lsa.202201778
45. Armani-Tourret M, Gao C, Hartana CA et al (2024) Selection of epigenetically privileged HIV-1 proviruses during treatment with panobinostat and interferon-alpha2a. Cell 187(5):1238–1254.e1214. https://doi.org/10.1016/j.cell.2024.01.037
46. Gisch DL, Brennan M, Lake BB et al (2024) The chromatin landscape of healthy and injured cell types in the human kidney. Nat Commun 15(1):433. https://doi.org/10.1038/s41467-023-44467-6
47. Innis SM, Cabot RA (2024) Chromatin profiling and state predictions reveal insights into epigenetic regulation during early porcine development. Epigenet Chromatin 17(1):16. https://doi.org/10.1186/s13072-024-00542-w
48. Gilbert G, Renaud Y, Teste C et al (2024) Drosophila TET acts with PRC1 to activate gene expression independently of its catalytic activity. Sci Adv 10(18):eadn5861. https://doi.org/10.1126/sciadv.adn5861
49. Baysoy A, Seddu K, Salloum T et al (2023) The interweaved signatures of common-gamma-chain cytokines across immunologic lineages. J Exp Med 220(7). https://doi.org/10.1084/jem.20222052
50. Janssens D, Henikoff J (2019) CUT&RUN: Targeted in situ genome-wide profiling with high efficiency for low cell numbers. protocols.io. https://doi.org/10.17504/protocols.io.zcpf2vn
51. Ahmad K (2018) CUT&RUN with Drosophila tissues V.1. protocols.io. https://doi.org/10.17504/protocols.io.umfeu3n
52. Jagani Z, Chenail G, Xiang K et al (2019) In-depth characterization and validation in

BRG1-mutant lung cancers define novelcatalytic inhibitors of SWI/SNF chromatin remodeling. https://doi.org/10.1101/812628
53. Martin BJE, Ablondi EF, Goglia C et al (2023) Global identification of SWI/SNF targets reveals compensation by EP400. Cell 186(24):5290–5307 e5226. https://doi.org/10.1016/j.cell.2023.10.006
54. Xiao L, Parolia A, Qiao Y et al (2022) Targeting SWI/SNF ATPases in enhancer-addicted prostate cancer. Nature 601(7893):434–439. https://doi.org/10.1038/s41586-021-04246-z
55. Farnaby W, Koegl M, Roy MJ et al (2019) BAF complex vulnerabilities in cancer demonstrated via structure-based PROTAC design. Nat Chem Biol 15(7):672–680. https://doi.org/10.1038/s41589-019-0294-6
56. Baker M (2015) Reproducibility crisis: blame it on the antibodies. Nature 521(7552):274–276. https://doi.org/10.1038/521274a
57. Polakiewicz RD (2015) Antibodies: the solution is validation. Nature 518(7540):483. https://doi.org/10.1038/518483b
58. Bradbury A, Pluckthun A (2015) Reproducibility: standardize antibodies used in research. Nature 518(7537):27–29. https://doi.org/10.1038/518027a
59. Shah RN, Grzybowski AT, Cornett EM et al (2018) Examining the roles of H3K4 methylation states with systematically characterized antibodies. Mol Cell 72(1):162–177.e167. https://doi.org/10.1016/j.molcel.2018.08.015
60. Radzisheuskaya A, Shliaha PV, Grinev VV et al (2021) Complex-dependent histone acetyltransferase activity of KAT8 determines its role in transcription and cellular homeostasis. Mol Cell 81(8):1749–1765.e1748. https://doi.org/10.1016/j.molcel.2021.02.012
61. EpiCypher (2024) Chromatin antibodies. https://chromatinantibodies.com/. Accessed Sept 2024.
62. Baker M (2015) Antibody anarchy: a call to order. Nature 527(7579):545–551. https://doi.org/10.1038/527545a
63. Miyamoto Y, Ikeuchi M, Noguchi H et al (2018) Long-term cryopreservation ofhuman and other mammalian cells at -80 degrees C for 8 years. Cell Med 10:2155179017733148. https://doi.org/10.1177/2155179017733148
64. Liu D, Li Q, Luo J et al (2023) An SPRI beads-based DNA purification strategy for flexibility and cost-effectiveness. BMC Genom 24:125. https://doi.org/10.1186/s12864-023-09211-w
65. Blanco E, Di Croce L, Aranda S (2021) SpikChIP: a novel computational methodology to compare multiple ChIP-seq using spike-in chromatin. NAR Genom Bioinform 3(3):lqab064. https://doi.org/10.1093/nargab/lqab064
66. Meers MP, Tenenbaum D, Henikoff S (2019) Peak calling by sparse enrichment analysis for CUT&RUN chromatin profiling. Epigenet Chromatin 12(1):42. https://doi.org/10.1186/s13072-019-0287-4
67. Zhu Q, Liu N, Orkin SH et al (2019) CUT&RUNTools: a flexible pipeline for CUT&RUN processing and footprint analysis. Genome Biol 20(1):192. https://doi.org/10.1186/s13059-019-1802-4
68. Sunitha Kumary VUN, Venters BJ, Raman K et al (2024) Emerging approaches to profile accessible chromatin from formalin-fixed paraffin-embedded sections. Epigenomes 8(2):20. Published 2024 May 12. https://doi.org/10.3390/epigenomes8020020
69. Landt SG, Marinov GK, Kundaje A et al (2012) ChIP-seq guidelines and practices of the ENCODE and modENCODE consortia. Genome Res 22(9):1813–1831. https://doi.org/10.1101/gr.136184.111
70. Small EC, Maryanski DN, Rodriguez KL et al (2021) Chromatin immunoprecipitation (ChIP) to study DNA-protein interactions. Methods Mol Biol 2261:323–343. https://doi.org/10.1007/978-1-0716-1186-9_20
71. Sabbattini P, Canzonetta C, Sjoberg M et al (2007) A novel role for the Aurora B kinase in epigenetic marking of silent chromatin in differentiated postmitotic cells. EMBO J 26(22):4657–4669. https://doi.org/10.1038/sj.emboj.7601875
72. Marunde MR, Fuchs HA, Burg JM et al (2024) Nucleosome conformation dictates the histone code. Elife 13. https://doi.org/10.7554/eLife.78866
73. de Bock CE, Demeyer S, Degryse S et al (2018) HOXA9 cooperates with activated JAK/STAT signaling to drive leukemia development. Cancer Discov 8(5):616–631. https://doi.org/10.1158/2159-8290.CD-17-0583
74. Larsen SB, Cowley CJ, Sajjath SM et al (2021) Establishment, maintenance, and recall of inflammatory memory. Cell Stem Cell 28(10):1758–1774.e1758. https://doi.org/10.1016/j.stem.2021.07.001
75. Liu N, Hargreaves VV, Zhu Q et al (2018) Direct promoter repression by BCL11A controls the fetal to adult hemoglobin switch. Cell 173(2):430–442.e417. https://doi.org/10.1016/j.cell.2018.03.016
76. Miao W, Porter DF, Lopez-Pajares V et al (2023) Glucose dissociates DDX21 dimers to regulate mRNA splicing and tissue differentiation. Cell 186(1):80–97.e26. https://doi.org/10.1016/j.cell.2022.12.004

77. Uyehara CM, McKay DJ (2019) Direct and widespread role for the nuclear receptor EcR in mediating the response to ecdysone in Drosophila. Proc Natl Acad Sci USA 116(20):9893–9902. https://doi.org/10.1073/pnas.1900343116
78. Chou J, Provot S, Werb Z (2010) GATA3 in development and cancer differentiation: cells GATA have it!. J Cell Physiol 222(1):42–49. https://doi.org/10.1002/jcp.21943
79. Liu D, Li Q, Luo J et al (2023) An SPRI beads-based DNA purification strategy for flexibility and cost-effectiveness. BMC Genomics 24(1):125. https://doi.org/10.1186/s12864-023-09211-w
80. Beckman Coulter I (2024) SPRIselect bead protocol for size selection. https://www.beckman.com/reagents/genomic/cleanup-and-size-selection/size-selection/spriselect-protocol.
81. Rouillard AD, Gundersen GW, Fernandez NF et al (2016) The harmonizome: a collection of processed datasets gathered to serve and mine knowledge about genes and proteins. Database. https://doi.org/10.1093/database/baw100

Check for updates

Chapter 6

CUT&Tag for Chromatin Profiling in Cancer Tissue

Yang Liu and Paul A. Wade

Abstract

Cancer initiation and progression are intricately linked with abnormal epigenetic modifications. The detection of these epigenetic alterations throughout tumorigenesis is essential for early cancer diagnosis and effective treatment. Cleavage Under Targets and Tagmentation (CUT&Tag) is a relatively new method for profiling diverse chromatin components. Here, we describe the genome-wide profiling of histone modifications by CUT&Tag in cancer tissues. Lightly cross-linked nuclei are immobilized on beads and incubated with a specific primary antibody against histone modifications. Then a secondary antibody is added, followed by protein AG-Tn5 transposase fusion protein. Tn5 transposase is activated to cleave the accessible chromatin proximal to antibody-bound target loci. PCR amplification is conducted to generate the next-generation sequencing (NGS) libraries.

Keywords Histone modifications, Cancer tissue, Genomics, CUT&Tag, Epigenetics, Tagmentation

1 Introduction

Chromatin proteins regulate gene expression through interaction with DNA, and aberrant DNA–protein interactions are closely associated with cancer susceptibility and tumor growth [1]. Mapping of chromatin features genome-wide is most commonly carried out using chromatin immunoprecipitation and high-throughput sequencing (ChIP-seq). However, the effective implementation of ChIP-seq is challenged by the low signal-to-noise ratio, the experimental complexity, and the requirement for a larger amount of starting material, typically in the range of 1–10 million cells per immunoprecipitation [2, 3]. Thus, several methods based on targeting of enzymatic activities to specific loci have been developed, including DNA adenine methyltransferase identification (DamID) [4], chromatin endogenous cleavage and high-throughput sequencing (ChEC-seq) [5], and Cleavage Under Targets & Release Using Nuclease (CUT&RUN) [6]. These techniques facilitate the exploration of dynamic alterations in DNA–protein interactions during

Archana Dhasarathy and Motoki Takaku (eds.), *Cancer Epigenetics*, Methods in Molecular Biology, vol. 3005, https://doi.org/10.1007/978-1-0716-5072-1_6,

tumorigenesis and cancer treatment, thereby deepening our understanding of the molecular mechanisms underlying cancer development and therapy.

The CUT&Tag assay, which builds on CUT&RUN, is an innovative immunotethering strategy for chromatin profiling first reported in 2019 [7]. It relies on antibody-directed tagmentation by a pAG-Tn5 transposase fusion protein. A major advantage of CUT&Tag is that the adaptor ligation and cleavage are completed simultaneously in intact nuclei, which simplifies the preparation of sequencing libraries. Notably, CUT&Tag uses more stringent high-salt washes than CUT&RUN to minimize nonspecific cleavage of exposed DNA, which may destabilize the interactions between loosely-bound factors and chromatin [8]. Here, we describe the CUT&Tag assay, based on the previous protocol [8], for histone modification profiling using snap-frozen tissues. The primary antibody, the secondary antibody, and the pAG-Tn5 fusion protein are used for the selective tagmentation of antibody-bound chromatin in intact nuclei. NGS libraries are prepared by PCR amplification of target chromatin fragments with primers that recognize the adapter-ligated DNA.

2 Materials

2.1 Biological Materials

Collect normal and cancer tissues, flash freeze in liquid nitrogen and store at −80°C for later use or keep fresh tissue on ice for immediate homogenization.

2.2 Buffers

Prepare the following buffers with nuclease-free water.

1. Binding buffer (~120 μL per reaction): 20 mM HEPES-KOH pH 7.9, 10 mM KCl, 1 mM $CaCl_2$ and 1 mM $MnCl_2$. Store the buffer at 4°C up to 6 months.
2. Nuclei extraction buffer (1 mL per tissue sample): 20 mM HEPES-KOH pH 7.9, 10 mM KCl, 0.5 mM spermidine, 0.1% Triton-X100, 20% glycerol, and EDTA-free protease inhibitor tablet. Store the buffer at 4°C up to 3 months.
3. Wash buffer (~2 mL per tissue sample and 0.5 mL per reaction): 20 mM HEPES-NaOH pH 7.5, 150 mM NaCl, 0.5 mM spermidine, and EDTA-free protease inhibitor tablet. Store the buffer at 4°C for up to 1 month.
4. Antibody buffer (50 μL per reaction): 0.1% BSA, 2 mM EDTA in wash buffer (*see* **Note 1**).

5. High-salt wash buffer (450 μL per reaction): 20 mM HEPES-NaOH pH 7.5, 300 mM NaCl, 0.5 mM spermidine, and EDTA-free protease inhibitor tablet. Store at 4°C for up to 1 month.
6. Tagmentation buffer (100 μL per reaction): 10 mM $MgCl_2$ in high-salt wash buffer (*see* **Note 2**).

2.3 Reagents

1. Primary antibodies: Histone H3K27ac mouse monoclonal antibody and H3K27me3 rabbit polyclonal antibody
2. Secondary antibodies: Guinea pig polyclonal anti-rabbit IgG and Rabbit Anti-Mouse IgG
3. pAG-Tn5 fusion protein
4. Concanavalin A (ConA) beads
5. NEBNext HiFi 2× PCR Master mix
6. MinElute PCR Purification Kit
7. AMPure XP reagent
8. Phosphate-buffered saline solution (Calcium & Magnesium Free, PBS-CMF)
9. 16% Formaldehyde solution (Methanol-free)
10. Glycine (2.5 M)
11. 0.5 M EDTA
12. 10% SDS
13. Proteinase K (20 mg/mL)
14. Indexed primers (listed in Table 1)

2.4 Equipment

1. Qubit Fluorometer
2. TapeStation 4200
3. 0.2 mL 8-strip PCR tubes
4. 1.7 mL low-retention microcentrifuge tube
5. ThermoMixer
6. Magnetic stand
7. 1 mL Dounce tissue grinders
8. Hemacytometer and light microscope
9. Nutator
10. 70 μM cell strainer
11. PCR thermal cycler

Table 1
i5 and i7 barcoded primer sequences

Name	Barcode	Sequence (5′ to 3′)
i5.2	CTCTCTAT	AATGATACGGCGACCACCGAGATCTACACCTCTCTATTCGTCGGCAGCGTCAGATGTGTAT
i5.3	TATCCTCT	AATGATACGGCGACCACCGAGATCTACACTATCCTCTTCGTCGGCAGCGTCAGATGTGTAT
i5.5	GTAAGGAG	AATGATACGGCGACCACCGAGATCTACACGTAAGGAGTCGTCGGCAGCGTCAGATGTGTAT
i5.6	ACTGCATA	AATGATACGGCGACCACCGAGATCTACACACTGCATATCGTCGGCAGCGTCAGATGTGTAT
i7.1	TAAGGCGA1	CAAGCAGAAGACGGCATACGAGATTCGCCTTAGTCTCGTGGGCTCGGAGATGTG
i7.2	CGTACTAG	CAAGCAGAAGACGGCATACGAGATCTAGTACGGTCTCGTGGGCTCGGAGATGTG
i7.3	AGGCAGAA	CAAGCAGAAGACGGCATACGAGATTTCTGCCTGTCTCGTGGGCTCGGAGATGTG
i7.4	TCCTGAGC	CAAGCAGAAGACGGCATACGAGATGCTCAGGAGTCTCGTGGGCTCGGAGATGTG
i7.5	GGACTCCT	CAAGCAGAAGACGGCATACGAGATAGGAGTCCGTCTCGTGGGCTCGGAGATGTG
i7.6	TAGGCATG	CAAGCAGAAGACGGCATACGAGATCATGCCTAGTCTCGTGGGCTCGGAGATGTG
i7.7	CTCTCTAC	CAAGCAGAAGACGGCATACGAGATGTAGAGAGGTCTCGTGGGCTCGGAGATGTG
i7.10	CGAGGCTG	CAAGCAGAAGACGGCATACGAGATCAGCCTCGGTCTCGTGGGCTCGGAGATGTG
i7.11	AAGAGGCA	CAAGCAGAAGACGGCATACGAGATTGCCTCTTGTCTCGTGGGCTCGGAGATGTG
i7.12	GTAGAGGA	CAAGCAGAAGACGGCATACGAGATTCCTCTACGTCTCGTGGGCTCGGAGATGTG

3 Methods

3.1 ConA Beads Activation

1. Resuspend the ConA magnetic beads slurry and transfer $N \times$ 5 μL of beads into a 1.7 mL low-retention microcentrifuge tube ($N =$ the final sample number).
2. Add at least ten volumes of binding buffer and place the tube on a magnetic stand to allow the mixture to clear.
3. Withdraw all the liquid and remove the tube from the magnetic stand.
4. Resuspend the beads with at least ten volumes of binding buffer and place the tube on the magnetic stand to allow the mixture to clear.
5. Discard all the liquid and remove the tube from the magnetic stand.
6. Resuspend the beads in $(N + 1) \times 10$ μL of binding buffer and hold on ice until nuclei are ready.

3.2 Nuclei Preparation

1. Homogenize 10–25 mg of snap-frozen or fresh tissues in 1 mL of ice-cold nuclei extraction buffer on ice using a loose pestle for 20 strokes with a Dounce homogenizer and incubate on ice for 10 min (*see* **Note 3**).
2. Transfer homogenized samples into a 1.7 mL low-retention microcentrifuge tube and then centrifuge for 5 min at 600 × *g* at 4°C.
3. Remove the supernatant and gently resuspend the nuclei in 1 mL of PBS-CMF containing 0.1% formaldehyde, then incubate at room temperature for 2 min (*see* **Note 4**).
4. Add 30 μL of 2.5 M glycine to quench the formaldehyde and terminate the cross-linking.
5. Centrifuge for 5 min at 800 × *g* at 4°C and then remove the supernatant.
6. Resuspend nuclei in 1 mL of wash buffer and filter through a 70 μM cell strainer.
7. Centrifuge for 5 min at 800 × *g* at 4°C and then remove the supernatant.
8. Resuspend nuclei in wash buffer to a final concentration of ~1 million nuclei/mL. Count nuclei with a hemacytometer using a light microscope (*see* **Note 5**).

3.3 Binding of Nuclei to Activated ConA Beads

1. Transfer 20,000–50,000 nuclei per sample in 0.2 mL 8-strip PCR tubes (*see* **Note 6**).
2. Add 10 μL of activated ConA beads and gently vortex.

3. Mix the nuclei and beads on the nutator for 10 min at room temperature.
4. After a quick spin (~100 × *g*, 1 s, room temperature), place PCR tubes on a magnetic stand for 1 min and discard all the liquid.

3.4 Binding of Primary Antibody and Secondary Antibody

1. Remove the tubes from the magnetic stand and resuspend nuclei in 50 μL of antibody buffer containing target primary antibody (1:100) (*see* **Note 7**).
2. Mix by flicking the tubes and then incubate 8-strip PCR tubes on the nutator at 4°C overnight or at room temperature for 2 h.
3. After a quick spin (~100 × *g*, 1 s, room temperature), place 8-strip PCR tubes on a magnetic stand for 1 min and discard all the liquid.
4. Remove the 8-strip PCR tubes from the magnetic stand, and then resuspend nuclei in 50 μL of wash buffer containing secondary antibody (1:100).
5. Mix by flicking the tubes and incubate 8-strip PCR tubes on a nutator for 1 h at room temperature.
6. After a quick spin (~100 × *g*, 1 s, room temperature), place 8-strip PCR tubes on a magnetic stand for 1 min and discard all the liquid.
7. With 8-strip PCR tubes on the magnetic stand, add 200 μL of cold wash buffer onto the beads and then remove all the liquid (*see* **Note 8**).
8. Repeat step 7 for a total of two washes.

3.5 Binding of pAG-Tn5 Adaptor Complex

1. Remove the 8-strip PCR tubes from the magnetic stand and add 50 μL of high-salt wash buffer (*see* **Note 9**).
2. Add 2.5 μL of pAG-Tn5 (20× stock) to each sample and mix by flicking.
3. Incubate 8-strip PCR tubes on a nutator for 1 h at room temperature.
4. After a quick spin (~100 × *g*, 1 s, room temperature), place 8-strip PCR tubes on a magnetic stand for 1 min and discard all the liquid.
5. With 8-strip PCR tubes on the magnetic stand, add 200 μL of cold high-salt wash buffer onto the beads and then remove all the liquid.
6. Repeat step 5 for a total of two washes.

3.6 Chromatin Tagmentation and DNA Purification

1. Remove the 8-strip PCR tubes from the magnetic stand and add 100 μL of tagmentation buffer.
2. Mix by flicking and incubate the 8-strip PCR tubes at 37°C for 1 h on a Thermomixer at 1000 rpm.

3. Remove the 8-strip PCR tubes from ThermoMixer and subject samples to a quick spin (~100 × *g*, 1 s, room temperature).
4. Add 4 μL of 0.5 M EDTA, 2 μL of 10% SDS, and 1 μL of 20 mg/mL Proteinase K, and mix by flicking the 8-strip PCR tubes.
5. Incubate the 8-strip PCR tubes at 58°C for 1 h on the ThermoMixer at 1000 rpm.
6. Remove the 8-strip PCR tubes from ThermoMixer. After a quick spin (~100 × *g*, 1 s, room temperature), place the 8-strip PCR tubes on the magnetic stand for 1 min.
7. Transfer the liquid into the new microcentrifuge tubes and purify the DNA with Qiagen MinElute PCR purification Kit following the manufacturer's manual.
8. Elute the DNA with 24 μL of EB buffer in the MinElute PCR purification Kit.

3.7 PCR Amplification and Cleanup

1. Transfer 21 μL of eluted DNA to fresh 8-strip PCR tubes and add 2 μL of 10 μM barcoded i5 primer, 2 μL of 10 μM barcoded i7 primer, and 25 μL of NEBNext HiFi 2× PCR Master mix for each sample (Table 1, *see* **Note 10**).
2. Mix by vortexing, quick spin (~100 × *g*, 1 s, room temperature), and then perform PCR with the following conditions:
   a. 58°C for 5 min (Gap fill-in)
   b. 72°C for 5 min (Extension)
   c. 98°C for 45 s (DNA melting)
   d. 98°C for 15 s (DNA melting)
   e. 63°C for 10 s (Primer annealing and short extension)
   f. Repeat steps d, e for a total of 12–16 cycles (*see* **Note 11**)
   g. 72°C for 1 min (Final extension)
   h. Hold at 4°C
3. Remove the 8-strip PCR from the Thermocycler and add 1.2 volume (60 μL) of AMPure XP beads. Follow the manufacturer's recommendations for DNA purification.
4. Elute DNA in 15 μL nuclease-free water and quantify the purified PCR product with Qubit following the manufacturer's instructions (*see* **Note 12**).
5. Determine the size distribution of libraries by TapeStation. Typical TapeStation results for H3K27ac and H3K27me3 CUT&Tag sequencing libraries are shown in Fig. 1 (*see* **Note 13**).
6. Pool libraries and perform paired-end Illumina sequencing as per the manufacturer's recommendations.

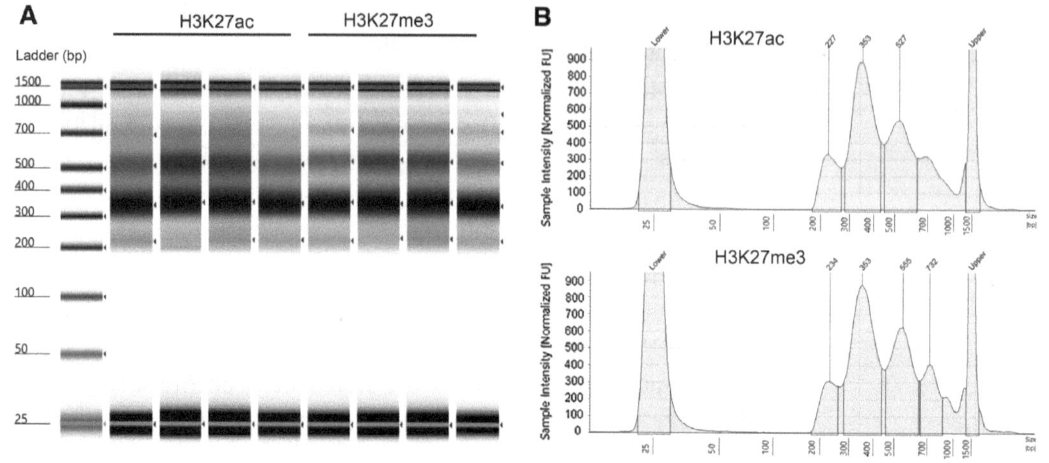

Fig. 1 TapeStation results for H3K27ac and H3K27me3 CUT&Tag libraries prepared with snap-frozen tissues. **A** Gel view of TapeStation results. **B** Electropherogram view of TapeStation results

4 Notes

1. Prepare the antibody buffer freshly each time.
2. Make tagmentation buffer fresh each time.
3. Nuclei can also be isolated using a commercial nuclei isolation kit (Invent Biotechnologies, SN-047). For samples rich in adipocytes, it is recommended to utilize the adipose tissue-specific kit (Invent Biotechnologies, AN-029) for nuclei isolation.
4. Make fresh 0.1% formaldehyde in PBS-CMF solution each time. Light cross-linking serves to maintain the integrity of nuclei and minimize clumping throughout the procedure.
5. Isolated nuclei may be transferred into cryogenic vials with 10% DMSO, then slow-frozen in Mr. Frosty container overnight at −80°C for long-term storage at −80°C.
6. The utilization of 8-strip PCR tubes can mitigate bead adhesion to the tube and facilitate a faster workflow with multichannel pipettes.
7. To evaluate the success of CUT&Tag, it is recommended to include a no-primary antibody negative control.
8. To prevent losing beads, withdraw the liquid carefully.
9. 300 mM NaCl is critical in the high-salt wash buffer to prevent the non-specific binding of pAG-Tn5 to exposed DNA.
10. Indexed primers were designed as previously reported [9] and purchased from IDT. Unique dual indexing allows us

to increase the number of samples sequenced per run and reduce per-sample cost.

11. PCR should be performed for at least 12 cycles to minimize the contribution of large DNA fragments.
12. The CUT&Tag assay does not have a "typical yield." However, comparing yields between positive controls (e.g., H3K27me3) and negative controls (e.g., no-primary antibody) can be informative. Generally, the yields for positive controls should exceed those of negative controls.
13. Predominant enrichment of mono-nucleosome size fragments (~330 bp, including sequencing adapters) is the best indicator of successful CUT&Tag assay prior to NGS.

Acknowledgments

We thank Dr. Yesenia Rodriguez at the NIEHS for critical reading of the manuscript. This work was supported by the Intramural Research Program of the NIEHS, NIH (Award ES101965 to P. A.W.).

References

1. Lu YJ, Chan YT, Tan HY, Li S, Wang N, Feng YB (2020) Epigenetic regulation in human cancer: the potential role of epi-drug in cancer therapy. Mol Cancer 19(1). https://doi.org/10.1186/s12943-020-01197-3
2. Policastro RA, Zentner GE (2018) Enzymatic methods for genome-wide profiling of protein binding sites. Brief Funct Genomics 17(2):138–145. https://doi.org/10.1093/bfgp/elx030
3. Akdogan-Ozdilek B, Duval KL, Meng FW, Murphy PJ, Goll MG (2022) Identification of chromatin states during zebrafish gastrulation using CUT&RUN and CUT&Tag. Dev Dyn 251(4):729–742. https://doi.org/10.1002/dvdy.430
4. van Steensel B, Delrow J, Henikoff S (2001) Chromatin profiling using targeted DNA adenine methyltransferase. Nat Genet 27(3):304–308. https://doi.org/10.1038/85871
5. Zentner GE, Kasinathan S, Xin B, Rohs R, Henikoff S (2015) ChEC-seq kinetics discriminates transcription factor binding sites by DNA sequence and shape in vivo. Nat Commun 6:8733. https://doi.org/10.1038/ncomms9733
6. Skene PJ, Henikoff S (2017) An efficient targeted nuclease strategy for high-resolution mapping of DNA binding sites. Elife 6. https://doi.org/10.7554/eLife.21856
7. Kaya-Okur HS, Wu SJ, Codomo CA, Pledgers ES, Bryson TD, Henikoff JG, Ahmad K, Henikoff S (2019) CUT&Tag for efficient epigenomic profiling of small samples and single cells. Nat Commun 10. https://doi.org/10.1038/s41467-019-09982-5
8. Kaya-Okur HS, Janssens DH, Henikoff JG, Ahmad K, Henikoff S (2020) Efficient low-cost chromatin profiling with CUT&Tag. Nat Protoc 15(10):3264–3283. https://doi.org/10.1038/s41596-020-0373-x
9. Buenostro JD, Wu BJ, Litzenburger UM, Ruff D, Gonzales ML, Snyder MP, Chang HY, Greenleaf WJ (2015) Single-cell chromatin accessibility reveals principles of regulatory variation. Nature 523(7561):486–U264. https://doi.org/10.1038/nature14590

Part III

Chromatin Accessibility and Transposable Element Analysis

Check for updates

Chapter 7

Assay for Transposase-Accessible Chromatin Sequencing (ATAC-seq) of Cancer Cells in Culture

Kaitlyn Gura, Trevor Long, Motoki Takaku, Archana Dhasarathy and Manu

Abstract

The Assay for Transposase-Accessible Chromatin followed by sequencing (ATAC-seq) is a powerful and widely-used technique for chromatin accessibility profiling due to its low input requirement, simple protocol, high sensitivity, and ability to study single cells. ATAC-seq is a powerful tool in cancer epigenetics research that has enabled the scrutiny of gene regulatory mechanisms, aided the identification of therapeutic targets, and in the exploration of anticancer immunity, thereby advancing our understanding of cancer biology. The protocol includes five main steps: cell isolation and nuclei preparation, Tn5 tagmentation, PCR amplification, library sequencing, and data analysis. Here, we have outlined a comprehensive ATAC-seq protocol that is utilized by our laboratories, along with detailed troubleshooting steps at each point, in the hopes that it will aid other researchers in utilizing this tool to the fullest capability. We also describe bioinformatics pipelines for processing the raw sequence data and identifying regulatory regions by differential accessibility analysis, as well as more advanced methods for the visualization of transcription factor binding sites.

Keywords Chromatin accessibility, Tagmentation, Tn5 transposase, ATAC-seq, Cancer, Transcription factor, *Cis*-regulation

1 Introduction

The textbook view of chromatin, the complex of DNA and proteins in the nuclei of eukaryotic cells, is that of open (euchromatin) vs. closed (heterochromatin), each with its own set of histone modifications [1–7]. This binary organization has been linked to function, with open chromatin states being more accessible to cellular machinery involved in gene expression, DNA replication, etc. This concept has been challenged in recent years due to the discovery of condensed

These authors contributed equally: Kaitlyn Gura, Trevor Long

Archana Dhasarathy and Motoki Takaku (eds.), *Cancer Epigenetics*, Methods in Molecular Biology, vol. 3005,
https://doi.org/10.1007/978-1-0716-5072-1_7,

domains even within euchromatin [8], and an MNase study that showed euchromatin has similar DNA accessibility as heterochromatin [9]. Nevertheless, several assays have sought to define chromatin accessibility to probe the genomic landscape, and how it is altered in disease states. To this end, applications such as DNase I hypersensitive sites sequencing (DNase-seq) [10], Formaldehyde-Assisted Isolation of Regulatory Elements sequencing (FAIRE-seq) [11], Chromatin Immunoprecipitation sequencing (ChIP-seq) [12, 13], and Micrococcal Nuclease sequencing (MNase-seq) [14, 15] were developed. Most if not all of these methods required large input volume of cells, and are complex and time-intensive.

The development of the Assay for Transposase-Accessible Chromatin followed by sequencing (ATAC-seq) [16, 17] has in the few short years since it was developed been widely adopted for chromatin accessibility studies for a number of reasons. Unlike other methods of probing the chromatin landscape, ATAC-seq is simpler, faster, requires no prior knowledge of DNA-bound proteins or epigenetic marks, and uses a relatively small number of input cells, allowing for the study of rare or limited cell types [18]. More recently, the adaptation of single cell [19, 20] and spatial [21] technologies to ATAC-seq brings the added advantage of studying accessibility at the level of single cells or tissue level. Further, several groups have successfully adapted this method for use with frozen or cryopreserved tissue [22–24], which enables querying chromatin accessibility of patient samples possible as well.

Some caveats to keep in mind for ATAC-seq include mitochondrial DNA contamination, Tn5 sequence bias, and potential for technical variability and batch effects, due to low number of cells used relative to the other methods [25]. Normalization methods can also affect ATAC-seq analysis and can introduce bias in interpretations of data [26]. Further, while ATAC-seq can identify open chromatin regions, integration with other data (ChIP-seq and/or RNA-seq) is required to determine the type of regulatory element (enhancer, silencer, promoter) present in those regions. That being said, deep-sequencing coupled with computational modeling methods has proven invaluable for prediction of transcription factor (TF) binding sites [27–31].

Despite the ATAC-seq workflow being relatively simple, it is a powerful tool that enables researchers to probe both nucleosome structure and transcription factor binding, in addition to detection of accessibility changes. Here, we will outline the fundamental ATAC-seq protocol (Fig. 1) including key troubleshooting tips and tricks. Although we are presenting the protocol as it applies to cancer cells in culture, the main points here are broadly applicable to and can be modified for cancer tissues as well as techniques such as single-cell and spatial ATAC-seq. We believe this protocol has the potential to aid other researchers in achieving a heightened level of precision and

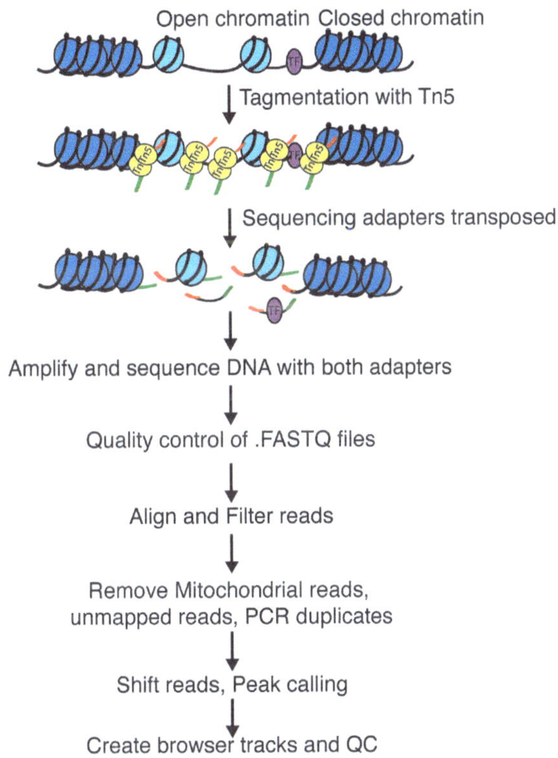

Fig. 1 Overview of ATAC-seq protocol

accuracy in their ATAC-seq techniques, and ultimately developing a deeper understanding of the complex mechanisms underlying the development and progression of cancer.

2 Materials

1. Cell culture reagents: Growth media specific for cells (e.g., DMEM/F-12 with 10% fetal bovine serum) and for spike-in control cells such as *Drosophila* S2 cells, if using; Trypsin-EDTA; Phosphate-buffered saline (calcium-magnesium free, sterile). Any additional supplements/antibiotics as needed.
2. Cell lysis: Prepare enough lysis buffer to add 1 mL to each sample consisting of one million cells. *See* **Note 1**.
3. Tn5 tagmentation: Tn5 Transposase and Tn5 buffer (Diagenode or other source).
4. DNA purification: 3M sodium acetate (pH 5.2) and Qiagen MinElute PCR purification kit (or similar).
5. PCR amplification: Sequencing primer indices (from Diagenode in this example), SYBR Gold (diluted 1000×), NEB 2× PCR master mix, and nuclease-free water.

6. DNA purification: SPRI (AMPure) beads, 80% ethanol, nuclease-free water, or elution buffer (10 mM Tris-HCl pH 8.0).

3 Methods

3.1 Cell Culture

1. Grow cancer cells in media suitable for the cancer cell type to around 80% confluence. Cells should be actively dividing for best results.
2. Trypsinize and isolate roughly 1×10^6 cells per sample. Centrifuge cells in 15 mL tube for 3 min at 300 × *g* in a swinging bucket rotor to minimize cell loss. Wash twice with 5 mL cold PBS (Calcium/Magnesium-free). *See* **Note 2**.
3. Add spike-in controls at this step, if using. *See* **Note 3**.

3.2 Cell Lysis

1. Add 1 mL of cell lysis buffer optimized to cell type of interest and incubate on ice for about 5 min.
2. Check nuclei under microscope by adding 5 μL of trypan blue to 5 μL of cell lysate on a glass slide and mixing gently by rotating the slide. Nuclei should appear to be intact and in good condition. *See* **Note 1** and Fig. 2.
3. Count and transfer 25 μL or 2.5×10^4 cells to a new 1.7 mL tube. If starting with fewer than 1 million cells, transfer appropriate volume corresponding to 2.5×10^4 cells.
4. Centrifuge at 500 × *g* at 4°C for 5 min and carefully discard supernatant. Leave ~3–5 μL of buffer in the tube to avoid cell loss.

3.3 Tn5 Tagmentation

1. Immediately resuspend pellet in 25 μL of Transposase solution. *See* **Note 4** and Fig. 3.
2. Incubate at 37°C for 30 min, while gently mixing solution by tapping the tubes with a finger every 10 min, or place in a heated tube mixer set to 37°C.

3.4 DNA Purification by MinElute PCR Purification Kit

1. Add 5 μL of 3M sodium acetate (pH 5.2) to each sample.
2. Immediately add 125 μL of Buffer PB (Qiagen MinElute kit) and mix well.
3. Place sample in a MinElute column in a 2 mL collection tube and centrifuge for 1 min at full speed (17,000 × *g*) in a mini-centrifuge at room temperature.
4. Discard flow-through and place column back in the same collection tube.
5. Add 700 μL Buffer PE (Qiagen MinElute kit) to the column and centrifuge for 1 min at 17,000 × *g* at room temperature.

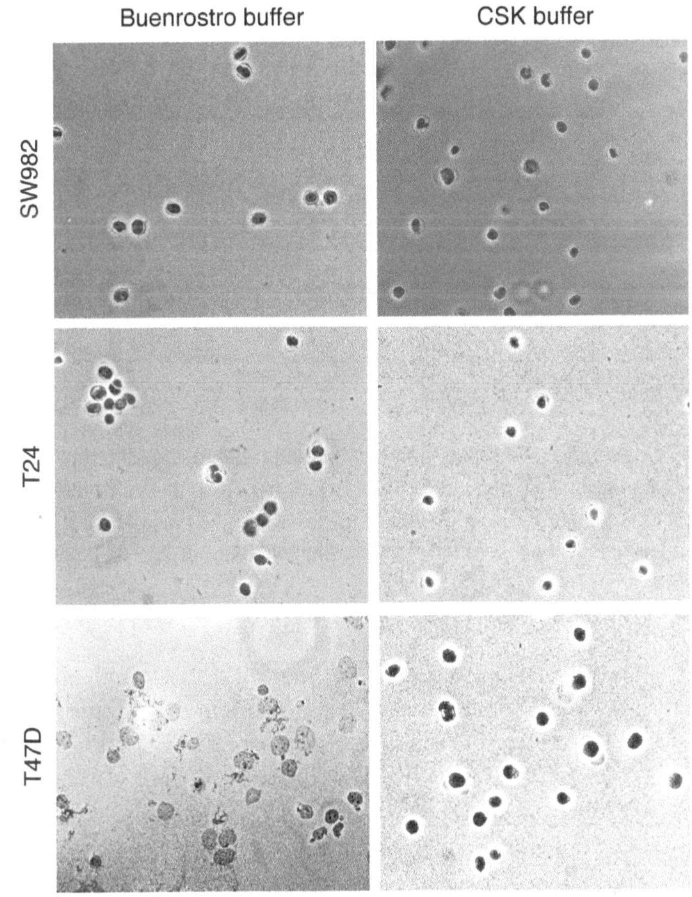

Fig. 2 Cell lysis. Human synovial sarcoma cells (SW982), urinary bladder epithelial transitional carcinoma cells (T24), and human breast cancer cells (T47D) were resuspended in either the original ATAC-seq (Buenrostro et al.) buffer (left panels) or cytoskeletal (CSK) buffer (right panels) for 5 min. Nuclei were imaged by addition of 4 μL of Trypan blue to 4 μL of cell resuspension, placed on a glass slide, and observed under a phase-contrast microscope at 20×. SW982 nuclei are visible with both buffers. T24 nuclei appear to be better with CSK (we can still observe whole cells in Buenrostro buffer after 5 min). T47D nuclei appear damaged with use of Buenrostro buffer but are intact after treatment with CSK buffer

6. Discard flow-through and place column back in the same collection tube.
7. Centrifuge column for 5 min to dry the membrane completely.
8. Discard flow-through and place column in a new 1.7 mL microcentrifuge tube.
9. Elute DNA fragments with 15 μL Buffer EB.
10. Let the column stand for 2–3 min at room temperature.
11. Centrifuge for 1 min at 10,000 × *g* at room temperature to elute DNA. At this point the DNA can be stored at −20°C.

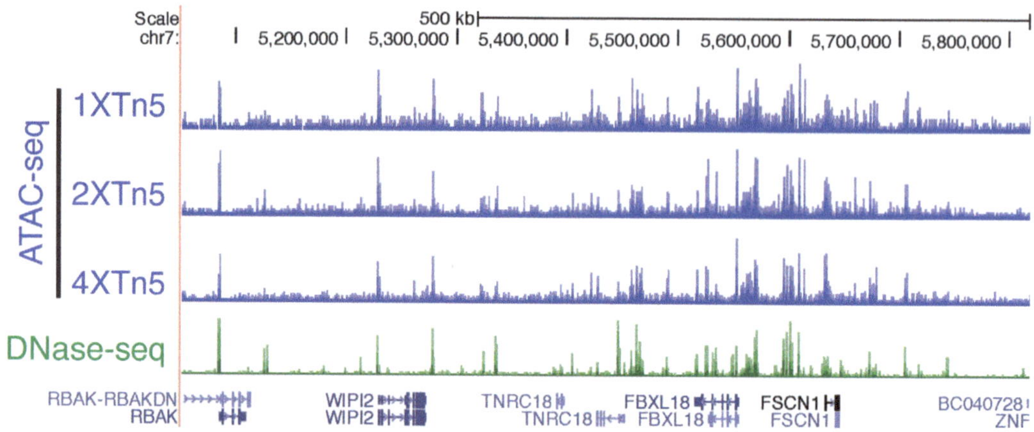

Fig. 3 Titration of Tn5 enzyme. ATAC-seq data were generated using different Tn5 concentrations in MDA-MB-231 cells. Specifically, 1.25 μL (1×), 2.5 μL (2×), or 5 μL (4×) of Tn5 Transposase was added to the tagmentation solution for a total of 25 μL, with nuclease-free water used to adjust the final volume. DNase-seq track is shown as a reference. The background signal is slightly higher in 1× Tn5 condition compared to 2× or 4× Tn5

3.5 PCR Amplification

1. To amplify the tagmented DNA, mix eluted DNA (15 μL), 1.25 μM final of each sequencing adapter (variable- e.g., use 1 μL if using Diagenode kit), NEB 2× PCR master mix (25 μL) and q.s. to 50 μL with nuclease-free water.
2. Prepare PCR mix and amplify in a standard thermal cycler under the following conditions:

```
Step 1: 72°C for 5min,
Step 2: 98°C for 30s,
Step 3: 98°C for 10s,
Step 4: 63°C for 30s,
Step 5: 72°C for 1min.
Repeat Steps 3-5 four times (not more than 5 total cycles)
Hold at 4°C.
```

3. Add 5 μL of the PCR amplified DNA to 5μL NEB 2× PCR master mix, 0.4 μM of each primer index (optional), 0.75 μL diluted SYBR gold, and q.s. to 15 μL with nuclease-free water. Keep the remaining 45 μL PCR on ice for use later in the protocol. The precise cycle number of cycles for library amplification will be determined by RT-qPCR (*see* **Note 5** and Fig. 4).
4. Amplify using a real-time PCR machine under the following conditions:

```
Step 1: 98°C for 30s
Step 2: 98°C for 10s,
Step 3: 63°C for 30s,
Step 4: 72°C for 1min.
Plate read.
```

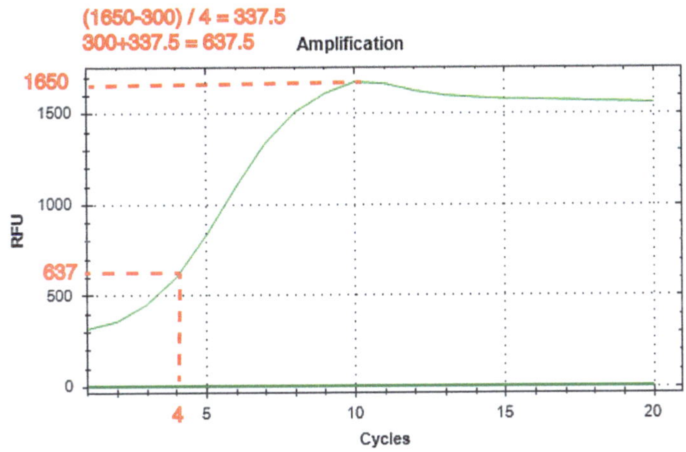

Fig. 4 Determination of PCR amplification cycles. Real-time qPCR of DNA product from initial qPCR amplification is used to calculate the number of cycles that correspond to 1/4 the maximum/saturated fluorescence intensity. Here the maximum fluorescent intensity was 1650 RFUs, but the amplification curve starts at 300. So (1650 – 300)/4 = 337.5 RFUs. If we add 300–337.5, this gives us 637.5, which corresponds to 4 more cycles to further amplify the library DNA before sequencing

```
Repeat Steps 2-4 19 times (20 total cycles)
Hold at 4°C.
```

5. Calculate the number of cycles corresponding to 1/4 the maximum saturated fluorescence intensity to determine the additional number of cycles needed [Buenrostro et al. (2015) Buenrostro, Wu, Chang, and Greenleaf]. *See* **Note 5** and Fig. 4.
6. Perform PCR amplification of the remaining 45 μL PCR product from the initial amplification with the calculated number of cycles.

```
Step 1: 98°C for 30s,
Step 2: 98°C for 10s,
Step 3: 63°C for 30s,
Step 4: 72°C for 1min.
Repeat Steps 2-5 (using the calculated number of
cycles from Section 3.5.5.)
Hold at 4°C.
```

3.6 DNA Purification

1. Transfer the PCR products to 8-well strip tubes or similar 0.2 mL tubes, and add 150 μL of SPRI magnetic (e.g., AMPure) beads.
2. Incubate for 15 min at room temperature.
3. Place on a magnetic stand for 5 min to allow the magnetic beads to adhere to the wall of the tube closest to the magnet, then carefully remove the supernatant.

4. Wash beads with 200 μL of freshly made 80% ethanol twice, then remove ethanol completely.
5. Briefly spin the tubes and remove any remnant ethanol with a thin pipette tip (P10 or P20). Let samples air dry at room temperature for about 30 s.
6. Elute the DNA from the beads by adding 30–50 μL of nuclease-free water or elution buffer (10 mM Tris-HCl pH 8.0). Leave tubes on magnetic stand and carefully transfer supernatant to new Eppendorf tubes. Repeat AMPure bead purification as needed. *See* **Note 6**.
7. Measure DNA concentration using the Qubit dsDNA HS Assay Kit or similar method that uses DNA intercalation. Check quality of library using gel electrophoresis/Tapestation/Bioanalyzer or similar method. *See* **Note 7** and Fig. 5.
8. Samples are now ready for sequencing. Standard ATAC-sequencing for mammalian cells is usually performed at about 40–50 million reads per sample [31]. However, deeper sequencing is required for TF footprinting (100–200 million reads or more) [30]. See Fig. 6 for examples of browser tracks.

3.7 Data Processing and Analysis Software

The data processing and analysis workflow steps up to identifying accessible genomic regions ("peak calling") utilize a heterogeneous

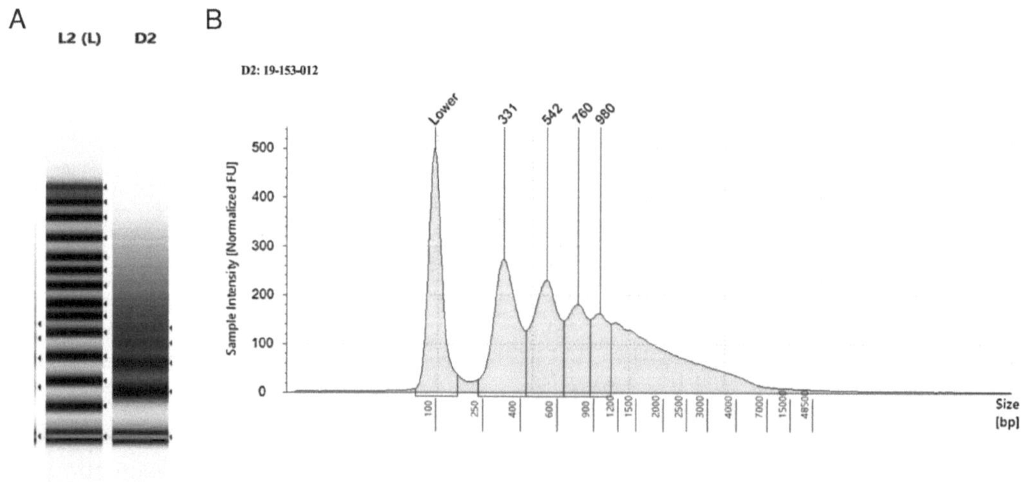

Fig. 5 ATAC-seq library quality assessment. (**A**) Gel image and (**B**) Electropherogram generated by the Agilent TapeStation for representative ATAC-seq libraries. For (B), the x-axis represents the fragment size in base pairs (bp), and the y-axis indicates the fluorescence intensity, which corresponds to the quantity of DNA fragments of each size. The nucleosomal ladder and distinct peaks indicate high-quality libraries with minimal adapter dimer contamination. The presence of a broad range of fragment sizes reflects successful tagmentation and library preparation

mixture of Unix software while subsequent steps are executed in the **R** software environment for statistical computing. In our implementation [27], the first phase of the workflow is automated using a workflow management system called `snakemake`. The command-line tools are installed into a compartmentalized environment using `conda`, an open-source package and environment manager, and mamba, a C++ based command-line interface (CLI) tool for conda. The following software is utilized in the workflow.

1. Linux/Terminal emulator: All the steps described here require the Linux operating system and a terminal emulator such as gnome-terminal or xterm. Any relatively recent Linux distribution, such as Ubuntu 22.04 LTS (Linux kernel 5.15) or newer would suffice.
2. Samtools [32]: Used to view and manipulate Sequence Alignment/Map (SAM) files that store information about the alignment of sequencing reads to a reference genome.
3. Picard Tools [33]: Used for manipulating and analyzing sequencing data.
4. FastQC: For quality control of next-generation sequencing (NGS) data.
5. Trimmomatic [34]: Used to remove sequencing adapter sequences ("trimming") and filter reads.
6. Hisat2 [35]: Aligns/maps NGS sequencing reads to a reference genome.
7. Macs2 [36]: Identifies genomic regions statistically enriched in signal either relative to background or to a control (peak calling).

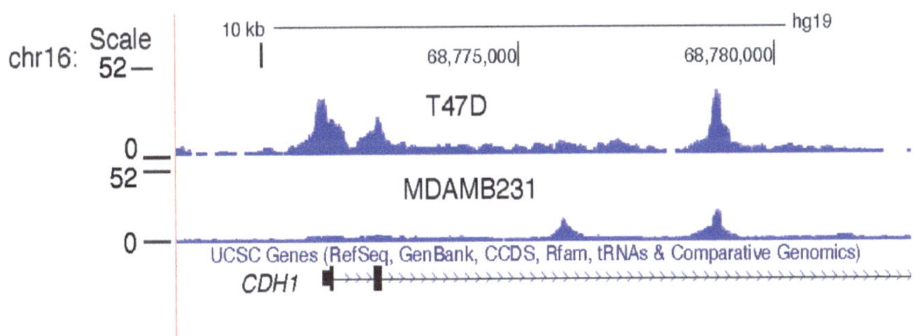

Fig. 6 Example tracks uploaded to the UCSC genome browser. Following ATAC-seq data analysis, tracks for T47D (luminal breast cancer cells) and MDA-MB-231 (basal breast cancer cells) were uploaded to the UCSC genome browser. Clear peaks of accessibility are visible over the E-cadherin (*CDH1*) gene promoter in the T47D luminal cells, which express this gene, while no peaks are visible in the MDA-MB-231 cell line, which possesses a highly repressed *CDH1* gene

8. bedGraphToBigWig: Converts the text *bedGraph* format files created by macs2 to the indexed, compressed, and binary *bigWig* format for efficient storage and access.
9. Conda: An environment manager used for tool compartmentalization.
10. Mamba: A command-line interpreter (CLI) tool to manage conda environments.
11. **R**: An open source programming language used for data analysis and visualization.
12. Rstudio: An **R**-based graphical user interface (GUI).
13. Snakemake [37]: A Python-based workflow management system.

3.8 Workflow

This protocol describes the workflow, also referred to as a *pipeline*, of ATAC-seq data processing, downstream analysis, and visualization. Here, we instruct the user through the basic usage of a snakemake workflow written to batch process ATAC-seq read data on the Linux command line. All the steps described below are to be executed on the command line in a terminal emulator such as `gnome-terminal` or `xterm`.

3.9 Obtaining the Snakemake Workflow, R Scripts, and Example Data

The `snakemake` workflow, **R** scripts, and example data used in this ATAC-seq analysis walk-through may be downloaded from https://github.com/mlekkha/ATACSeqWorkFlow by clicking the Code dropdown menu and choosing "Download ZIP". Uncompressing the ZIP file will yield a directory, `ATACSeqWorkFlow-master`, which can be placed in a location of the user's choosing. *The code below assumes that the directory is located in the user's home directory. In the commands below, a → indicates a continuation of the line without any white space.*

3.10 Conda and Mamba Installation

1. Download and install conda and mamba.

```
wget "https://github.com/conda-forge/mini-
forge/releases/latest/download/Mini-
forge3-$(uname)-$(uname -m).sh"
bash Miniforge3-$(uname)-$(uname -m).sh
```

2. Answer "yes" to the question

```
Do you wish to update your shell profile to automati-
cally initialize conda?
```

to allow the installer to modify the shell environment configuration files. To initialize `conda`, exit the terminal with the `exit` command and open a new one. More information and detailed installation instructions may be found at https://github.com/conda-forge/miniforge/.

3.11 Creating a Conda Environment for ATAC-seq Data Processing

1. The conda environment, which is a list of all the packages required for the pipeline, has been specified in a *yet another markup language* (YAML) file located in the Files directory of the repository. This step needs to be run only once.

```
conda activate
cd ~/ATACSeqWorkFlow-master/Files/
mamba env create --file condaenv.yaml
```

2. Activate the ATAC_Analysis environment. This must be repeated each time a new terminal is launched in order to execute the workflow or interact with the software tools.

```
conda activate ATAC_Analysis
```

3.12 Downloading the Hisat2 Genomic Index

The sequence alignment steps of the example workflows require the indexed human reference genome sequence. Download and unzip the Hisat2 index for the hg38 release of the human reference genome. This step needs to be run only once. *The workflow assumes that the genome index is located in the* genome_index *subdirectory.*

```
cd ~/ATACSeqWorkFlow-master/genome_index/
wget  https://genome-idx.s3.amazonaws.com/hisat/
hg38_genome.tar.gz
tar -xvf hg38_genome.tar.gz
rm hg38_genome.tar.gz
```

3.13 Obtaining the Trimmomatic Adapters

For adapter trimming, trimmomatic requires a list of adapter sequences. These sequences are included with trimmomatic. Copy the adapter sequence files into the adapters subdirectory for ease of use. This step needs to be run only once.

```
cd ~/ATACSeqWorkFlow-master/adapters/
cp ~/miniforge3/envs/ATAC_Analysis/share/trimmo-
matic/adapters/* .
```

3.14 Running the Workflow

Note: all the commands below have been written assuming that the repository directory is the current working directory. This can be accomplished by changing the working directory.

```
cd ~/ATACSeqWorkFlow-master/
```

3.14.1 The Configuration File

The first step of running the workflow is to write a configuration file that specifies details such as input file names and the location of the reference genome (Table 1) in the YAML format. We have provided two datasets and corresponding configuration files as examples.

3.14.2 Small Test Dataset

This dataset has only 10^5 reads and should run in a few minutes. It allows users to quickly test whether the repository has been set up correctly. The test subdirectory includes the data files and the corresponding configuration file.

Table 1
Snakemake configuration file variables

Variable	Description	Type
indir	<path_to_sequence_files>	String
samples	Define the sample names, which are used to name output files, and the names of files containing the read 1 and read 2 sequences. The sequence file names should not include the extension, which is defined separately.	String
file_extension	The sequence file extension.	String
threads_used	The number of threads to use in parallel computing. This should not exceed the total cores of the user's system.	Integer
trim_phred_encoding	The ASCII encoding of the base quality phred scores in the sequence file. phred33 encoding is the most common.	String
trim_steps	The parameters utilized by trimmomatic.	String
hisat2_index	<path_to_hisat2Index>	String
hisat2_args	Arguments passed to hisat2.	String
sorting_memory	The maximum memory utilized per thread by samtools while sorting alignments.	String
macs2_args	Arguments passed to macs2 callpeak.	String
chromInfo	<path_to_chromSizesFile>	String

1. Change current working directory to `test`.

```
cd test
```

2. Create a subdirectory in the repository directory to hold the intermediate and final outputs of the workflow and change the working directory to it.

```
mkdir test_out
cd test_out
```

3. Copy the provided test data configuration file to the output directory.

```
cp ../Files/config-test.yaml.
```

4. Execute the workflow.

```
snakemake --snakefile ~/ATACSeqWorkFlow-master/
Scripts/ATACSeq_Preprocessing_Snakefile.yaml
--configfile config-test.yaml --cores 1
```

3.14.3 Breast Cancer Dataset

ATAC-seq of MDA-MB-231 cells overexpressing either GFP (control) or GATA3 (accession number GSE72141) [38]. Process this dataset to reproduce the analyses presented in this chapter. The dataset is ~42GB in size and the intermediate and final output files require an additional ~200GB of storage.

1. Change working directory to the repository.

```
cd ~/ATACSeqWorkFlow-master/
```

2. Download the dataset. Run the `download-exampleData.sh` shell script provided in the `example/Data` directory. The sample SRR numbers are provided in the accompanying `Sample_Acc_List.txt` file. *Note: The download may take several hours depending on the internet speed.*

```
cd example/Data
nice -n 19 ./download-exampleData.sh Sample_Acc_
List.txt
```

3. Create a subdirectory in `example` to hold the intermediate and final outputs of the workflow and change the working directory to it.

```
cd ..
mkdir bc_out
cd bc_out
```

4. Copy the provided breast cancer data configuration file to the output directory.

```
cp ../Files/config-bc.yaml .
```

5. Execute the workflow. *See* **Note 8** to change the number of CPU cores the workflow runs on. *See* **Note 9** to run the workflow on only a subset of the samples. *See* **Note 10** to rerun the workflow on a previously processed dataset.

```
snakemake --snakefile ../../Scripts/ATACSeq_Pre-
processing_Snakefile.yaml --configfile config-bc.
yaml --cores 1
```

3.14.4 User Datasets

Our `snakemake` workflow can also be used to process user datasets after customizing the configuration file appropriately.

1. Change working directory to the repository.

```
cd ~/ATACSeqWorkFlow-master/
```

2. Create directory to hold the user project and its subdirectory to hold the user dataset. Change working directory to the user project data directory.

```
mkdir -p user_project/Data
cd user_project/Data
```

3. Copy the user dataset into the `Data` subdirectory of the project directory.
4. Create a subdirectory in `user_project` to hold the intermediate and final outputs of the workflow and change the working directory to it.

```
cd ..
mkdir up_out
cd up_out
```

5. A template for the configuration file has been provided in the `Files` subdirectory. Copy the template to the output directory.

```
cp ../../Files/config-atacseq.yaml .
```

6. Edit the configuration file using a text editor to adapt it to the user dataset. While many of the entries can be copied from the breast cancer dataset configuration file, the `samples` section, which specifies the input and output file names, must be edited.

```
[Format of the samples section of the configura-
tion file]
samples:
      <sample1name>:
          read1: <read1filename>
          read2: <read2filename>
      <sample2name>:
          read1: <read1filename>
          read2: <read2filename>
```

Above, `<read1filename>` and `<read2filename>` should be replaced with the names of the read1 and read2 files *without their extension*. `<sample1name>`, `<sample2name>` etc. should be replaced with a descriptive name of the sample. The intermediate and final output files generated by the workflow are named using the sample names specified here. For example, the `bigWig` files used to plot ATAC-seq signal are named `<sample1name>.bw`, `<sample2name>.bw`, and so on.

7. Run the workflow.

```
snakemake --snakefile ../../Scripts/ATACSeq_
Preprocessing_Snakefile.yaml --configfile con-
fig-atacseq.yaml --cores 1
```

Table 2
Outputs of the snakemake workflow

Directory	Generated by	Description
initialQC	FastQC	Raw read QC
trimmed	Trimmomatic	Trim statistics
trimmedQC	FactQ	Trimmed read QC
marked_reads	Picard Tools	Alignments annotated with PCR and optical duplicates
peaks/<sample_name>_peaks	macs2	Information about regions enriched in mapped reads compared to background ("peaks")

3.15 Outputs of the Snakemake Workflow

Successful completion of the snakemake pipeline results in the creation of new subdirectories in the output directory (Table 2) containing the intermediate and final outputs which may be used in downstream analysis.

3.16 Downstream Analyses

Analysis downstream of peak calling depend on the goals of the study. Here we provide instructions and include software for (1) identifying regions where the accessibility changes between two samples in a statistically significant manner ("differential accessibility analysis"), (2) determining the distribution of accessible regions according to their genomic location ("peak annotation"), (3) visualization of the fine-scale accessibility profile within accessible regions, (4) identification of TFs potentially binding protected regions, and (5) visualizing the genome-wide aggregate footprint of TFs ("aggregate TF footprinting"). We have provided examples of these analyses using the breast cancer dataset.

1. All these downstream analyses are performed in **R**. Download and install **R** and RStudio from https://posit.co/download/rstudio-desktop/ by following the instructions provided there. Launch RStudio.

2. **R** packages for bioinformatics are available from Bioconductor. Install Bioconductor. Determine the latest version number from https://www.bioconductor.org/install/ and replace `<version_number>` with it below. This step needs to be performed only once.

```
if (!require("BiocManager", quietly = TRUE))
    install.packages("BiocManager")
BiocManager::install(version = "<version_number>")
```

3. Install required **R** packages. This steps needs to be executed only once.

```
install.packages(c("reshape2", "ggplot2", "zoo",
    "gridExtra", "dplyr", "MatrixGenerics",
    "stringr", "Cairo"))

BiocManager::install(c("DiffBind", "ChIPseeker",
    "TxDb.Hsapiens.UCSC.hg38.knownGene", "org.
    Hs.eg.db", "GenomicRanges", "ATACseqQC",
    "BSgenome.Hsapiens.UCSC.hg38", "MotifDb",
    "GenomicAlignments", "BSgenome", "GenomeIn-
    foDb", "Rsamtools", "GenomicFeatures", "TFB-
    STools", "Biostrings", "ChIPpeakAnno"))
```

4. All the paths below are relative to the repository directory. Set the working directory in **R**

```
setwd(" ~/ATACSeqWorkFlow-master")
```

3.17 Differential Accessibility Analysis

We will utilize the **R** package DiffBind [39] to identify regions where accessibility changes significantly between two samples.

1. Load the `DiffBind` package.

```
require(DiffBind)
```

2. `DiffBind` requires a CSV sample sheet with eight columns, "SampleID", "Cell", "Factor", "Condition", "Replicate", "bamReads", "Peaks", and "PeakCaller", and one row per sample. We have provided an example sample sheet in the `example/Files/` subdirectory of the repository. Read this sample sheet into a data frame (`samples`).

```
samples <- read.csv("example/Files/DiffBind_Sam-
plesheet.csv")
```

3. The first step of analysis by `DiffBind` is to read in the peaks identified by MACS to create a DBA object (`data`). `DiffBind` requires a "score" metric for each peak that indicates the statistical confidence in or the degree of accessibility of the peak. Several different metrics are available in the `XLS` file generated by `macs2`, which is located in the `peaks/<sample_name>_peaks` subdirectory of the output directory. The column containing the metric to be utilized is specified with the scoreCol argument to dba. Here, the fold enrichment of each peak, contained in the eighth columns, is utilized as the score.

```
data <- dba(sampleSheet = samples, scoreCol = 8)
```

4. Generate the count matrix, containing the number of reads in all the peaks identified across the samples. *This will take a couple hours depending on the available RAM*

```
data$config$RunParallel <- FALSE
data <- dba.count(data)
```

5. After counting, the number of consensus peaks and associated Fraction of Reads in Peaks (FRiP) can be found by examining the DBA object.

```
print(data)
```

6. Normalize the counts with the default parameters. By default, `DiffBind` utilizes the standard `DESeq2` [40] normalization method.

```
data <- dba.normalize(data)
```

7. Establish a statistical model with the appropriate comparisons. Below, the samples are being compared between conditions.

```
data <- dba.contrast(data, categories = DBA_FAC-
TOR, minMembers = 2)
```

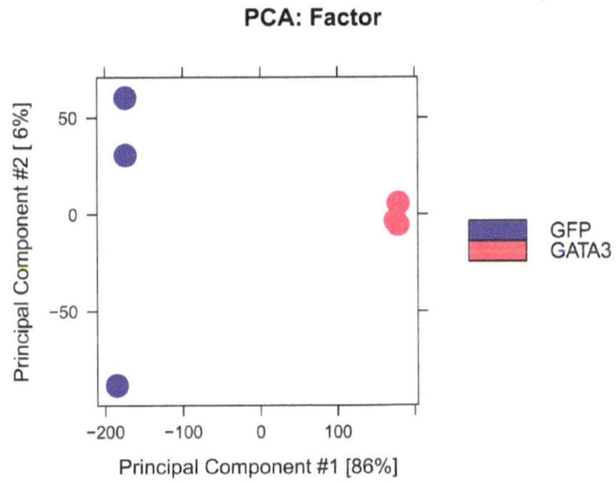

Fig. 7 PCA analysis of GFP- and GATA3-expressing MDA-MB-231 cells. PCA of the accessibility signal in regions having statistically significant differences between GFP-expressing (blue) and GATA3-expressing (pink) MDA-MB-231 cells [38]. GATA3 expression account for the bulk of the variation in the data

8. Conduct differential accessibility analysis. `DiffBind` can utilize common methods such as `DESeq2` and `EdgeR` [41]. Here, both methods are being utilized.

```
data <- dba.analyze(data, method = DBA_ALL_METHODS)
```

9. A summary showing the number of regions with statistically significant differences in accessibility at a false discovery rate (FDR) of 0.05 can be printed out.

```
dba.show(data, bContrasts = TRUE)
```

10. Retrieve the regions having differential accessibility, determined by edgeR in this example, into a data frame (`edger_df`).

```
res_edger <- dba.report(data, method = DBA_EDGER)
edger_df <- as.data.frame(res_edger)
```

11. PCA can be performed to determine the largest effects in the accessibility data (Fig. 7).

```
dba.plotPCA(data, method = DBA_EDGER, attri-
butes = DBA_FACTOR)
```

12. Similarity of samples to each other can be evaluated by computing the pairwise Pearson correlation between the accessibility of the consensus peakset between all pairs of samples (Fig. 8).

```
plot(data)
```

13. Visualize the number of accessible regions shared between or unique to samples using Venn diagrams (Fig. 9).

```
dba.plotVenn(data, contrast = 1, bGain = TRUE,
    bLoss = TRUE, bAll = FALSE, method = DBA_ALL_
    METHODS, main = "Peak Accessibility Chang-
    es", sub = "GATA3 vs. GFP")
```

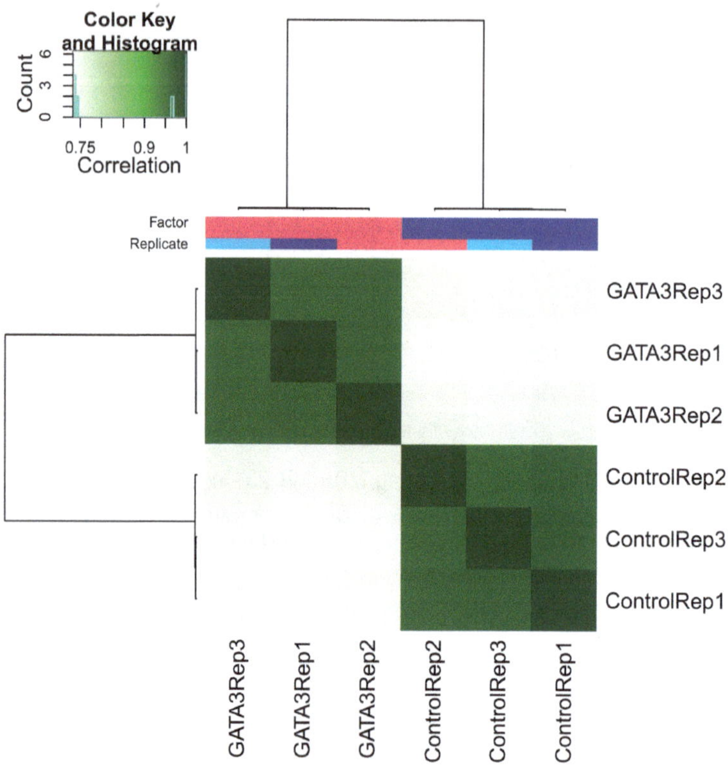

Fig. 8 Similarity between samples determined by pairwise Pearson correlation. Pearson correlation between the accessibility of the consensus peakset was computed for all pairs of samples and is displayed as a heatmap. The histogram (inset) shows the distribution of Pearson correlation. Sample are named according to treatment (Control/GATA3) and replicate number

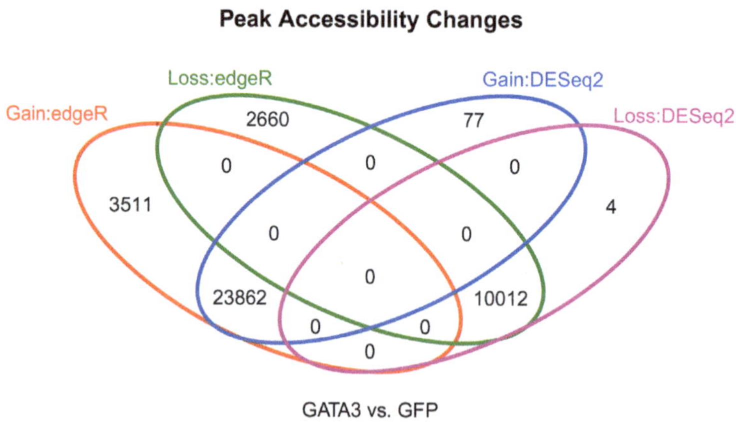

Fig. 9 Venn diagram of regions with statistically significant changes in accessibility between treatment groups. More regions increase in accessibility than decrease in it. EdgeR detects more significant regions than DESeq2

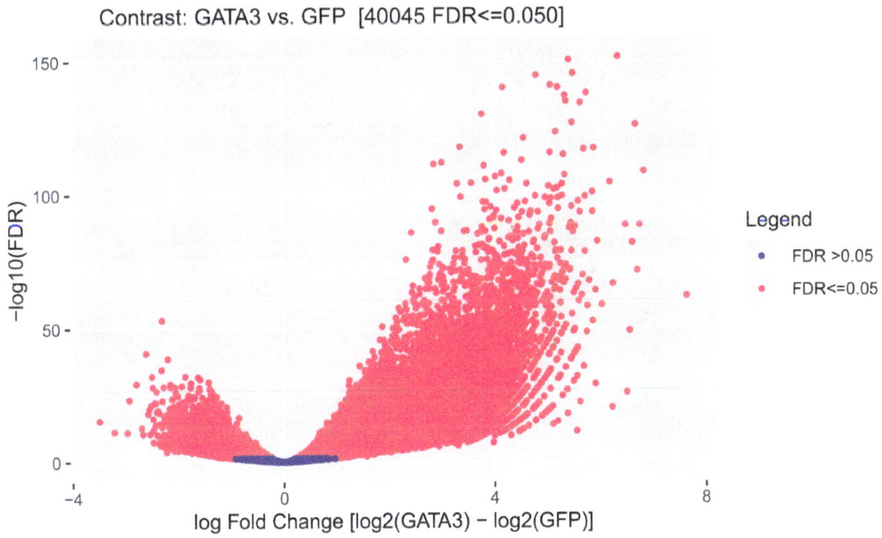

Fig. 10 Volcano plot of differentially accessible regions. *x*-axis is the $\log_2$ fold change in the mean Tn5 signal between the treatment groups. *y*-axis is $-\log_{10}(\text{FDR})$. Statistically significant (FDR $\leq$ 0.05) and not significant (FDR$>$0.05) regions are magenta and blue, respectively

14. Visualize the relationship between the statistical significance and fold change of accessible regions using volcano plots (Fig. 10).

```
dba.plotVolcano(data, method = DBA_EDGER)
```

3.18 Peak Annotation

1. Once a set of peaks ("peakset") of regions differing in accessibility between a set of conditions is obtained, determining their distribution in different genomic regions can provide insight into functional differences between conditions [40, 41]. This is accomplished by annotating each peak according to its location, intergenic, intronic, exonic, or promoter, relative to the nearest gene and then plotting this distribution. Use the `annotatePeak()` function from the `ChIPseeker` package [42] to annotate peaks. The first argument is a `GRanges` object containing peak information (`res_edger`), the second one defines the promoter region, while the last two furnish annotation databases. The required **R** packages are loaded beforehand.

```
require(ChIPseeker)
require(TxDb.Hsapiens.UCSC.hg38.knownGene)
require(org.Hs.eg.db)
require(GenomicRanges)
txdb <- TxDb.Hsapiens.UCSC.hg38.knownGene
annot.peaks <- annotatePeak(res_edger, tssRegion
   = c(-1000,1000), TxDb = txdb, annoDb = "org.
   Hs.eg.db")
```

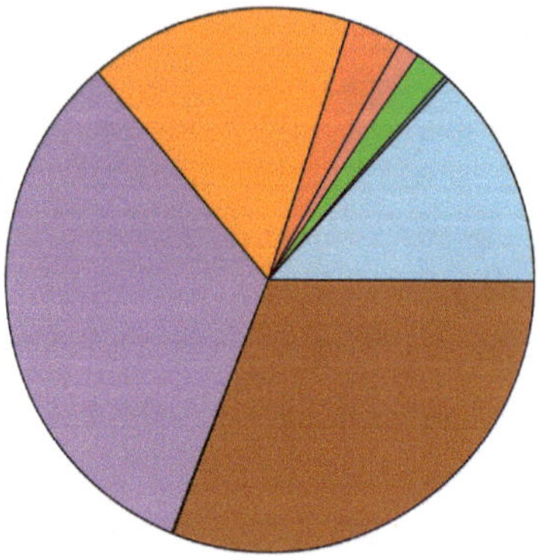

Fig. 11 The distribution of the locations of differentially accessible regions. Peaks are assigned to the nearest gene. 79.95% of the regions changing in accessibility are either intronic or intergenic, suggesting that most of the regulation occurs at enhancers

2. Visualize the distribution of the differentially accessible peaksets (Fig. 11).

```
plotAnnoPie(annot.peaks)
```

3. The annotation of individual differentially accessible peaks may be inspected by converting the `annot.peaks csAnno` object to a dataframe.

```
df.annot.peaks <- as.data.frame(annot.peaks)
```

3.19 Visualizing the Fine-Scale Accessibility Profile Within Accessible Regions

The fine-scale accessibility profile at the resolution of individual nucleotides contains information about the chromatin structure of regulatory regions. In order to visualize the fine-scale accessibility profile, it is essential to count individual Tn5 cleavage events ("cuts") rather than computing the coverage—the number of reads mapping to a nucleotide—which is effectively an average of the Tn5 cuts over the mean fragment length. During Tn5 transposition, each sequenceable fragment is generated by two cuts that occur at each end of the fragment. Given that Tn5 leaves a 9bp overhang, simply counting the cuts at the ends of each fragment results in the same cut corresponding to two different locations depending on whether the read maps to the + or the – strand of the reference genome (Fig. 12). Shifting the cut location by +4bp or −4bp if the read maps to the + or – strand respectively improves the genomic resolution of the accessibility profile. Another issue impacting the accuracy of accessibility profiles is that Tn5 does not cleave in a uniformly random manner and displays preference for certain DNA sequences [29, 43]. It is important to correct this bias, which is accomplished by determining the cut frequency of every k-mer in pure genomic DNA and adjusting the number of cuts in transposed chromatic accordingly [27].

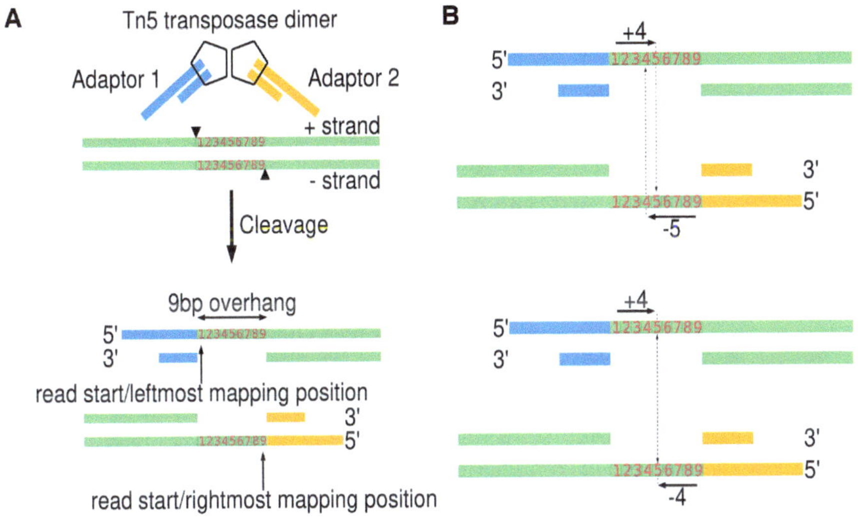

Fig. 12 Strategy for aligning Tn5 cut positions recorded on the + and – strands. (**A**) Schematic demonstrating the insertion of adaptor 1 (blue) and adaptor 2 (gold) into genomic DNA (green) during transposition. Tn5 leaves a 9bp overhang during adaptor insertion. Simply recording the start of a read as the position of the Tn5 cut results in recording two different positions for the same cut depending on whether the read maps to the + or – strand. (**B**) Schematic showing two methods of shifting the Tn5 cut position during data processing to reconcile the difference between reads mapping to different strands. The standard (+4/–5 bp) shift [16] (top) does not align the cut positions observed on the two strand whereas a +4/–4 bp shift (bottom) aligns them correctly. Reproduced from ref. [27]

1. We provide an **R** library, `ATACSeq_DataAnalysis.R`, containing a function, `createATACDataTable()` [27] that calculates the Tn5 cuts within a genomic region and corrects the significant values for Tn5 sequence biases. Source the library and call the `createATACDataTable()` function with a samplesheet (`samplesheet`) and the genomic region (`chrom`, `regionstart`, `regionend`) as arguments. See `Count_Samplesheet.csv` in the `Files` subdirectory of the repository for an example of the samplesheet. In the example below, we have chosen a differentially accessible region with a high average accessibility in the breast cancer dataset.

```
source("Scripts/ATACSeq_DataAnalysis.R")
counts <- createATACDataTable(
        samplesheet = "example/Files/Count_Sample
        sheet.csv",
        chrom = "chr6",
        regionstart = 43013859,
        regionend = 43014259,
        biascorrect = TRUE,
        combinedbiastable = "Files/Tn5_hexamer-
        Bias.txt", genome = "hg38")
```

2. `createATACDataTable()` returns a dataframe (`counts`) with one row per nucleotide and columns for the raw and corrected Tn5 cuts, among other statistics, for each sample. Plot

the total Tn5 cuts across all samples at each nucleotide to visualize the fine-scale accessibility profile (*see* **Note 12**).

```
ggplot(counts, aes(x=Location, y=correctedcuts))+
geom_col() + ylab("Pooled Tn5 Cuts")
```

3.20 Identification of TFs Potentially Binding Protected Regions

ATAC-seq is a protection assay and deeply sequenced libraries have the potential to reveal TF binding sites. Potential TF binding sites are detectable as 10–50bp protected regions having very few Tn5 cuts in an otherwise accessible region [27]. We provide the function `IdentifyFootprints()` which takes a genomic region of interest (`chrom, regionstart, regionend`) as arguments and returns a data frame, `binding_summary` in the example below, containing footprints and TFs potentially binding to them. Footprints are identified as regions in which each nucleotide has fewer Tn5 cuts than a threshold (Fig. 13). The threshold is set to the mean number of Tn5 cuts in the region of interest by default but can also be set manually (*see* **Note 13**).

1. Call the `IdentifyFootprints()` function.

```
binding_summary <- IdentifyFootprints(counts,
                                      genome = "hg38",
                                      chrom = "chr6",
                                      regionstart = 43013859,
                                      regionend = 43014259,
```

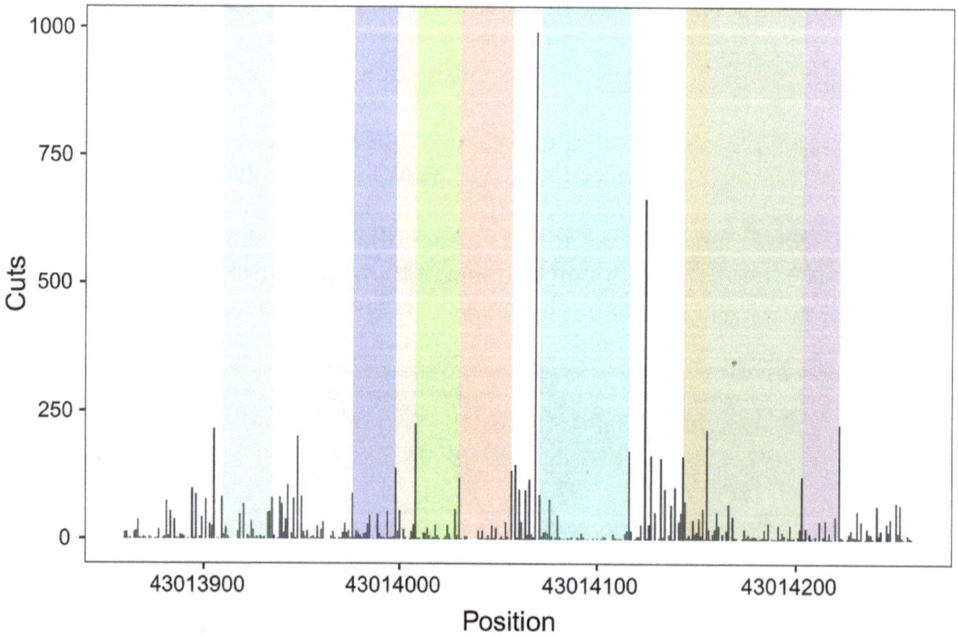

Fig. 13 Fine-scale accessibility profile of a differentially accessible region. The sum of Tn5 cuts across samples at each nucleotide in a 401 bp differentially accessible region. Potential TF footprints have been identified as spans of low signal flanked by nucleotides having high signal and highlighted as shaded regions

```
                    minWidth = 10,
                    maxWidth = 50)
```

2. Inspect the output of `IdentifyFootprints()` to view the protected regions.

```
print(binding_summary)
```

3. Finally, potential TFs binding to the protected regions can be identified by scoring the sequence with position weight matrices (PWMs) [44]. We provide `MatchJASPARMotifs()`, a wrapper function that utilizes the `matchPWM()` function from the `TFBSTools` package [45], to score a sequence (sequence argument) with a list of PWMs (`JASPARmatrix`). The function returns the TFs that score more than `min.score` on the test sequence as well as the positions of their binding sites. Call the `MatchJASPARMotifs()` to identify candidate TFs. *Note:* `prior.params` *specifies the background frequency of the nucleotides in the reference genome and should be changed according to the species. The values below are from the human genome.*

```
MatchJASPARMotifs(JASPARmatrix = "Files/JASPAR_
AllVertebrates_2024_MotifList.txt", sequence = "
TCCTCGGTCTCGACACATATTTGCACATGCGCAGAATCTCTCCTAC",
min.score = "90%",prior.params = c(A=0.303,
C=0.199, T=0.303, G=0.195))
```

3.21 Aggregate TF Footprinting

One common task is to determine whether a genome-wide signal of protection, an aggregate footprint, is detectable for a given TF. Here, we use the `factorFootprints()` function from the `ATACSeqQC` package [46] to plot the *CTCF* footprint from chromosome 14 separately in the GATA3 samples.

1. Load the `ATACseqQC`, `MotifDb` [47], and the hg38 human reference genome packages.

```
require(ATACseqQC)
require(MotifDb)
require(BSgenome.Hsapiens.UCSC.hg38)
```

2. As was done with fine-scale accessibility profiling (see Sect. 3.19), Tn5 cuts must be centered before computing the footprint. We provide a function, `shiftBAMs`, a wrapper for `ATACseqQC::shiftGAlignmentsList()`, to shift the read positions in the alignment BAM files. Shift read positions by calling `shiftBAMs()` with the name of the directory containing the alignment files (`indir`), the name of a directory to save the shifted alignments (`outdir`), and the number of nucleotides to shift by (`positiveShift, negativeShift`)

as arguments. *The procedure can be restricted to a single chromosome with the* `seqlev` *argument.*

```
shiftBAMs(indir = "example/bc_out/marked_reads/",
          outdirname = "shifted/",
          seqlev = "chr14",
          genome = "hg38",
          positiveShift = 4L,
          negativeShift = -4L)
```

3. Read in the list of shifted BAM files.

```
bamList <- list.files("example/bc_out/marked_
reads/shifted/",
                      pattern = ".bam$",
                      full.names = T)
```

4. Subset the list of samples for which aggregate footprints are desired.

```
GATA3 <- bamList[4:6]
```

5. Retrieve the PWM.

```
PWM <- query(MotifDb, "CTCF")
```

6. Call `factorFootprints()` with the BAM file list, the PWM (`pfm`), and the genome sequence (`genome`) as arguments (Fig. 14). Optionally, the minimum PWM score for a potential site to be included (`min.score`) and the window surrounding the

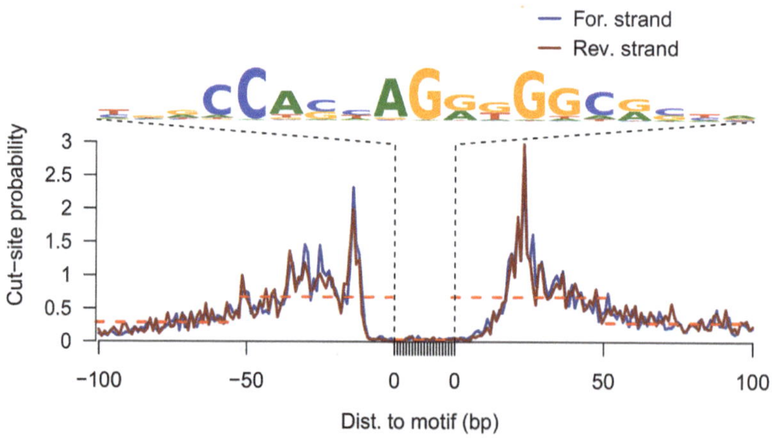

Fig. 14 *CTCF* aggregate footprint in the GATA3-expressing samples. Aggregate *CTCF* footprint calculated on chromosome 14 of the GATA3-expressing MDA-MB-231 samples. The *CTCF* PWM was used to score the entire length of the chromosome and positions scoring more than 95% of the consensus sequence score and occurring on the same strand were aligned. Tn5 cut probability was computed across all such sites in a 200bp region centered at the sites and is plotted separately for + and − strand sites. There is a strong depletion near the binding sites, demonstrating a chromosome-wide pattern of protection by *CTCF*

potential site (`upstream`, `downstream`) can be adjusted and the analysis can be restricted to a particular chromosome (`seqlev`).

```
factorFootprints(GATA3,  pfm=PWM[["Hsapiens-jas-
par2018-CTCF-MA0139.1"]],
    → genome=BSgenome.Hsapiens.UCSC.hg38,  se-
    qlev = "chr14", min.score="
    → 95%", upstream=100, downstream=100)
```

4 Notes

1. The use of CSK buffer or Hypotonic Greenleaf (Buenrostro) buffer can be optimized for the specific cell type used. We recommend testing both buffers and using Trypan blue to determine which buffer yields more intact nuclei (Fig. 2). Prepare Greenleaf buffer (10 mM Tris-HCl pH 7.4, 10 mM NaCl, 3 mM MgCl2 and 0.1% NP-40). Prepare CSK buffer (10 mM PIPES pH 6.8, 100 mM NaCl, 300 mM sucrose, 3 mM MgCl2 and 0.1% Triton X-100). In our experience, different cell types react differently to these buffers. Some cell types form robust nuclei in both buffers, while others are damaged by one or the other (Fig. 2).
2. We recommend starting with at least a million cells to minimize cell loss during washes, even if ultimately using a much lower cell number for the experiment.
3. Spike-in controls: Prepare nuclei from a different species (e.g., *Drosophila* if your primary samples are from mouse or human). Ensure these spike-in nuclei are prepared in the same way as your experimental nuclei. Count the spike-in nuclei accurately to make sure that you add a consistent amount to each sample. Typically, you should aim for 1–2% of the total experimental nuclei.
4. We are using Tn5 from Diagenode in this example. The amount of enzyme will need to be optimized for each commercial source/lot of enzyme/different cell numbers. For 2.5×10^4 cells, we use 2.5 μL of Tn5 Transposase to 12.5 μL of 2XTn5 buffer and 10 μL of nuclease-free water to obtain 25 μL final solution (*see* Fig. 3).
5. Method for determining number of PCR amplification cycles: *see* Fig. 4.
6. Repeat purification helps to remove adapter dimers but can also result in some loss of sample.
7. For DNA concentration, we prefer Qubit for better quantitation relative to Nanodrop, Tapestation or Bioanalyzer. For DNA fragment size, load at least ~30 ng by gel electrophoresis

using SYBR GOLD (0.5× TBE, 1.5% Agarose gel), or use the Bioanalyzer or TapeStation if available (Fig. 5).

8. The maximum number of CPU cores the workflow runs on can be changed with the `--cores <numcores>` argument to the `snakemake` command.

9. It is possible to run the workflow on only a subset of the samples

```
snakemake
--snakefile    ../Scripts/ATACSeq_Preprocessing_
Snakefile.yaml
--cores  1  peaks/{<sample1name>,<sample2name>
,...}.bw
```

Above, `<sample1name>, <sample2name>` etc. should be replaced with the names of the samples specified in the samples section of the configuration file, highlighted in red below.
An example `samples` section of the configuration file

```
samples:
   ControlRep1:
        read1: G2024-GSM1855976-ContRep1-R1
        read2: G2024-GSM1855976-ContRep1-R2
   ControlRep2:
        read1: G2024-GSM1855977-ContRep2-R1
        read2: G2024-GSM1855977-ContRep2-R2
```

10. `Snakemake` performs automatic checkpointing so that, upon rerunning the workflow, it resumes from last completed step instead of the start. If all the steps have already been completed, then `snakemake` will exit without performing any steps. It is possible to ATAC-seq of cancer cells force `snakemake` to run again on a previously processed dataset using the `--forceall` command-line switch.

```
snakemake  --snakefile  Scripts/ATACSeq_Prepro-
cessing_Snakefile.yaml --
   → configfile  test/config-test.yaml  --cores  1
   --forceall
```

11. The memory usage of `dba.count()` scales with the number of CPU cores and can be as high as 2TB. Here, we have turned off parallel processing to prevent unexpected behavior. Multiple CPUs can be utilized for faster processing at the cost of increased memory utilization.

```
data$config$RunParallel <- TRUE
data$config$cores <- <num_cores>
data <- dba.count(data)
```

<num_cores> is the number of cores to be utilized.

12. As written, plotting the Tn5 counts per individual nucleotide will pool all samples together. While this will increase the depth of coverage in the region and provide clearer patterns of accessibility, it is important to check each sample individually to ensure information is not being lost via pooling.

```
ggplot(counts[counts$Sample == <SampleName>,],
aes(x=Location, y=
    → correctedcuts)) + geom_col() +
ylab("Pooled Tn5 Cuts")
```

<SampleName> is the name of the sample the user wishes to plot.

13. While IdentifyFootprints() automatically sets the threshold for identifying nucleotides of high accessibility, one may be interested in manually setting the threshold. This can be done by utilizing the in-built argument manualThreshold.

```
binding_summary <- IdentifyFootprints(counts,
                      genome = "hg38",
                      chrom = "chr6",
                      regionstart = 43013859,
                      regionend = 43014259,
                      minWidth = 10,
                      maxWidth = 50,
                      manualThreshold = <numCuts>)
```

<numCuts> is the number of cuts of which the threshold will be set.

Acknowledgments

This work was supported by NIH grants P20GM104360 (MT, AD), R01GM148729 (MT), the IDeA Networks of Biomedical Research Excellence (INBRE) grant UND0026601 (AD), a congressionally directed medical research program (CDMRP) grant W81XWH-20-1-0168 (AD), and National Science Foundation (NSF) grant 1942471 (M). KG is supported by an NSF-GRFP fellowship (NSF FAIN# 2334433). M.T. is a Research Scholar of the American Cancer Society (ACS Research Scholar Award, RSG-23-645952-01-DMC). We thank the UND Genomics Core for assistance with ATAC-seq sample sequencing and analysis. The Core is supported by the NIH awards U54GM128729, P20GM104360, and P30GM154634.

References

1. Allshire RC, Madhani HD (2018) Ten principles of heterochromatin formation and function. Nat Rev Mol Cell Biol 19(4):229–244. https://doi.org/10.1038/nrm.2017.119
2. Grewal SI, Jia S (2007) Heterochromatin revisited. Nat Rev Genet 8(1):35–46. https://doi.org/10.1038/nrg2008
3. Huisinga KL, Brower-Toland B, Elgin SC (2006) The contradictory definitions of heterochromatin: transcription and silencing. Chromosoma 115(2):110–122. https://doi.org/10.1007/s00412-006-0052-x
4. Jin C, Zang C, Wei G, Cui K, Peng W, Zhao K, Felsenfeld G (2009) H3.3/H2A.Z double variant-containing nucleosomes mark 'nucleosome-free regions' of active promoters and other regulatory regions. Nat Genet 41(8):941–945. https://doi.org/10.1038/ng.409
5. Penagos-Puig A, Furlan-Magaril M (2020) Heterochromatin as an important driver of genome organization. Front Cell Dev Biol 8:579137. https://doi.org/10.3389/fcell.2020.579137
6. Strahl BD, Allis CD (2000) The language of covalent histone modifications. Nature 403(6765):41–45. https://doi.org/10.1038/47412
7. Venkatesh S, Workman JL (2015) Histone exchange, chromatin structure and the regulation of transcription. Nat Rev Mol Cell Biol 16(3):178–189. https://doi.org/10.1038/nrm3941
8. Maeshima K, Iida S, Shimazoe MA, Tamura S, Ide S (2024) Is euchromatin really open in the cell? Trends Cell Biol 34(1):7–17. https://doi.org/10.1016/j.tcb.2023.05.007
9. Schwartz U, Nemeth A, Diermeier S, Exler JH, Hansch S, Maldonado R, Heizinger L, Merkl R, Langst G (2019) Characterizing the nuclease accessibility of DNA in human cells to map higher order structures of chromatin. Nucleic Acids Res 47(3):1239–1254. https://doi.org/10.1093/nar/gky1203
10. Song L, Crawford GE (2010) DNase-seq: a high-resolution technique for mapping active gene regulatory elements across the genome from mammalian cells. Cold Spring Harb Protoc 2010(2):pdb prot5384. https://doi.org/10.1101/pdb.prot5384
11. Giresi PG, Kim J, McDaniell RM, Iyer VR, Lieb JD (2007) Faire (formaldehyde-assisted isolation of regulatory elements) isolates active regulatory elements from human chromatin. Genome Res 17(6):877–885. https://doi.org/10.1101/gr.5533506
12. Johnson DS, Mortazavi A, Myers RM, Wold B (2007) Genome-wide mapping of in vivo protein-DNA interactions. Science 316(5830):1497–1502. https://doi.org/10.1126/science.1141319
13. Ren B, Robert F, Wyrick JJ, Aparicio O, Jennings EG, Simon I, Zeitlinger J, Schreiber J, Hannett N, Kanin E, Volkert TL, Wilson CJ, Bell SP, Young RA (2000) Genome-wide location and function of DNA binding proteins. Science 290(5500):2306–2309. https://doi.org/10.1126/science.290.5500.2306
14. Cui K, Zhao K (2012) Genome-wide approaches to determining nucleosome occupancy in metazoans using mnase-seq. Methods Mol Biol 833:413–419. https://doi.org/10.1007/978-1-61779-477-3_24
15. Schones DE, Cui K, Cuddapah S, Roh TY, Barski A, Wang Z, Wei G, Zhao K (2008) Dynamic regulation of nucleosome positioning in the human genome. Cell 132(5):887–898. https://doi.org/10.1016/j.cell.2008.02.022
16. Buenrostro JD, Giresi PG, Zaba LC, Chang HY, Greenleaf WJ (2013) Transposition of native chromatin for fast and sensitive epigenomic profiling of open chromatin, DNA-binding proteins and nucleosome position. Nat Methods 10(12):1213–1218. https://doi.org/10.1038/nmeth.2688
17. Buenrostro JD, Wu B, Chang HY, Greenleaf WJ (2015) ATAC-seq: a method for assaying chromatin accessibility genome-wide. Curr Protoc Mol Biol 109:21291–21299. https://doi.org/10.1002/0471142727.mb2129s109
18. Cooper M, Ray A, Bhattacharya A, Dhasarathy A, Takaku M (2022) ATAC-seq optimization for cancer epigenetics research. J Vis Exp https://doi.org/10.3791/64242
19. Buenrostro JD, Wu B, Litzenburger UM, Ruff D, Gonzales ML, Snyder MP, Chang HY, Greenleaf WJ (2015) Single-cell chromatin accessibility reveals principles of regulatory variation. Nature 523(7561):486–490. https://doi.org/10.1038/nature14590
20. Cusanovich DA, Daza R, Adey A, Pliner HA, Christiansen L, Gunderson KL, Steemers FJ, Trapnell C, Shendure J (2015) Multiplex single cell profiling of chromatin accessibility by combinatorial cellular indexing. Science 348(6237):910–914. https://doi.org/10.1126/science.aab1601
21. Llorens-Bobadilla E, Zamboni M, Marklund M, Bhalla N, Chen X, Hartman J, Frisen J, Stahl PL (2023) Solid-phase capture and profiling of open chromatin by spatial atac. Nat Biotechnol 41(8):1085–1088. https://doi.org/10.1038/s41587-022-01603-9
22. Carraro C, Bonaguro L, Srinivasa R, van Uelft M, Isakzai V, Schulte-Schrepping J, Gambhir

P, Elmzzahi T, Montgomery JV, Hayer H, Li Y, Theis H, Kraut M, Mahbubani KT, Aschenbrenner AC, Konig I, Fava E, Fried HU, De Domenico E, Beyer M, Saglam A, Schultze JL (2023) Chromatin accessibility profiling of targeted cell populations with laser capture microdissection coupled to ATAC-Seq. Cell Rep Methods 3(10):100598. https://doi.org/10.1016/j.crmeth.2023.100598
23. Fujiwara S, Baek S, Varticovski L, Kim S, Hager GL (2019) High quality ATAC-Seq data recovered from cryopreserved breast cell lines and tissue. Sci Rep 9(1), 516. https://doi.org/10.1038/s41598-018-36927-7
24. Corces MR, Trevino AE, Hamilton EG, Greenside PG, Sinnott-Armstrong NA, Vesuna S, Satpathy AT, Rubin AJ, Montine KS, Wu B, Kathiria A, Cho SW, Mumbach MR, Carter AC, Kasowski M, Orloff LA, Risca VI, Kundaje A, Khavari PA, Montine TJ, Greenleaf WJ, Chang HY (2017) An improved ATAC-Seq protocol reduces background and enables interrogation of frozen tissues. Nat Methods 14(10):959–962. https://doi.org/10.1038/nmeth.4396
25. Yan F, Powell DR, Curtis DJ, Wong NC (2020) From reads to insight: a hitchhiker's guide to ATAC-Seq data analysis. Genome Biol 21(1), 22. https://doi.org/10.1186/s13059-020-1929-3
26. Reske JJ, Wilson MR, Chandler RL (2020) ATAC-Seq normalization method can significantly affect differential accessibility analysis and interpretation. Epigenet Chromatin 13(1):22. https://doi.org/10.1186/s13072-020-00342-y
27. Long T, Bhattacharyya T, Repele A, Naylor M, Nooti S, Krueger S, Manu (2023) The contributions of DNA accessibility and transcription factor occupancy to enhancer activity during cellular differentiation. G3 Bethesda. https://doi.org/10.1093/g3journal/jkad269
28. Thibodeau A, Khetan S, Eroglu A, Tewhey R, Stitzel ML, Ucar D (2021) CoRE-ATAC: a deep learning model for the functional classification of regulatory elements from single cell and bulk ATAC-Seq data. PLoS Comput Biol 17(12):e1009670. https://doi.org/10.1371/journal.pcbi.1009670
29. Li Z, Schulz MH, Look T, Begemann M, Zenke M, Costa IG (2019) Identification of transcription factor binding sites using ATAC-Seq. Genome Biol 20(1), 45. https://doi.org/10.1186/s13059-019-1642-2
30. Cazares TA, Rizvi FW, Iyer B, Chen X, Kotliar M, Bejjani AT, Wayman JA, Donmez O, Wronowski B, Parameswaran S, Kottyan LC, Barski A, Weirauch MT, Prasath VBS, Miraldi ER (2023) maxATAC: genome-scale transcription-factor binding prediction from ATAC-Seq with deep neural networks. PLoS Comput Biol 19(1):e1010863. https://doi.org/10.1371/journal.pcbi.1010863
31. Karabacak Calviello A, Hirsekorn A, Wurmus R, Yusuf D, Ohler U (2019) Reproducible inference of transcription factor footprints in ATAC-seq and DNase-seq datasets using protocol-specific bias modeling. Genome Biol 20(1), 42. https://doi.org/10.1186/s13059-019-1654-y
32. Danecek P, Bonfield JK, Liddle J, Marshall J, Ohan V, Pollard MO, Whitwham A, Keane T, McCarthy SA, Davies RM, Li H (2021) Twelve years of samtools and bcftools. GigaScience 10. https://doi.org/10.1093/gigascience/giab008
33. Picard toolkit (2019) Picard toolkit. https://broadinstitute.github.io/picard/
34. Bolger AM, Lohse M, Usadel B (2014) Trimmomatic: a flexible trimmer for illumina sequence data. Bioinformatics 30:2114–2120. https://doi.org/10.1093/bioinformatics/btu170
35. Kim D, Paggi JM, Park C, Bennett C, Salzberg SL (2019) Graph-based genome alignment and genotyping with hisat2 and hisat-genotype. Nat Biotechnol 37:907–915. https://doi.org/10.1038/s41587-019-0201-4
36. Zhang Y, Liu T, Meyer CA, Eeckhoute J, Johnson DS, Bernstein BE, Nusbaum C, Myers RM, Brown M, Li W, Liu XS (2008) Model-based analysis of ChIP-Seq (MACS). Genome Biol 9, R137. https://doi.org/10.1186/gb-2008-9-9-r137
37. Mölder F, Jablonski KP, Letcher B, Hall MB, Tomkins-Tinch CH, Sochat V, Forster J, Lee S, Twardziok SO, Kanitz A, Wilm A, Holtgrewe M, Rahmann S, Nahnsen S, Köster J (2021) Sustainable data analysis with snakemake. F1000Research 10:33. https://doi.org/10.12688/f1000research.29032.2
38. Takaku M, Grimm SA, Shimbo T, Perera L, Menafra R, Stunnenberg HG, Archer TK, Machida S, Kurumizaka H, Wade PA (2016) Gata3-dependent cellular reprogramming requires activation-domain dependent recruitment of a chromatin remodeler. Genome Biol 17, 36. https://doi.org/10.1186/s13059-016-0897-0
39. Stark R, Brown G (2011) DiffBind: differential binding analysis of ChiP-Seq peak data. http://bioconductor.org/packages/release/bioc/vignettes/DiffBind/inst/doc/DiffBind.pdf
40. Love MI, Huber W, Anders S (2014) Moderated estimation of fold change and dispersion for RNA-Seq data with DESeq2. Genome Biol 15, 550. https://doi.org/10.1186/s13059-014-0550-8
41. Robinson MD, McCarthy DJ, Smyth GK (2010) edgeR: a bioconductor package for differential expression analysis of digital gene expression data. Bioinform 26(1):139–140. https://doi.org/10.1093/BIOINFORMATICS/BTP616
42. Yu G, Wang LG, He QY (2015) ChiPseeker: an R/bioconductor package for chip peak

annotation, comparison and visualization. Bioinformatics 31(2382):2383. https://doi.org/10.1093/bioinformatics/btv145
43. Bentsen M, Goymann P, Schultheis H, Klee K, Petrova A, Wiegandt R, Fust A, Preussner J, Kuenne C, Braun T, Kim J, Looso M (2020) ATAC-seq footprinting unravels kinetics of transcription factor binding during zygotic genome activation. Nat Commun 11(1), 4267. https://doi.org/10.1038/s41467-020-18035-1
44. Stormo GD (2000) DNA binding sites: representation and discovery. Bioinformatics 16(16):23
45. Tan G, Lenhard B (2016) TFBSTools: an R/bioconductor package for transcription factor binding site analysis. Bioinformatics 32:1555–1556. https://doi.org/10.1093/bioinformatics/btw024
46. Ou J, Liu H, Yu J, Kelliher MA, Castilla LH, Lawson ND, Zhu LJ (2018) ATACseqQC: a bioconductor package for post-alignment quality assessment of ATAC-seq data. BMC Genom 19, 169. https://doi.org/10.1186/s12864-018-4559-3
47. Shannon P, Richards M (2025) MotifDb: An annotated collection of protein–DNA binding sequence motifs (R package version 1.50.0). Bioconductor. https://doi.org/10.18129/B9.bioc.MotifDb

Check for updates

Chapter 8

Investigating the Physiological Regulation of Transposable Elements In Vivo: Nuclei Preparation from Primary Mouse Tissues for Epigenomic Analyses and Computational Approaches to Interrogate the Repeat Genome

Sajad A. Wani and Amanda R. Wasylishen

Abstract

Investigating the physiological regulation of the epigenome using genetically engineered mouse models, with a focus on heterochromatic regions of DNA, presents two major challenges: (i) the isolation and preparation of high-quality nuclei from tissue for analysis, (ii) the computational considerations for repetitive regions of the genome. Here we outline the experimental methods for tissue procurement and cryopreservation used in our laboratory, and the steps for nuclei isolation, which can be used for several downstream epigenomic and transcriptomic applications, including ATAC-seq, CUT&RUN, and single-nucleus RNA-sequencing. We then describe the computational considerations and bioinformatics pipelines for interrogating the repeat genome, with a particular focus on transposable elements.

Keywords Genetically engineered mouse models, Nuclei isolation, Heterochromatin, Transposable elements, ATAC-seq, ChIP-seq, CUT&RUN, RNA-seq

1 Introduction

Cancer genome sequencing has both highlighted and emphasized the impact of epigenetic dysregulation on the pathogenesis of disease. Across diverse cancer types there is a high frequency of genetic alterations in genes that regulate the epigenome [1, 2], which has contributed to nonmutational epigenetic reprogramming being recognized as an enabling characteristic of cancer more generally [3]. Notable examples include mutations in the *DNMT3A* DNA methyltransferase in myeloid leukemias [4, 5], recurrent mutations in the histone chaperones *DAXX* and *ATRX* in pancreatic neuroendocrine tumors [6], and even mutations in genes encoding histones in pediatric and adult brain tumors [7–9].

As both epigenomes and transcriptomes are profoundly impacted by biological context, physiologically relevant models are critical

Archana Dhasarathy and Motoki Takaku (eds.), *Cancer Epigenetics*, Methods in Molecular Biology, vol. 3005, https://doi.org/10.1007/978-1-0716-5072-1_8,

for understanding both normal biology and the downstream and potentially pathogenic consequences of genetic or environmental perturbations. Genetically engineered mouse models (GEMMs) provide a powerful platform to understand the physiological function(s) of genes and proteins. Additionally, many genetic tools exist to allow for both spatial and temporal control of genetic perturbations. Despite the many advantages of GEMMs and other experimental models in vivo, additional experimental and practical considerations are required to obtain high quality tissue for downstream epigenomic and transcriptomic analyses.

Only a small fraction of the genome, approximately 1–2%, encodes for proteins and the non-coding genome is important for regulating gene expression. Additionally, a significant portion of the genome is comprised of repetitive sequences, which impact genome structure, function and stability. For example, telomeres are tandem repeats located at the ends of chromosomes which protect the DNA from degradation [10]. Interspersed repeats, which are identical or very similar sequences found in different locations throughout the genome, comprise approximately 45% of the human genome and 38% of the mouse genome and are primarily comprised of transposable elements (TEs). TEs, which are also referred to as "jumping genes", are mobile DNA sequences that can move from one location in the genome to another. For a comprehensive description of TEs and their history, we direct readers to several comprehensive resources [11–15]. Epigenetic mechanisms are critical for silencing TEs in the genome, including DNA methylation, and histone modifications, including the H3K9me3 heterochromatin mark [11]. Additionally, the histone 3 variant, H3.3, also contributes to the silencing of TEs, specifically endogenous retroviruses, when deposited by a chaperone complex containing DAXX and ATRX [16–19]. Loss or weakening of these silencing mechanisms can result in the acquisition of active histone marks, such as H3K4me1/2/3 and H3K27ac, and can lead to de-repression of TEs. These de-repressed TEs can function as cis-regulatory elements [20–26], acting as alternative promoters or enhancers that influence gene expression, and in some cases, participate in long-range chromatin interactions that reshape transcriptional programs [27–31]. TEs have been implicated in a number of disease states including L1 and Alu insertions in the factor VIII gene causing hemophilia A, and the onco-exaptation of TE-derived promoters to drive oncogene expression in cancer [12, 32–35]. Given the potential roles of the TEs to contribute to disease states, it is important to consider these when evaluating epigenetic and transcriptional mechanisms. Despite the growing appreciation for TEs, they present methodological challenges, owing in large part to their abundance and repetitive nature. Recent and evolving methods now allow these sequences to be considered when evaluating genomic datasets and are providing novel and important insights.

Herein, we provide experimental methods for the isolation of high-quality nuclei from mouse tissues for downstream epigenomic analysis, and the computation considerations and framework for interrogating the repeat genome.

2 Materials

2.1 Laboratory Experiments

2.1.1 Tissue Collection

1. 70% ethanol solution
2. Standard dissecting tools
3. Cryotubes
4. Liquid nitrogen

2.1.2 Nuclei Isolation

1. Mortar and pestle
2. Spatula
3. Dounce homogenizer
4. 6× Homogenization Buffer—Stable (30 mM $CaCl_2$, 18 mM $Mg(Ac)_2$, 60 mM Tris pH=7.8)
5. 6× Homogenization Buffer—Unstable, prepared fresh (6× Homogenization Buffer—Stable, 0.1 mM PMSF, 1 mM β-mercaptoethanol)
6. 1 M sucrose
7. 1× Homogenization Buffer prepared fresh (1× Homogenization Buffer—Unstable, 320 mM sucrose, 0.1 mM EDTA, 0.1% NP40)
8. Cell strainers (35 μm)
9. Refrigerated benchtop microcentrifuge

2.1.3 Nuclei Purification (Optional)

1. 60% Iodixanol
2. 6× Homogenization Buffer—Unstable
3. 1M sucrose solution
4. 50% iodixanol solution (1× Homogenization Buffer—Unstable, 50% iodixanol)
5. 29% iodixanol solution (1× Homogenization Buffer—Unstable, 160 mM sucrose, 29% iodixanol)
6. 35% iodixanol solution (1× Homogenization Buffer—Unstable, 160 mM sucrose, 35% iodixanol)
7. Refrigerated centrifuge with swinging bucket rotor

2.2 Computational Tools

1. FastQC [36]
2. MultiQC [37]

3. TrimGalore [38]
4. Bowtie2 [39]
5. MACS2 [40]
6. SEACR
7. featureCounts (Subread) [42]
8. ChiPSeeker [43]
9. deepTools [44]
10. HOMER [45]
11. BEDTools [46]
12. STAR [47]
13. DESeq2 [48]
14. TEtranscripts [49]
15. SQuIRE [50]

3 Methods

3.1 Laboratory Experiments

3.1.1 Tissue Collection

1. Euthanize mouse following procedures approved by Institutional Animal Care and Use Committee (IACUC) protocol (*see* **Note 1**).
2. Position the mouse on dissecting tray, immobilize, and spray 70% ethanol solution. Use standard dissecting tools (sharp scissors and forceps) to expose the appropriate body cavity and isolate required tissue. Larger tissues, such as liver, may be cut into small pieces prior to freezing. Some tissues are more susceptible to autolysis and may require additional considerations (*see* **Note 2**).
3. Immediately place tissue in a labeled cryovial and flash freeze in liquid nitrogen. Use appropriate safety procedures when working with liquid nitrogen (*see* **Note 3**). If needed, prior to freezing, a portion of the tissue can be isolated for other applications, including formalin fixation for histopathology evaluation.
4. Once tissue is frozen, cryovial can be transferred to a liquid nitrogen storage container or to a –80°C freezer.

3.1.2 Nuclei Isolation

The protocol for nuclei isolation from frozen tissues is based on published methodology from the Greenleaf and Chang laboratories developed for ATAC-seq [51]. To prevent over-lysis and ensure sample integrity, the best practice is to ensure all reagents are cold, samples are maintained on ice, and all centrifugation steps are conducted at 4°C.

Create Frozen Tissue Powder

There are several ways to homogenize tissue by pulverization in liquid nitrogen. Here we outline a simple procedure using a mortar and pestle, but other devices are also available and can be used.

1. Precool a mortar, pestle and a spatula with liquid nitrogen (*see* **Note 3**).
2. Pour a small amount of liquid nitrogen into the mortar and place frozen tissue piece in the liquid nitrogen. If you are processing multiple samples, *see* **Note 4**.
3. Wait approximately 1 min for tissue to reduce in temperature and become brittle.
4. Grind tissue into a fine powder using chilled pestle.
5. With precooled spatula, transfer tissue into a pre-chilled microcentrifuge tube on dry ice.
6. Clean the mortar, pestle and spatula and repeat with additional tissue samples.

Homogenize and Lyse Tissue

1. Retrieve frozen tissue powder and add to a prechilled Dounce Homogenizer (2 mL) with cold 1x Homogenization Buffer (2 mL). The amount of tissue powder required will be influenced by the tissue itself and the downstream application. In our experience 50–100 mg of pancreas is sufficient for downstream ATAC-seq and CUT&RUN applications. If processing tissue for single-nucleus RNA-sequencing analysis or other RNA applications, consider the addition of RNase inhibitors starting at this step (*see* **Note 5**).
2. Allow frozen tissue to thaw.
3. Dounce with pestle A (loose) for 10 strokes, or until resistance is no longer evident
4. Dounce with pestle B (tight) for 20 strokes. The number of strokes may be modified if there is evidence of over or under lysis.
5. Preclear to remove larger pieces and connective tissue by short centrifugation at 100 RCF for 1 minute at 4°C.
6. Collect supernatant, avoiding any pelleted connective tissue.
7. 400 μL of supernatant can be collected and used for Sect. 3.1.3 Nuclei purification below, or nuclei can be pelleted at 500 RCF for 5 min at 4°C. Nuclei can then be resuspended in PBS + 0.1% BSA or other buffer appropriate for downstream application. If needed, nuclei suspension can be passed through a cell strainer (35 μm) to remove any debris or aggregates.

Evaluate Nuclei Quality and Quantity

1. Stain nuclei with trypan blue. An aliquot of the nuclei suspension can be mixed with an equal volume of 0.4% trypan blue solution. Isolated nuclei will be positive for trypan blue stain and appear dark blue.

2. Transfer 10 μL of nuclei-trypan blue mixture to a hemocytometer or appropriate counting slide.
3. View under a brightfield microscope or cell counter and assess nuclei number and quality (*see* **Note 6**). Representative images and good and poor quality nuclei preparations are provided in Fig. 1.

3.1.3 Nuclei Purification (Optional)

Nuclei can be purified using an iodixanol (Optiprep) gradient. Best practice is to perform centrifugation using a swinging bucket centrifuge to ensure layers are maintained.

1. Add an equal volume of 50% iodixanol solution and mix by pipetting. The final concentration of iodixanol will be 25%.
2. Layer 600 μL of 29% iodixanol solution under the 25% layer, avoiding mixing.
3. Layer 600 μL of 35% iodixanol solution under the 29% layer, avoiding mixing.
4. Spin at 3000 RCF for 20 minutes with the brake off to prevent mixing of layers
5. Nuclei will be at the interface of the 29% and 35% layers. Aspirate the top layers to within approximately 300 μL of the nuclei band.
6. Aiming for a volume of approximately 200 μL, collect the nuclei band and transfer to a clean tube. If needed, nuclei can

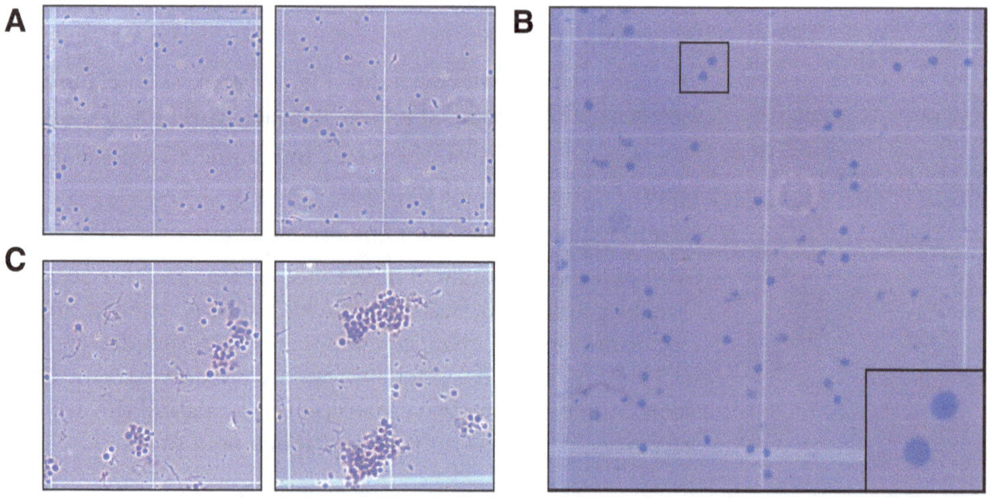

Fig. 1 Examples from nuclei preparations. **A, B** Example of a good-quality nuclei preparation. The majority of the nuclei are intact and well-distributed in the sample. **C** Example of a poor-quality nuclei preparation with clumping and debris present. Nuclei are stained with trypan blue, loaded on a hemocytometer and visualized. Gridlines from the hemocytometer form 0.25 mm × 0.25 mm boxes, with four boxes represented in each image

be diluted with an appropriate buffer (e.g., PBS + 0.1% BSA or ATAC-Resuspension Buffer + 0.1% Tween 20).

7. Evaluate and count nuclei as in Sect. 3.1.2.3 above.

3.1.4 Potential Applications

Isolated nuclei can be used for several downstream applications. We have successfully isolated and used iodixanol-purified nuclei for ATAC-sequencing from mouse pancreas [18]. Additionally, isolated nuclei have undergone additional wash steps (10× Genomics Wash and Resuspension Buffer) and successfully been used for single nucleus RNA-sequencing from the pancreas, in which 18 cell clusters were identified representing all major cell types in the pancreas, including epithelial (endocrine and exocrine), stromal and immune cells (unpublished). Finally, we have also used isolated nuclei as input for CUT&RUN analysis of H3K4me3 (Wani SA and Wasylishen AR, unpublished).

3.2 Library Preparation for Sequencing

Library preparation for sequencing is primarily performed using sequencing core facilities; however, several considerations must be followed to optimize results. For studies focusing on TE, ribosomal ribonucleic acid (rRNA) depletion should be prioritized over poly-A selection to retain non-polyadenylated TE transcripts and antisense RNA species. The use of paired-end sequencing is highly recommended, as it enhances read alignment accuracy for TEs. Opting for a read length of over 75 base pairs (bp) is beneficial as it enhances the uniqueness of the mapped fragments, thereby minimizing ambiguities during the alignment process. As supported by benchmarking studies [52], the use of paired-end RNA-seq libraries improves TE read mapping by up to 30% for young TE families. Shorter read lengths, such as 50 bp and 75 bp, often fail to fully cover many TEs, leading to incomplete spanning of these elements [53, 54]. While utilizing longer reads (≥100 bp) can greatly diminish confusion in regions abundant with repetitive sequences by increasing the likelihood of covering unique sequences adjacent to the TEs [55].

3.3 Computational Approaches Focused on Transposable Elements

One of the main methodological challenges when studying TEs is their highly repetitive nature. Standard pipelines whether for histone mark profiling (CUT&RUN/ChIP-seq) or for gene expression analysis (RNA-seq) focus on uniquely mapped reads to define peaks or measure transcript abundance and discard multi-mapped reads. For instance, the H3K9me3 histone mark, which is typically enriched in constitutive heterochromatin, a region abundant in repeat elements, may be underreported if only uniquely mapped reads are considered, leading to skewed estimates of enrichment and background levels [56]. For transcription factor (TF) ChIP-seq, which typically yields narrow peaks (around 50–150 nucleotides), the situation is more challenging. There are often too few uniquely mapped reads surrounding the narrow binding sites to reliably inform the statistical redistribution of multi-mapped reads [57]. Moreover, TE-derived regulatory elements constitute 40% of TF binding sites [58]. Also, the standard approach can underestimate TE measurements because many TE-derived reads map to multiple loci especially for young TE families, which have high sequence similarity and often lack uniquely mappable regions [59]. Thus, studies focusing on TEs require specialized analytical approaches beyond standard pipelines.

In this section, we describe strategies tailored to TE-rich regions, highlighting the nuanced balance between retaining multi-mapped reads to capture genuine TE signals and controlling for inflated false positives. There is a trade-off between sensitivity and specificity. For example, retaining multi-mapped reads increases sensitivity to detect TE expression or enrichment, but it reduces specificity since it becomes unclear whether a read originated from a specific locus or another identical copy elsewhere in the genome. Conversely, focusing solely on uniquely mapped reads ensures specificity but sacrifices sensitivity, leading to incomplete biological conclusions regarding TE. Although many tools offer different strategies for dealing with multi-mapping, the problem is still not solved. We believe there is a necessity for specialized algorithms capable of accurately determining the origin of reads that map to multiple locations, an essential step for reliable downstream analysis.

3.3.1 Initial Data Processing

For both ChIP-seq/CUT&RUN and RNA-seq, rigorous quality control prior to alignment is crucial to maintain data integrity, regardless of whether the focus is on genes or TEs.

1. Use tools such as FastQC or MultiQC to assess read quality, examining metrics including base quality scores, GC content distribution, and identifying potential adapter contamination [37].
2. Apply trimming and filtering procedures using tools like Trim Galore [38] to remove low-quality bases and adapter sequences, thus reducing technical artifacts that could affect downstream analyses.
3. Use the resulting trimmed and filtered FASTQ files as input for subsequent ChIP-seq/CUT&RUN and RNA-seq alignment and downstream analysis steps.

3.3.2 Mapping Tools to Analyze Histone Modifications Around TEs

There are various aligners available for ChIP-seq data, including BWA, Bowtie, Bowtie2, and Subread [39, 42, 60, 61]; however, Bowtie2 is used more frequently because it seamlessly integrates fast full-text indexing with single-instruction multiple-data accelerated dynamic programming, delivering both speed and high accuracy for gapped alignments across diverse read lengths and sequencing platforms. Bowtie2 offers three distinct reporting modes default mode, -k mode and -a mode as explained in **Note 7**. Although Bowtie2 offers multiple reporting modes, recent publications have demonstrated that its default mode strikes the best balance between sensitivity and specificity in TE focused ChIP-seq data analysis [21, 25, 26, 62–64]. Nevertheless, STAR's multimapping mode and SQuIRE-based pipelines have also been adapted as explained in **Notes 8 and 9** [20, 65]. For ChIP-seq and CUT&RUN analyses, we adhere to Bowtie2's default mode, which reports only the best alignment for multi-mapped reads (Fig. 2).

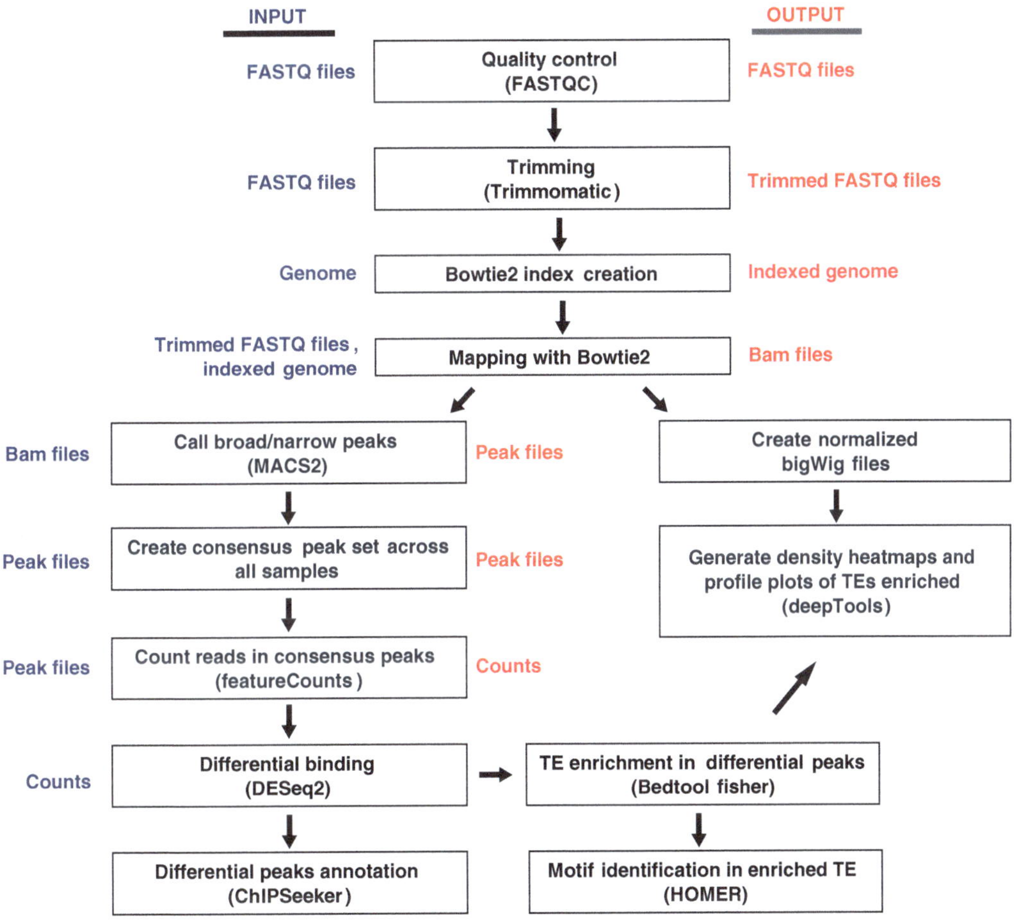

Fig. 2 ChIP-seq data analysis with a TE focus. Starting from raw FASTQ files, reads undergo quality control (FastQC/MultiQC) and adapter-quality trimming (Trim Galore) before alignment to the reference genome using Bowtie2. Mitochondrial and unassigned reads are removed, keeping only concordantly mapped pairs. Peaks are identified with MACS2 (narrow or broad calls depending on the mark), annotated and merged across samples to form a consensus peak set. FeatureCounts extracts read counts per consensus peak, and differential enrichment between conditions is assessed with DESeq2. HOMER is used for motif discovery on differentially enriched peaks, deepTools generates heatmaps of ChIP signal intensity over peak regions, and BEDTools Fisher's exact test evaluates TE-specific enrichment within the peaks

1. Begin by utilizing Bowtie2's default alignment mode, which reports only the single best alignment for multi-mapped reads. If multiple equally good alignments exist, Bowtie2 randomly selects one, reducing bias in repetitive genomic regions and balancing sensitivity and specificity.
2. Bowtie2 defaults to end-to-end alignment, aligning the entire read length. If partial alignments or soft clipping are desired (e.g., detecting partial TE insertions or chimeric TE-gene transcripts), use the – local parameter to allow soft clipping at one or both read ends.
3. Filter out reads mapping to mitochondrial DNA or unassigned genomic sequences. For paired-end data, retain only concordantly aligned pairs to ensure proper read pairing.
4. Consider using the—very-sensitive preset in Bowtie2 to enhance alignment stringency. This setting allows more exhaustive searches for valid alignments, which is beneficial when analyzing highly similar TE subfamilies.
5. Apply mapping quality filters (e.g., MAPQ ≥ 1 or MAPQ ≥ 30) to adjust sensitivity and specificity. For TE analyses, note that a strict threshold (MAPQ ≥ 30) may discard too many genuine reads from evolutionarily young repeats with near-identical sequences. Thus, a lower threshold (e.g., MAPQ ≥ 1) may be preferable [66–68]. MAPQ is a Phred-based score indicating alignment confidence.

Ultimately, the decision to keep or discard multi-mapped reads depends on balancing the need to capture genuine TE signals against minimizing false positives. Retaining all multi-mappers can inflate TE coverage with spurious signals, while discarding them outright can underestimate TE enrichment, particularly in TE subfamilies lacking uniquely mappable regions.

3.3.3 Peak Calling and Downstream Analysis in TE Studies

For peak calling MACS2 and SEACR are commonly used for ChIP-seq and CUT&RUN, respectively, each typically expecting a set of uniquely mapped reads [40, 41].

1. Use MACS2 for ChIP-seq datasets, as it effectively detects narrow and broad peaks using a robust, model-based approach suitable for typical ChIP-seq background signals. Use SEACR for CUT&RUN datasets, which is specifically optimized for sparse and low-background signals common in CUT&RUN.
2. Use BAM files generated from Bowtie2 alignments as input for peak callers, ensuring these files primarily include uniquely mapped reads to maintain high-quality peak calling.
3. Select narrow or broad peak detection based on the specific histone marks or TFs studied. Histone marks produce broad

or narrow peaks depending on their biological function, while TFs typically generate narrow peaks.

4. Annotate identified peaks by associating them with genomic features (e.g., genes, promoters, TEs), providing biological context and facilitating downstream interpretation.
5. Construct a consensus peak set by merging peaks identified across multiple samples or replicates, establishing a unified framework for comparative analysis.
6. Quantify read coverage within the consensus peak regions using tools such as featureCounts, generating count matrices for subsequent quantitative analyses.
7. Conduct clustering analysis and differential binding analysis to compare peak intensities or enrichments between different biological groups or experimental conditions, identifying statistically significant differences relevant to TE biology.

Several other strategies have been developed as explained in **Note 10** to manage ambiguously mapped reads in repetitive genomic regions. Overall, the chosen strategy depends on factors such as the dataset complexity, read length, and biological goals, with each approach offering different trade-offs in accuracy, resolution, and computational demands. Many recent TE-focused studies adopt Bowtie2's default single-best-mapping approach to handle multi-mapped reads, and we likewise follow this convention to ensure balanced sensitivity and specificity. There remains a considerable opportunity to improve ChIP-seq analysis in repetitive genomic regions.

3.3.4 Enrichment of TE Regions in Histone Modification Peaks

Enrichment of histone modifications at TE loci is crucial for understanding their regulatory potential; by determining whether TEs exhibit active enhancer marks or repressive chromatin signatures, researchers can decipher how these sequences contribute to gene regulation and influence cellular identity and development. Enrichment analyses of TE regions in ChIP-seq and CUT&RUN data have evolved to employ a diverse suite of computational tools, each tailored to capture different aspects of TE epigenetic regulation. We adhere to, TEenrich-FET which uses Fisher's Exact Test on TE coordinates to assess subfamily-specific enrichment [64, 69, 70]. Other tools used for studying enrichment of TEs in peaks are explained in **Note 11**. While these computational methods offer valuable predictions on histone modification enrichment at TE loci, experimental validation remains essential to confirm their true biological functions.

3.3.5 Mapping Tools to Analyze RNA-Seq Data to Measure Expression of TEs

There are various aligners available for RNA-Seq data like STAR, HISAT2, TopHat2, Subread, Bowtie2, and pseudoalignment based methods like Salmon [39, 42, 47, 71–73]; however, STAR is used more frequently because it offers exceptionally fast, accurate, and splice-aware alignments, making it highly effective for detecting exon-exon junctions. These aligners generate BAM files that are then used for quantification via tools like featureCounts, RNA-Seq by Expectation Maximization (RSEM), HTSeq-count, or in some cases tools like Salmon and kallisto directly produce the count data [71]. The resulting counts are subsequently employed in differential expression analyses using methods such as DESeq2, edgeR, LIMMA, EBSeq [48, 74–76], particularly in standard quantification workflows for coding genes. Studies focusing on TEs require specialized analytical approaches beyond these standard pipelines.

Regarding the expression of TEs, STAR has been the recommended choice within the TE research field [20, 26, 62, 63, 77–79]; however, its parameters must be carefully tailored to specifically address the unique challenges presented by TE sequences [52]. STAR offers three reporting modes unique, random, and multi each suited to different alignment needs as explained in **Note 12**.

1. Increase the default threshold (--outFilterMultimapNmax) to a higher value (e.g., 5000 instead of the default of 10) to effectively capture abundant multi-mapping TE-derived reads.
2. Use --outMultimapperOrder Random to minimize alignment assignment bias among reads mapping equally well to multiple genomic locations.
3. Tailor STAR alignment parameters specifically according to the genome composition and the nature of TEs studied, ensuring optimal accuracy and specificity.
4. Adjust parameters such as --chimSegmentMin and --chimJunctionOverhangMin to facilitate accurate identification of spliced junctions and chimeric transcripts. This is crucial for detecting novel transcript isoforms involving TE-gene fusion events, or TEs acting as alternative promoters or enhancers.
5. Moreover, the alignment parameters must be carefully tailored to the genome and its specific TE content [52].

Standard quantification tools, primarily designed for gene expression analysis, often underestimate TE transcripts due to the inherently repetitive nature of TEs. Numerous specialized tools have been developed for TE expression quantification as mentioned in **Note 13**. However, we adhere to TEtranscripts (Fig. 3) and SQuIRE (Fig. 4), with TEtranscripts being the most widely used tool for subfamily-level analysis and SQuIRE functioning as an end-to-end pipeline for detailed locus-specific quantification. These TE-focused tools are better suited to capture the complex transcriptional landscape of TEs including polymorphic insertions and pervasive co-transcription with host genes that standard methods often miss. None of these tools alone provides a complete solution; the choice of methods

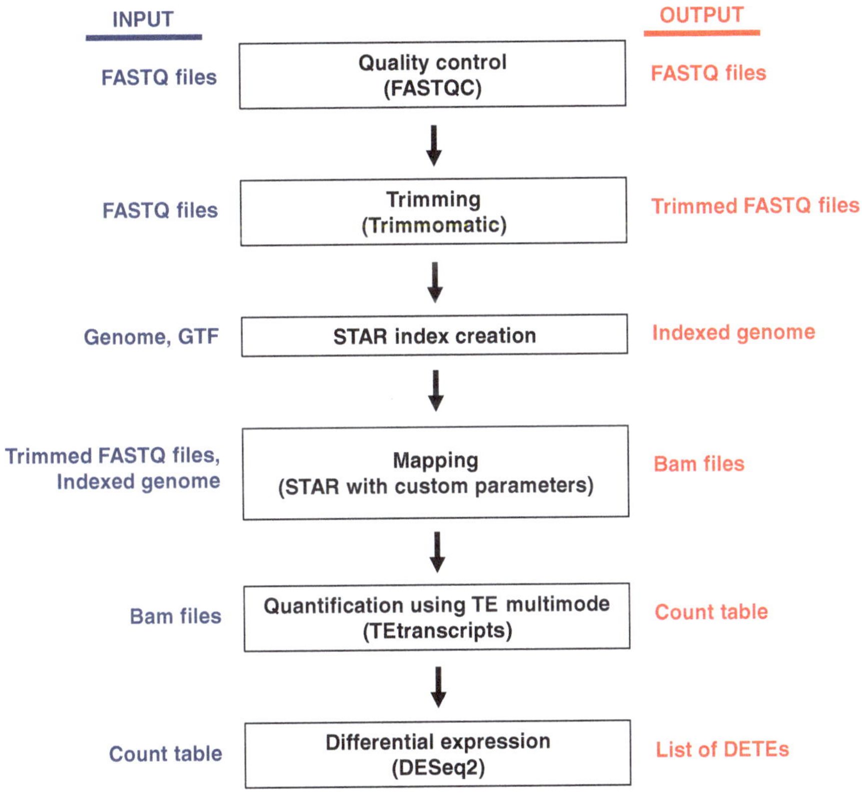

Fig. 3 Sub-family analysis of TE expression using TEtranscripts. Raw FASTQ reads are first quality-checked (FastQC/MultiQC) and adapter-/quality-trimmed (Trim Galore). Reads are mapped to the genome with STAR using repeat-optimized parameters. TEtranscripts assigns uniquely mapped reads directly and uses an expectation–maximization step to apportion multi-mapped reads across TE subfamilies, producing subfamily-level count tables that are normalized and analyzed for differential expression with DESeq2

should be driven by your specific research questions, often requiring complementary approaches.

TEtranscripts:

1. Prepare input BAM files from aligners such as STAR.
2. Download the necessary TE annotation GTF files from the TEtranscripts website.
3. Employ TEtranscripts in "multi" mode, using the expectation–maximization (EM) algorithm, to iteratively allocate multimapping reads among closely related TE subfamilies. This step helps to mitigate biases against younger TEs with fewer uniquely mappable regions.
4. Combine unique-read counts with optimized multi-read abundances to produce final TE transcript abundances and summarize counts at TE subfamily (element) level.
5. Conduct differential expression using DESeq2.

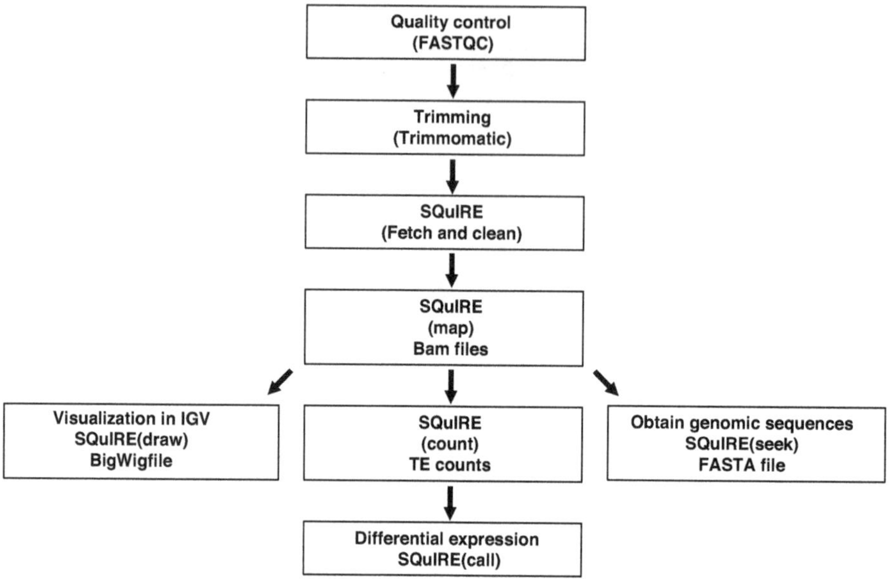

Fig. 4 End-to-end pipeline for loci level expression of TEs using SQuIRE. Preparation: Fetch reference FASTA/GTF (genes + RepeatMasker TEs), build STAR index, and format TE GTF. Map: Align RNA-seq with STAR (multi-mode, discordant allowed). Count: Run SQuIRE Count to assign unique and multi-mapped reads at individual TE loci using an EM algorithm, producing locus-level counts and FPKM. Call: Execute SQuIRE Call (DESeq2) for differential locus-level TE expression. Follow-up: Generate bedgraphs for visualization (Draw) and extract TE sequences (Seek) for downstream analyses

SQuIRE:

1. Fetch and Clean: Obtain genome annotation files (RepeatMasker, RefSeq) and create STAR alignment indices. Reformat TE annotations into BED format for analysis.
2. Map: Align RNA-seq data using STAR with optimized parameters for multi-mapping TE-derived reads.
3. Count: Quantify expression levels at locus-specific and subfamily levels using unique and multi-mapped reads via an expectation-maximization (EM) algorithm.
4. Call: Perform differential expression analysis for TEs and genes using DESeq2.
5. Draw: Generate bedgraph files for visualization in genome browsers like IGV or UCSC.
6. Seek: Retrieve FASTA sequences for user-specified genomic regions or TE loci.

3.3.6 Evolving Best Practices

Despite the distinct goals of ChIP-seq and RNA-seq, evolving best practices increasingly converge on a shared need to handle multi-mapped reads with care, particularly in regions enriched for TEs.

In both histone mark profiling (CUT&RUN/ChIP-seq) and gene expression analysis (RNA-seq), retaining multi-mapped reads without robust modeling can lead to inflated signals in repetitive regions, whereas discarding them outright often underestimates true biological variation. A hybrid approach such as Bowtie2's default single-best-mapping mode for ChIP-seq and STAR's multimapping parameters for RNA-seq helps balance sensitivity and specificity by retaining enough multi-mapped reads to capture true TE events while minimizing false positives. This synergy is further enhanced by specialized bioinformatic tools (e.g., TEtranscripts, SQuIRE) that employ expectation–maximization algorithms to distribute ambiguous reads more accurately. Through these combined strategies, researchers can glean nuanced epigenetic and transcriptomic insights from TE-rich regions, while adhering to widely accepted pipelines and parameter settings for both ChIP-seq and RNA-seq.

3.3.7 Long-Read Sequencing: A Promise for Accurate TE Mapping

Recent advances in long-read sequencing technologies, such as Pacific Biosciences and Oxford Nanopore platforms, offer significant advantages for TE research by overcoming many limitations of short-read sequencing in TEs. Long-read methods can span entire TE or even multiple adjacent repeats, thereby reducing alignment ambiguity, improving de novo assembly, and facilitating the study of full-length TE insertions and their flanking regions. Moreover, these techniques enable single-molecule resolution of epigenetic marks, a feature especially relevant for dissecting complex TE regulation and its interplay with chromatin structure. For instance, newly developed methods like DiMeLo-seq leverage long-read, single-molecule sequencing to map protein-DNA interactions and endogenous DNA modifications simultaneously, even within highly repetitive domains [80]. As these strategies continue to mature and become more cost-effective, they will substantially enhance the accuracy of TE discovery, quantification, and functional characterization.

4 Conclusions

TEs challenge standard sequencing approaches due to their repetitive nature, necessitating special consideration of multi-mapped reads and careful parameter tuning to avoid skewed results. While gene-centric studies often default to uniquely mapped reads, TE-focused research shows that ignoring multi-mapped reads risks overlooking critical regulatory elements and underestimating TE expression. On the other hand, retaining every multi-mapper without reliable models inflates false positives and obscures genuine biological signals. Adapting established tools such as using Bowtie2 with TE-friendly settings for CUT&RUN data, leveraging STAR's multi-hit parameters for RNA-seq, and applying expectation-maximization-based methods for subfamily-level quantification enables researchers to unlock the functional complexity of TEs. Although many computational approaches attempt to address multi-mapping challenges, there is no one-size-fits-all solution, and no single pipeline can fully resolve all biases. Ultimately, researchers must determine which assumptions and trade-offs best suit their specific goals, accepting certain limitations while refining their analyses to capture the most accurate representation of TE regulatory activity. As emerging long-read technologies and novel algorithms continue to evolve, the ability to pinpoint TE contributions in both normal physiology and disease will only grow stronger, offering further insights into the role of TEs in the genome.

5 Notes

1. All activities involving the use of live vertebrate animals must be reviewed and approved by the Institutional Animal Care and Use Committee (IACUC) prior to the commencement of any research. Animal euthanasia must be performed by appropriately trained personnel and using methods approved by an IACUC protocol. The American Veterinary Medical Association Guidelines for the Euthanasian of Animals is a valuable resource.
2. Some tissues are more subject to rapid autolysis than others. When multiple tissues are being collected these should be isolated first. For example, tissues of the gastrointestinal tract, including the pancreas and intestine, are sensitive to autolysis. Additionally, if you are collecting tissue from a large group of animals, to limit the potential for autolysis it is best practice to euthanize animals one at a time and rapidly collect tissue.
3. Liquid nitrogen is extremely cold (−196 °C) and can cause severe burns. Appropriate safety precautions should be taken, including personal protective equipment, such as gloves and eye protection.
4. To facilitate the processing of multiple samples, tissue may also be wrapped in a small piece of heavy-duty aluminum foil (pre-cooled) prior to placing in mortar. This keeps tissue powder contained, simplifying clean-up between tissues and limits potential cross-sample contamination.
5. If intending to use nuclei for single nuclei RNA-sequencing, RNase inhibitors should be included. RNase inhibitors successfully applied to this protocol are Protector RNase Inhibitor (#50-100-3383, Roche) and SUPERase-In RNase Inhibitor (#AM2696, Invitrogen), 30 μL of each added per 1 mL of 6× Homogenization Buffer–Unstable.

6. Good quality intact nuclei should be round and smooth with a continuous and intact membrane. Compromised nuclei may show evidence of blebbing and the membrane will not appear smooth. In addition to evaluating the integrity of the nuclei, suspensions should also be determined to be free of debris or nuclei clumping. Nuclei clumping can be a sign of over lysis, as the DNA released from the damaged nuclei can cause clumping. Consider reducing the number of strokes with the Dounce Homogenizer if clumping is detected. Additionally, nuclei clumping can also occur if the reagents being used are warmer than ice cold, ensuring all reagents are kept on ice can reduce clumping.
7. In its default mode, Bowtie2 searches for multiple valid alignments for each read but reports only one best alignment and selected randomly among equally scoring options to minimize bias in repetitive regions. With the -k mode, users can specify a maximum number of distinct alignments per read (for example, -k 2 will report up to two alignments), while the -a mode reports all valid alignments without an upper limit, listing them in descending order of score. Alternative modes (-k and -a) offer flexibility in capturing multiple mapping events, though they necessitate careful filtering to prevent inflated signals and false-positive peak calls.
8. There is a report that has adapted STAR's multimapping mode (explained in detail in Sect. 3.3.5) [20], typically used for RNA-seq, for ChIP-seq and CUT&RUN analyses. This approach allows for increased sensitivity in repetitive regions by enabling multimapping reads to be retained (e.g., – outFilterMultimapNmax 5000) while ensuring that each read is assigned to only one random location (– outSAMmultNmax 1). While STAR's multimapping mode enhances sensitivity in TE analysis for ChIP-seq and CUT&RUN, its application requires careful interpretation, as retaining multiple alignments may introduce challenges in distinguishing true binding events from background noise at peak calling.
9. Another group leveraged a SQuIRE-based pipeline (explained in Sect. 3.3.5) designed for RNA-seq TE quantification to process CUT&RUN data [65]. This pipeline offers a complementary strategy to traditional peak-calling methods by providing locus-specific quantitative measurements of histone marks at repetitive regions. In this method, CUT&RUN fastq files are first aligned to the reference genome, generating BAM files that contain mapped reads. These BAM files are then input into SQuIRE's counting module, which uses curated TE annotations to assign reads to individual TE loci. The resulting counts are aggregated at the TE subfamily

level, normalized using DESeq2's rlog transformation and finally visualized as heatmaps.

10. Strategies can be broadly categorized as coverage-based, enrichment-based, mapping-quality-based, or minimal/no-assumption methods with each making different assumptions about how ambiguous reads should be allocated [53]. Coverage-based or uni-mapper-driven approaches reassign ambiguous reads according to unique-read coverage but can underrepresent young TEs [81]. Enrichment-based methods presume that genuinely bound regions outcompete random coverage, risking misallocation if regions are equally enriched [82]. Most strategies assume that multi-mappers and uniquely mapping reads are similarly distributed across the genome, but multi-mappers are concentrated into a few regions of the genome [83]. Mapping-quality filter method incorporates alignment-quality scores into an iterative Bayesian reweighting algorithm, enabling proportional assignment of multi-mappers rather than their exclusion [83]. Specifically, this method maps reads with Bowtie2, retains and assigns initial weights to multi-mappers based on alignment quality, iteratively refines these weights using a Bayesian approach with efficient data structures, and finally uses the combined weighted signal for peak calling and further downstream analyses. Its utility is limited by a fixed maximum alignment threshold, potential over-refinement with excessive iterations, and significant memory overhead from its use of binary-indexed trees. Many pipelines favor minimal or no-assumption approaches for handling multimapping reads. For instance, some methods randomly assign a multi-mapper to one of its possible locations (as in Bowtie2's default), while others, such as T3E and Crunch, distribute the weight of a multimapping read among all its potential alignments, typically by inversely weighting according to the number of mappings [84, 85]. However, T3E computes enrichment from ChIP-seq data at the level of transposable element (TE) families/subfamilies rather than distinguishing among individual copies. In contrast, MapRRCon adopts a consensus-based strategy by first mapping reads to the full genome, then extracting those falling within annotated repetitive regions, and finally re-aligning these reads to a consensus sequence (or pseudogenome) for the TE subfamily [86]. However, this strategy aggregates signal across all copies of a TE subfamily rather than distinguishing individual insertions. LONUT reassigns non-unique tags to their most likely binding sites by calculating an empirical score based on distance to uniquely mapped peaks and peak enrichment, thereby enhancing detection of enriched regions in ChIP-seq-data [87].

11. T3E leverages simulated background distributions generated from the input control to estimate the expected random mapping of reads across the genome and compares this to the actual ChIP-seq read counts for TE families/subfamilies [84, 88]. Tools such as regioneR implement customizable permutation tests to statistically assess and characterize associations between genomic regions [89]. Tools like giggle compute enrichment scores by comparing the overlap between TE coordinates and ChIP-seq-peaks against a randomized background model, thereby quantifying the statistical significance of histone modification enrichment at specific TE families [90, 91]. Recently, a web-based approach, Transposable Element ENrichment Analyzer (TEENA) has been developed to improve speed and flexibility in TE enrichment analysis, utilizing a Sweeping Line algorithm for rapid overlap analysis and offering customizable parameters to exclude genomic gaps and promoter regions for specialized applications [92].
12. The unique mode reports only uniquely mapped reads for high-confidence alignments, while the random mode selects one alignment at random when a read maps equally well to multiple locations. In multi-mode, STAR reports all valid alignments up to a specified limit. However, care must be taken when reporting all multi-mappers; without proper filtering, specificity can be compromised, leading to inaccurate results in regions dense with TEs.
13. Numerous specialized tools have been developed for TE expression quantification as reviewed including TEtranscripts, SQuIRE, RepEnrich, TEtools, SalmonTE, ERVmap, TEcandidates, REdiscoverTE, TeXP, Telescope, and SoloTE [49, 50, 93–102]. SoloTE is the only tool for measuring locus-specific TE expression in single-cell RNA-Seq.

Acknowledgments

Work reported in this publication was supported by the National Cancer Institute and the National Institute of Diabetes and Digestive and Kidney Diseases of the National Institutes of Health under award numbers K22CA258674 and R01DK136632 to A.R.W. and a Steven Goldman Memorial Pancreatic Cancer Research Grant to A.R.W.

References

1. Shen H, Laird PW (2013) Interplay between the cancer genome and epigenome. Cell 153(1):38–55. https://doi.org/10.1016/j.cell.2013.03.008
2. You JS, Jones PA (2012) Cancer genetics and epigenetics: two sides of the same coin? Cancer Cell 22(1):9–20. https://doi.org/10.1016/j.ccr.2012.06.008

3. Hanahan D (2022) Hallmarks of cancer: new dimensions. Cancer Discov 12(1): 31–46. https://doi.org/10.1158/2159-8290.CD-21-1059
4. Ley TJ, Ding L, Walter MJ, McLellan MD, Lamprecht T, Larson DE, Kandoth C, Payton JE, Baty J, Welch J, Harris CC, Lichti CF, Townsend RR, Fulton RS, Dooling DJ, Koboldt DC, Schmidt H, Zhang Q, Osborne JR, Lin L, O'Laughlin M, McMichael JF, Delehaunty KD, McGrath SD, Fulton LA, Magrini VJ, Vickery TL, Hundal J, Cook LL, Conyers JJ, Swift GW, Reed JP, Alldredge PA, Wylie T, Walker J, Kalicki J, Watson MA, Heath S, Shannon WD, Varghese N, Nagarajan R, Westervelt P, Tomasson MH, Link DC, Graubert TA, DiPersio JF, Mardis ER, Wilson RK (2010) DNMT3A mutations in acute myeloid leukemia. N Engl J Med 363(25):2424–2433. https://doi.org/10.1056/NEJMoa1005143
5. Yan XJ, Xu J, Gu ZH, Pan CM, Lu G, Shen Y, Shi JY, Zhu YM, Tang L, Zhang XW, Liang WX, Mi JQ, Song HD, Li KQ, Chen Z, Chen SJ (2011) Exome sequencing identifies somatic mutations of DNA methyltransferase gene DNMT3A in acute monocytic leukemia. Nat Genet 43(4):309–315. https://doi.org/10.1038/ng.788
6. Jiao Y, Shi C, Edil BH, de Wilde RF, Klimstra DS, Maitra A, Schulick RD, Tang LH, Wolfgang CL, Choti MA, Velculescu VE, Diaz Jr.LA, Vogelstein B, Kinzler KW, Hruban RH, Papadopoulos N (2011) DAXX/ATRX, MEN1, and mTOR pathway genes are frequently altered in pancreatic neuroendocrine tumors. Science 331(6021):1199–1203. https://doi.org/10.1126/science.1200609
7. Schwartzentruber J, Korshunov A, Liu XY, Jones DT, Pfaff E, Jacob K, Sturm D, Fontebasso AM, Quang DA, Tonjes M, Hovestadt V, Albrecht S, Kool M, Nantel A, Konermann C, Lindroth A, Jager N, Rausch T, Ryzhova M, Korbel JO, Hielscher T, Hauser P, Garami M, Klekner A, Bognar L, Ebinger M, Schuhmann MU, Scheurlen W, Pekrun A, Fruhwald MC, Roggendorf W, Kramm C, Durken M, Atkinson J, Lepage P, Montpetit A, Zakrzewska M, Zakrzewski K, Liberski PP, Dong Z, Siegel P, Kulozik AE, Zapatka M, Guha A, Malkin D, Felsberg J, Reifenberger G, von Deimling A, Ichimura K, Collins VP, Witt H, Milde T, Witt O, Zhang C, Castelo-Branco P, Lichter P, Faury D, Tabori U, Plass C, Majewski J, Pfister SM, Jabado N (2012) Driver mutations in histone H3.3 and chromatin remodelling genes in paediatric glioblastoma. Nature 482(7384):226–231. https://doi.org/10.1038/nature10833
8. Wu G, Broniscer A, McEachron TA, Lu C, Paugh BS, Becksfort J, Qu C, Ding L, Huether R, Parker M, Zhang J, Gajjar A, Dyer MA, Mullighan CG, Gilbertson RJ, Mardis ER, Wilson RK, Downing JR, Ellison DW, Zhang J, Baker SJ, St. Jude Children's Research Hospital-Washington University Pediatric Cancer Genome P (2012) Somatic histone H3 alterations in pediatric diffuse intrinsic pontine gliomas and non-brainstem glioblastomas. Nat Genet 44(3):251–253. https://doi.org/10.1038/ng.1102
9. Sturm D, Witt H, Hovestadt V, Khuong-Quang DA, Jones DT, Konermann C, Pfaff E, Tonjes M, Sill M, Bender S, Kool M, Zapatka M, Becker N, Zucknick M, Hielscher T, Liu XY, Fontebasso AM, Ryzhova M, Albrecht S, Jacob K, Wolter M, Ebinger M, Schuhmann MU, van Meter T, Fruhwald MC, Hauch H, Pekrun A, Radlwimmer B, Niehues T, von Komorowski G, Durken M, Kulozik AE, Madden J, Donson A, Foreman NK, Drissi R, Fouladi M, Scheurlen W, von Deimling A, Monoranu C, Roggendorf W, Herold-Mende C, Unterberg A, Kramm CM, Felsberg J, Hartmann C, Wiestler B, Wick W, Milde T, Witt O, Lindroth AM, Schwartzentruber J, Faury D, Fleming A, Zakrzewska M, Liberski PP, Zakrzewski K, Hauser P, Garami M, Klekner A, Bognar L, Morrissy S, Cavalli F, Taylor MD, van Sluis P, Koster J, Versteeg R, Volckmann R, Mikkelsen T, Aldape K, Reifenberger G, Collins VP, Majewski J, Korshunov A, Lichter P, Plass C, Jabado N, Pfister SM (2012) Hotspot mutations in H3F3A and IDH1 define distinct epigenetic and biological subgroups of glioblastoma. Cancer Cell 22(4):425–437. https://doi.org/10.1016/j.ccr.2012.08.024
10. Blackburn EH (2001) Switching and signaling at the telomere. Cell 106(6): 661–673. https://doi.org/10.1016/s0092-8674(01)00492-5
11. Bourque G, Burns KH, Gehring M, Gorbunova V, Seluanov A, Hammell M, Imbeault M, Izsvak Z, Levin HL, Macfarlan TS, Mager DL, Feschotte C (2018) Ten things you should know about transposable elements. Genome Biol 19(1):199. https://doi.org/10.1186/s13059-018-1577-z
12. Payer LM, Burns KH (2019) Transposable elements in human genetic disease. Nat Rev Genet 20(12):760–772. https://doi.org/10.1038/s41576-019-0165-8
13. Fueyo R, Judd J, Feschotte C, Wysocka J (2022) Roles of transposable elements in the regulation of mammalian transcription. Nat Rev Mol Cell Biol 23(7):481–497. https://doi.org/10.1038/s41580-022-00457-y

14. Ravindran S (2012) Barbara McClintock and the discovery of jumping genes. Proc Natl Acad Sci USA 109(50):20198–20199. https://doi.org/10.1073/pnas.1219372109
15. Wells JN, Feschotte C (2020) A field guide to eukaryotic transposable elements. Annu Rev Genet 54:539–561. https://doi.org/10.1146/annurev-genet-040620-022145
16. Elsasser SJ, Noh KM, Diaz N, Allis CD, Banaszynski LA (2015) Histone H3.3 is required for endogenous retroviral element silencing in embryonic stem cells. Nature 522(7555):240–244. https://doi.org/10.1038/nature14345
17. Goldberg AD, Banaszynski LA, Noh KM, Lewis PW, Elsaesser SJ, Stadler S, Dewell S, Law M, Guo X, Li X, Wen D, Chapgier A, DeKelver RC, Miller JC, Lee YL, Boydston EA, Holmes MC, Gregory PD, Greally JM, Rafii S, Yang C, Scambler PJ, Garrick D, Gibbons RJ, Higgs DR, Cristea IM, Urnov FD, Zheng D, Allis CD (2010) Distinct factors control histone variant H3.3 localization at specific genomic regions. Cell 140(5): 678–691. https://doi.org/10.1016/j.cell.2010.01.003
18. Wasylishen AR, Sun C, Moyer SM, Qi Y, Chau GP, Aryal NK, McAllister F, Kim MP, Barton MC, Estrella JS, Su X, Lozano G (2020) Daxx maintains endogenous retroviral silencing and restricts cellular plasticity in vivo. Sci Adv 6(32):eaba8415. https://doi.org/10.1126/sciadv.aba8415
19. He Q, Kim H, Huang R, Lu W, Tang M, Shi F, Yang D, Zhang X, Huang J, Liu D, Songyang Z (2015) The Daxx/Atrx complex protects tandem repetitive elements during DNA hypomethylation by promoting H3K9 trimethylation. Cell Stem Cell 17(3):273–286. https://doi.org/10.1016/j.stem.2015.07.022
20. Enriquez-Gasca R, Gould PA, Tunbak H, Conde L, Herrero J, Chittka A, Beck CR, Gifford R, Rowe HM (2023) Co-option of endogenous retroviruses through genetic escape from TRIM28 repression. Cell Rep 42(6):112625. https://doi.org/10.1016/j.celrep.2023.112625
21. Wu F, Liufu Z, Liu Y, Guo L, Wu J, Cao S, Qin Y, Guo N, Fu Y, Liu H, Li Q, Shu X, Pei D, Hutchins AP, Chen J, He J (2022) Species-specific rewiring of definitive endoderm developmental gene activation via endogenous retroviruses through TET1-mediated demethylation. Cell Rep 41(11):111791. https://doi.org/10.1016/j.celrep.2022.111791
22. Karttunen K, Patel D, Xia J, Fei L, Palin K, Aaltonen L, Sahu B (2023) Transposable elements as tissue-specific enhancers in cancers of endodermal lineage. Nat Commun 14(1):5313. https://doi.org/10.1038/s41467-023-41081-4
23. Bourque G, Leong B, Vega VB, Chen X, Lee YL, Srinivasan KG, Chew JL, Ruan Y, Wei CL, Ng HH, Liu ET (2008) Evolution of the mammalian transcription factor binding repertoire via transposable elements. Genome Res 18(11):1752–1762. https://doi.org/10.1101/gr.080663.108
24. Goke J, Lu X, Chan YS, Ng HH, Ly LH, Sachs F, Szczerbinska I (2015) Dynamic transcription of distinct classes of endogenous retroviral elements marks specific populations of early human embryonic cells. Cell Stem Cell 16(2):135–141. https://doi.org/10.1016/j.stem.2015.01.005
25. Pal D, Patel M, Boulet F, Sundarraj J, Grant OA, Branco MR, Basu S, Santos SDM, Zabet NR, Scaffidi P, Pradeepa MM (2023) H4K16ac activates the transcription of transposable elements and contributes to their cis-regulatory function. Nat Struct Mol Biol 30(7):935–947. https://doi.org/10.1038/s41594-023-01016-5
26. Fernandes LP, Enriquez-Gasca R, Gould PA, Holt JH, Conde L, Ecco G, Herrero J, Gifford R, Trono D, Kassiotis G, Rowe HM (2022) A satellite DNA array barcodes chromosome 7 and regulates totipotency via ZFP819. Sci Adv 8(43):eabp8085. https://doi.org/10.1126/sciadv.abp8085
27. Calvet S, Sallis S, Saksouk N, Rebouissou C, Teyssier C, Lesne A, Cammas F, Forne T (2022) Endogenous retroviral sequences behave as putative enhancers controlling gene expression through HP1-regulated long-range chromatin interactions. Cells 11 (15). https://doi.org/10.3390/cells11152392
28. Fuentes DR, Swigut T, Wysocka J (2018) Systematic perturbation of retroviral LTRs reveals widespread long-range effects on human gene regulation. Elife 7. https://doi.org/10.7554/eLife.35989
29. Raviram R, Rocha PP, Luo VM, Swanzey E, Miraldi ER, Chuong EB, Feschotte C, Bonneau R, Skok JA (2018) Analysis of 3D genomic interactions identifies candidate host genes that transposable elements potentially regulate. Genome Biol 19(1):216. https://doi.org/10.1186/s13059-018-1598-7
30. Zhang Y, Li T, Preissl S, Amaral ML, Grinstein JD, Farah EN, Destici E, Qiu Y, Hu R, Lee AY, Chee S, Ma K, Ye Z, Zhu Q, Huang H, Fang R, Yu L, Izpisua Belmonte JC, Wu J, Evans SM, Chi NC, Ren B (2019) Transcriptionally active HERV-H retrotransposons demarcate topologically associating domains in human pluripotent stem cells.

Nat Genet 51(9):1380–1388. https://doi.org/10.1038/s41588-019-0479-7
31. Choudhary MNK, Quaid K, Xing X, Schmidt H, Wang T (2023) Widespread contribution of transposable elements to the rewiring of mammalian 3D genomes. Nat Commun 14(1):634. https://doi.org/10.1038/s41467-023-36364-9
32. Burns KH (2017) Transposable elements in cancer. Nat Rev Cancer 17(7):415–424. https://doi.org/10.1038/nrc.2017.35
33. Kazazian Jr.HH, Wong C, Youssoufian H, Scott AF, Phillips DG, Antonarakis SE (1988) Haemophilia A resulting from de novo insertion of L1 sequences represents a novel mechanism for mutation in man. Nature 332(6160):164–166. https://doi.org/10.1038/332164a0
34. Schneider AM, Duffield AS, Symer DE, Burns KH (2009) Roles of retrotransposons in benign and malignant hematologic disease. Cellscience 6(2):121–145
35. Babaian A, Mager DL (2016) Endogenous retroviral promoter exaptation in human cancer. Mob DNA 7:24. https://doi.org/10.1186/s13100-016-0080-x
36. Andrews S (2010) FastQC: a quality control tool for high throughput sequence data. http://www.bioinformatics.babraham.ac.uk/projects/fastqc.
37. Ewels P, Magnusson M, Lundin S, Kaller M (2016) MultiQC: summarize analysis results for multiple tools and samples in a single report. Bioinformatics 32(19):3047–3048. https://doi.org/10.1093/bioinformatics/btw354
38. Krueger F, James F, Ewels P, Afyounian E, Weinstein M, Schuster-Boeckler B, Hulselmans G, Sclamons. (2023) FelixKrueger/TrimGalore: v0.6.10 - add default decompression path (0.6.10). Zenodo. https://doi.org/10.5281/zenodo.7598955.
39. Langmead B, Salzberg SL (2012) Fast gapped-read alignment with Bowtie 2. Nat Methods 9(4):357–359. https://doi.org/10.1038/nmeth.1923
40. Zhang Y, Liu T, Meyer CA, Eeckhoute J, Johnson DS, Bernstein BE, Nusbaum C, Myers RM, Brown M, Li W, Liu XS (2008) Model-based analysis of ChIP-Seq (MACS). Genome Biol 9(9):R137. https://doi.org/10.1186/gb-2008-9-9-r137
41. Meers MP, Tenenbaum D, Henikoff S (2019) Peak calling by sparse enrichment analysis for CUT&RUN chromatin profiling. Epigenet Chromatin 12(1):42. https://doi.org/10.1186/s13072-019-0287-4
42. Liao Y, Smyth GK, Shi W (2013) The Subread aligner: fast, accurate and scalable read mapping by seed-and-vote. Nucleic Acids Res 41(10):e108. https://doi.org/10.1093/nar/gkt214
43. Yu G, Wang LG, He QY (2015) ChIPseeker: an R/Bioconductor package for ChIP peak annotation, comparison and visualization. Bioinformatics 31(14):2382–2383. https://doi.org/10.1093/bioinformatics/btv145
44. Ramirez F, Ryan DP, Gruning B, Bhardwaj V, Kilpert F, Richter AS, Heyne S, Dundar F, Manke T (2016) deepTools2: a next generation web server for deep-sequencing data analysis. Nucleic Acids Res 44(W1):W160–W165. https://doi.org/10.1093/nar/gkw257
45. Heinz S, Benner C, Spann N, Bertolino E, Lin YC, Laslo P, Cheng JX, Murre C, Singh H, Glass CK (2010) Simple combinations of lineage-determining transcription factors prime cis-regulatory elements required for macrophage and B cell identities. Mol Cell 38(4):576–589. https://doi.org/10.1016/j.molcel.2010.05.004
46. Quinlan AR, Hall IM (2010) BEDTools: a flexible suite of utilities for comparing genomic features. Bioinformatics 26(6):841–842. https://doi.org/10.1093/bioinformatics/btq033
47. Dobin A, Davis CA, Schlesinger F, Drenkow J, Zaleski C, Jha S, Batut P, Chaisson M, Gingeras TR (2013) STAR: ultrafast universal RNA-seq aligner. Bioinformatics 29(1):15–21. https://doi.org/10.1093/bioinformatics/bts635
48. Love MI, Huber W, Anders S (2014) Moderated estimation of fold change and dispersion for RNA-seq data with DESeq2. Genome Biol 15(12):550. https://doi.org/10.1186/s13059-014-0550-8
49. Jin Y, Tam OH, Paniagua E, Hammell M (2015) TEtranscripts: a package for including transposable elements in differential expression analysis of RNA-seq datasets. Bioinformatics 31(22):3593–3599. https://doi.org/10.1093/bioinformatics/btv422
50. Yang WR, Ardeljan D, Pacyna CN, Payer LM, Burns KH (2019) SQuIRE reveals locus-specific regulation of interspersed repeat expression. Nucleic Acids Res 47(5):e27. https://doi.org/10.1093/nar/gky1301
51. Corces MR, Trevino AE, Hamilton EG, Greenside PG, Sinnott-Armstrong NA, Vesuna S, Satpathy AT, Rubin AJ, Montine KS, Wu B, Kathiria A, Cho SW, Mumbach MR, Carter AC, Kasowski M, Orloff LA, Risca VI, Kundaje A, Khavari PA, Montine TJ, Greenleaf WJ, Chang HY (2017) An improved ATAC-seq protocol reduces background and enables interrogation of frozen tissues. Nat Methods 14(10):959–962. https://doi.org/10.1038/nmeth.4396

52. Teissandier A, Servant N, Barillot E, Bourc'his D (2019) Tools and best practices for retrotransposon analysis using high-throughput sequencing data. Mob DNA 10:52. https://doi.org/10.1186/s13100-019-0192-1
53. Almeida da Paz M, Warger S, Taher L (2024) Disregarding multimappers leads to biases in the functional assessment of NGS data. BMC Genomics 25(1):455. https://doi.org/10.1186/s12864-024-10344-9
54. Sexton CE, Han MV (2019) Paired-end mappability of transposable elements in the human genome. Mob DNA 10:29. https://doi.org/10.1186/s13100-019-0172-5
55. Amarasinghe SL, Su S, Dong X, Zappia L, Ritchie ME, Gouil Q (2020) Opportunities and challenges in long-read sequencing data analysis. Genome Biol 21(1):30. https://doi.org/10.1186/s13059-020-1935-5
56. Zhang T, Cooper S, Brockdorff N (2015) The interplay of histone modifications - writers that read. EMBO Rep 16(11):1467–1481. https://doi.org/10.15252/embr.201540945
57. O'Neill K, Brocks D, Hammell MG (2020) Mobile genomics: tools and techniques for tackling transposons. Philos Trans R Soc Lond B Biol Sci 375(1795):20190345. https://doi.org/10.1098/rstb.2019.0345
58. Sundaram V, Cheng Y, Ma Z, Li D, Xing X, Edge P, Snyder MP, Wang T (2014) Widespread contribution of transposable elements to the innovation of gene regulatory networks. Genome Res 24(12):1963–1976. https://doi.org/10.1101/gr.168872.113
59. Deschamps-Francoeur G, Simoneau J, Scott MS (2020) Handling multi-mapped reads in RNA-seq. Comput Struct Biotechnol J 18:1569–1576. https://doi.org/10.1016/j.csbj.2020.06.014
60. Li H, Durbin R (2009) Fast and accurate short read alignment with Burrows-Wheeler transform. Bioinformatics 25(14):1754–1760. https://doi.org/10.1093/bioinformatics/btp324
61. Langmead B, Trapnell C, Pop M, Salzberg SL (2009) Ultrafast and memory-efficient alignment of short DNA sequences to the human genome. Genome Biol 10(3):R25. https://doi.org/10.1186/gb-2009-10-3-r25
62. He J, Fu X, Zhang M, He F, Li W, Abdul MM, Zhou J, Sun L, Chang C, Li Y, Liu H, Wu K, Babarinde IA, Zhuang Q, Loh YH, Chen J, Esteban MA, Hutchins AP (2019) Transposable elements are regulated by context-specific patterns of chromatin marks in mouse embryonic stem cells. Nat Commun 10(1):34. https://doi.org/10.1038/s41467-018-08006-y
63. Torre D, Fstkchyan YS, Ho JSY, Cheon Y, Patel RS, Degrace EJ, Mzoughi S, Schwarz M, Mohammed K, Seo JS, Romero-Bueno R, Demircioglu D, Hasson D, Tang W, Mahajani SU, Campisi L, Zheng S, Song WS, Wang YC, Shah H, Francoeur N, Soto J, Salfati Z, Weirauch MT, Warburton P, Beaumont K, Smith ML, Mulder L, Villalta SA, Kessenbrock K, Jang C, Lee D, De Rubeis S, Cobos I, Tam O, Hammell MG, Seldin M, Shi Y, Basu U, Sebastiano V, Byun M, Sebra R, Rosenberg BR, Benner C, Guccione E, Marazzi I (2023) Nuclear RNA catabolism controls endogenous retroviruses, gene expression asymmetry, and dedifferentiation. Mol Cell 83(23):4255–4271 e4259. https://doi.org/10.1016/j.molcel.2023.10.036
64. Yang J, Cook L, Chen Z (2024) Systematic evaluation of retroviral LTRs as cis-regulatory elements in mouse embryos. Cell Rep 43(3):113775. https://doi.org/10.1016/j.celrep.2024.113775
65. Fang Y, Barrows D, Dabas Y, Carroll TS, Tap WD, Nacev BA (2023) ATRX guards against aberrant differentiation in mesenchymal progenitor cells. bioRxiv. https://doi.org/10.1101/2023.08.08.552433
66. Chaouch A, Berlandi J, Chen CCL, Frey F, Badini S, Harutyunyan AS, Chen X, Krug B, Hebert S, Jeibmann A, Lu C, Kleinman CL, Hasselblatt M, Lasko P, Shirinian M, Jabado N (2021) Histone H3.3 K27M and K36M mutations de-repress transposable elements through perturbation of antagonistic chromatin marks. Mol Cell 81(23):4876–4890 e4877. https://doi.org/10.1016/j.molcel.2021.10.008
67. Hoffmann FT, Kim M, Beh LY, Wang J, Vo PLH, Gelsinger DR, George JT, Acree C, Mohabir JT, Fernandez IS, Sternberg SH (2022) Selective TnsC recruitment enhances the fidelity of RNA-guided transposition. Nature 609(7926):384–393. https://doi.org/10.1038/s41586-022-05059-4
68. Forestan C, Farinati S, Aiese Cigliano R, Lunardon A, Sanseverino W, Varotto S (2017) Maize RNA PolIV affects the expression of genes with nearby TE insertions and has a genome-wide repressive impact on transcription. BMC Plant Biol 17(1):161. https://doi.org/10.1186/s12870-017-1108-1
69. Yu M, Hu X, Pan Z, Du C, Jiang J, Zheng W, Cai H, Wang Y, Deng W, Wang H, Lu J, Sun MA, Cao B (2023) Endogenous retrovirus-derived enhancers confer the transcriptional regulation of human trophoblast syncytialization. Nucleic Acids Res 51(10):4745–4759. https://doi.org/10.1093/nar/gkad109

70. Du C, Jiang J, Li Y, Yu M, Jin J, Chen S, Fan H, Macfarlan TS, Cao B, Sun MA (2023) Regulation of endogenous retrovirus-derived regulatory elements by GATA2/3 and MSX2 in human trophoblast stem cells. Genome Res 33(2):197–207. https://doi.org/10.1101/gr.277150.122
71. Kim D, Paggi JM, Park C, Bennett C, Salzberg SL (2019) Graph-based genome alignment and genotyping with HISAT2 and HISAT-genotype. Nat Biotechnol 37(8):907–915. https://doi.org/10.1038/s41587-019-0201-4
72. Kim D, Pertea G, Trapnell C, Pimentel H, Kelley R, Salzberg SL (2013) TopHat2: accurate alignment of transcriptomes in the presence of insertions, deletions and gene fusions. Genome Biol 14(4):R36. https://doi.org/10.1186/gb-2013-14-4-r36
73. Patro R, Duggal G, Love MI, Irizarry RA, Kingsford C (2017) Salmon provides fast and bias-aware quantification of transcript expression. Nat Methods 14(4):417–419. https://doi.org/10.1038/nmeth.4197
74. Robinson MD, McCarthy DJ, Smyth GK (2010) edgeR: a Bioconductor package for differential expression analysis of digital gene expression data. Bioinformatics 26(1):139–140. https://doi.org/10.1093/bioinformatics/btp616
75. Ritchie ME, Phipson B, Wu D, Hu Y, Law CW, Shi W, Smyth GK (2015) limma powers differential expression analyses for RNA-sequencing and microarray studies. Nucleic Acids Res 43(7):e47. https://doi.org/10.1093/nar/gkv007
76. Leng N, Dawson JA, Thomson JA, Ruotti V, Rissman AI, Smits BM, Haag JD, Gould MN, Stewart RM, Kendziorski C (2013) EBSeq: an empirical Bayes hierarchical model for inference in RNA-seq experiments. Bioinformatics 29(8):1035–1043. https://doi.org/10.1093/bioinformatics/btt087
77. Pabis K, Barardo D, Sirbu O, Selvarajoo K, Gruber J, Kennedy BK (2024) A concerted increase in readthrough and intron retention drives transposon expression during aging and senescence. Elife 12. https://doi.org/10.7554/eLife.87811
78. Wu MJ, Kondo H, Kammula AV, Shi L, Xiao Y, Dhiab S, Xu Q, Slater CJ, Avila OI, Merritt J, Kato H, Kattel P, Sussman J, Gritti I, Eccleston J, Sun Y, Cho HM, Olander K, Katsuda T, Shi DD, Savani MR, Smith BC, Cleary JM, Mostoslavsky R, Vijay V, Kitagawa Y, Wakimoto H, Jenkins RW, Yates KB, Paik J, Tassinari A, Saatcioglu DH, Tron AE, Haas W, Cahill D, McBrayer SK, Manguso RT, Bardeesy N (2024) Mutant IDH1 inhibition induces dsDNA sensing to activate tumor immunity. Science 385(6705):eadl6173. https://doi.org/10.1126/science.adl6173
79. He J, Babarinde IA, Sun L, Xu S, Chen R, Shi J, Wei Y, Li Y, Ma G, Zhuang Q, Hutchins AP, Chen J (2021) Identifying transposable element expression dynamics and heterogeneity during development at the single-cell level with a processing pipeline scTE. Nat Commun 12(1):1456. https://doi.org/10.1038/s41467-021-21808-x
80. Altemose N, Maslan A, Smith OK, Sundararajan K, Brown RR, Mishra R, Detweiler AM, Neff N, Miga KH, Straight AF, Streets A (2022) DiMeLo-seq: a long-read, single-molecule method for mapping protein-DNA interactions genome wide. Nat Methods 19(6):711–723. https://doi.org/10.1038/s41592-022-01475-6
81. Chung D, Kuan PF, Li B, Sanalkumar R, Liang K, Bresnick EH, Dewey C, Keles S (2011) Discovering transcription factor binding sites in highly repetitive regions of genomes with multi-read analysis of ChIP-Seq data. PLoS Comput Biol 7(7):e1002111. https://doi.org/10.1371/journal.pcbi.1002111
82. Newkirk D, Biesinger J, Chon A, Yokomori K, Xie X (2011) AREM: aligning short reads from ChIP-sequencing by expectation maximization. J Comput Biol 18(11):1495–1505. https://doi.org/10.1089/cmb.2011.0185
83. Shah RN, Ruthenburg AJ (2021) Sequence deeper without sequencing more: Bayesian resolution of ambiguously mapped reads. PLoS Comput Biol 17(4):e1008926. https://doi.org/10.1371/journal.pcbi.1008926
84. Almeida da Paz M, Taher L (2022) T3E: a tool for characterising the epigenetic profile of transposable elements using ChIP-seq data. Mob DNA 13(1):29. https://doi.org/10.1186/s13100-022-00285-z
85. Berger S, Pachkov M, Arnold P, Omidi S, Kelley N, Salatino S, van Nimwegen E (2019) Crunch: integrated processing and modeling of ChIP-seq data in terms of regulatory motifs. Genome Res 29(7):1164–1177. https://doi.org/10.1101/gr.239319.118
86. Sun X, Wang X, Tang Z, Grivainis M, Kahler D, Yun C, Mita P, Fenyo D, Boeke JD (2018) Transcription factor profiling reveals molecular choreography and key regulators of human retrotransposon expression. Proc Natl Acad Sci USA 115(24):E5526–E5535. https://doi.org/10.1073/pnas.1722565115
87. Wang R, Hsu HK, Blattler A, Wang Y, Lan X, Wang Y, Hsu PY, Leu YW, Huang TH, Farnham PJ, Jin VX (2013) LOcating non-unique matched tags (LONUT) to improve the detection of the enriched regions for ChIP-seq

data. PLoS One 8(6):e67788. https://doi.org/10.1371/journal.pone.0067788
88. Trovato M, Bunina D, Yildiz U, Fernandez-Novel Marx N, Uckelmann M, Levina V, Perez Y, Janeva A, Garcia BA, Davidovich C, Zaugg JB, Noh KM (2024) Histone H3.3 lysine 9 and 27 control repressive chromatin at cryptic enhancers and bivalent promoters. Nat Commun 15(1):7557. https://doi.org/10.1038/s41467-024-51785-w
89. Gel B, Diez-Villanueva A, Serra E, Buschbeck M, Peinado MA, Malinverni R (2016) regioneR: an R/Bioconductor package for the association analysis of genomic regions based on permutation tests. Bioinformatics 32(2):289–291. https://doi.org/10.1093/bioinformatics/btv562
90. Layer RM, Pedersen BS, DiSera T, Marth GT, Gertz J, Quinlan AR (2018) GIGGLE: a search engine for large-scale integrated genome analysis. Nat Methods 15(2):123–126. https://doi.org/10.1038/nmeth.4556
91. Buttler CA, Ramirez D, Dowell RD, Chuong EB (2023) An intronic LINE-1 regulates IFNAR1 expression in human immune cells. Mob DNA 14(1):20. https://doi.org/10.1186/s13100-023-00308-3
92. Li Y, Lyu R, Chen S, Wang Y, Sun MA (2024) TEENA: an integrated web server for transposable element enrichment analysis in various model and non-model organisms. Nucleic Acids Res 52(W1):W126–W131. https://doi.org/10.1093/nar/gkae411
93. Lanciano S, Cristofari G (2020) Measuring and interpreting transposable element expression. Nat Rev Genet 21(12):721–736. https://doi.org/10.1038/s41576-020-0251-y
94. Criscione SW, Zhang Y, Thompson W, Sedivy JM, Neretti N (2014) Transcriptional landscape of repetitive elements in normal and cancer human cells. BMC Genomics 15: 583. https://doi.org/10.1186/1471-2164-15-583
95. Lerat E, Fablet M, Modolo L, Lopez-Maestre H, Vieira C (2017) TEtools facilitates big data expression analysis of transposable elements and reveals an antagonism between their activity and that of piRNA genes. Nucleic Acids Res 45(4):e17. https://doi.org/10.1093/nar/gkw953
96. Jeong HH, Yalamanchili HK, Guo C, Shulman JM, Liu Z (2018) An ultra-fast and scalable quantification pipeline for transposable elements from next generation sequencing data. Pac Symp Biocomput 23:168–179
97. Tokuyama M, Kong Y, Song E, Jayewickreme T, Kang I, Iwasaki A (2018) ERVmap analysis reveals genome-wide transcription of human endogenous retroviruses. Proc Natl Acad Sci USA 115(50):12565–12572. https://doi.org/10.1073/pnas.1814589115
98. Valdebenito-Maturana B, Riadi G (2018) TEcandidates: prediction of genomic origin of expressed transposable elements using RNA-seq data. Bioinformatics 34(22):3915–3916. https://doi.org/10.1093/bioinformatics/bty423
99. Kong Y, Rose CM, Cass AA, Williams AG, Darwish M, Lianoglou S, Haverty PM, Tong AJ, Blanchette C, Albert ML, Mellman I, Bourgon R, Greally J, Jhunjhunwala S, Chen-Harris H (2019) Transposable element expression in tumors is associated with immune infiltration and increased antigenicity. Nat Commun 10(1):5228. https://doi.org/10.1038/s41467-019-13035-2
100. Navarro FC, Hoops J, Bellfy L, Cerveira E, Zhu Q, Zhang C, Lee C, Gerstein MB (2019) TeXP: deconvolving the effects of pervasive and autonomous transcription of transposable elements. PLoS Comput Biol 15(8):e1007293. https://doi.org/10.1371/journal.pcbi.1007293
101. Bendall ML, de Mulder M, Iniguez LP, Lecanda-Sanchez A, Perez-Losada M, Ostrowski MA, Jones RB, Mulder LCF, Reyes-Teran G, Crandall KA, Ormsby CE, Nixon DF (2019) Telescope: characterization of the retrotranscriptome by accurate estimation of transposable element expression. PLoS Comput Biol 15(9):e1006453. https://doi.org/10.1371/journal.pcbi.1006453
102. Rodriguez-Quiroz R, Valdebenito-Maturana B (2022) SoloTE for improved analysis of transposable elements in single-cell RNA-Seq data using locus-specific expression. Commun Biol 5(1):1063. https://doi.org/10.1038/s42003-022-04020-5

Part IV

Single-Cell Chromatin Profiling and Imaging in Cancer Cells

Check for updates

Chapter 9

Single Molecule Tracking: Visualizing Transcription of Endogenous Loci in Living Cells

Pelin Yaşar, Christopher R. Day and Joseph Rodriguez

Abstract

Live-cell imaging of RNA provides mechanistic insights into the spatiotemporal organization and regulation of transcription. The most widely used method for visualizing transcription in living cells is the MS2 aptamer system. During nascent RNA synthesis of an MS2-labeled gene, a transcribed array of MS2 RNA stem-loops is bound by fluorescently labeled MS2 coat proteins to enable the visualization and tracking of single RNA molecules. Here, we present a protocol for visualizing nascent RNA transcription at an endogenous gene locus. Given the importance of the native chromatin context in regulating transcription, using this approach is crucial for uncovering mechanisms and understanding gene expression dynamics.

Keywords Single-molecule tracking, MS2, Single-cell, Transcription

1 Introduction

Live-cell imaging is a method used to observe cells in real-time through advanced microscopy. This technique often employs fluorescence and confocal microscopy to capture high-resolution images and videos of living cells. Unlike traditional methods that rely on fixed and stained samples, live-cell imaging captures cellular activities in their natural state, offering a dynamic view of biological and molecular processes. Over the last 20 years, the MS2 aptamer system has been used to directly label RNA in living cells. This system relies on the binding of fluorescently labeled bacteriophage MS2 coat proteins (MCPs) to an array of RNA stem-loops (Fig. 1) [1–3]. The enrichment of fluorescent MCP on single RNAs manifests as bright punctate spots on a fluorescence microscope . This approach has been utilized to investigate RNA biology broadly from synthesis and decay to RNA localization and transport [4].

One of the most frequent uses of MS2 is to investigate gene regulation. Transcription occurs in short bouts of nascent RNA synthesis known as transcriptional bursts. Given that live-cell MS2 imaging

Archana Dhasarathy and Motoki Takaku (eds.), *Cancer Epigenetics*, Methods in Molecular Biology, vol. 3005, https://doi.org/10.1007/978-1-0716-5072-1_9,

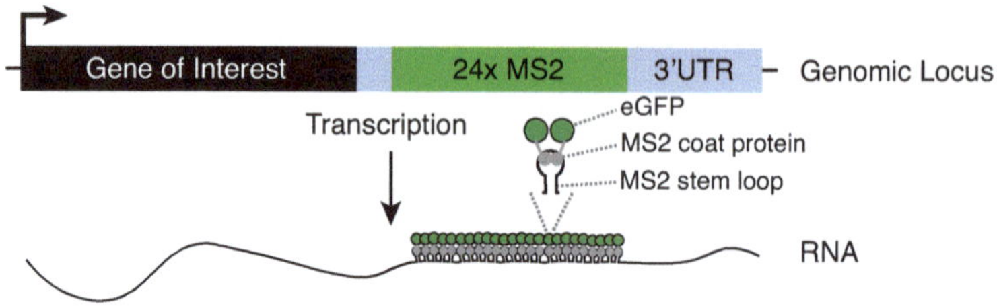

Fig. 1 Schematics of RNA visualization using the MS2/MCP-GFP imaging system. MS2 coat protein fused to GFP reporter recognizes and binds to newly transcribed MS2 stem loops

provides high temporal resolution, this approach is ideal for extracting spatial and kinetic information of this dynamic process. For example, the expression of the *Drosophila even-skipped (eve)* gene has acutely illustrated the spatiotemporal patterning of *eve* transcriptional bursting in developing embryos [5]. Moreover, the kinetic steps of RNA synthesis (burst initiation, elongation, splicing, and cleavage) were evaluated using dual-color live-cell imaging of a transgene labeled with MS2 and the orthologous system PP7 [6]. Furthermore, the development of CRISPR has opened new avenues by enabling the investigation of transcription dynamics from endogenous loci of interest, such as enhancers, long non-coding RNAs and coding genes [7].

The key advance of this method is the ability to monitor transient expression and molecular processes of endogenous genes over time. Since transcriptional heterogeneity gives rise to resistant cancer cells, it is critical to investigate mechanisms of transcriptional dynamics in single cells. Bulk cell assays fail to measure this heterogeneity in response to signaling pathways or therapeutic agents. In contrast, live-cell RNA imaging allows the quantification of the variation in transcriptional responses at the single cell level [8]. In addition, this method facilitates the interrogation of cancer-associated mutations of epigenetic factors and their effect on transient gene expression. In summary, live-cell MS2 imaging is a powerful tool that enables the direct and temporal observation of transcriptional activity in response to stimuli. This chapter focuses on preparing, optimizing, and utilizing the MS2 system for visualizing transcriptional activation of a specific gene in live cells.

2 Materials

2.1 Plasmid Cloning

1. Reporter plasmid containing 24×MS2 stem loops from Addgene (plasmid #31865).
2. Sequence of 2× MS2 cassette, stem loops are underlined: GATCCTACGGTACTTATTGCCAAGAAAGCAC-GAGCATCAGCCGTGCCTCCAGGTCGAATCT-TCAAACGACGACGATCACGCGTCGCTCCAGTATTC-CAGGGTTCATCAG

3. Protein expression plasmid containing GFP fused tandem MS2 Coat Protein (MS2-GFP plasmid, Addgene #61764).
4. Restriction enzymes for cloning.
5. Quick-Load Taq 2× Master Mix.
6. Gel DNA Recovery Kit, DNA Clean & Concentrator.
7. Screening Primers specific to the gene of interest (see Sect. 3.1 for the screening primer design).
8. Competent *E. coli* cells.
9. QIAprep Spin Miniprep Kit.
10. NEB Gibson Assembly Cloning Kit.
11. Luria Bertani Broth supplemented with 100 μg/mL ampicillin.

2.2 Tissue Culture and Lentivirus

1. MS2-GFP Lentivirus generated from the MS2-GFP plasmid (Addgene #61764).
2. Cell line: MCF-7 (ATCC #HTB-22).
3. Tissue culture dishes 10 cm.
4. Tissue culture 6-well and 96-well plates.
5. Glass bottom dishes for live-cell imaging (Nunc Lab-Tek II Chambered Coverglass No. 1.5).
6. Cell strainer for cell sorting, 40 μm nylon.
7. Polybrene (a cationic polymer that increases the infection efficiency).
8. Lipofectamine 3000 transfection reagent.
9. Puromycin for antibody selection of infected cells.
10. 17-beta estradiol.
11. Genomic DNA isolation kit.
12. Quick-DNA 96 Kit.

2.3 Media

1. Tissue culture maintenance media: MEM with Earle's Salts and without L-Glutamine supplemented with 1% L-glutamine, 1% Penicillin–Streptomycin, and 10% Fetal Bovine Serum.
2. Tissue culture hormone depletion media: MEM with Earle's Salts, without L-glutamine and Phenol Red-free, supplemented with 1% L-glutamine, 1% Penicillin–Streptomycin, and 10% Charcoal-Dextran Stripped FBS.
3. Tissue culture antibiotic selection media: MEM with Earle's Salts and without L-Glutamine supplemented with 1% L-glutamine, 1% Penicillin–Streptomycin, 10% Fetal Bovine Serum, and optimal concentration of puromycin (*see* Sect. 3.2.1).
4. Trypsin-EDTA (0.25%), with Phenol Red.

5. Sorted cells were collected in tissue culture maintenance media supplemented with 5% Penicillin–Streptomycin.
6. Live-cell imaging was performed in hormone-depletion media supplemented with 1 nM E2.
7. Media for transfection: Opti-MEM.

2.4 Fluorescent Activated Cell Sorting (FACS)

1. Cells: MS2-GFP coat protein synthesizing MCF7 cells.
2. Trypsin-EDTA (0.25%).
3. Hormone depletion media.
4. Falcon Round-Bottom Polystyrene Test Tubes.

2.5 Single-Molecule Fluorescence In Situ Hybridization (smFISH)

1. Quasar 670 labeled intron probes were ordered from LGC Biosearch Technologies.
2. MS2-L1 probe: ATGAACCCTGGAATACTGGAG/3Cy3Sp/ MS2-L2 probe: TTTGAAGATTCGACCTGGAG/3Cy3Sp/.
3. Fixation solution: 4% PFA. Mix 1.5 mL 32% Paraformaldehyde and 10.5 mL PBS.
4. Permeabilization solution: 0.5% Triton X-100. 125 μL Triton X-100 and 25 mL PBS.
5. Prehybridization/Wash Buffer: 5 mL 20× SSC, 5 mL deionized formamide, 40 mL ultrapure DEPC-treated water. Prepare fresh.
6. Hybridization buffer: Dissolve 1 g dextran sulfate in 7 mL ultrapure DEPC-treated water by rotating (may take 1–2 h). Once dissolved, add 1 mL 20× SSC and 1 mL deionized formamide, and fill up with ultrapure DEPC-treated water to 10 mL. Prepare fresh or aliquot in 1 mL and store at −20°C.
7. Nuclei staining: ProLong Gold Antifade Mountant with DNA Stain DAPI.
8. 18 mm #1.5 thickness coverslips.

2.6 Live-Cell Imaging

A widefield or confocal microscope equipped with a 40× or 60× 1.4 NA objective and an EMCCD camera is sufficient for live-cell imaging.

Example Microscope:

Andor Dragonfly in spinning disk confocal mode equipped with:

1. 1.49 NA, CFI Apo TIRF 60x objective (Nikon).
2. iXon EMCCD.
3. 40 μm pinhole.
4. Drift Stabilization.
5. Environmental control system CO_2 and 37°C (Cage Incubator, Okolab).

2.7 Data Analysis

1. ImageJ FIJI. https://imagej.net/software/fiji/downloads
2. Localize. (9) https://github.com/CBIIT/Larson-Lab-CCR-NCI
3. HMM trace fit (8, 10) https://sites.psu.edu/smbiophysics/hmm-code-and-manual/

3 Methods

The overall strategy to enable live-cell transcription imaging of endogenous genes is to integrate approximately 1.4 kb of 24× MS2 loops into a non-coding region of the gene locus using CRISPR homologous repair. To efficiently achieve this task at moderately expressed loci, using a promoter-driven resistance gene is recommended. However, this promoter resistance gene must be removed prior to using the cell line for imaging. An outline of the major steps is provided in Fig. 2.

3.1 Plasmid Design and Cloning to Label Gene of Interest with 24× MS2 Stem Loops

The primary goal of this section is to generate a CRISPR donor vector containing 24× MS2 stem loops, a promoter-driven puromycin resistance gene with polyA tail signal, which is flanked by flippase recombination target sites (FRT) to edit the genomic locus of the target gene. The 5′ and 3′ homology arms will be cloned around this editing cassette to ensure the MS2 stem loops are inserted to the desired region on the genome after homologous recombination (Fig. 3).

Screening is the most critical step of the gene editing process, and the strategy described here requires the incorporation of the screening primers into the donor vectors. These screening primers must be designed outside of the homology arms. Since the screening primers are incorporated into the donor construct, PCR amplification creates double-band amplicons from a single reaction if the integration is correct at the targeted locus.

1. Identify the insertion site for MS2 loops with the following considerations:
   a. Locate guide RNA binding sites near the desired insertion site of the MS2 stem loops (*see* **Note 1**).
   b. Ensure that insertion of the MS2 stem loops does not disrupt key regulatory elements of the RNA.
2. Define 5′ and 3′ homology arms between 500 and 1000 bp in length.
3. Identify PCR screening primers around the homology arms, which will amplify unedited alleles (screening primers will be designed based on the homology arm sequence, Fig. 3). Test primers using PCR of the locus using genomic DNA. For the

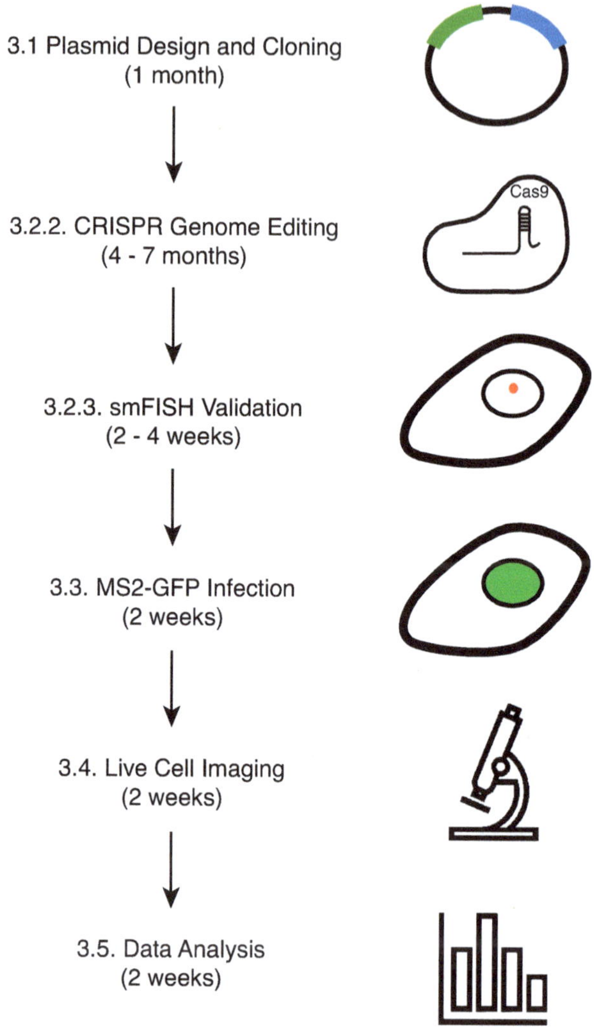

Fig. 2 Schematics endogenous editing workflow. Approximate durations of each step are provided based on MCF-7 cells

screening strategy to function properly, the distances between the primers and the insertion site should differ by at least 200 bp.

4. Clone a donor construct with 5′ homology arm, 24× MS2 stem loops and a PGK-Puromycin-polyA resistance gene flanked by FLP recognition target (FRT) sites (*see* **Note 2**) and ending with the 3′ homology arm. Insert reverse screening primer sequence (RSP) at the end of the 5′ homology arm and insert forward screening primer sequence (FSP) at the beginning of the 3′ homology arm and to aid in PCR screening of insertions (Fig. 3).
5. Clone guide RNA into the pX330 backbone using the standard Zhang Lab protocol (*see* **Note 3**).

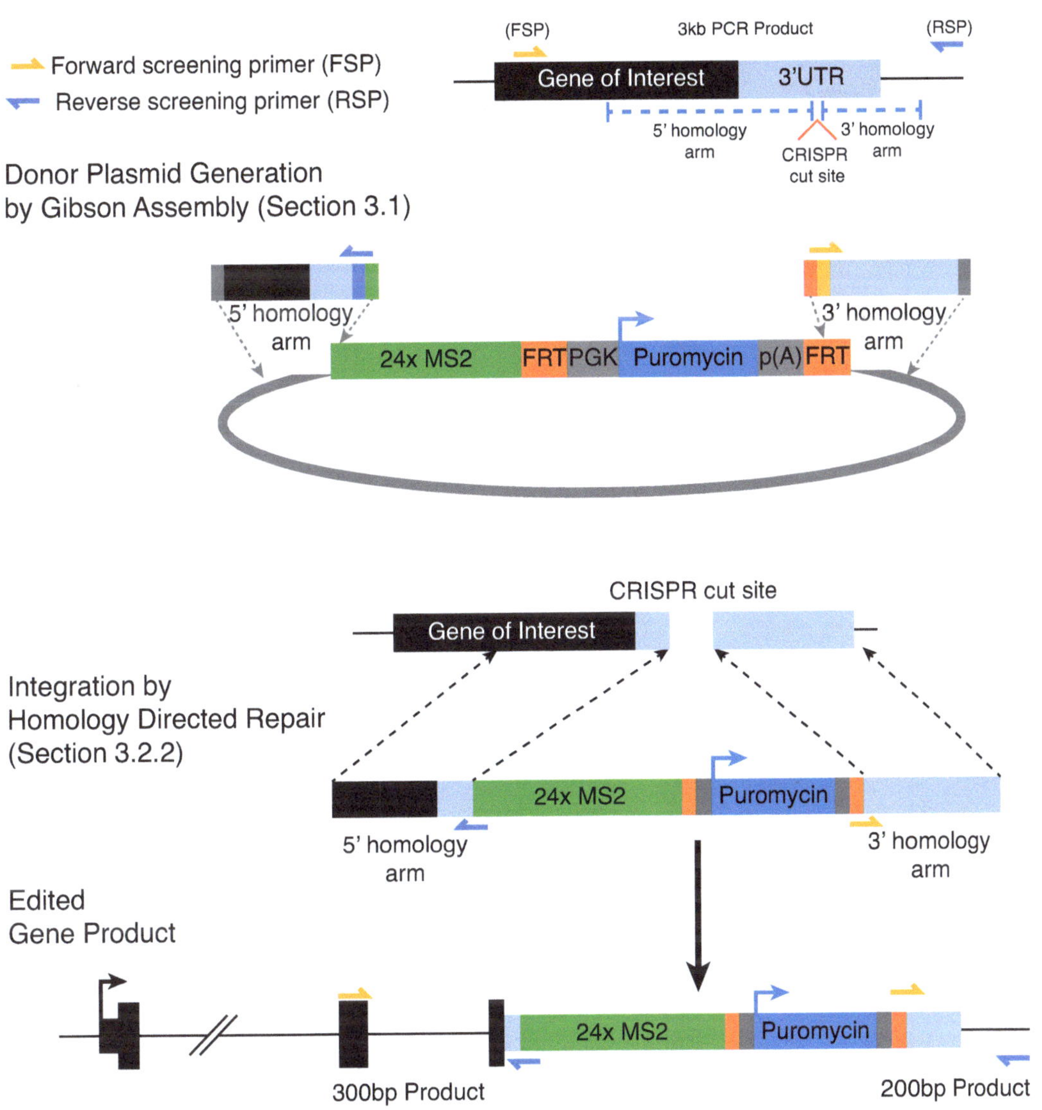

Fig. 3 Schematics of designing the CRISPR donor vector and 24× MS2 stem loop insertion. We generated a CRISPR donor vector with 5′ and 3′ homology arms to insert MS2 stem loops and puromycin resistance gene at the 3′ UTR of the gene of interest. A guide RNA is designed for targeting the insertion site and cloned into the pX330 vector. The cassette is then inserted through homology-directed repair when co-transfected with the pX330 plasmid, which contains the gRNA

3.2 Generation of Cells Expressing MS2-labeled RNA for Visualization

This section covers gene editing and screening steps. Since resistance will be used for selection, an optimal puromycin concentration for antibiotic selection is determined by generating a kill curve. The goal is to allow the cells enough time to recover and express the resistance gene from the edited loci before selection.

3.2.1 Puromycin Kill Curve for Antibiotic Selection of CRISPR-Edited Cells

1. Plate cells in a 6-well plate at an appropriate density in 2 mL maintenance media and grow for 2 days—plate two replicates for each concentration of puromycin. Cells should reach 70–80% confluence on the day antibiotic treatment begins.
2. Replace the tissue culture maintenance media with 2 mL media supplemented with increasing doses of puromycin (0, 0.05, 0.1, 0.2, 0.5, 1, 2 μg/mL). Change the antibiotic-supplemented media every 2 days.
3. Culture cells for 7–10 days and observe cells under the light microscope every day, and record the percentage of cell confluence.
4. On day 10 (*see* **Note 4**), determine the concentration of puromycin that kills all the cells (in both replicates). Control cells maintained in tissue culture maintenance medium (0 μg/mL) should be overconfluent by the time.
5. Generate kill curve graphs based on confluency data collected over this 10-day period.

3.2.2 CRISPR Gene Editing of 24× MS2 Stem Loops at the Endogenous Location

1. Plate cells in a 6-well plate with 2 mL tissue culture maintenance media.
2. Allow cells to reach 80% confluency, replacing media every 2 days.
3. Co-transfect cells with plasmids containing 5′ and 3′ homology arms with MS2 cassette and pX330 containing the guide RNA using Lipofectamine 3000 Transfection Reagent.
4. Maintain cells 10–14 days after transfection, expanding into a 10 cm dish if necessary (*see* **Note 5**).
5. Begin 10-day puromycin treatment with a dose identified from the kill curve.
6. After selection, allow cells to recover. When cells are confluent, pass 1/8 into a new dish. Pellet the remaining cells and isolate gDNA from excess cells using a Genomic DNA isolation Kit.
7. Confirm integration by performing the screening PCR using primers FSP and RSP (Fig. 4A). The cells with correct integration must exhibit amplification of two different band sizes.
8. After on-target integration is confirmed, remove the resistance gene and polyA site by treating cells with flippase to induce recombination at the FRT sites.
9. When cells reach confluency, pass the cells with a 1/8 ratio. Generate a single cell suspension by pipetting up and down with a P1000 tip.
10. Seed cells in a 96-well plate at a density of 0.8 cells per well to generate monoclones (*see* **Note 6**).

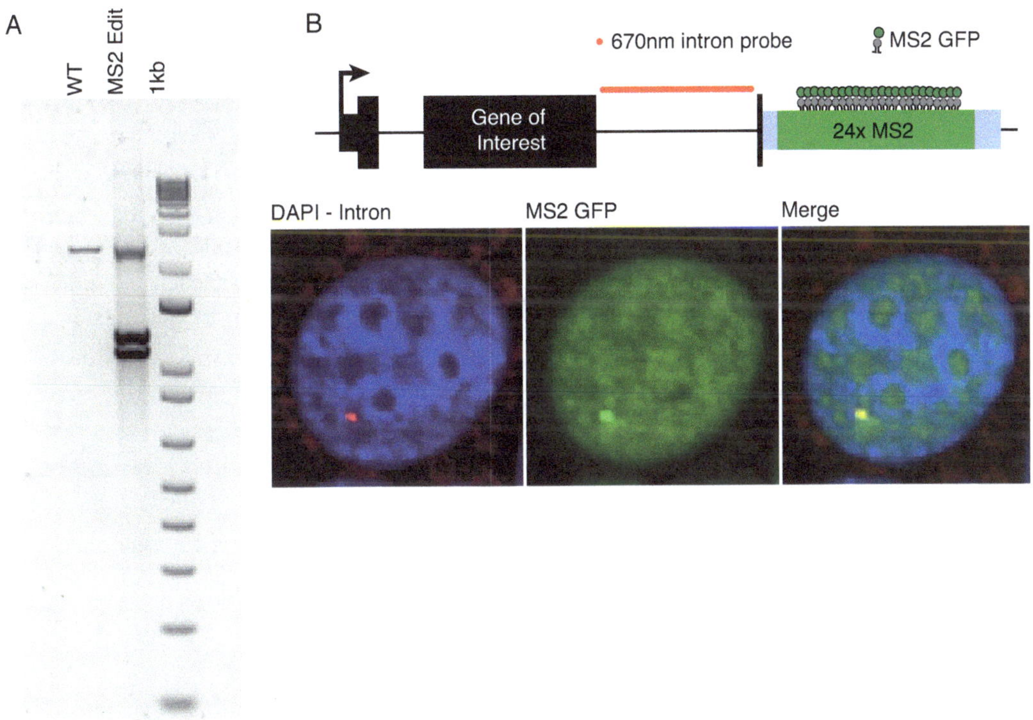

Fig. 4 (**A**) Representative double band PCR confirming MS2 integration at the endogenous locus. PCR reaction was performed using Forward Screening (FPS) and Reverse Screening Primers (RSP) on gDNAs from Wild Type (WT, parental cells) and CRISPR edited (MS2 Edit) cells. As expected in edited cells along with the WT size band, two bands appear due to the integration of the MS2 caseate carrying additional FSP and RSP. (**B**) A representative image of an edited MS2-GFP cell with a burst and smFISH for the intron of the endogenous gene. Colocalization of the two segments of RNA can be observed, validating integration at the expected locus

11. When the monoclones are confluent (*see* **Note 7**), pass the cells into replica plates with one plate receiving ¼ of the cells for maintenance and the rest of the cells seeded into the other plate for gDNA isolation.
12. When the second copy plate is confluent, isolate gDNA using the 96-well Genomic DNA isolation Kit to isolate the gDNA.
13. Use isolated gDNA from each clone, validate the removal of the Puromycin gene and repeat the double band PCR to finally confirm the selected monoclones that have the integration at the right location.

3.2.3 Screening of Monoclones with the Correct Integration Using smFISH

1. Reconstitute the probes to a 12.5 μM final concentration using TE.
2. Add an 18 mm #1.5 thickness coverslip to each well of a 12-well plate. Next, seed 65,000 cells per well.
3. Allow cells to reach 80–90% confluency, changing media every 2 days.

4. Fix cells with 4% PFA for 10 min, remove excess PFA with a PBS wash.
5. Permeabilize cells with 0.5% Triton X-100 for 5 min (*see* **Note 8**).
6. Prehybridize cells using wash buffer (*see* **Note 9**).
7. For each coverslip, mix 0.4 μL of each probe set, for a final concentration of 0.1 μM, with 50 μL of hybridization buffer. Prepare a volume sufficient for each coverslip to receive 50 μL of probe and hybridization buffer master mix.
8. Place a 50 μL drop of probe and hybridization on the surface of a chamber lined with parafilm. Next, place the coverslip cell side down on top of each droplet.
9. Incubate cells with probes suspended in hybridization buffer for 4 h at 37°C.
10. Remove coverslips, placing them back in a 12-well plate with 1 mL wash buffer per well for 30 min at 37°C.
11. Repeat 30 min wash at 37°C.
12. Rince cells with 1 mL 2× SSC.
13. Wash cells for 5 min at RT with 1 mL PBS.
14. Place 20 μL ProLong Gold Antifade Mountant with DNA Stain DAPI on a glass slide. Remove coverslips from PBS, air drying them for 2 min, before placing them cell surface down on the droplet of ProLong Gold. Allow the mounting media to cure for at least 24 h in the dark.
15. Using a microscope equipped with a 40× water objective and an EMCCD camera, capture 10 Fields of View (FOV) per coverslip.
16. For each FOV, collect z sections that span the nuclear volume, approximately 10 μm, with a spacing of 0.5 μm for the 405 (DAPI), 488 (MS2-GFP) and 640 (RNA FISH) channels.
17. Using max projections, identify a clone where 100% of the MS2 spots colocalize with intron spots (Fig. 4B).

3.3 Infecting Cell Lines With MS2-GFP Coat Protein Synthesizing Cell Line for Live-Cell Imaging

Optimizing coat protein levels is critical to obtaining good signal-to-noise and observing transcription sites. Infect positive integrant monoclones with the mammalian expression vector containing GFP fused MS2-coat protein to generate a stable MS2-GFP expressing cell line. Following the infection, a polyclonal cell line with heterogeneous MS2-GFP expressing cells is generated. FACS sorts these cells to obtain homogeneous MS2-GFP expressing cells with optimal levels of the protein.

1. Plate cells in a 6-well plate with 2 mL maintenance media.
2. When cells reach 70% confluency, infect the cells with the MS2-GFP coat protein virus with a Multiplicity of Infection (MOI) of 1, following viral safety protocols (*see* **Note 10**).
3. Allow cells to recover for 3 days.
4. Trypsinize and plate 65,000 cells on a Nunc Lab-Tek II Chambered Coverglass dish in tissue culture maintenance media.
5. Transfer the rest of the cells to a 10 cm dish in tissue culture maintenance media for growing as a polyclonal cell line.
6. Transfer cells to a phenol red-free tissue culture maintenance media and use a fluorescent microscope with a 40–60× objective to assess expression levels of the MS2-GFP coat protein in the population (*see* **Note 11**).
7. Perform FACS to isolate a population of cells with optimal, medium levels of MS2-GFP (Fig. 5A).

3.4 Live-Cell Imaging of MS2 Labeled Transcription Sites

At this point, a monoclonal cell line with MS2 stem loops inserted at an endogenous allele should have been generated and validated. This cell line can be used to examine transcription kinetics and RNA dynamics at the endogenous allele using time-lapse imaging. For the study of bursting dynamics, longer intervals of up to 100 s are used to observe burst durations and intervals between bursts. Additionally, imaging should take place over long periods of time, up to or longer than 14 h, to allow for the observation of many bursts and to generate an accurate estimate of the maximal intervals between bursts. However, careful consideration should be given to laser power and exposure times to minimize photobleaching.

1. Plate approximately 50,000 cells that have successful integration of MS2 stem loops in a Nunc Lab-Tek II Chambered Coverglass dish.
2. Allow cells to recover for 2 days in maintenance media.
3. Replace maintenance media with hormone depletion media supplemented with 1 nM Estrogen.
4. Place the Chambered Coverglass dish on the microscope with an incubator at 37°C and 5% CO_2.
   - An Andor Dragonfly spinning disk confocal equipped with a 60× Nicon objective, an Andor iXon EMCCD camera, a 40 μm pinhole and an Okolab Cage incubator was used to generate example data.
5. Place a generous amount of oil on the objective so as not to run out of oil during imaging.
6. Set the 488 nm laser to 6% power, 100 ms exposures and a gain of 100.
7. Find up to 10 FOVs of interest (*see* **Note 12**).

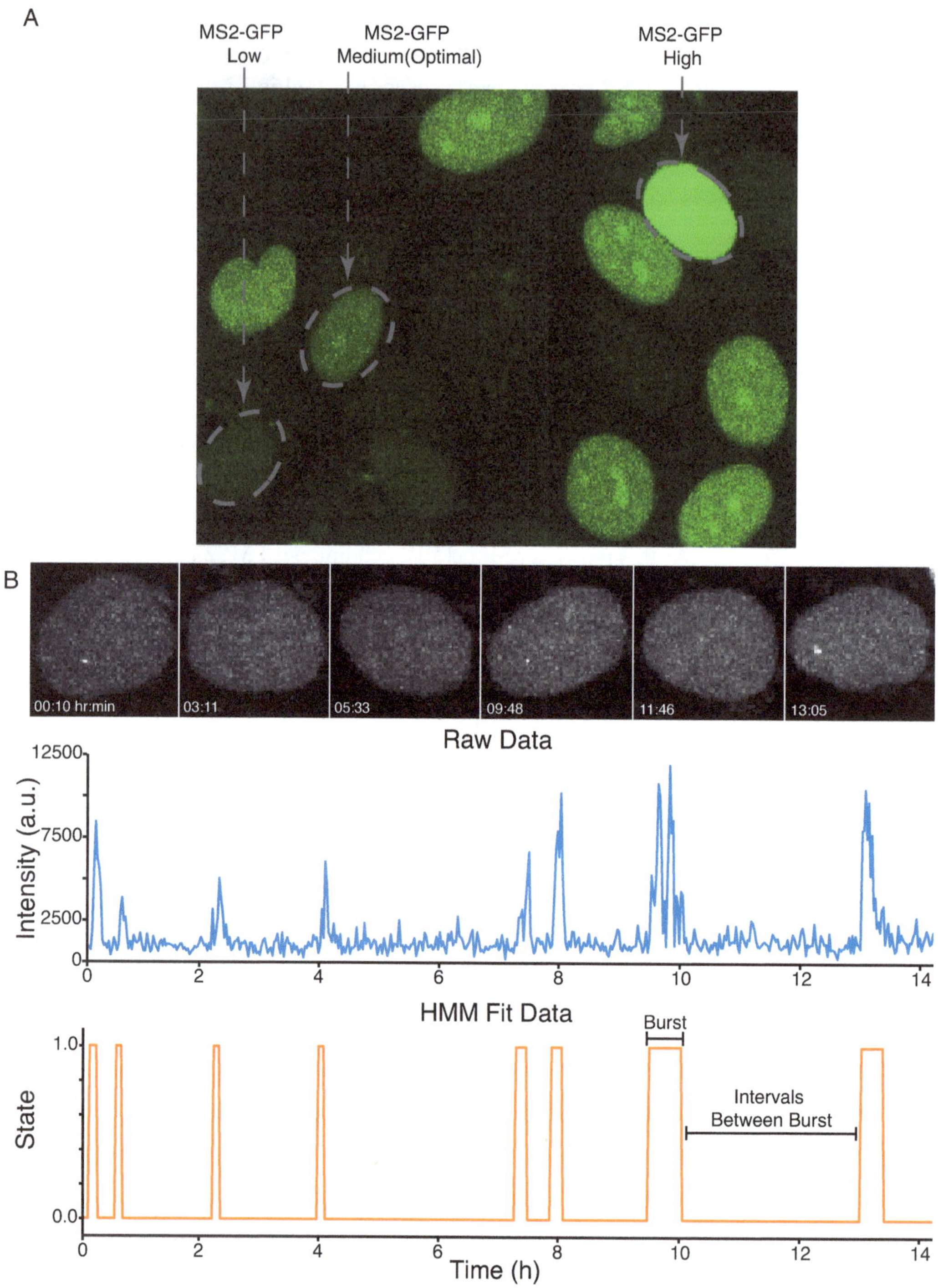

Fig. 5 (**A**) Polyclonal MS2-GFP MCF-7 cells. Representative image of MS2-GFP cells showing the heterogeneous levels of the coat protein expression with high, medium (optimal), or low levels. (**B**) Visualization of the bursting of the MS2MS2-labeled endogenous gene. The fluorescence time trace extracted from this allele and the fit of the HMM are displayed

8. When FOVs are identified, set up to 7 z steps which span 10 μm and center the z stack for each FOV.
9. Engage drift stabilization and check the focal plane for each FOV.
10. Set up a time lapse with 512 intervals spaced 100 s apart, which makes it a total of 14.2 h of imaging.

3.5 Transcription Site Tracking Analysis

The data generated by the cell lines can be used to extract kinetic information about bursting. Bursts in the time-lapse acquisitions are observed as diffraction-limited spots. Burst duration and interval can be determined by counting the number of frames the burst is present and the number of frames between bursts. To aid in this, a semiautomated pipeline is described below. This pipeline uses the software Localize to identify and track bursts. It then extracts the fluorescence intensity as a time trace. The intensity trace is fit to Hidden Markov models (HMM) to extract burst durations.

1. Open one FOV timelapse and use ImageJ FIJI to generate a maximum intensity projection. In the menu Image > Stacks > Z project.
2. Save the max projections as .tif files.
3. Observe each cell in the maximum intensity projection and identify cells with transcription sites (TS, Fig. 5B).
4. To reduce the number of TS in the FOV while tracking bursts, crop the nuclei with TS from the FOV (*see* **Note 13**).
5. To help initialize tracking of bursts while using the Localize software, copy the first frame with an active TS and append it to the beginning of the time course.
6. Using Localize with the FCS function activ,e extract intensities of TS and the background between bursts [9] (Fig. 5B, *see* **Note 14**).
7. Using .trk files generated by Localize, apply an HHM to extract on and off periods of the gene (Fig. 5B).
8. After this process is repeated for 30–50 alleles, a distribution of burst durations and intervals between bursts can be generated.

4 Notes

1. Online tools are available for gRNA design, including through IDT. https://www.idtdna.com/site/order/designtool/index/CRISPR_CUSTOM
2. LoxP sites and Blasticidin are also suitable for this application. Newer versions of MS2 stem loops are available from

Addgene. Versions 5–7 have different coat protein binding efficiencies and have been shown to generate fewer coat protein aggregates.

3. Zhang Lab CRISPR vector cloning protocol: https://media.addgene.org/data/plasmids/62/62988/62988-attachment_i-jdFt6Gm-ft.pdf

4. Higher doses of puromycin with 3–5 days of selection can also be used. However, lower doses of puromycin with longer selection times are preferred to decrease the antibiotic cytotoxicity and increase the yield of cells that are truly antibiotic resistant. The faster selection method has a caveat that causes the selection of transfected cells instead of cells with integrations.

5. The level of cell death due to the transfection varies from cell line to cell line. Therefore, it is important to observe cells during the recovery period after transfection and not let them become overconfluent.

6. We seeded 0.8 cells per well to prevent seeding more than one cell into a single well. It is important to generate true monoclones that originated from a single cell at this step. It is advisable to set up 10–20× 96-well plates at this stage. 30–60 clones are recovered from each plate, and low integration efficiency may result in the need to screen many clones.

7. We pass monoclonal cells at least they reach 80–90% confluency. Passaging before this confluency could result in the loss of the monoclones during trypsinization. Depending on the cell line, the cell proliferation rates may change, so observe the monoclones under a light microscope to decide the right time for passaging.

8. Ethanol is generally used to permeabilize cells; however, to preserve the eGFP signal, an incubation with 0.5% Triton X-100 in PBS for 5 min is used instead of 70% Ethanol.

9. Stellaris RNA FISH protocol for adherent cells: https://biosearch-cdn.azureedge.net/assetsv6/bti_stellaris_protocol_adherent_cell.pdf

10. A value of MOI = 1 means that, on average, every host cell will be receiving one virus particle; therefore, host cells will have one integrant of MS2-GFP. To prevent excess MS2-GFP coat protein expression in cells, we used an MOI of 1.

11. Extensive heterogeneity can be observed in the expression levels of the MS2-GFP coat protein. Determining the optimum level of MCP is critical for a robust signal-to-noise ratio (SNR) when bursts are active. Low levels of coat protein make it difficult to visualize transcription sites because of the low concentration of coat protein. On the other hand, high

levels of the coat protein result in masking of the transcription sites because the signal of the coat protein is not above the background.

12. Spacing FOVs close together reduces travel time between each FOV, allowing for more to be captured within 100 s, and reduces the chance of drift in the focal plane.
13. The ImageJ plugin Trackmate is an excellent tool for tracking and extracting nuclei.
14. Extracting the background between bursts allows the HMM model to better detect bursts.

Acknowledgments

This research was supported in part by the Intramural Research Program of the NIH ZIA ES103331 (J.R.). Use of the Andor Dragonfly was provided by the Fluorescence Microscopy and Imaging Center (NIEHS), and the MCP virus was generated by the Viral Vector Core (NIEHS).

References

1. Fouts DE, True HL, Celander DW (1997) Functional recognition of fragmented operator sites by R17/MS2 coat protein, a translational repressor. Nucleic Acids Res 25(22):4464–4473. https://doi.org/10.1093/nar/25.22.4464
2. Valegard K et al (1997) The three-dimensional structures of two complexes between recombinant MS2 capsids and RNA operator fragments reveal sequence-specific protein-RNA interactions. J Mol Biol 270(5):724–738. https://doi.org/10.1006/jmbi.1997.1144
3. Bertrand E et al (1998) Localization of ASH1 mRNA particles in living yeast. Mol Cell 2(4):437–445. https://doi.org/10.1016/s1097-2765(00)80143-4
4. Tutucci E, Livingston NM, Singer RH, Wu B (2018) Imaging mRNA In Vivo, from Birth to Death. Annu Rev Biophys 47:85–106. https://doi.org/10.1146/annurev-biophys-070317-033037
5. Bothma JP et al (2014) Dynamic regulation of eve stripe 2 expression reveals transcriptional bursts in living Drosophila embryos. Proc Natl Acad Sci USA 111(29):10598–10603. https://doi.org/10.1073/pnas.1410022111
6. Coulon A et al (2014) Kinetic competition during the transcription cycle results in stochastic RNA processing. Elife 3. https://doi.org/10.7554/eLife.03939
7. Spille JH et al (2019) A CRISPR/Cas9 platform for MS2-labelling of single mRNA in live stem cells. Methods 153:35–45. https://doi.org/10.1016/j.ymeth.2018.09.004
8. Rodriguez J et al (2019) Intrinsic dynamics of a human gene reveal the basis of expression heterogeneity. Cell 176(1–2):213–226 e218. https://doi.org/10.1016/j.cell.2018.11.026
9. Larson DR et al (2011) Real-time observation of transcription initiation and elongation on an endogenous yeast gene. Science 332(6028):475–478. https://doi.org/10.1126/science.1202142
10. Lee TH (2009) Extracting kinetics information from single-molecule fluorescence resonance energy transfer data using hidden Markov models. J Phys Chem B 113(33):11535–11542. https://doi.org/10.1021/jp903831z

Chapter 10

Advanced High-Resolution and Super-Resolution Microscopy Techniques for Investigating Chromatin Structure and Epigenetics

Yu-Lin Chen and Aussie Suzuki

Abstract

Fluorescence microscopy has emerged as an indispensable tool for investigating chromatin structure and epigenetic states. This technique not only enables super-resolution imaging to unveil detailed localizations and structures but also offers precise quantitative capabilities for comparing variations across multiple conditions. In this chapter, we present a comprehensive guide to immunofluorescence methods, detailing two primary fixation protocols. Additionally, we describe the application of super-resolution spinning disc microscopy, emphasizing its utility in detecting nuclear and epigenetic markers.

Keywords Immunofluorescence, Fluorescence Microscopy, Confocal microscopy, Super-resolution microscopy, Chromatin, Epigenetics

1 Introduction

Immunofluorescence (IF) staining is a technique for elucidating the subcellular localization, functional attributes, or intermolecular interactions of proteins. With recent advancements in microscopy, antibody engineering, and analysis software, the application of IF staining has significantly expanded. This versatile technique is now employed in various fields, from basic cell biology to translational medicine, including epigenetics, facilitating discoveries and advances in our understanding of biological processes and disease mechanisms [1, 2]. One advantage of IF methods is the ability to fluorescently label multiple targets within a cell using specific dyes and antibodies conjugated with fluorophores. This allows for the determination of the spatiotemporal dynamics of labeled targets [3], such as tracking epigenomic changes during development [4], during cell cycle progression [5, 6], and quantitatively comparing their amounts and locations between different samples, such as studying differences in epigenomic status between cancer and non-cancer cells [7, 8]. Ideal

Archana Dhasarathy and Motoki Takaku (eds.), *Cancer Epigenetics*, Methods in Molecular Biology, vol. 3005,
https://doi.org/10.1007/978-1-0716-5072-1_10,

fluorescence microscopy can achieve approximately 250 nm lateral and 600 nm axial resolution, known as the diffraction limit [9, 10]. Furthermore, super-resolution microscopy offers even greater resolution [10–12, 13], enabling the visualization of relatively small extrachromosomal DNA (ecDNA) in cancer cells [14] and the evolution of higher-order chromatin folding during carcinogenesis [15]. For accurate detection and quantification with fluorescence microscopy, it is crucial to capture images with a high signal-to-noise ratio (S/N), achievable through optimizations in IF methods [3, 9, 16].

The standard IF staining process begins with sample preparation, including chemical fixation to preserve cellular architecture integrity and permeabilization to facilitate antibody access to cellular antigens. We present two different chemical fixation methods: paraformaldehyde (PFA) and cold methanol. It is important to note that some antibodies are compatible only with one of these methods, but PFA fixation generally ensures better preservation of cellular architecture, including the nucleus, compared to methanol fixation [17, 18]. Following fixation, specimens are labeled with fluorescently tagged dyes and/or antibodies. In this protocol, we outline a comprehensive workflow for IF staining and imaging using a non-transformed retinal pigment epithelial RPE1 cell line, fluorescently labeled with DNA and an epigenetic histone marker (Histone H3K4me3 or Histone H3K27me3 as an example). Histone H3K4me3 is known as a marker for active euchromatin, whereas Histone H3K27me3 is a marker for chromatin with transcriptional silencing [19, 20]. This protocol encompasses the imaging process of spinning disc confocal microscopy and spinning disc-based super-resolution microscopy (Yokogawa CSU SoRa) [9, 21], achieving super-resolution inherently, without the need for post-image processing.

2 Materials

Prepare all solutions using DI water (18.2 MΩ-cm).

1. 3% PFA (500 mL): Mix 15 g of paraformaldehyde with about 300 mL of DI water. Mix and heat the solution on the hotplate stirrer at 65°C, avoiding exposure to light and preventing overheating (boiling). Gradually add 5N NaOH until the solution becomes clear. Add 125 mL of 1 M HEPES buffer (Sigma, H4034, pH 7.5). Make up to 500 mL with DI water, store at 4°C (for up to 1 month, avoiding exposure to light and temperature fluctuations) (*see* **Note 1**).
2. 500 mL of 1× PBS (Phosphate-Buffered Saline): 137 mM NaCl, 10 mM Na_2HPO_4, 2.7 mM KCl, 1.8 mM KH_2PO_4, pH 7.4.
3. Permeabilization buffer (50 mL): 0.5% NP-40 in 1× PBS.

4. Blocking buffer (50 mL): 0.5% of BSA (Bovine Serum Albumin, Sigma, A3311) in 1× PBS.
5. DAPI (4',6-Diamidino-2-phenylindole dihydrochloride) solution (10 mL): Mix 2 μL of 1 mg/mL DAPI (Sigma, D9542) in 10 mL 1× PBS.
6. Mounting medium (100 mL): 20 mM Tris (Sigma, T6066), pH 8.0, 0.5% N-propyl gallate (Sigma, 02370), 90% Glycerol (Sigma, 1.04095). Store at 4°C (avoiding exposure to light).
7. Coverslips with a thickness of #1.5 (*see* **Note 2**).
8. Microscope Slides (Corning, CLS294875x25).
9. Any fluorescence microscope: an example setup used in this protocol includes a Nikon Ti2 stand equipped with a Yokogawa CSU SoRa spinning disc confocal, Yokogawa Uniformizer, Piezo Z-stage (Prior), and lasers from Nikon with wavelengths of 405, 488, 561, and 640 nm, each at 100 mW. The system also included a Hamamatsu Flash V3 sCMOS camera and a 100× NA 1.45 oil objective. The entire setup was controlled by the Nikon Element software (NIS v5.30).
10. Antibodies: Histone H3K4me3 (Active Motif, 39060), Histone H3K27me3 (Cell Signaling, 9733S), Anti-centromere antibody (ACA) (Antibodies Incorporated, 15-234), Donkey anti-Rabbit IgG Alexa 488 (min X, JacksonImmuno Research, 711-546-152), Donkey anti-human IgG Rhodamine Red X (min X, 709-296-149) (*see* **Note 3**).

3 Methods

3.1 Acid-Wash Coverslips (see Note 4)

1. Heat coverslips in 1 M HCl at 60°C for 4–16 h.
2. Cool to room temperature.
3. Rinse the coverslips with DI water.
4. Sonicate coverslips with DI water in a water bath for 30 min.
5. Repeat washing and sonication (twice).
6. Rinse and sonicate with 50% ethanol in a water bath for 30 min.
7. Rinse and sonicate with 70% ethanol in a water bath for 30 min.
8. Rinse and sonicate with 95% ethanol in a water bath for 30 min.
9. Rinse and sonicate with 100% ethanol in a water bath for 30 min.
10. Store coverslips in 70% ethanol at room temperature.

3.2 Cell Culture

Acid-washed coverslips should be rinsed with 1× PBS (cell culture grade) before use. Cells should be plated on these acid-washed coverslips in a 35-mm or 24-well dish at least 1 day prior to fixation. In this protocol, we used the immortalized retinal pigment epithelial (RPE1) cell line as an example. It is essential to maintain the cells in an incubator under appropriate conditions (typically, 5% CO_2 at 37°C) until the time of fixation (*see* **Note 4**).

3.3 PFA Fixation (see Note 5)

1. Warm up 3% PFA in a 42°C water bath before use. Add 1 mL of 3% PFA to a new 24-well plate. Fix the cells by gently moving the coverslips into the wells containing 3% PFA and incubating at 37°C for 15 min.
2. Rinse fixed cells three times with 1× PBS.
3. Permeabilize the cells with 0.5% NP-40 for 15 min at room temperature.
4. Rinse the cells with 1× PBS three times.
5. Place a piece of parafilm on a solid surface. Add 15 μL of prepared primary antibody mix (anti-Histone H3K4me3 with ACA, or Histone H3K27me3 with ACA), and place the coverslip with the cell side down on the antibody droplet.
6. Incubate the cells at 37°C for 2.5 h in a humidity chamber.
7. Transfer coverslips into a 24-well plate (with cell side facing up) and rinse the cells three times with 1× PBS.
8. Prepare new parafilm, add 15 μL of prepared secondary antibody mix (Donkey anti-Rabbit IgG Alexa 488 and Donkey anti-human IgG Rhodamine Red X), and place the coverslip with the cell side down on the antibody droplet.
9. Incubate the cells at 37°C for 1.5 h in a dark, humidified chamber to avoid light exposure.
10. Transfer coverslips into a 24-well plate and rinse the cells three times with 1× PBS.
11. Incubate the cells with 500 μL of DAPI solution at room temperature for 15 min.
12. Rinse the cells three times with 1× PBS.
13. Mount the coverslip on a slide with 4.5 μL of mounting medium (*see* **Note 6**).
14. Remove the extra mounting medium by blotting the slide with filter paper or tissue paper (*see* **Note 7**).

3.4 Cold Methanol Fixation (see Note 5)

1. Chill the 100% methanol at −20°C before use. Add 2.5 mL of methanol to a new 24-well plate and store it at −20°C (*see* **Note 8**).
2. Incubate the cells with cold methanol at −20°C for 20 min by gently moving the coverslips into the wells containing the cold methanol.

3. Rinse fixed cells three times with 1× PBS.
4. Place a piece of parafilm in a humidity chamber or plate. Add 15 μL of prepared primary antibody mix (anti-Histone H3K4me3 with ACA, or Histone H3K27me3 with ACA), and place the coverslip with the cell side down on the antibody droplet.
5. Incubate the cells at room temperature for 1.5 h (*see* **Note 9**).
6. Transfer coverslips into a 24-well plate and rinse the cells three times with 1× PBS.
7. Prepare new parafilm, add 15 μL of prepared secondary antibody mix (Donkey anti-Rabbit IgG Alexa 488 and Donkey anti-human IgG Rhodamine Red X), and place the coverslip with the cell side down on the antibody droplet.
8. Incubate the cells at room temperature for 1 h in a dark, humidified chamber to avoid light exposure (*see* **Note 9**).
9. Transfer coverslips into a 24-well plate and rinse the cells with 1× PBS three times.
10. Incubate the cells with 500 μL of DAPI solution at room temperature for 15 min.
11. Rinse the cells three times with 1× PBS.
12. Mount the coverslip on a slide with 4.5 μL of mounting medium (*see* **Note 6**).
13. Remove the extra mounting medium by blotting the slide with filter paper or tissue paper (*see* **Note 7**).

3.5 High- and Super-Resolution Spinning Disc Confocal Microscopy Imaging

1. Set up the SoRa super-resolution microscopy system by turning on all components, including the camera, Yokogawa SoRa spinning disc system, microscope stand, lasers, and computer.
2. Launch the Nikon NIS Elements software and verify the connectivity of all microscopy components.
3. Clean the surface of the coverslips that you have stained.
4. Apply Nikon Type F immersion oil to a 100× oil objective lens with a numerical aperture (NA) of 1.49 (*see* **Note 10**).
5. Place a stained slide on the stage.
6. Carefully adjust the objective lens to make contact with the oil's surface (*see* **Note 11**).
7. In the Nikon NIS Elements software, select the CSU-W1 50 μm disc and set magnification to 1× (Fig. 1) (*see* **Note 12**).
8. Activate Live mode using the 405 nm laser wavelength and adjust the focus to locate the cells.
9. Once the cells are located, fine-tune the laser power and camera exposure time for all wavelengths (405, 488, and 561 nm).

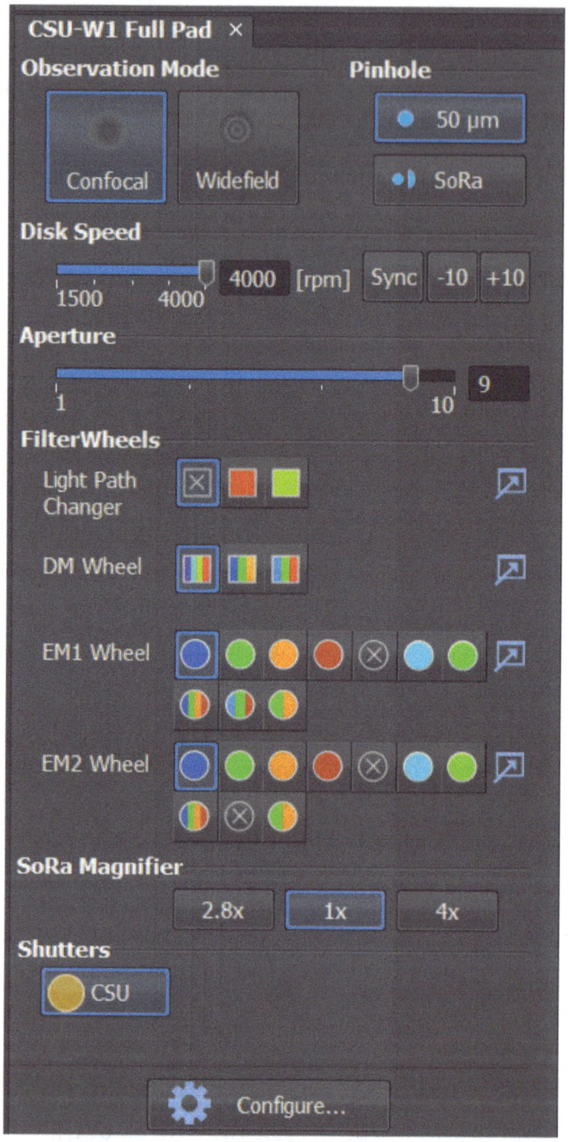

Fig. 1 NIS-Elements Software Confocal Setup: Select the "50 μm" Pinhole and "1×" SoRa Magnifier to achieve high-resolution confocal imaging

10. Determine the *z*-axis limits (bottom and top) of the cells and choose a z-step size of less than 200 nm.
11. Proceed with image acquisition.
12. Repeat steps 9 and 10 at a different location on the slide.
13. For super-resolution imaging, identify cells for imaging under the conditions described above.
14. Exit Live mode, then select the SoRa disc and set magnification to 2.8× in the Nikon NIS Elements software (Fig. 2).

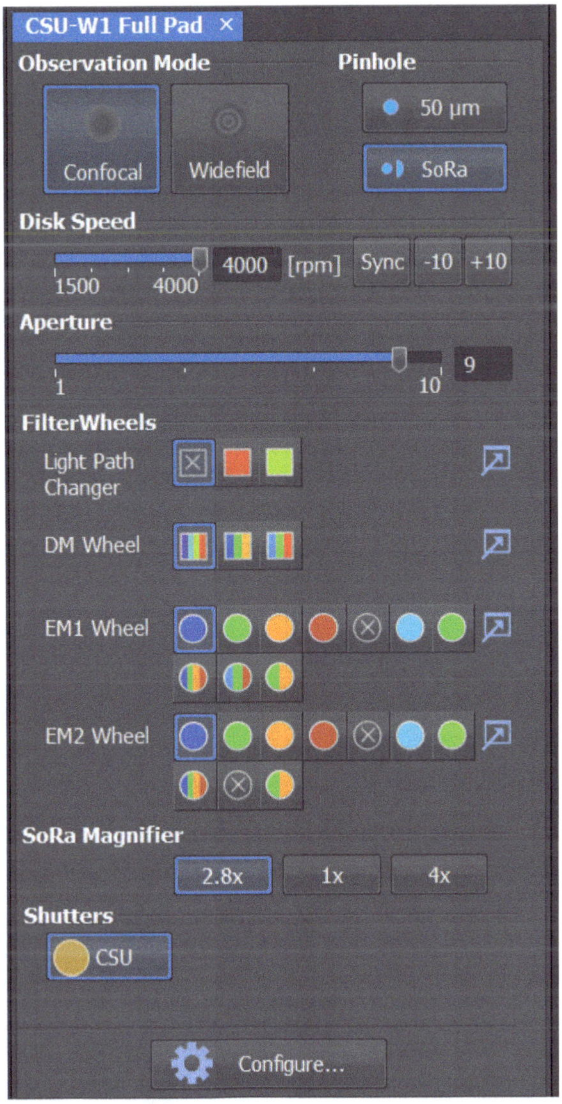

Fig. 2 NIS-Elements Software Super-Resolution SoRa Setup: Choose the "SoRa" Pinhole and "2.8×" SoRa Magnifier for super-resolution confocal microscopy imaging. A magnification of "2.8×" is suitable for 100× objectives, while a magnification of "4×" is recommended for 60× and lower magnification objectives

15. Re-enter Live mode using the 405 nm laser wavelength and adjust the focus.
16. Adjust the laser power and camera exposure time for all wavelengths (405, 488, and 561 nm).
17. Set the *z*-axis limits (bottom and top) of the cells and select a z-step size of less than 200 nm.

18. Execute image acquisition.
19. Continue the process outlined in steps 17 and 18 at a new location on the slide.

3.6 Review Images

1. The obtained images can be viewed for further quantification and visualization using Nikon NIS Elements, ImageJ (an open-source software), or other imaging software.
2. In our experience, most antibodies exhibited significantly brighter fluorescence with methanol fixation compared to paraformaldehyde (PFA) fixation (*see* Fig. 3). However, PFA fixation preserved cellular structures more effectively.
3. High-resolution confocal images of Histone H3 lysine 4 trimethylation (H3K4me3) with the centromere and Histone H3 lysine 27 trimethylation (H3K27me3) with the centromere are presented in Fig. 4. H3K4me3 staining was more uniformly distributed throughout the entire interphase nucleus, whereas H3K27me3 was predominantly localized in heterochromatin regions.
4. Figure 5 illustrates examples of both high-resolution confocal and super-resolution Structured Illumination Microscopy (SoRa) images of the same cell. The super-resolution SoRa image demonstrats significantly enhanced resolution.

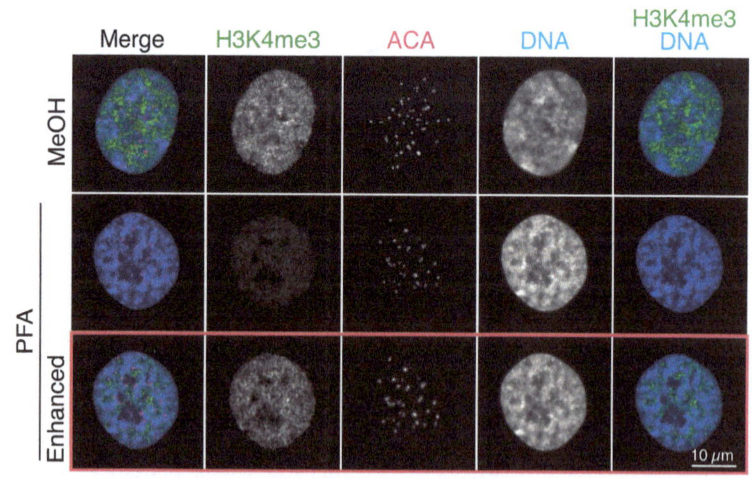

Fig. 3 Confocal Imaging of MeOH and PFA Fixed Samples: RPE1 cells were fixed using either methanol (MeOH) or paraformaldehyde (PFA) and subjected to immunofluorescence staining with anti-Histone H3K4me3 and ACA antibodies. DNA was visualized with DAPI staining. The top panels show cells fixed with MeOH, while the middle panels depict PFA-fixed cells. All images within these two sets were adjusted to the same contrast levels. The bottom panels display images identical to the middle set but with enhanced contrast for clearer visualization

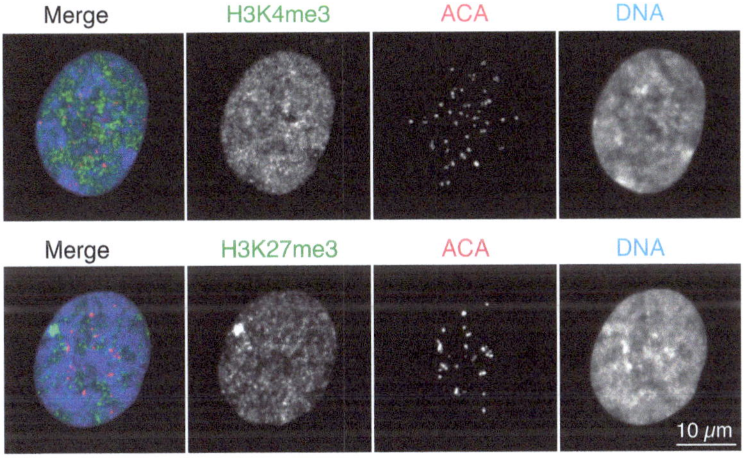

Fig. 4 Confocal Images Stained with Histone Antibodies: Samples were stained with either anti-Histone H3K4me3 or anti-Histone H3K27me3 antibodies, alongside labeling with ACA antibody and DAPI for additional structural and nuclear context. The top panels highlight staining with Histone H3K4me3, whereas the bottom panels feature Histone H3K27me3 staining, illustrating differential histone modifications

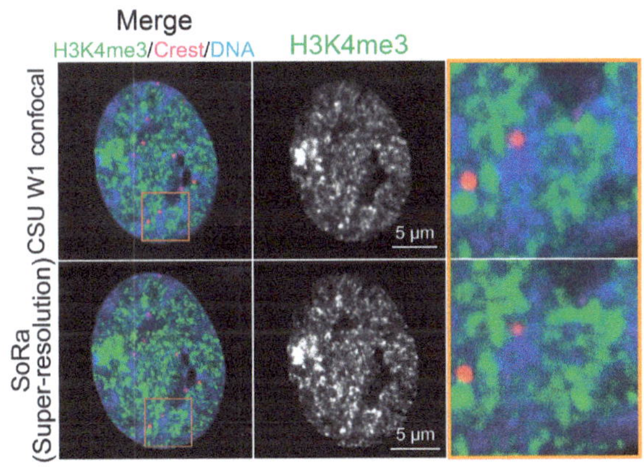

Fig. 5 Confocal and Super-Resolution SoRa Imaging of Histone H3K27me3: Samples stained with the anti-Histone H3K27me3 antibody were further labeled with ACA antibody and DAPI. Top images were captured using conventional confocal microscopy, and the bottom images were acquired through super-resolution SoRa microscopy, featuring the same cell for direct comparison. The SoRa images demonstrate significantly enhanced resolution, providing finer structural details

4 Notes

1. PBS is not recommended for a fixation solution. HEPES or PHEM buffer preserves cellular architecture better.
2. Any size of coverslips with #1.5 thickness can be used. However, most objectives for fluorescence microscopy are optimized for use with coverslips that are 0.170 mm thick.

As such, #1.5 coverslips are considered the optimal choice. However, it's important to note that the thickness of #1.5 coverslips can vary between 0.160 mm and 0.190 mm. For more consistent results, we recommend using #1.5H (high-performance) or #1.5HP (high precision) coverslips, which offer a thickness of 0.170 ± 0.005 mm. These specifications ensure the best and most consistent performance for fluorescence microscopy applications.

3. Although any vendor's secondary antibodies can be used, it is critical to use min X (minimal cross-reactivity) antibodies for multi-labeled immunofluorescence.
4. We recommend plating cells on acid-washed coverslips at least 1 day before fixation. For those interested in studying cell cycle progression, it is advisable to grow cells on coverslips for two consecutive nights.
5. Although samples fixed with PFA cannot be stored, samples fixed with methanol can be stored in methanol at −20°C for at least 1 week.
6. Any commercial mounting medium may be employed; however, the use of mounting media containing DAPI is discouraged due to the potential for increased background interference. Furthermore, caution is advised when considering hardening mounting mediums, as they are known to cause shrinkage of cellular components. In such cases, it is recommended to proceed with imaging immediately after mounting, before the medium hardens.
7. Should the coverslip shift, nail polish can be applied to seal the edges of the coverslip, thereby preventing further movement.
8. It is crucial to utilize methanol at −20°C. Higher temperatures may lead to further distortion of cellular structures.
9. Avoid incubating primary and secondary antibodies at 37°C in methanol-fixed samples, as this can occasionally cause serious structural distortions.
10. Resolution is primarily determined by the Numerical Aperture (NA) of an objective. We recommend using an objective lens with a higher NA, ideally greater than 1.4.
11. Be careful when moving an objective upwards, to avoid bumping into coverslips.
12. This setup is equivalent to a normal Yokogawa CSU W1 spinning disc confocal system.

Acknowledgments

We would like to thank Mr. Yoshitaka Sekizawa, Nikon USA, and Yokogawa Electric Corporation for critical equipment and technical support. Part of this work was supported by the Wisconsin Partnership Program, Research Forward initiative from the University of Wisconsin-Madison Office of the Vice Chancellor for Research with funding from the Wisconsin Alumni Research Foundation, Big Data Challenge from the Wisconsin Alumni Research Foundation, start-up funding from University of Wisconsin-Madison SMPH, UW Carbone Cancer Center, and McArdle Laboratory for Cancer Research, and NIH grant R35GM147525 (to A.S.).

References

1. Hickey SM et al (2021) Fluorescence microscopy-an outline of hardware, biological handling, and fluorophore considerations. Cells 11(1). https://doi.org/10.3390/cells11010035
2. Lichtman JW, Conchello JA (2005) Fluorescence microscopy. Nat Methods 2(12):910–919. https://doi.org/10.1038/nmeth817
3. Yu-Chia, Chen Ece, Kilic Evelyn, Wang Will, Rossman Aussie, Suzuki (2025) CENcyclopedia: dynamic landscape of kinetochore architecture throughout the cell cycle Nature Communications 16(1) 10.1038/s41467-025-62316-6
4. Ma Y, Buttitta L (2017) Chromatin organization changes during the establishment and maintenance of the postmitotic state. Epigenet Chromatin 10(1):53. https://doi.org/10.1186/s13072-017-0159-8
5. Yu-Lin, Chen Yu-Chia, Chen Aussie, Suzuki (2024) ImmunoCellCycle-ID – a high-precision immunofluorescence-based method for cell cycle identification ABSTRACT Journal of Cell Science 137(22) 10.1242/jcs.263414
6. Yu-Lin, Chen Syon, Reddy Aussie, Suzuki (2025) (2025) Reversible and effective cell cycle synchronization method for studying stage-specific processes Life Science Alliance 8(5) e202403000-10.26508/lsa.202403000
7. Pierzynska-Mach A et al (2023) Imaging-based study demonstrates how the DEK nanoscale distribution differentially correlates with epigenetic marks in a breast cancer model. Sci Rep 13(1), 12749. https://doi.org/10.1038/s41598-023-38685-7
8. Cheng Y et al (2019) Targeting epigenetic regulators for cancer therapy: mechanisms and advances in clinical trials. Signal Transduct Target Ther 4:62. https://doi.org/10.1038/s41392-019-0095-0
9. Loi J et al (2023) Semi-automated 3D fluorescence speckle analyzer (3D-Speckler) for microscope calibration and nanoscale measurement. J Cell Biol 222(4). doi:https://doi.org/10.1083/jcb.202202078
10. Liu WJ et al (2018) Breaking the axial diffraction limit: a guide to axial super-resolution fluorescence microscopy. Laser Photon Rev 12(8). https://doi.org/10.1002/lpor.201700333
11. Sanderson MJ et al (2014) Fluorescence microscopy. Cold Spring Harb Protoc 2014(10):pdb top071795. https://doi.org/10.1101/pdb.top071795
12. Vangindertael J et al (2018) An introduction to optical super-resolution microscopy for the adventurous biologist. Methods Appl Fluoresc 6(2):022003. https://doi.org/10.1088/2050-6120/aaae0c
13. Roshan X., Norman Yu-Chia, Chen Emma E., Recchia Jonathan, Loi Quincy, Rosemarie Sydney L., Lesko Smit, Patel Nathan, Sherer Motoki, Takaku Mark E., Burkard Aussie, Suzuki (2025) One step 4× and 12× 3D-ExM enables robust super-resolution microscopy of nanoscale cellular structures Journal of Cell Biology 224(2) 10.1083/jcb.202407116
14. Wu S et al (2019) Circular ecDNA promotes accessible chromatin and high oncogene expression. Nature 575(7784):699–703. https://doi.org/10.1038/s41586-019-1763-5
15. Xu J et al (2020) Super-resolution imaging reveals the evolution of higher-order chromatin folding in early carcinogenesis. Nat commun 11(1). https://doi.org/10.1038/s41467-020-15718-7
16. Aussie, Suzuki Sarah K, Long Edward D, Salmon (2018) An optimized method for 3D fluorescence co-localization applied to human kinetochore protein architecture eLife 7. https://doi.org/10.7554/eLife.32418

17. Hua K, Ferland RJ (2017) Fixation methods can differentially affect ciliary protein immunolabeling. Cilia 6:5. https://doi.org/10.1186/s13630-017-0045-9
18. Whelan DR, Bell TD (2015) Image artifacts in single molecule localization microscopy: why optimization of sample preparation protocols matters. Sci Rep 5, 7924. https://doi.org/10.1038/srep07924
19. Lawrence M et al (2016) Lateral thinking: how histone modifications regulate gene expression. Trends Genet 32(1):42–56. https://doi.org/10.1016/j.tig.2015.10.007
20. Millan-Zambrano G et al (2022) Histone post-translational modifications—cause and consequence of genome function. Nat Rev Genet 23(9):563–580. https://doi.org/10.1038/s41576-022-00468-7
21. Azuma T, Kei T (2015) Super-resolution spinning-disk confocal microscopy using optical photon reassignment. Opt Express 23(11):15003–15011. https://doi.org/10.1364/OE.23.015003

Chapter 11

A Framework for Analyzing the Epigenetic and Transcriptomic Landscape of Cancers Using Single-Cell Multiomic Data

Avinash Veerappa and Chittibabu Guda

Abstract

Single-cell approaches to study gene regulation using multi-modal data (sc-Multiome), such as genomes, transcriptomes, and chromatin accessibility of tumor cells, offer great insights into the development trajectory. By enabling high-resolution barcoding, single-cell isolation facilitates the detection of subpopulations, uncovers molecular mechanisms, and characterizes cell types to reveal cellular heterogeneity within complex tissues. The pipeline for analyzing multiome data includes four main steps: processing Single-Cell Multiome ATAC+Gene Expression sequencing data, gene expression analysis using Seurat, chromatin accessibility analysis and the joint embedding of the expression and accessibility data using SnapATAC2. Here, we outline a comprehensive protocol for joint embedding of chromatin accessibility and transcriptomic data using best practices developed in our laboratories, along with detailed parameter tuning at each step to fully leveraging this pipeline. This approach enhances our understanding of the intricate tumor microenvironment and aids in determining cellular landscape. This chapter focuses on the integration of single-cell multi-omics methodologies, emphasizing their utility in cancer research.

Keywords Multiome, Gene expression, Chromatin accessibility, ATAC, Cancer, Seurat, SnapATAC2, Joint embedding

Abbreviations

sc-Multiome	Single-cell Multiome
ATAC	Assay for Transposase-Accessible Chromatin
sc-RNAseq	Single-cell RNA Sequencing
sc-ATACseq	Single-cell Assay for Transposase-Accessible Chromatin Sequencing
TF	Transcription Factor
GEX	Gene Expression
TEA-seq	Transcriptome, Epigenome, and Antigensequencing
ASAP-seq	Assay for Single-cell ATAC-seq and Protein-seq

Archana Dhasarathy and Motoki Takaku (eds.), *Cancer Epigenetics*, Methods in Molecular Biology, vol. 3005,
https://doi.org/10.1007/978-1-0716-5072-1_11,

CITE-seq	Cellular Indexing of Transcriptomes and Epitopes by Sequencing
REAP-seq	RNA Expression and Protein-sequencing
sci-CAR	Single-cell Combinatorial Indexing Chromatin Accessibility and mRNA
SNARE-seq	Single-nucleus Chromatin Accessibility and RNA Expression Sequencing
SHARE-seq	Simultaneous High-throughput ATAC and RNA Expression Sequencing
ICICLE-seq	Integrated Chromatin, Immune Cell Indexing, and Chromatin Landscape Epigenomics Sequencing
DOGMA-seq	DNA, RNA, and Global Measurement of Activity Sequencing
RAM	Random Access Memory
CPUs	Central Processing Units
GPUs	Graphics Processing Units
SSD	Solid-State Drive
GB	Gigabyte
UMAP	Uniform Manifold Approximation and Projection
UMIs	Unique Molecular Identifiers
GTF	Gene Transfer Format
TSS	Transcription Start Site
TSSE	Transcription Start Site Enrichment
CRAN	Comprehensive R Archive Network
CLR	Centered Log-Ratio
PCA	Principal Component Analysis
tSNE	t-distributed Stochastic Neighbor Embedding
vst	Variance Stabilizing Transformation
KNN	K-Nearest Neighbors
MTX	Matrix Market

1 Introduction

Understanding the biology of a cell population through single-omics approaches, such as transcriptomics or epigenomics, provides only a one-dimensional perspective of cellular function, as it focuses on gene regulation using only one data modality at a time. In contrast, multi-omics assays such as single-cell multiome (sc-RNAseq/sc-ATACseq) allow for the capture of multi-dimensional information on the same cell and integrate data from various data modalities. This comprehensive approach enhances our understanding of gene regulation and molecular mechanisms and provides a more accurate and nuanced view of cell function, leading to increased confidence in biological insights [1, 2].

Biology of cancer is inherently complex, characterized by intricate interactions among and within multimolecular complexes that form extensive networks in both healthy and diseased states [3]. This complexity is evident at both genetic, transcriptomic and epigenetic levels, where chromatin remodeling complexes play vital roles in regulating chromatin accessibility and transcriptional events associated with oncogenesis and cancer progression [4]. Understanding these complex interactions and mechanisms is essential for developing more effective cancer therapies and improving patient outcomes.

Multiomics approaches integrate multiple layers of omics data to provide a comprehensive understanding of gene regulation in cancer [5, 6]. By combining genomics, transcriptomics, chromatin accessibility and transcriptional factor (TF) footprinting, cancer diagnosis and subtyping have been improved [6, 7]. Multi-omics analyses offer unprecedented opportunities to classify cancers into precise subtypes, enhance survival predictions, and uncover critical pathophysiological processes [5]. This integration is shifting cancer research from a single-parametric to multi-parametric analytical domain, enabling a more holistic view of disease mechanisms [7]. Despite challenges in data processing and standardization, multi-omic data integration holds great promise for personalized and adaptable cancer treatments [8]. As bioinformatics technologies advance, multi-omics approaches are becoming increasingly feasible for clinical applications, paving the way for tailored therapies based on individual genomic and metabolomic profiles [8].

Various single-cell based single-omic and multi-omic platforms exist, each designed for specific types of molecules being assayed, which include 10× Multiome for GEX+ATAC [9], TEA-seq for GEX+Protein Epitopes+ATAC [10], ASAP-seq for Proteins+ATAC [11], and CITE-seq for GEX+Protein [12]. These high-throughput single-cell multi-omic technologies have evolved from single-omic transcriptome measurements to bimodal platforms to tri-modal, simultaneously capturing transcriptomes and cell surface protein epitopes in CITE-seq and REAP-seq [12, 13], transcriptomes and chromatin accessibility in sci-CAR [14], SNARE-seq [15], SHARE-seq [16], and chromatin accessibility with cell surface protein epitopes in ASAP-seq [11], and ICICLE-seq [10]. Recently, trimodal platforms have been developed to simultaneously measure transcriptomes, cell surface protein epitopes, and chromatin accessibility in DOGMA-seq [11], and TEA-seq [10], providing even greater insights into cellular function and complexity.

In this study, we utilized the 10× Genomics multiome platform to assay bi-modal GEX+ATAC molecules, providing a comprehensive evaluation of astrocytoma, a type of brain tumor. We outline best-practice approaches and key parameter selections to perform joint embedding of GEX+ATAC and enhance reproducibility and accuracy in single-cell analysis. We believe this framework offers significant potential for advancing the understanding of the complex mechanisms underlying cancer development and progression.

2 Computation Requirements/Materials

For analyzing a minimal set of two 10× Chromium Multiome samples (both sc-RNAseq and sc-ATACseq data) using Seurat and SnapATAC2, substantial computational resources are required due to the high-dimensional nature of single-cell multi-omics data. Below is a detailed breakdown of the computational requirements for processing and analyzing these datasets efficiently.

2.1 Operating System

- Linux (Ubuntu or CentOS recommended) or macOS.

2.2 Hardware Requirements

2.2.1 Memory (RAM)

- Recommended: At least 128 GB of RAM for efficient processing.
- Minimum: 64 GB for smaller datasets, but may lead to slower processing

2.2.2 CPUs (Processing Power)

- Recommended: 16-32 CPU cores to parallelize computational tasks
- Minimum: 8 CPU cores

2.2.3 GPUs (Optional)

- Recommended: 1 or 2 high-performance GPUs (e.g., NVIDIA Tesla V100 or A100).

2.2.4 Disk Storage

- Recommended: 1–2 TB of SSD storage for faster data access.
- Minimum: 500 GB is required but can be restrictive.

2.3 Software Requirements

2.3.1 Operating System

- Linux (Ubuntu or CentOS recommended) or macOS.

2.3.2 R and Python Installations

- R: Version 4.0 or higher (required for *Seurat*).
- *Python*: Version 3.8 or higher (required for *SnapATAC2*).

2.3.3 Required Libraries

- *Seurat*: Version ≥4.1 or higher
- *SnapATAC2*: ≥2.7
- *Harmony*: For batch effect correction and data integration within Seurat.
- *zellkonverter*: For converting single-cell data formats when working with 'SingleCellExperiment' objects.
- *SingleCellExperiment*: Provides a flexible structure for single-cell data and compatibility with other single-cell tools.

- *Matrix*: For efficient handling of sparse matrices, particularly with large datasets.
- *openai*: Provides access to OpenAI's API
- *GPTCelltype*: Package for cell type annotation using GPT models.

2.3.4 Other R Packages

- For handling ATACseq data with Seurat: 'Signac'
- For data manipulation and visualization: 'ggplot2', 'dplyr', 'patchwork'

2.3.5 Python Packages

- For single-cell data processing: 'anndata', 'scanpy'
- For matrix operations and data handling: 'scipy', 'numpy', 'pandas'
- For dimensionality reduction and clustering: 'umap-learn', 'scikit-learn'
- Required for GPU-accelerated UMAP in SnapATAC2: CUDA (for GPU support)

3 Methods

The R and Python scripts used in this multiome analysis walk-through are available for download from the GitHub repository at https://github.com/gudalab/multiome. To access the files, navigate to the repository and click on the "Code" dropdown menu, selecting "Download ZIP" from the options. Once downloaded, extract the contents of the ZIP file to a directory of your choice, which will be named "Multiome-pipeline-master."

3.1 Sample Cohort

For this walkthrough, we used a multiome sample of astrocytoma, a type of brain tumor. Single-cell data from IDH-mutant astrocytoma samples generated using the 10× Genomics Multiome of Gene Expression (GEX) and Assay for Transposase-accessible chromatin (ATAC) platform were obtained from NCBI GEO with the accession number GSE230389.

3.2 Data Preprocessing

10× Genomics' Cell Ranger *Count* pipeline should be used to process FASTQ files with the GRCh38p14 reference, to perform alignment, de-duplication, filtering, feature-barcode matrix generation, and gene expression and chromatin accessibility analysis. Sample sheets in the below format should be prepared and saved as a Comma Separated Value (.csv) file format, which will be used as an input file under --libraries when executing the 'cellranger-arc count' command.

```
fastqs,sample,library_type
/home/jdoe/runs/HNGEXSQXXX/outs/fastq_path,exam-
ple,Gene Expression
/home/jdoe/runs/HNATACSQXX/outs/fastq_path,exam-
ple,Chromatin Accessibility
```

Next, 'Cell Ranger-ARC Count' should be used as shown in the below code block to align with GRCh38p14, perform filtering, barcode counting, peak calling, and counting of both ATAC and GEX molecules.

```
cellranger-arc count --id=sample345 \
                     --reference=/opt/refdata-cell-
                     ranger-arc-GRCh38-2020-A-2.0.0 \
                     --libraries=/home/jdoe/runs/
                     libraries.csv \
                     --localcores=16 \
                     --localmem=64
```

Upon completion of the cellranger-arc count pipeline, it outputs three types of files: features, barcode, matrix among several others as shown in the below code block. In sc-RNAseq dataset, you will have Gene Expression data in the features files, while in sc-ATACseq dataset, the features file will show the accessible Peaks. For the Gene Expression feature each element in the matrix is the number of UMIs (Unique molecular identifiers) associated with the corresponding feature (row) and barcode (column). For a Peaks feature each element in the matrix is the number of cut sites associated with the corresponding feature (row) and barcode (column).

```
$ cd /home/jdoe/runs/sample345/outs

$ tree filtered_feature_bc_matrix

filtered_feature_bc_matrix

├── barcodes.tsv.gz

├── features.tsv.gz

└── matrix.mtx.gz
```

The first and second columns of the features.tsv.gz file store the gene ID and name as defined in the reference Gene transfer format file (GTF), respectively. The third column identifies the type of feature, which will be one of Gene Expression or Peaks, depending on the feature type. The fourth, fifth and sixth columns store the chromosome, start and end positions of the cellranger-arc determined transcription start site (TSS) for this gene in 0-based bed-format.

The following represents a simplified sample of a features.tsv.gz file, typically found in sc-RNAseq or multiome datasets. This example showcases data for three genes, illustrating the basic structure of a gene expression feature file.

```
$ zcat filtered_feature_bc_matrix/features.tsv.gz |
head -n 5
ENSG00000139687    RB1    Gene Expression    chr13
48303725          48303747
ENSG00000141510    TP53  Gene Expression    chr17
7675492          7687550
ENSG00000012048    BRCA1  Gene Expression    chr17
43125314          43125483
```

Below is a simple example of features.tsv.gz file from the features file of sc-ATACseq/multiome dataset showing data collected for 2 peaks

```
chr13:48301972-48306754  chr13:48301972-48306754
Peaks
chr13  48301972  48306754
chr17:76751362-76755140  chr17:76751362-76755140
Peaks
chr17  76751362  76755140
```

Barcode sequences correspond to column indices.

```
$ zcat filtered_feature_bc_matrices/barcodes.tsv.gz
| head -n 5
AAACAGCCAAATATCC-1
AAACAGCCAGGAACTG-1
AAACAGCCAGGCTTCG-1
AAACCAACACCTGCTC-1
AAACCAACAGATTCAT-1
```

Below is the illustration showing the basic structure of a gene expression matrix file.

```
1. $ zcat matrix.mtx.gz | head -n 10
2. %%MatrixMarket matrix coordinate integer general
3. %metadata_json: {"software_version": "cellrang-
   er-arc-2.0.2", "format_version": 2}
4. 66770 5983 13567892
5. 82 1 1
6. 261 1 1
 7. 527 1 1
 8. 651 1 1
 9. 694 1 1
10. 1048 1 1
11. 1139 1 1
```

The output represents the first 10 lines of a Matrix Market (MTX) file, which is commonly used in single-cell genomics to store sparse matrices, such as gene expression or chromatin accessibility data. The header '%%MatrixMarket matrix coordinate integer general' indicates that the file is in Matrix Market format, representing a sparse matrix in coordinate format (COO) with integer values and no specific symmetry. The following metadata line, '%metadata_json: {"software_version": "cellranger-arc-2.0.2", "format_version": 2}', provides additional details, showing that the file was generated using Cell Ranger ARC version 2.0.2 and follows format version 2. The line '66770 5983 13567892' specifies the dimensions of the matrix: 66,770 rows (representing features, such as genes or peaks), 5,983 columns (representing cells), and 13,567,892 non-zero entries, indicating the sparsity of the data. The subsequent lines, such as '82 1 1', represent non-zero entries in the matrix, where the format 'row_index column_index value' is used. For instance, '82 1 1' indicates that in row 82 (feature 82) and column 1 (cell 1), the value is 1. This pattern is repeated for other entries, highlighting the sparse nature of the data where only non-zero values are stored.

Next, we will use Seurat [17] for analyzing GEX data, while SnapATAC2 [18] will be employed to analyze chromatin accessibility (ATAC) data and to characterize and annotate cell clusters by integrating GEX and ATAC. For the sc-RNAseq analysis before integration with sc-ATACseq data, we employed a detailed workflow encompassing data quality control, clustering, and annotation processes, utilizing the Seurat [17].

3.3 Creating Environment, Installing Packages and Required Dependencies for Transcriptome Analysis

3.3.1 Step 1: Create the Conda Environment

```
conda create -n seurat_env r-base=4.1.2 -y # Specify
R version compatible with Seurat and Bioconductor
conda activate seurat_env
```

3.3.2 Step 2: Install Required Dependencies in Conda

```
conda install -c conda-forge r-seurat r-ggplot2
r-dplyr r-matrix -y
```

3.3.3 Step 3: Install BiocManager and Bioconductor Packages in R

Now, open R from the Conda environment:

```
R
```

The next step is to install some packages in R; for this, you will need to use a combination of CRAN and Bioconductor repositories. The below code block shows how you can install each package:

First, make sure you have the latest version of R installed, Open R or RStudio and Run the following commands:

```
# Install BiocManager
if (!requireNamespace("BiocManager", quietly = TRUE))
   install.packages("BiocManager")

# Install CRAN packages
install.packages(c("Seurat",   "ggplot2",   "dplyr",
"Matrix"))

# Install Bioconductor packages
BiocManager::install(c("SingleCellExperiment",
"zellkonverter"))

# Install harmony
install.packages("harmony")

# Install openai
install.packages("openai")

# Install GPTCelltype from GitHub (assuming it's not
on CRAN)
if (!requireNamespace("devtools", quietly = TRUE))
   install.packages("devtools")
devtools::install_github("xmc811/GPTCelltype")
```

After all the essential packages are installed, it is time to import them using the library( ) function.

3.4 Single-Cell Transcriptome Analysis Using Seurat Package

```
library(Seurat)
library(harmony)
library(zellkonverter)
library(SingleCellExperiment)
library(ggplot2)
library(dplyr)
library(Matrix)
library("openai")
library("GPTCelltype")
```

3.4.1 Import Data Files

Once all the packages and their dependencies are installed and imported, the next step is to read and process features, matrix and barcode data using the Read10X function. The Read10X function combines the information from the three files (barcodes, features, and matrix) to create a single data structure that represents the gene expression matrix. Each row typically represents a gene, each column represents a cell, and the values in the matrix represent the number of transcripts detected for each gene in each cell.

```
barcode_file <- "/path/to/file/barcodes.tsv.gz"
features_file <- "/path/to/file/features.tsv.gz"
matrix_file <- "/path/to/file/matrix.mtx.gz"

expression_matrix <- Read10X(data.dir = "/path/to/
file/filtered_feature_bc_matrix")
```

3.4.2 Creating Seurat Object

The next step is to create a Seurat object, which is a fundamental data structure in the Seurat package for sc-RNAseq analysis. CreateSeuratObject( ) is a function from the Seurat package that initializes a new Seurat object. counts = expression_matrix is an expression_matrix variable, which was created in the previous code line, is passed as the counts parameter. This tells the function to use the gene expression matrix as the initial data for the Seurat object. The result of this line of code is that a new Seurat object is created and stored in the seurat_object variable. This Seurat object will contain the gene expression data, as well as various metadata and analysis results that will be generated as the analysis progresses. The Seurat object is the central data structure used throughout the Seurat workflow for tasks such as data normalization, dimensionality reduction, clustering, and differential expression analysis.

```
seurat_object <- CreateSeuratObject(counts = ex-
pression_matrix, project = "snRNAseq")
```

The next step is to set the default assay in the Seurat object to "RNA". An "assay" refers to a specific type of data that is stored within the Seurat object.

```
DefaultAssay(seurat_object) <- "RNA"
```

3.4.3 Normalizing Gene Expression Data

In the next step, we perform data normalization on the Seurat object using the "CLR" (Center Log-Ratio) normalization method.

```
seurat_object <- NormalizeData(seurat_object,
normalization.method = "CLR", assay = "RNA")
```

```
> seurat_object <- NormalizeData(seurat_object, normalization.method = "CLR", assay = "RNA")
Normalizing layer: counts.Gene Expression
Normalizing across features
  |++++++++++++++++++++++++++++++++++++++++++++++++++| 100% elapsed=04s
Normalizing layer: counts.Peaks
Normalizing across features
  |++++++++++++++++++++++++++++++++++++++++++++++++++| 100% elapsed=03s
```

3.4.4 Identifying Cell Cycle-Associated Genes to Regress Out the Cell Cycle Effect

The next step is to assign the "s.genes" (S-phase genes) from the cc.genes.updated.2019 dataset to the variable s.genes. The rationality behind this is to identify cell cycle-associated genes, which can be used to regress out the cell cycle effect during the sc-RNAseq analysis. The cell cycle stage of each cell can be a significant source of unwanted variation, as it can affect the gene expression profile. By identifying and removing this cell cycle-associated variation, the downstream analysis can focus on other biologically relevant sources of heterogeneity, such as cell type or disease state.

```
s.genes <- cc.genes.updated.2019$s.genes
```

Similarly, in the next step, we assign the "g2m.genes" (G2/M-phase genes) from the cc.genes.updated.2019 dataset to the variable g2m.genes. The principle behind this is similar to the previous step of assigning the "s.genes" (S-phase genes): it is part of the process of identifying cell cycle-associated genes, which can be used to regress out the cell cycle effect during the sc-RNAseq analysis.

```
g2m.genes <- cc.genes.updated.2019$g2m.genes
```

The next step involves intersecting the s.genes and g2m.genes variables with the gene names (row names) present in the "RNA" assay of the Seurat object. The principle behind this is to ensure that the cell cycle gene lists (s.genes and g2m.genes) only contain genes that are actually present in the sc-RNAseq data.

```
s.genes  <-  intersect(rownames(seurat_object[["R-
NA"]]), s.genes)
g2m.genes <- intersect(rownames(seurat_object[["R-
NA"]]), g2m.genes)
```

3.4.5 Extracting RNA and Peak Data

The next step extracts the available data layers from the "RNA" assay of the Seurat object and stores them in the rna_layers variable. The principle behind this is to understand the different types of data that are available within the Seurat object. The "RNA" assay in a Seurat object can have multiple data layers, which can include:

- The primary gene expression matrix
- Normalized expression values
- Scaled expression values
- Dimensionality reduction coordinates (e.g., PCA, UMAP, tSNE)
- Cell cycle scores
- Any other custom data layers added during the analysis

By storing the available data layers in the rna_layers variable, you can later use this information to access and work with the specific data layers you need for downstream analyses, such as the normalized expression values or the dimensionality reduction coordinates.

```
rna_layers <- Layers(seurat_object[["RNA"]])
print(rna_layers)
```

```
> s.genes <- cc.genes.updated.2019$s.genes
> s.genes <- intersect(rownames(seurat_object[["RNA"]]), s.genes)
> g2m.genes <- intersect(rownames(seurat_object[["RNA"]]), g2m.genes)
> rna_layers <- Layers(seurat_object[["RNA"]])
> print(rna_layers)
[1] "counts.Gene Expression" "counts.Peaks"          "data.Gene Expression"
[4] "data.Peaks"
>
```

In the next step, we will extract the primary gene expression data from the "RNA" assay of the Seurat object and store it in the data_gene_expression variable. The reason behind this is to access the core gene expression data that was initially loaded into the Seurat object, which is essential for further analysis and visualization.

```
data_gene_expression <- GetAssayData(seurat_object,
assay = "RNA", layer = "data.Gene Expression")
```

The next step extracts the "counts.Peaks" data layer from the "RNA" assay of the Seurat object and stores it in the counts_peaks variable. The principle behind this is to access the peak count data, which is typically associated with single-cell ATACseq data that has been integrated into the Seurat object.

```
counts_peaks <- GetAssayData(seurat_object, assay =
"RNA", layer = "counts.Peaks")
```

The next step involves extracting the "data.Peaks" layer from the "RNA" assay of the Seurat object and storing it in the data_peaks variable. The principle behind this step is to access a version of the peak data that has potentially been preprocessed or normalized. This step is part of a multi-omics analysis approach, allowing you to work with both gene expression and chromatin accessibility data in parallel, providing a more comprehensive view of cellular states and regulatory mechanisms.

```
data_peaks <- GetAssayData(seurat_object, assay =
"RNA", layer = "data.Peaks")
```

In the next step, the "counts.Gene Expression" layer from the "RNA" assay of the Seurat object is stored in the 'rna_data' variable. This step is crucial as it provides access to the most fundamental form of the gene expression data. It allows for maximum flexibility in downstream analyses and ensures that you can always return to the original data if needed.

```
rna_data <- GetAssayData(seurat_object[["RNA"]],
layer = "counts.Gene Expression")
```

In the next step, the function takes two arguments, data (A matrix or data frame of gene expression data) and features (A vector of gene names or features to consider), and performs a selection of rows from data corresponding to the specified features, calculates the column means of this subset using colMeans() followed by returning the mean values. The principle behind this function is to calculate a "module score" for a set of genes (the module) across all cells.

```
calculate_module_score <- function(data, features)
{
avg_expression <- colMeans(data[features, , drop =
FALSE])
return(avg_expression)
}
```

3.4.6 Calculating Module Score for Cell Cycle-Associated Genes Across All Cells

In the next step, we use the calculate_module_score function we just defined and use the raw gene expression data (rna_data) and the list of S-phase genes (s.genes) as inputs. The result is stored in the s_score variable. Using the rna_data (raw gene expression count data we extracted earlier) and s.genes (list of S-phase genes we defined earlier), the function calculates the average expression of the S-phase genes for each cell. The principle behind this step is to quantify the activity of the S-phase of the cell cycle in each cell. By calculating the average expression of S-phase genes, we are essentially creating a "score" for how active the S-phase of the cell cycle is in each cell. Cells with high S-phase scores are likely to be in the DNA synthesis phase of the cell cycle. These scores can be used later to regress out cell cycle effects from the gene expression data.

```
s_score <- calculate_module_score(rna_data, s.genes)
```

Similar to the previous step, the next step calculates the score for the G2/M phase of the cell cycle.

```
g2m_score <- calculate_module_score(rna_data, g2m.
genes)
```

The next set of lines is a nested set of code lines that add the calculated S-phase and G2/M-phase scores to the Seurat object as metadata.

```
seurat_object <- AddMetaData(seurat_object, metada-
ta = s_score, col.name = "S.Score")
seurat_object <- AddMetaData(seurat_object, metada-
ta = g2m_score, col.name = "G2M.Score")
```

The next step is to assign a discrete cell cycle phase label to each cell based on its S and G2M scores. This is important for several reasons: (1) Simplified Cell Cycle Classification: It reduces the continuous scores into a binary classification, making it easier to interpret and use in downstream analyses. (2) Cell State Characterization: It provides a clear label for each cell's likely cell cycle state, which can be used for various analyses and visualizations. (3) Population Analysis: It allows you to easily quantify the proportion of cells in different cell cycle phases in your sample. (4) Differential Analysis: You can use these labels to perform differential expression analysis between cells in different cycle phases. (5) Quality Control: Unusual distributions of cell cycle phases might indicate technical issues or interesting biological phenomena.

```
seurat_object$Phase <- ifelse(seurat_object$S.Score
> seurat_object$G2M.Score, "S", "G2M")
```

The next line uses the 'FindVariableFeatures' function from the Seurat package to identify the most variable genes in the dataset. 1. 'FindVariableFeatures()' is a Seurat function, which identifies features (genes) that are highly variable across cells. 2. 'assay = "RNA"': Specifies that we are working with the RNA assay data. 3. 'layer = "counts.Gene Expression"': Indicates that are using the raw count data for gene expression. 4. 'selection.method = "vst"': Specifies the method for selecting variable features. "vst" stands for "variance stabilizing transformation". 5. 'nfeatures = 2000': Sets the number of variable features to select (in this case, 2000). The principle behind this step is to identify genes that show high cell-to-cell variation in the dataset. This is crucial for several reasons:

1. Dimensionality reduction: By focusing on the most variable genes, we can reduce the dimensionality of the data while retaining most of the relevant biological information.
2. Computational efficiency: Analyzing thousands of genes across thousands of cells is computationally intensive. Focusing on the most variable genes makes downstream analyses more manageable.
3. Noise reduction: Many genes show little variation across cells and may primarily represent technical noise. Focusing on highly variable genes helps to filter out this noise.
4. Biological signal: Genes with high cell-to-cell variation are often the most informative about biological differences between cells.
5. Preparation for clustering: Variable genes are typically used as input for clustering algorithms to identify cell types or states.
6. Differential expression: These genes are often the most likely to show differential expression between cell types or conditions.
7. Trajectory analysis: Variable genes are crucial for methods that reconstruct developmental or differentiation trajectories. The "vst" method used here is particularly good at identifying variable genes across a range of expression levels, not just highly expressed genes.

3.4.7 Dimensionality Reduction Using 'vst' to Identify the Most Variable Genes Using 'FindVariableFeatures' Function

```
seurat_object <- FindVariableFeatures(seurat_object, assay = "RNA", layer = "counts.Gene Expression", selection.method =vst", nfeatures = 2000)
```

```
> counts_gene_expression <- GetAssayData(seurat_object, assay = "RNA", layer = "counts.Gene Expression")
> data_gene_expression <- GetAssayData(seurat_object, assay = "RNA", layer = "data.Gene Expression")
> counts_peaks <- GetAssayData(seurat_object, assay = "RNA", layer = "counts.Peaks")
> data_peaks <- GetAssayData(seurat_object, assay = "RNA", layer = "data.Peaks")
> rna_data <- GetAssayData(seurat_object[["RNA"]], layer = "counts.Gene Expression")
> calculate_module_score <- function(data, features) {
+ avg_expression <- colMeans(data[features, , drop = FALSE])
+ return(avg_expression)
+ }
> s_score <- calculate_module_score(rna_data, s.genes)
> g2m_score <- calculate_module_score(rna_data, g2m.genes)
> seurat_object <- AddMetaData(seurat_object, metadata = s_score, col.name = "S.Score")
> seurat_object <- AddMetaData(seurat_object, metadata = g2m_score, col.name = "G2M.Score")
> seurat_object$Phase <- ifelse(seurat_object$S.Score > seurat_object$G2M.Score, "S", "G2M")
> seurat_object <- FindVariableFeatures(seurat_object, assay = "RNA", layer = "counts.Gene Expression", selection.method = "vst", nfeatures = 2000)
Finding variable features for layer counts.Gene Expression
Calculating gene variances
0%   10   20   30   40   50   60   70   80   90   100%
[----|----|----|----|----|----|----|----|----|----|
**************************************************|
Calculating feature variances of standardized and clipped values
0%   10   20   30   40   50   60   70   80   90   100%
[----|----|----|----|----|----|----|----|----|----|
**************************************************|
> 
```

3.4.8 Scaling and Centering Gene Expression Data

In the next step, we use the ScaleData function from the Seurat package to scale and center the gene expression data, while also regressing out the effects of cell cycle scores. This is to prepare the data for downstream analysis by standardizing gene expression levels and removing unwanted sources of variation.

```
seurat_object <- ScaleData(seurat_object, assay =
"RNA", layer = "counts.Gene Expression", vars.to.
regress = c("S.Score", "G2M.Score"))
```

```
> seurat_object <- ScaleData(seurat_object, assay = "RNA", layer = "counts.Gene Expression", vars.to.regress = c("S.Score", "G2M.Score
Regressing out S.Score, G2M.Score
  |======================================================================| 100%
Centering and scaling data matrix
  |======================================================================| 100%
>
```

3.4.9 Performing Principal Component Analysis on the Scaled Data

The next line is to perform Principal Component Analysis (PCA) on the Seurat object. RunPCA() function performs PCA on the scaled data, while the assay = "RNA" specifies that we're using the RNA assay data for PCA. This step is to reduce the dimensionality of the data while retaining most of the important variation. This is crucial for Dimensionality Reduction, since sc-RNAseq data is high-dimensional (thousands of genes). PCA reduces this to a smaller number of dimensions that capture the most variation. Further, to perform Noise Reduction, PCA helps to separate signal from noise by focusing on the directions of greatest variation in the data.

```
seurat_object <- RunPCA(seurat_object, assay = "RNA")
head(seurat_object@meta.data)
```

```
> seurat_object <- RunPCA(seurat_object, assay = "RNA")
PC_ 1
Positive:  RBFOX1, SNHG14, NRG3, CACNA1B, DLGAP2, KCNIP4, CAMK1D, RYR2, SYT1, MEG3
           KALRN, CHD5, KCNMA1, CACNA1C, GALNT17, SPTBN4, GRIN2A, RALYL, LRRC7, R3HDM1
           KCNQ5, CABP1, RBFOX3, ATP8A2, DOCK3, DNM1, GRIN1, KSR2, MYT1L, PRKCB
Negative:  RRM2, SH3TC1, SLC11A1, C3, RHBDF2, CCNE2, STAB1, APBB1IP, CSF2RA, CSF1R
           SERPINE1, VSIG4, C1QA, TYMP, FGD2, ENSG00000269514, CARD11, P2RY8, SLC2A5, CD83
           ENSG00000289021, EGR3, SRGN, RPL36AL, CCL4L2, SIGLEC10, ENSG00000249293, SELENOV, RGS1, EZH2
PC_ 2
Positive:  CACNA1A, MEG3, SRRM3, DSCAML1, FGF12, KCNIP4, HECW1, MYT1L, CACNA1B, PRKCB
           CHD5, OPCML, RYR2, GRIN1, LHFPL3, SEZ6L, RBFOX1, RBFOX3, KSR2, ATCAY
           CELF4, DSCAM, CACNA1C, ATP8A2, KAZN, XKR4, RAPGEF4, SYT1, TPTEP1, GALNT17
Negative:  OBI1-AS1, KCNN3, SLC39A11, ATP1A2, ADGRV1, LINC01088, ENSG00000237844, STOX1, RFX4, AHNAK
           FADS2, PPP1R9A-AS1, GLIS3, DAAM2, SORBS1, TIMP3, SLC1A3, AMZ2, ITPR2, HPSE2
           PAPLN, RASSF4, HIF3A, CLEC16A, ENTREP1, ARHGEF4, SLC25A18, NT5DC3, LPL, ENSG00000287704
PC_ 3
Positive:  TNR, LHFPL3, DSCAML1, BCAN, DPP6, VCAN, DSCAM, TPTEP1, MEGF11, OPCML
           DLGAP1, CACNA1A, FERMT1, ENSG00000274422, TNK2, SMOC1, FYN, C2orf27A, KAZN, CHD7
           ENSG00000286364, FGF12, ASIC4, POLR2F, GRIK3, ACAP3, LINC00689, CEROX1, NKD1, SEZ6L
Negative:  SYT1, RYR2, KCNQ3, GALNT17, PLXDC2, RBFOX1, TRPM2, CHD5, FKBP5, CAMK2A
           GRIN2A, CAMK1D, FRMPD4, SEC14L5, NRGN, TMEM132D, CABP1, SH3PXD2A, CHRM3, RGS6
           SHISA9, CRACDL, MYH9, C3, DLGAP2, NELL2, DENND3, EPIC1, CSF2RA, SLC17A7
PC_ 4
Positive:  PLXDC2, FKBP5, TBXAS1, PALD1, INPP5D, ST6GAL1, CSF2RA, SLCO2B1, C3, APBB1IP
           RHBDF2, DOCK8, CARD11, CTSB, SRGN, MT-CO2, SLC11A1, MT-ND4, MT-ATP6, MT-CYB
           MT-CO3, DENND3, TMEM156, MT-ND5, STAB1, MT-ND3, IRAK2, A2M, IRAK3, MT-CO1
Negative:  NRG3, MEG3, CHD5, GRIN1, SNHG14, MIAT, KALRN, CABP1, CACNA1B, PDIA2
           ADGRB2, NRGN, DNM1, RYR2, RBFOX1, RAPGEFL1, KNDC1, RIN1, DLGAP2, SEC14L5
           SPTBN4, RBFOX3, DPP10, MMP17, CAMK2A, CNTNAP1, LINC00632, SLC17A7, KCNT1, TNNT2
PC_ 5
Positive:  PLXDC2, FKBP5, C3, SLC11A1, CTSB, PALD1, RHBDF2, CSF2RA, KCNQ3, ST6GAL1
           APBB1IP, DENND3, INPP5D, RNASET2, EPB41L2, SAT1, STAB1, SLCO2B1, NEAT1, TBXAS1
           SH3TC1, DAGLB, TRPM2, PDGFB, OLR1, ZFP36, SPP1, SLC2A5, SRGN, LINC02798
Negative:  MT-ATP6, MT-CYB, MT-CO2, MT-ND4, MT-ND3, MT-ND5, MT-CO1, MT-CO3, MT-ND1, MT-ND2
           ENSG00000289474, ENSG00000289901, MT-ND4L, MT-ATP8, MT-ND6, GPC6, LINC02055, DISC1FP1, NELL1, GALNTL6
           LINC03000, EMCN, ENSG00000283692, GPC5, GNGT1, ANO3, CD36, PWRN1, ENSG00000225421, LINC01090
> head(seurat_object@meta.data)
                   orig.ident nCount_RNA nFeature_RNA    S.Score  G2M.Score
AAACAGCCAAACTGCC-1   snRNAseq        498          408 0.00000000 0.00000000
AAACAGCCAAGTGAAC-1   snRNAseq       3469         2102 0.00000000 0.00000000
AAACAGCCACTGACCG-1   snRNAseq       1400          934 0.02325581 0.01851852
AAACAGCCATGATTGT-1   snRNAseq       2377         1467 0.04651163 0.00000000
AAACATGCAAACTAAG-1   snRNAseq       3032         2088 0.06976744 0.05555556
AAACATGCACAGGGAC-1   snRNAseq       4723         2997 0.02325581 0.00000000
                   Phase
AAACAGCCAAACTGCC-1   G2M
AAACAGCCAAGTGAAC-1   G2M
AAACAGCCACTGACCG-1     S
AAACAGCCATGATTGT-1     S
AAACATGCAAACTAAG-1     S
AAACATGCACAGGGAC-1     S
>
```

3.4.10 Running UMAP to Further Reduce the Dimensionality

In the next step, we run UMAP (Uniform Manifold Approximation and Projection) on the Seurat object.

```
seurat_object <- RunUMAP(seurat_object, assay =
"RNA", dims = 1:20)
```

```
> seurat_object <- RunUMAP(seurat_object, assay = "RNA", dims = 1:20)
Warning: The default method for RunUMAP has changed from calling Python UMAP via reticulate to the R-native UWOT using the cosine metric
To use Python UMAP via reticulate, set umap.method to 'umap-learn' and metric to 'correlation'
This message will be shown once per session
12:03:10 UMAP embedding parameters a = 0.9922 b = 1.112
12:03:10 Read 5983 rows and found 20 numeric columns
12:03:10 Using Annoy for neighbor search, n_neighbors = 30
12:03:10 Building Annoy index with metric = cosine, n_trees = 50
0%   10   20   30   40   50   60   70   80   90   100%
[----|----|----|----|----|----|----|----|----|----|
**************************************************|
12:03:11 Writing NN index file to temp file /tmp/Rtmp7mfw6B/file3ae05c7c633f9
12:03:11 Searching Annoy index using 1 thread, search_k = 3000
12:03:12 Annoy recall = 100%
12:03:13 Commencing smooth kNN distance calibration using 1 thread with target n_neighbors = 30
12:03:14 Initializing from normalized Laplacian + noise (using RSpectra)
12:03:14 Commencing optimization for 500 epochs, with 277096 positive edges
Using method 'umap'
0%   10   20   30   40   50   60   70   80   90   100%
[----|----|----|----|----|----|----|----|----|----|
**************************************************|
12:03:20 Optimization finished
>
```

3.4.11 Constructing a K-nearest Neighbor Graph Using FindNeighbors Function

The next code line uses the FindNeighbors function from the Seurat package to construct a K-nearest neighbor (KNN) graph based on the Euclidean distance in PCA space. This is to identify cells that are similar to each other based on their gene expression profiles. This is essential for several reasons:

1. **Preparation for Clustering:** The KNN graph serves as the basis for many graph-based clustering algorithms, including the one used by Seurat (Louvain algorithm).
2. **Community Detection:** By identifying neighborhoods of similar cells, we can start to understand the community structure of the data.
3. **Data Structure:** The KNN graph captures the local and global structure of the data useful for downstream analyses.
4. **Dimensionality Handling:** Using the PCA dimensions aids in capturing the most important variations in the data.
5. **Noise Reduction:** Using a limited number of PCs (20 in this case) helps to focus on the most significant biological signals while reducing the impact of noise. This step is typically performed after dimensionality reduction (PCA and UMAP) and before clustering. It is a critical step in the Seurat workflow, as it forms the foundation for identifying cell types or states in the next steps.

```
seurat_object <- FindNeighbors(seurat_object, assay
= "RNA", dims = 1:20)
```

This next step is to use the FindClusters function from the Seurat package to perform graph-based clustering on the cells to group cells into distinct clusters based on their gene expression similarities. Clusters often correspond to distinct cell types or states, helping to identify the different cell populations in the sample. Clusters serve as the basis for identifying genes that are differentially expressed between cell types or states. Unexpected clusters might reveal novel cell types or states.

3.4.12 Graph-Based Cell Clustering Using the FindClusters Function

```
seurat_object <- FindClusters(seurat_object, reso-
lution = 0.2)
```

```
> seurat_object <- FindNeighbors(seurat_object, assay = "RNA", dims = 1:20)
Computing nearest neighbor graph
Computing SNN
> seurat_object <- FindClusters(seurat_object, resolution = 0.2)
Modularity Optimizer version 1.3.0 by Ludo Waltman and Nees Jan van Eck

Number of nodes: 5983
Number of edges: 264206

Running Louvain algorithm...
0%   10   20   30   40   50   60   70   80   90   100%
[----|----|----|----|----|----|----|----|----|----|
**************************************************|
Maximum modularity in 10 random starts: 0.9136
Number of communities: 6
Elapsed time: 0 seconds
>
```

3.4.13 Setting Environment Variable for Performing Cell Type Annotation

The next code line is to set an environment variable in your R session for performing cell type annotation which is a fundamental step in sc-RNAseq analysis. This process is often laborious and time-consuming. While manual annotation using marker genes remains widespread, the process relies heavily on subjective interpretation. Therefore, we will use automated cell type annotation methods by utilizing marker gene information.

In the below code, we use Sys.setenv( ): which is an R function used to set environment variables for the current R session.

1. OPENAI_API_KEY: This is the name of the environment variable being set.
2. 'Your-API-Key-Here': This is a placeholder for the actual API key you would use.

The principle behind this step is to securely store and access your OpenAI API key. This is important for several reasons:

1. Authentication: The API key is used to authenticate your requests to the OpenAI API, allowing you to use their services.
2. Security: By using an environment variable, you avoid hard-coding the API key directly in your script, which is a security best practice.
3. API Access: This key allows you to programmatically interact with OpenAI's services, such as GPT-4 or GPT-4o, from within your R environment.

4. Integration: It prepares your R environment to use OpenAI's services in conjunction with your single-cell analysis, potentially for tasks like automated cell type annotation or generating descriptions of cell clusters.
5. Reproducibility: By setting the environment variable in your script, you ensure that anyone running your code (with their own API key) can reproduce your results.

```
Sys.setenv(OPENAI_API_KEY = 'Your-API-Key-Here')
```

3.4.14 Combining Different Data into a Single Assay

In the next step, we will use the 'JoinLayers' function from the Seurat package to combine different data (like raw counts, normalized data, scaled data) layers in the Seurat object into a single assay.

```
seurat_object <- JoinLayers(seurat_object)
table(Idents(seurat_object))
```

```
> seurat_object <- JoinLayers(seurat_object)
> table(Idents(seurat_object))

   0    1    2    3    4    5
1688 1569 1204 1011  340  171
>
```

3.4.15 Identifying Marker Genes for Each Cluster

The next step after joining layers is to use the 'FindAllMarkers' function from the Seurat package to identify marker genes for each cluster in the Seurat object. 'FindAllMarkers()' function performs differential expression analysis to identify genes that are significantly up-regulated in each cluster compared to all other clusters. The results are stored in the 'markers' variable, which will be a data frame containing information about marker genes for each cluster.

```
markers <- FindAllMarkers(seurat_object)
```

```
> table(Idents(seurat_object))

   0    1    2    3    4    5
1688 1569 1204 1011  340  171
> markers <- FindAllMarkers(seurat_object)
Calculating cluster 0
For a (much!) faster implementation of the Wilcoxon Rank Sum Test,
(default method for FindMarkers) please install the presto package
--------------------------------------------
install.packages('devtools')
devtools::install_github('immunogenomics/presto')
--------------------------------------------
After installation of presto, Seurat will automatically use the more
efficient implementation (no further action necessary).
This message will be shown once per session
  |++++++++++++++++++++++++++++++++++++++++++++++++++| 100% elapsed=03m 12s
Calculating cluster 1
  |++++++++++++++++++++++++++++++++++++++++++++++++++| 100% elapsed=02m 55s
Calculating cluster 2
  |++++++++++++++++++++++++++++++++++++++++++++++++++| 100% elapsed=03m 05s
Calculating cluster 3
  |++++++++++++++++++++++++++++++++++++++++++++++++++| 100% elapsed=03m 04s
Calculating cluster 4
  |++++++++++++++++++++++++++++++++++++++++++++++++++| 100% elapsed=02m 58s
Calculating cluster 5
  |++++++++++++++++++++++++++++++++++++++++++++++++++| 100% elapsed=03m 13s
>
```

Using the next code line, we will save the markers into a CSV file.

```
write.csv(markers, file = "normal_cluster_markers.
csv", row.names = FALSE)
```

Alternatively, one can also use the below code line to identify markers for all clusters. Furthermore, by specifying Seurat's differential gene expression functions, you can use the min.pct and logfc.threshold parameters to filter genes based on their expression patterns and fold change.

```
markers <- FindAllMarkers(seurat_object, only.pos =
TRUE, min.pct = 0.25, logfc.threshold = 0.25)
```

To compare between two specific clusters:

```
cluster_markers <- FindMarkers(seurat_object, ident.1
= "cluster_1", ident.2 = "cluster_2", min.pct = 0.25,
logfc.threshold = 0.25)
```

To view the top markers for each cluster, one can use the below code:

```
top_markers <- markers %>% group_by(cluster) %>%
top_n(n = 10, wt = avg_log2FC)
```

This will give you the top 10 markers for each cluster based on the average log2 fold change.

3.4.16 Cell Type Annotation and Composition

The next step is to perform cell type annotation using the cluster-level markers. Cell annotation is conducted using the GPTCelltype framework as described by Hou et al. (https://github.com/Winnie09/GPTCelltype) [19]. For each cell cluster, the top ten differentially expressed genes are selected as input features to guide the annotation process. Models GPT-4o (13 June 2023 version) and GPT-3.5 (13 June 2023 version) can be employed. Cell type annotation reproducibility can be used as a factor while deciding to select the right model. The annotation process should be iterated 15–20 times to ensure high-confidence predictions, allowing for robust identification of cell types across the dataset. This iterative approach provides consistent and reliable cell type assignments, leveraging the power of large language models in biological data interpretation.

```
res <- gptcelltype(markers, model = 'gpt-3.5-turbo-
0125') #gpt-3.5-turbo-0125, gpt-4o-2024-05-13. gpt-4o

res
```

```
> res <- gptcelltype(markers, model = 'gpt-3.5-turbo-0125')
[1] "Note: OpenAI API key found: returning the cell type annotations."
[1] "Note: It is always recommended to check the results returned by GPT-4 in case of\n AI hallucination, before going to down-stream analysis."
> res
                                         0
                        "1. Oligodendrocyte"
                                         1
                                "2. Neuron"
                                         2
"3. Proliferating cell (possibly cancer cell)"
                                         3
                            "4. Macrophage"
                                         4
                        "5. Oligodendrocyte"
                                         5
                                "6. Neuron"
```

In the next step, we assign the cell type labels to cells in the Seurat object. The string seurat_object@meta.data$celltype creates a column named "celltype" in the metadata of the Seurat object for each cell is stored.

```
seurat_object@meta.data$celltype <- as.factor(res[as.
character(Idents(seurat_object))])
```

In the next step, we utilize the 'DimPlot' function in Seurat to visualize single-cell data by plotting cells in a reduced-dimensional space (e.g., PCA or UMAP) to highlight clusters. Setting 'group. by = 'celltype'' colors cells by their assigned cell type, helping reveal similarities and differences among them. This approach aids in understanding cell population structure and diversity.

```
DimPlot(seurat_object, group.by = 'celltype')
```

This next line captures and exports visualizations. The 'ggsave' function saves the most recent plot (from 'last_plot()') as a JPEG file at the specified path.

```
ggsave("/path/to/file/GSE230389_mutant_astrocytoma_
dimplot_seurat.jpeg", plot = last_plot(), device =
"jpeg", width = 10, height = 8, units = "in")
```

The UMAP plot shows single-cell data points clustered by cell type, with each color representing a different cell type (Fig. 1). There are two clusters of oligodendrocytes (labeled as 1 and 5) in red and light blue, indicating potential heterogeneity within this cell type. Similarly, two distinct clusters of neurons are visible (labeled as 2 and 6) in yellow and pink, possibly suggesting different neuron subtypes. Proliferating Cells (possibly cancer cells)/astrocytomas are in the green cluster (labeled 3) representing actively dividing cells, potentially cancerous and the blue cluster (labeled 4) represents immune cells involved in the tumor microenvironment. The UMAP plot shows how these cell types spatially separate based on gene expression, providing insights into cellular diversity and interactions in the astrocytoma sample (Fig. 1).

The code block in the next step creates a frequency table of cells by their assigned cell types in the 'seurat_object'. 'table(seurat_object$celltype)' counts the number of cells for each cell type, and 'cell_counts' stores this result. This summary helps in understanding the cell type composition in the dataset, which is essential for assessing cell population distribution.

```
cell_counts <- table(seurat_object$celltype)

cell_counts
```

The below is printed when the 'cell_counts' is executed describing the cell population distribution. Based on the output, there are ~2028 oligodendrocyte cells (including 1688 and 340 in different subtypes) and around 1740 neuron cells (including 1569 and 171 in different subtypes). Additionally, about 1204 cells were identified as proliferating cells (possibly cancerous), and ~1011 cells were classified as macrophages.

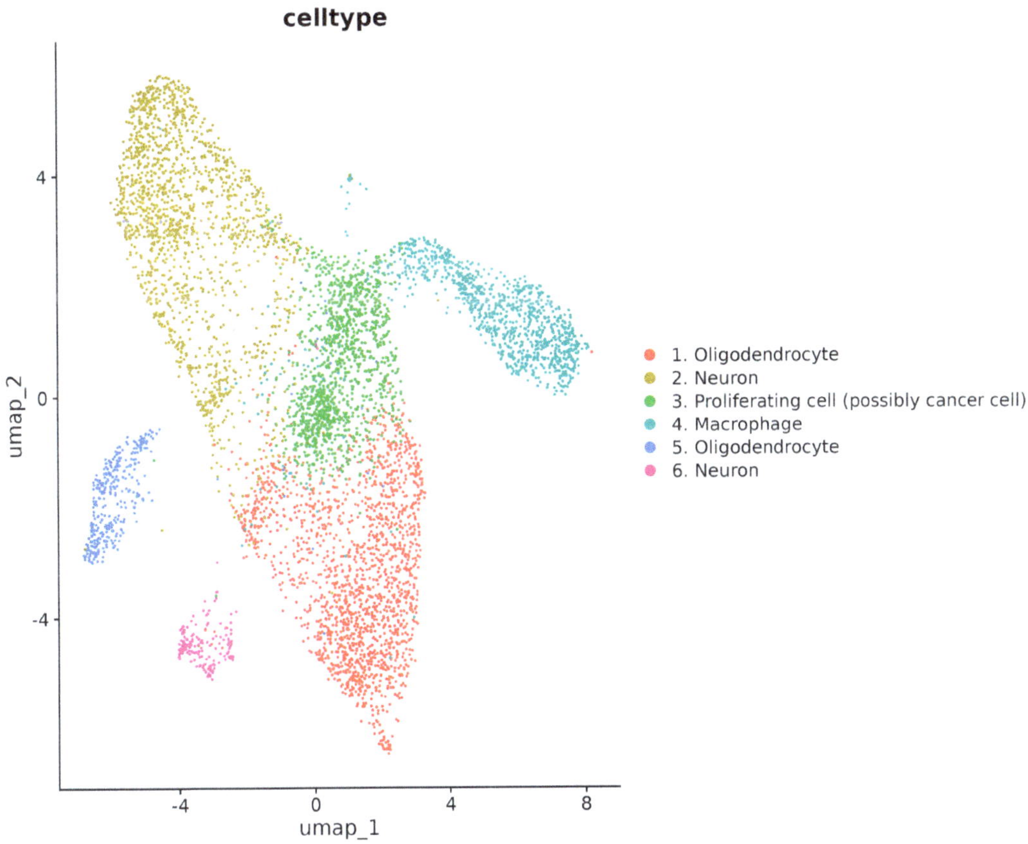

Fig. 1 The UMAP plot illustrates single-cell data points clustered by cell type, with each color corresponding to a specific cell type. Two subtypes of oligodendrocytes (labeled 1 and 5) are highlighted in red and light blue, suggesting potential heterogeneity within this cell type. Similarly, two distinct subtypes of neurons (labeled 2 and 6) appear in yellow and pink, possibly indicating different neuron subtypes. The green cluster (labeled 3) represents astrocytoma/proliferating cells, potentially cancerous and actively dividing, while the blue cluster (labeled 4) corresponds to immune cells involved in the tumor microenvironment. This spatial separation of cell types based on gene expression reveals insights into the cellular diversity and interactions present in the astrocytoma sample

cells were classified as macrophages.

```
> cell_counts <- table(seurat_object$celltype)
> cell_counts

                          1. Oligodendrocyte
                                        1688
                                   2. Neuron
                                        1569
3. Proliferating cell (possibly cancer cell)
                                        1204
                               4. Macrophage
                                        1011
                          5. Oligodendrocyte
                                         340
                                   6. Neuron
                                         171

>
```

3.4.17 File Saving Options

In the next line, we convert a Seurat object ('seurat_object') into a SingleCellExperiment (SCE) object, storing it in 'sce'. The 'as.SingleCellExperiment' function facilitates this conversion, allowing Seurat data to be used with tools and analyses specifically designed for SingleCellExperiment objects. This interoperability enables to leverage complementary analysis workflows across R packages.

```
sce <- as.SingleCellExperiment(seurat_object)
```

The next code line saves the 'sce' SingleCellExperiment object as an '.h5ad' file named '"GSE230389_mutant_astrocytoma.h5ad". The 'writeH5AD' function creates an HDF5-based file format commonly used in Python's 'scanpy' library, allowing the data to be shared and analyzed across R and Python ecosystems. This facilitates interoperability, enabling integration with other tools for further analysis.

```
writeH5AD(sce, "GSE230389_mutant_astrocytoma.h5ad")
```

This code saves the 'seurat_object' as an '.rds' file named '"GSE230389_mutant_astrocytoma.rds" using 'saveRDS'. The '.rds' format is specific to R, preserving the Seurat object structure for easy reloading and continued analysis within R. Saving in this format ensures data integrity and reproducibility, allowing future sessions to access the object exactly as saved.

```
saveRDS(seurat_object,file=("/path/to/file/GSE230389_
mutant_astrocytoma.rds")
```

The 'save.image' function saves the entire R workspace, including all objects, to a file named '"GSE230389_mutant_astrocytoma.RData" at the specified path. This command captures the current analysis state, allowing all objects to be reloaded exactly as they were for future sessions. Saving the workspace ensures that the analysis can be resumed without re-running previous code, enhancing reproducibility and efficiency.

```
save.image('/storage2/work/aveerappa/cancer_multi-
ome/GSE230389_mutant_astrocytoma/GSE230389_mutant_
astrocytoma.RData')
```

The next set of lines in the code block extracts barcodes (cell identifiers) and cell types from the metadata of a Seurat object, combines them into a data frame ('barcode_celltype_df'), and saves it as a CSV file named '"astrocytoma_barcodes_and_celltypes.csv". This process creates a standalone file linking cell barcodes with their assigned cell types, which can be useful for downstream analyses.

```
barcodes <- rownames(seurat_object@meta.data)
celltypes <- seurat_object@meta.data$celltype
barcode_celltype_df <- data.frame(barcode = bar-
codes, CellType = celltypes)
write.csv(barcode_celltype_df, file = "astrocytoma_
barcodes_and_celltypes.csv", row.names = FALSE)
```

3.5 Data Preprocessing and Quality Control of snATACseq Using SnapATAC2

3.5.1 Importing Libraries and Environment Setup

The below code creates a new Conda environment named 'snapatac2_env' with Python version 3.9. The '-y' flag automatically confirms the installation, streamlining the process.

```
# Create a new conda environment named snapatac2_env
conda create -n snapatac2_env python=3.9 -y
```

The next step is to activate the newly created environment.

```
# Activate the newly created environment
conda activate snapatac2_env
```

```
(base) aveerappa@lambda-scalar:/storage2/work/aveerappa/cancer_multiome/astrocytoma$ conda activate snapatac2_env
(snapatac2_env) aveerappa@lambda-scalar:/storage2/work/aveerappa/cancer_multiome/astrocytoma$ python -c "import snapatac2; print(snapatac2.__version
2.7.0
```

Followed by activation, we will then begin installing snapatac2 located here https://kzhang.org/SnapATAC2/install.html. We will use [recommend] to install all recommended dependencies.

```
# Install SnapATAC2 and its dependencies
pip install snapatac2[recommend]
```

Further, we will continue to install recommended dependencies for SnapATAC2.

```
# Install recommended dependencies for SnapATAC2
conda install -c bioconda samtools bedtools -y
conda install -c conda-forge hdf5 -y
pip install matplotlib numpy pandas scipy scikit-
learn anndata h5py
```

The next code installs PyTorch, a popular deep learning library, either for CPU-only usage or with GPU (CUDA 11.7) support. The CPU version is installed with 'pip install torch', while the GPU version is installed with specific CUDA-compatible versions of 'torch', 'torchvision', and 'torchaudio'. Installing the appropriate version based on hardware accelerates computations, making PyTorch models more efficient, especially for large datasets or complex tasks.

```
# Install PyTorch (CPU or GPU version depending on
your setup)
# CPU version
pip install torch
# GPU version (CUDA 11.7)
pip install torch==1.13.1+cu117 torchvision==0.14.1+
cu117
torchaudio==0.13.1 --extra-index-url https://down-
load.pytorch.org/whl/cu117
```

This code installs 'umap-learn' and 'scanpy', Python packages for dimensionality reduction and single-cell data. Adding these tools supports detailed and efficient visualization and analysis workflows in single-cell research.

```
# Install additional visualization tools
pip install umap-learn scanpy
```

In the next step, we verify if the SnapATAC2 program was installed successfully by importing it and printing the installed version.

```
# Verify the installation
python -c "import snapatac2; print(snapatac2.__ver-
sion__)"
```

The next step is to import all the libraries into the environment as shown below.

```
import snapatac2 as snap

import scanpy as sc

import matplotlib.pyplot as plt

import pandas as pd
```

3.5.2 Preprocessing by Importing Fragment Files

Data Preprocessing and Quality Control of snATACseq using SnapATAC2

The generated fragment files from the processed bam files obtained from the cell ranger output contain chromatin accessibility data, which is imported using snap.pp.import_data(). This function computed basic quality control (QC) metrics, such as counts of fragment size, TSS enrichment and unique fragments per cell, while storing the data in an AnnData object. We analyzed the fragment size distribution to identify key features: nucleosome-free regions, mono-nucleosomes, and di-nucleosomes using snap.pl.frag_size_distr() (Fig. 2). TSS enrichment scores were computed for each cell using snap.metrics.tsse(), enabling the identification of high-quality cells. Based on TSS enrichment and fragment counts, we filtered cells that did not meet the predefined quality thresholds using snap.pp.filter_cells(), specifying minimum and maximum counts of fragments and a threshold for TSS enrichment. The next code loads ATACseq fragment data using the SnapATAC package:

1. It checks the SnapATAC version and defines the path to the fragment file ('atac_fragments.tsv') and chromosome sizes for the hg38 genome.
2. 'snap.pp.import_data' imports the fragment data, preparing it for analysis by mapping each fragment to its genomic position.
3. This enables single-cell ATACseq data organized for downstream processing, helping identify accessible chromatin regions across the genome.

```
# Step 1: Load ATAC fragment data
print(snap.__version__)
fragment_file = "/storage2/work/aveerappa/cancer_
multiome/astrocytoma/atac_fragments.tsv"
```

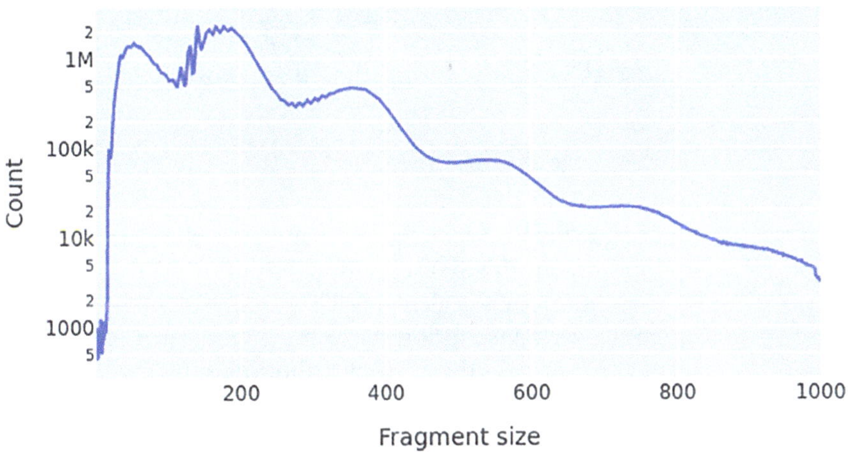

Fig. 2 The fragment size distribution plot shows a typical ATACseq pattern, with a high count of smaller fragments around 50–200 base pairs, likely representing mononucleosomal fragments associated with single nucleosomes . Secondary peaks at 400–600 base pairs suggest dinucleosomal or multinucleosomal fragments, indicating accessibility around multiple nucleosomes. The gradual decline in the count of larger fragments is also expected, as larger fragments are generally less frequent in ATACseq data. This pattern, especially the presence of distinct nucleosome-associated peaks, indicates good data quality and effective capture of accessible chromatin regions

```
chrom_sizes = snap.genome.hg38
data = snap.pp.import_data(
fragment_file,
chrom_sizes=chrom_sizes,
sorted_by_barcode=False,
)
print(data)
```

```
[>>> import snapatac2 as snap
[>>> import scanpy as sc
[>>> import matplotlib.pyplot as plt
[>>> import pandas as pd
[>>> print(snap.__version__)
 2.7.0
[>>> fragment_file = "/storage2/work/aveerappa/cancer_multiome/astrocytoma/atac_fragments.ts
[>>> chrom_sizes = snap.genome.hg38
[>>> data = snap.pp.import_data(
[... fragment_file,
[... chrom_sizes=chrom_sizes,
[... sorted_by_barcode=False,
[... )
 · Processed 178,548 barcodes in 40s (4,411.0305/s) ...
```

3.5.3 Calculate Quality Control Metrics

The next code generates and saves a plot of sc-ATACseq fragment size distribution (Fig. 2). snap.pl.frag_size_distr calculates and displays the fragment size distribution from the data. The y-axis is set to a logarithmic scale to better visualize differences in fragment frequencies (Fig. 2).

```
# Step 2: Plot fragment size distribution and save it
snap.pl.frag_size_distr(data, interactive=False)
fig = snap.pl.frag_size_distr(data, show=False)
fig.update_yaxes(type="log")
fig.write_image("astrocytoma_fragment_size_distribu-
tion.png")
```

3.5.4 Calculating the Size Distribution of Fragments

The next code calculates, plots and saves the Transcription Start Site Enrichment (TSSE) score distribution for ATACseq data (Fig. 3). snap.metrics.tsse calculates TSSE scores, which reflect the accessibility near transcription start sites, an indicator of data quality. snap.pl.tsse visualizes the TSSE distribution, showing how accessible TSS regions are across cells (Fig. 3).

```
# Step 3: Calculate TSSE metrics and plot
snap.metrics.tsse(data, chrom_sizes)
snap.pl.tsse(data, interactive=False)
snap.pl.tsse(data, interactive=False, show=False,
out_file="astrocytoma_tsse_distribution.png")
```

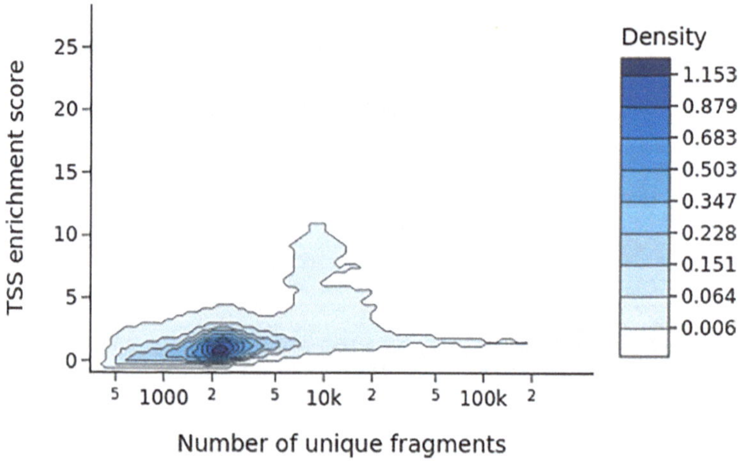

Fig. 3 The plot displays the relationship between the Transcription Start Site Enrichment (TSSE) score and the number of unique ATACseq fragments per cell, with density contours indicating cell distribution. The x-axis represents the number of unique fragments per cell, reflecting chromatin accessibility or sequencing depth, while the y-axis shows the TSSE score, a key indicator of data quality in ATACseq experiments. Most cells cluster at lower TSSE scores and fragment counts, suggesting lower-quality data or sparse chromatin accessibility. A smaller subset of cells has higher TSSE scores and fragment counts, indicating higher-quality data with strong TSS enrichment—these cells are more suitable for downstream analyses. This plot is crucial for assessing overall data quality, helping identify well-profiled cells based on their TSSE scores and fragment counts

3.5.5 Performing Feature Selection

The next set of code lines calculates summary statistics for 'n_fragment' (fragment counts per cell) and 'tsse' from an ATACseq dataset. For each metric, it computes the minimum, maximum, mean, standard deviation, median, and 25th and 75th percentiles. These statistics are stored in dictionaries ('count_stats' for fragment counts and 'tsse_stats' for TSSE) providing a quick overview of data distribution. This step aids in assessing data quality and variability, helping identify any anomalies or trends in fragment counts and TSSE scores across cells.

```
# Calculate summary statistics for n_fragment
(counts) and tsse
count_stats = {
    "min": np.min(data.obs['n_fragment']),
    "max": np.max(data.obs['n_fragment']),
    "25th_percentile": np.percentile(data.obs['n_
    fragment'], 25),
    "50th_percentile": np.median(data.obs['n_frag-
    ment']),
    "75th_percentile": np.percentile(data.obs['n_
    fragment'], 75),
    "mean": np.mean(data.obs['n_fragment']),
    "std": np.std(data.obs['n_fragment'])
}
tsse_stats = {
    "min": np.min(data.obs['tsse']),
    "max": np.max(data.obs['tsse']),
    "25th_percentile": np.percentile(data.obs['tsse'],
    25),
    "50th_percentile": np.median(data.obs['tsse']),
    "75th_percentile": np.percentile(data.obs['tsse'],
    75),
    "mean": np.mean(data.obs['tsse']),
    "std": np.std(data.obs['tsse'])
}
# Print the statistics
print("Count Metrics (n_fragment):", count_stats)
print("TSSE Metrics:", tsse_stats)
```

```
>>> print("Count Metrics (n_fragment):", count_stats)
Count Metrics (n_fragment): {'min': 200, 'max': 341552, '25th_percentile': 1553.0, '50th_percentile': 2181.0, '75th_percentile': 2859.0, 'mean': 4048.187162200442, 'std'
738.162600293947}
>>> print("TSSE Metrics:", tsse_stats)
TSSE Metrics: {'min': 0.0, 'max': 27.484143763213527, '25th_percentile': 0.5681818181818181, '50th_percentile': 1.0291595197255574, '75th_percentile': 1.6317016317016315
ean': 1.456627756611715, 'std': 1.9198480030994365}
```

Based on the above summary statistics, some suggested thresholds for 'min_counts', 'min_tsse', and 'max_counts' can be decided as follows:

1. min_counts: The 75th percentile for 'n_fragment' (count) is 2859, with a mean of 4048 and a standard deviation of 12,738. Setting 'min_counts' at around the 75th percentile would include cells with reasonably high fragment counts. Therefore, setting 2859 or slightly higher, such as 3000 would be ideal.

2. min_tsse: The 75th percentile for TSSE is 1.63, with a mean of 1.45. Since a TSSE threshold of 4 would be higher than both the 75th percentile and mean, it would be quite stringent. If you prefer a moderately high TSSE filter, 2.0 might be appropriate to capture cells with strong signal enrichment.
3. max_counts: The maximum count is 341,552, but the mean is much lower (4048), suggesting that a few cells may have very high fragment counts, which could skew the analysis. A value around 50,000 or 75,000 would include most cells with high counts while excluding extreme outliers.

Based on these suggestions, the next code line filters cells in the dataset based on fragment counts ('min_counts' and 'max_counts') and TSSE scores ('min_tsse'). snap.pp.filter_cells removes cells with fewer than 3000 fragments, fewer than a TSSE score of 2.0, or more than 50,000 fragments, retaining only high-quality cells. Filtering based on these thresholds ensures that low-quality cells with insufficient data or outliers are excluded from further analysis. This step is crucial for improving the accuracy and reliability of downstream analyses by focusing on well-profiled cells.

```
# Step 4: Filter cells based on counts and TSSE
snap.pp.filter_cells(data, min_counts=3000, min_
tsse=2.0, max_counts=50000)
print(data)
```

```
>>> print(data)
AnnData object with n_obs × n_vars = 6919 × 0
    obs: 'n_fragment', 'frac_dup', 'frac_mito', 'tsse'
    uns: 'reference_sequences', 'frag_size_distr', 'library_tsse', 'frac_overlap_TSS', 'TSS_profil
    obsm: 'fragment_paired'
>>> 
```

In the next code, the 'n_features' parameter in snap.pp.select_features refers to the number of top variable features (peaks) that will be selected based on variability, and it should be set according to the number of cells and the complexity of the data. Based on the metrics (see above) and considering the common practices in single-cell ATACseq data analysis, here are some guidelines; For Broad Coverage, 'n_features= 100,000' might be reasonable, especially if analyzing a large, complex dataset. However, 100,000 features can be quite high and may lead to memory or computational challenges, especially if your system has limited resources. For moderate selection, around 30,000–50,000 features are often more practical and still provides a good representation of variability in data. This range is typically enough for most analyses. Since the Cell Count is 6919 cells, selecting 30,000–50,000 features should work well without overwhelming the dataset. Based on this rationale, we can adjust 'n_features' as below. This selection should capture the most informative features without excessive computational cost.

```
# Step 5: Add tile matrix and select features
snap.pp.add_tile_matrix(data)
snap.pp.select_features(data, n_features=100000)
```

3.5.6 Applying Scrublet Algorithm to Identify and Filter Potential Doublets

The next code removes potential doublets from the dataset using the Scrublet algorithm. snap.pp.scrublet identifies cells that may contain mixed genetic material from two or more cells, known as doublets, which can confound analyses. Removing doublets improves data quality by ensuring each cell represents a single unique cell profile. This step is essential for accurate single-cell analysis, as doublets can skew results and reduce the interpretability of cell-type-specific patterns.

```
# Step 6: Remove doublets with Scrublet
snap.pp.scrublet(data)
```

3.5.7 Calculating Dimension Reduction Using Spectral Embedding

This next code performs spectral clustering and UMAP dimensionality reduction on the ATACseq dataset. snap.tl.spectral applies spectral clustering, which groups cells based on similarities in chromatin accessibility, helping to identify cell types or states. snap.tl.umap reduces the data to two dimensions using UMAP, creating a visualization that captures cell relationships and cluster structure. These steps help reveal the underlying structure and diversity in the data, facilitating interpretation of cell populations and their interactions.

```
# Step 7: Perform spectral clustering and UMAP
snap.tl.spectral(data)
snap.tl.umap(data)
```

3.5.8 Using UMAP on the Lower Dimensional Representation to Embed the Cells

The next code performs K-Nearest Neighbors (KNN) graph construction and Leiden clustering on the ATACseq dataset. snap.pp.knn builds a KNN graph to identify neighboring cells based on chromatin accessibility similarity, providing a foundation for clustering. snap.tl.leiden applies the Leiden algorithm on this graph to detect clusters, grouping similar cells into distinct communities. These steps help uncover cell-type clusters, enhancing the resolution and interpretability of cell populations in single-cell analysis.

```
# Step 8: KNN and Leiden clustering
snap.pp.knn(data)
snap.tl.leiden(data)
```

3.5.9 Performing Graph-Based Clustering to Identify Cell Clusters

The next code generates and saves UMAP plots of the ATACseq dataset, coloring cells by Leiden clusters. snap.pl.umap creates a UMAP visualization where cells are colored based on their Leiden cluster assignment, highlighting group structure. The plots are saved in different formats as output files.

```
# Step 9: Plot ATAC UMAP with Leiden clusters
snap.pl.umap(data,   color='leiden',   interactive=
False, height=500)
snap.pl.umap(data,   color='leiden',   interactive=
False, height=500, show=False, out_file="astrocyto-
ma_umap_leiden_clusters.png")
snap.pl.umap(data, color='leiden', show=False, out_
file="astrocytoma_umap.pdf", height=500)
```

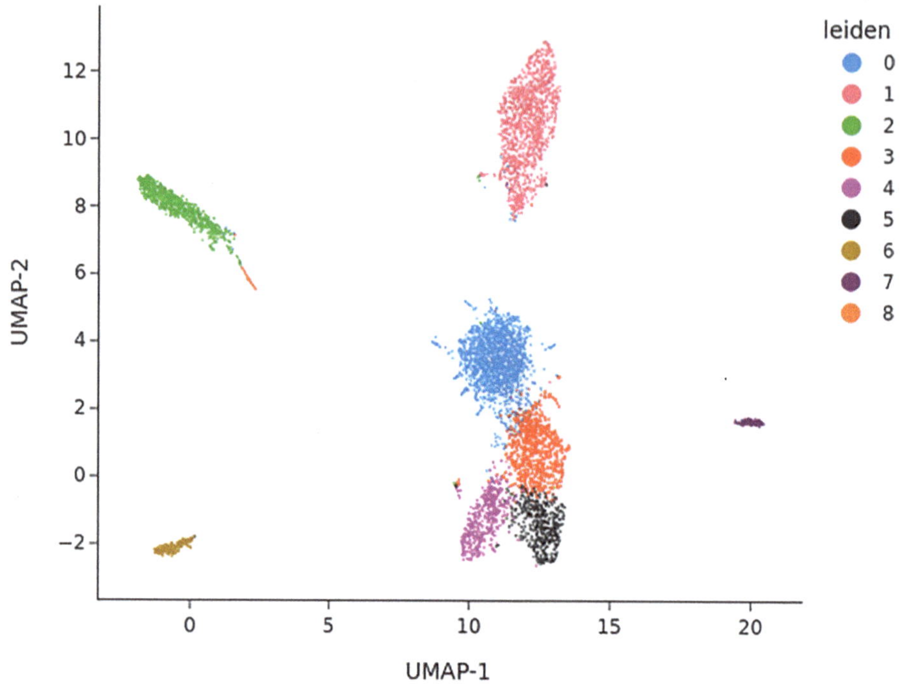

Fig. 4 The UMAP plot displays clusters of cells from the ATACseq dataset, with each color representing a distinct Leiden cluster (labeled from 0 to 8). This UMAP visualization provides an overview of the cell heterogeneity within the dataset, highlighting distinct chromatin accessibility patterns across different cell populations. The clusters are well-separated, indicating distinct cell populations, as identified by the Leiden clustering algorithm. Each cluster likely represents a unique cell type or subpopulation based on chromatin accessibility profiles. Clusters occupy different regions in the UMAP space, which reflects differences in chromatin accessibility patterns between groups. Cluster labels show identifiers (0 through 8), with each cluster visually differentiated by color

The UMAP plot displays clusters of cells from the ATACseq dataset, with each color representing a distinct Leiden cluster (labeled from 0 to 8) (Fig. 4). This UMAP visualization provides an overview of the cell heterogeneity within the dataset, highlighting distinct chromatin accessibility patterns across different cell populations. The clusters are well-separated, indicating distinct cell populations, as identified by the Leiden clustering algorithm. Each cluster likely represents a unique cell type or subpopulation based on chromatin accessibility profiles (Fig. 4). Clusters occupy different regions in the UMAP space, which reflects differences in chromatin accessibility patterns between groups. Cluster labels show identifiers (0 through 8), with each cluster visually differentiated by color (Fig. 4).

3.5.10 Cell Cluster Annotation

Create the Cell by Gene Activity Matrix

The next code block creates and preprocesses a gene accessibility matrix from ATACseq data, where each row represents a gene, and each column represents a cell. First, it generates a matrix by mapping accessible chromatin regions to nearby genes, allowing for gene-level insights. It then filters out genes accessible in fewer than 150 cells to focus on those with meaningful signal across the dataset. Next, it normalizes the counts across cells to correct for differences in sequencing depth, and finally, applies a log transformation to reduce data skewness. This processed matrix can be used to analyze gene accessibility patterns, offering insights into regulatory elements and active genes across different cell types or states.

Gene Filtering, Data Normalization, and Data Transformation

The 'min_cells' parameter in sc.pp.filter_genes defines the minimum number of cells in which a gene must be expressed (or detected) for it to be retained. This parameter is crucial for filtering out lowly expressed genes that might add noise to the analysis. To determine an appropriate value for 'min_cells', consider the following: since we have 6919 cells after filtering, we would want a 'min_cells' threshold that retains informative genes but filters out sparsely detected ones. Setting 'min_cells' too high could exclude valuable genes, while too low might retain too much noise. For single-cell data, setting 'min_cells' to include genes present in about 1–5% of cells are common. Therefore, for the dataset comprising of 6919 cells, 1% of cells would be ~70 cells, while 5% of cells would be ~350 cells. Based on this, we can use 'min_cells' value around 100–200 (2%-3%) to balance retaining genes with meaningful expression across cells while reducing noise.

This should retain genes that are expressed in at least 150 cells, which is a reasonable threshold for identifying informative genes without overwhelming the dataset with noise. One can adjust this based on the resulting gene count or experiment with different values (e.g., 100 or 200) to find the balance that best suits your analysis needs.

```
# Step 10: Generate gene matrix from ATAC data and
preprocess
gene_matrix = snap.pp.make_gene_matrix(data, chrom_
sizes)
sc.pp.filter_genes(gene_matrix, min_cells=150)
sc.pp.normalize_total(gene_matrix)
sc.pp.log1p(gene_matrix)
```

This next code block applies MAGIC imputation to the gene accessibility matrix and integrates the results with a UMAP embedding for visualization. MAGIC Imputation sc.external.pp.magic(gene_matrix, solver="approximate") applies MAGIC (Markov Affinity-based Graph Imputation of Cells) to the gene accessibility matrix, smoothening noisy single-cell data and inferring missing values. This method enhances signal by imputing likely gene activity, helping reveal gene expression patterns that may be obscured by technical noise. gene_matrix.obsm["X_umap"] = data.obsm["X_umap"] aligns the imputed gene accessibility data with the UMAP coordinates from the original data. This allows cells to be visualized in the UMAP space based on imputed gene accessibility values, preserving the spatial relationships.

Performing Imputation Using the MAGIC Algorithm and Data Smoothing

```
# Step 11: MAGIC imputation and UMAP integration
sc.external.pp.magic(gene_matrix, solver="approxi-
mate")
gene_matrix.obsm["X_umap"] = data.obsm["X_umap"]
```

3.5.11 Finding Marker Regions

The next code plots marker gene expression on a UMAP embedding to help interpret cell clusters in an astrocytoma dataset. By selecting specific marker genes ('GFAP', 'UGT8', 'MBP', 'CNP', 'PLP1', 'IDH1', and 'IDH2'), the plot highlights genes relevant to different cell types or tumor characteristics. Cells are visualized by their Leiden cluster and the expression of each marker gene. This enables identification of cell types or tumor states within the dataset, as marker gene expression patterns help distinguish cell populations.

```
# Step 12: Load marker genes and plot UMAP with gene
expression
marker_genes = ['GFAP', 'UGT8', 'MBP', 'CNP', 'PLP1',
'IDH1', 'IDH2']
sc.pl.umap(gene_matrix, use_raw=False, color=
["leiden"] + marker_genes)
plt.savefig("astrocytoma_umap_marker_genes.png")
gene_matrix.write("astrocytoma_gene_matrix.h5ad",
compression='gzip')
```

The UMAP plots display the expression levels of marker genes (GFAP, UGT8, MBP, CNP, PLP1, IDH1, and IDH2) across Leiden clusters in an astrocytoma dataset (Fig. 5). Each plot shows how these genes, associated with specific cell types, are distributed among clusters. For instance, GFAP expression suggests astrocytic cells, while UGT8, MBP, CNP, and PLP1 indicate oligodendrocyte lineage cells. The IDH1 and IDH2 markers, commonly altered in gliomas, highlight tumor-associated clusters (Fig. 5). These visualizations help characterize the cellular diversity within the astrocytoma sample, identifying potential cell types and revealing tumor heterogeneity (Fig. 5).

The next step is to save the gene matrix into a h5ad file format using the below code.

```
#Save the gene matrix file to h5ad file gene_matrix.
write("astrocytoma_atac_gene_mat.h5ad", compression=
'gzip')
```

3.6 Integrating sc-RNAseq Data with sc-ATACseq Data

3.6.1 Importing the Annotated Reference sc-RNAseq Data

In the next step, the code loads an RNA-seq dataset from a file named .h5ad and prepares it for integration with ATACseq data by mapping cell type labels. This integration step links RNA expression profiles with chromatin accessibility data across the same cell types, allowing for a comprehensive analysis of gene expression and regulatory elements within the astrocytoma sample. By combining both data types, we can gain deeper insights into cell-type-specific gene regulation and tumor-specific characteristics, enhancing our

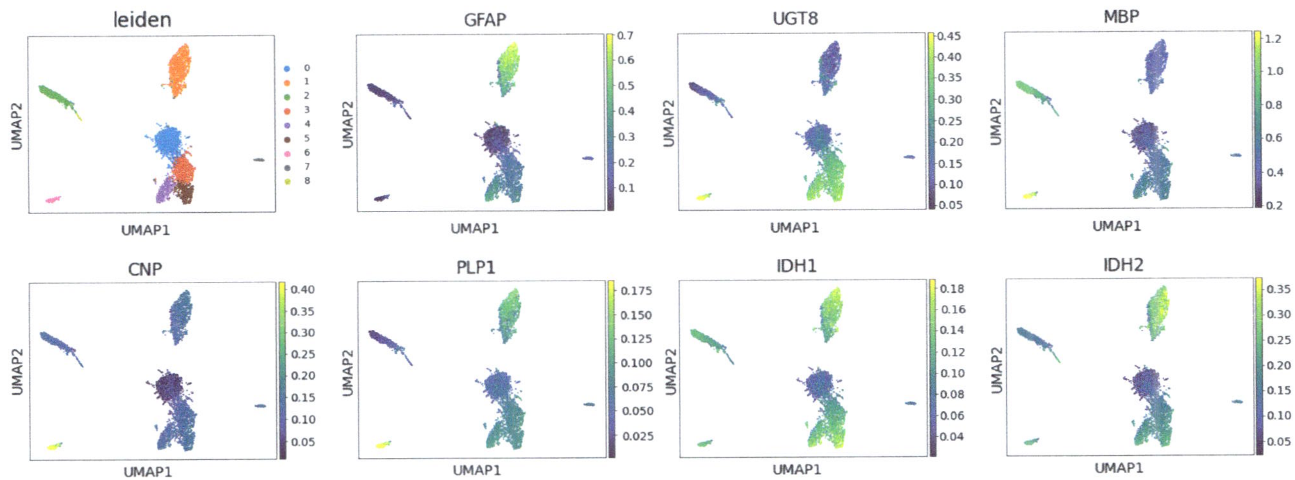

Fig. 5 The UMAP plots illustrate the expression patterns of marker genes (GFAP, UGT8, MBP, CNP, PLP1, IDH1, and IDH2) across Leiden clusters in the astrocytoma sample. Each plot highlights the distribution of these genes, which are linked to specific cell types, among the clusters. For example, GFAP expression points to astrocytic cells, while UGT8, MBP, CNP, and PLP1 are indicative of oligodendrocyte lineage cells. The markers IDH1 and IDH2, frequently mutated in gliomas, emphasize tumor-associated clusters. These visualizations provide valuable insights into the cellular composition of the astrocytoma sample, shedding light on potential cell types and tumor heterogeneity

understanding of cellular diversity and regulatory mechanisms in the tumor environment.

```
# Step 13: Load RNA data and map CellType labels
(This is the key integration step)
rna = sc.read_h5ad("astrocytoma.h5ad")
print(rna)
```

The next code is to load a CSV file containing cell barcodes and corresponding cell type labels into a pandas DataFrame in order to associate each cell with its identity. By reading this file, we can enable mapping of cell type labels based on barcodes, facilitating data integration. This step is crucial for unifying different data types, enabling a cohesive analysis of cellular composition and gene regulatory mechanisms across identified cell types in the astrocytoma sample.

```
# Load the CSV with barcodes and CellType labels
celltype_labels =
pd.read_csv("/path/to/your/GSE230389_mutant_astro-
cytoma_barcodes_and_celltypes.csv ")
```

The next code line sets the 'barcode' column as the index of 'celltype_labels', aligning it with 'rna.obs_names' for easy merging with the RNA dataset. This enables efficient label mapping based on cell barcodes.

```
# Set the 'barcode' column as the index to match it
with rna.obs_names
celltype_labels.set_index('barcode', inplace=True)
```

The next code maps cell type labels from 'celltype_labels' to the 'rna' dataset based on matching barcodes, adding a 'CellType' column to 'rna.obs'. This enables analysis of RNA data by cell type.

```
# Map the CellType labels to the RNA data based on
the barcodes
rna.obs['CellType'] = celltype_labels['CellType'].
reindex(rna.obs_names).astype('category')
```

In the next step, we verify the cell type mapping in the 'rna' dataset: rna.obs['CellType'].unique() prints the unique cell types, confirming that all expected cell types were mapped correctly, while rna.obs[['CellType']].head() displays the first few rows of the 'CellType' column, allowing a quick check of the mapping accuracy. This verification step ensures that cell type labels have been correctly assigned to each cell in the RNA data, confirming successful integration.

```
# Verify the CellType mapping
print(rna.obs['CellType'].unique())
print(rna.obs[['CellType']].head())
```

```
>>> celltype_labels = pd.read_csv("GSE230389_mutant_astrocytoma_barcodes_and_celltypes.csv")
>>> celltype_labels.set_index('barcode', inplace=True)
>>> rna.obs['CellType'] = celltype_labels['CellType'].reindex(rna.obs_names).astype('category')
>>> print(rna.obs['CellType'].unique())
['2. Neuron', '1. Oligodendrocyte', '4. Macrophage', '3. Proliferating cell (possibly cancer cell)', '6. Neuron', '5. Oligodendrocyte
Categories (6, object): ['1. Oligodendrocyte', '2. Neuron',
                         '3. Proliferating cell (possibly cancer cell)', '4. Macrophage', '5. Oligodendrocyte', '6. Neuron']
>>> print(rna.obs[['CellType']].head())
                                                       CellType
AAACAGCCAAACTGCC-1                                    2. Neuron
AAACAGCCAAGTGAAC-1                           1. Oligodendrocyte
AAACAGCCACTGACCG-1                                4. Macrophage
AAACAGCCATGATTGT-1                                    2. Neuron
AAACATGCAAACTAAG-1  3. Proliferating cell (possibly cancer cell)
```

The code preprocesses RNA data by selecting highly variable genes, normalizing expression values, and applying a log transformation. First, it identifies the top 3000 most variable genes, narrowing the analysis to those with meaningful expression differences across cells. The data is then normalized to a uniform total count per cell (10,000) to standardize expression levels, and cells with non-zero expression are retained. Finally, log transformation is applied to reduce skewness, making the data more suitable for downstream analyses. This preprocessing ensures a focus on biologically relevant genes and comparability across cells, improving the robustness of subsequent analysis steps.

```
# Step 14: Process the RNA data (select highly vari-
able genes and normalize)
sc.pp.highly_variable_genes(rna, flavor='seurat_v3',
n_top_genes=3000)
rna = rna[:, rna.var.highly_variable]
sc.pp.normalize_total(rna, target_sum=1e4)
rna = rna[:, rna.var.highly_variable].copy()
rna = rna[rna.X.sum(axis=1) > 0, :]
sc.pp.normalize_total(rna, target_sum=1e4)
sc.pp.log1p(rna)
```

The UMAP plot displays expression data from the astrocytoma sample, with cells grouped by type, such as oligodendrocytes, neurons, proliferating cells (potentially cancerous)/astrocytomas, and macrophages (Fig. 6). Each cell type forms distinct clusters, indicating differences in gene expression profiles. The separate clustering of proliferating cells suggests tumor-related activity, while macrophages highlight immune cell presence within the tumor environment (Fig. 6). This visualization reveals the sample's cellular diversity, including both normal and potentially cancerous cells, offering a foundation for studying cell interactions and the unique gene expression patterns associated with different cell populations in the tumor (Fig. 6).

The next code applies spectral and UMAP embeddings to the RNA data, creating a lower-dimensional representation for visualization. First, the code converts expression matrix format rna.X = rna.X.tocsr( ) to CSR format, optimizing it for efficient computation. Followed by that, it performs spectral embedding snap.tl.spectral, a technique that groups similar cells by reducing data dimensions

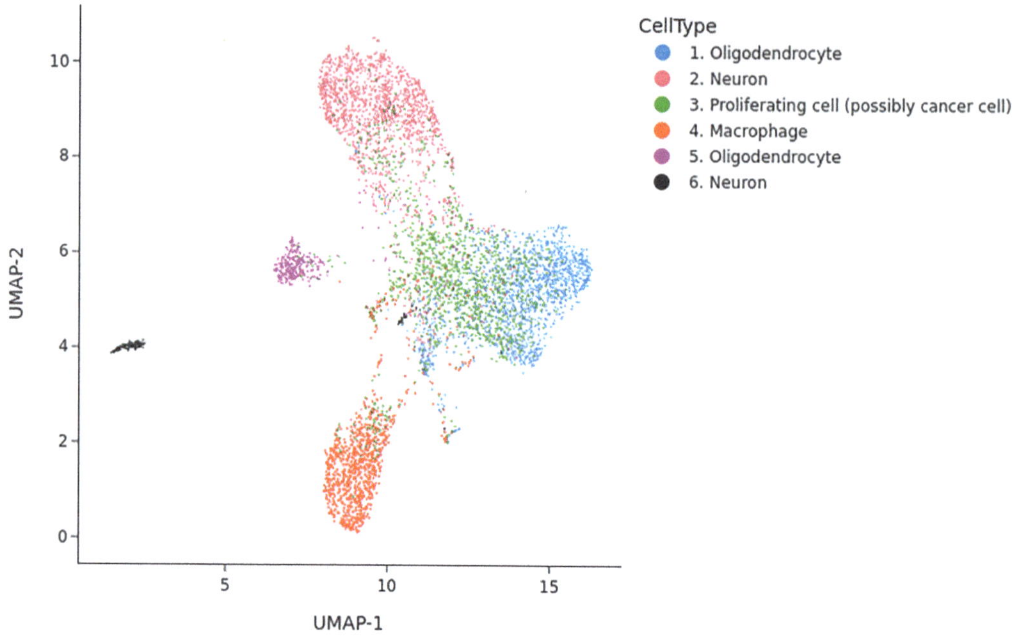

Fig. 6 The UMAP plot visualizes expression data from the astrocytoma sample, clustering cells by type, including oligodendrocytes, neurons, astrocytomas/proliferating cells (potentially cancerous), and macrophages. Distinct clusters for each cell type reflect variations in gene expression profiles. The clustering of astrocytomas/proliferating cells suggests tumor activity, while macrophages indicate the presence of immune cells within the tumor microenvironment. This visualization highlights the cellular diversity of the sample, encompassing both normal and astrocytoma cells, and provides a foundation for exploring cell interactions and the unique gene expression patterns associated with different cell populations in the tumor

based on graph structure. Finally, UMAP Embedding and Plotting is performed by applying snap.tl.umap for further dimensionality reduction, while snap.pl.umap plots cells colored by cell type.

```
# Step 15: Spectral and UMAP embedding on RNA
rna.X = rna.X.tocsr()
snap.tl.spectral(rna, features=None)
snap.tl.umap(rna)
snap.pl.umap(rna, color='CellType', interactive=False,
show=False, out_file='astrocytoma_rna_umap.png')
```

The UMAP plot displays expression data from the astrocytoma sample, with distinct clusters representing different cell types, such as oligodendrocytes, neurons, proliferating cells (likely cancerous)/astrocytoma, and macrophages (Fig. 7). This visualization reveals the cellular diversity within the sample, highlighting the presence of both normal and tumor-associated cells, which is crucial for studying cellular interactions and gene expression patterns in astrocytoma (Fig. 7).

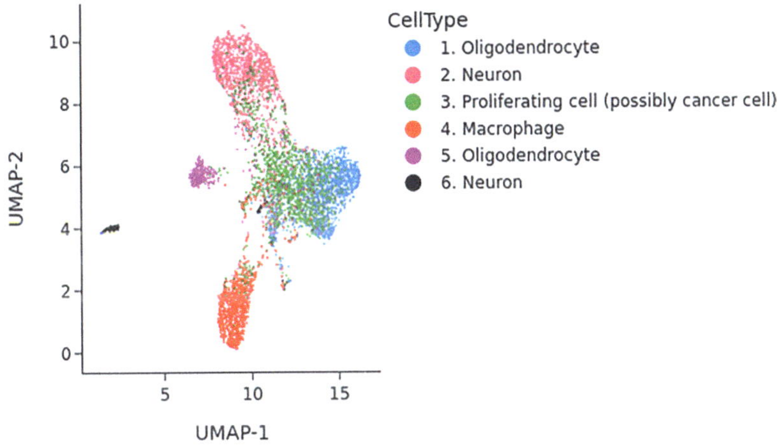

Fig. 7 The UMAP plot showcases expression data from the astrocytoma sample, with distinct clusters corresponding to various cell types, including oligodendrocytes, neurons, astrocytomas/proliferating cells (likely cancerous), and macrophages. This visualization emphasizes the sample's cellular diversity, capturing both normal and tumor-associated cells

3.6.2 Importing the Cell-by-Bin Matrix of sc-ATACseq Data

The next code loads an ATAC gene matrix and converts it to a sparse matrix format for efficient processing. We load ATAC matrix using sc.read_h5ad("astrocytoma_atac_gene_mat.h5ad") and loads the ATACseq gene matrix into the 'atac' variable. Secondly, we convert it to Sparse Matrix using csr_matrix(atac.X) which converts the data to CSR (Compressed Sparse Row) format, reducing memory usage by storing only non-zero values. Performing these two steps optimizes data storage and computation, particularly useful for large datasets, as sparse matrices save memory and speed up analyses.

```
# Step 16: Read ATAC gene matrix and handle sparse
matrix conversion for spectral embedding
atac = sc.read_h5ad("astrocytoma_atac_gene_mat.h5ad")
from scipy.sparse import csr_matrix
atac.X = csr_matrix(atac.X)
```

The next code performs spectral clustering and UMAP dimensionality reduction on the ATAC data. Spectral Clustering is performed using snap.tl.spectral and groups similar cells based on chromatin accessibility patterns, identifying clusters within the ATAC data, while UMAP Embedding snap.tl.umap reduces data to two dimensions for visualization, preserving cluster structure and relationships. These two steps help uncover and visualize cell-type-specific chromatin accessibility profiles, enabling insights into regulatory diversity within the dataset.

```
# Step 17: Perform spectral clustering and UMAP on
ATAC data
snap.tl.spectral(atac, features=None)
snap.tl.umap(atac)
```

The next set of code lines identifies cells present in both the RNA and ATAC datasets. Firstly, it will find common cells using rna.obs_names.intersection(atac.obs_names) and retrieves cell identifiers that are shared between the two datasets, storing them in 'common_cells'. Secondly, it will count common cells using print(f"-Number of common cells: {len(common_cells)}") and displays the number of overlapping cells. This step enables integration of RNA and ATAC data for the same cells, facilitating a combined analysis of gene expression and chromatin accessibility to understand regulatory mechanisms at a single-cell level.

```
# Step 18: Find common cells after mapping CellType
common_cells = rna.obs_names.intersection(atac.obs_
names)
print(f"Number of common cells: {len(common_cells)}")
```

```
>>> atac = sc.read_h5ad("astrocytoma_atac_gene_mat.h5ad")
>>> from scipy.sparse import csr_matrix
>>> atac.X = csr_matrix(atac.X)
>>> snap.tl.spectral(atac, features=None)
>>> snap.tl.umap(atac)
/home/aveerappa/.conda/envs/snapatac2_env/lib/python3.9/site-packages/umap/umap_.py:1945: UserWarnin

n_jobs value 1 overridden to 1 by setting random_state. Use no seed for parallelism.

>>> common_cells = rna.obs_names.intersection(atac.obs_names)
>>> print(f"Number of common cells after removing prefix: {len(common_cells)}")
Number of common cells after removing prefix: 5003
```

The next code subsets the RNA and ATAC datasets to include only the common cells and maps cell type labels from RNA to ATAC. Firstly, it performs subset for common cells using 'rna[common_cells]' and 'atac[common_cells]' and creates subsets of RNA and ATAC data that contain only the cells present in both datasets. Secondly, it performs mapping of cell type labels using atac_common.obs['CellType'] = rna_common.obs['CellType'] and assigns cell type labels from the RNA dataset to the ATAC dataset. Finally, the UMAP plot snap.pl.umap is used to visualize the ATAC data with cell types. This step enables a unified analysis of chromatin accessibility data with cell type labels, facilitating cross-validation and deeper insights into cell-type-specific regulatory features.

```
# Step 19: Subset RNA and ATAC for common cells, map
CellType labels to ATAC data
if len(common_cells) > 0:
rna_common = rna[common_cells]
atac_common = atac[common_cells]
atac_common.obs['CellType'] = rna_common.obs['Cell-
Type']
snap.pl.umap(atac_common, color='CellType',
interactive=False, show=False,
out_file='astrocytoma_atac_umap.png')
else:
print("No common cells found after preprocessing.")
```

3.6.3 Perform Joint Embedding

This next code integrates RNA and ATAC data using joint spectral and UMAP embeddings. Firstly, it will perform multi-modality spectral analysis using snap.tl.multi_spectral([rna_common, atac_common]) and aligns RNA and ATAC data, creating a combined embedding that captures shared information across both modalities. Followed by this, it will then perform joint UMAP embedding using snap.tl.umap(atac_common, use_rep='X_joint') and then perform UMAP, visualizing both data types in a unified space. Finally, snap.pl.umap is utilized to display the joint UMAP with cell type labels. This method enables a cohesive analysis of gene expression and chromatin accessibility, providing insights into how these features interact across cell types in a shared visual context.

```
# Step 20: Multi-modality spectral analysis and joint UMAP
embedding = snap.tl.multi_spectral([rna_common, atac_common], features=None)[1]
atac_common.obsm['X_joint'] = embedding
snap.tl.umap(atac_common, use_rep='X_joint')
snap.pl.umap(atac_common, color="CellType", interactive=False,
show=False, out_file='astrocytoma_RNA_atac_joint_umap.png')
```

```
>>> if len(common_cells) > 0:
...     rna_common = rna[common_cells]
...     atac_common = atac[common_cells]
...     atac_common.obs['CellType'] = rna_common.obs['CellType']
...     snap.pl.umap(atac_common, color='CellType', interactive=False, show=False, out_file='astrocytoma_atac_umap.png')
... else:
...     print("No common cells found after preprocessing.")
...
<stdin>:4: ImplicitModificationWarning:

Trying to modify attribute `.obs` of view, initializing view as actual.

>>> embedding = snap.tl.multi_spectral([rna_common, atac_common], features=None)[1]
2024-10-31 11:02:13 - INFO - Compute normalized views...
2024-10-31 11:05:55 - INFO - Compute embedding...
>>> atac_common.obsm['X_joint'] = embedding
>>> snap.tl.umap(atac_common, use_rep='X_joint')
/home/aveerappa/.conda/envs/snapatac2_env/lib/python3.9/site-packages/umap/umap_.py:1945: UserWarning:

n_jobs value 1 overridden to 1 by setting random_state. Use no seed for parallelism.

>>> snap.pl.umap(atac_common, color="CellType", interactive=False, show=False, out_file='astrocytoma_RNA_atac_joint_umap.pn
>>> 
```

The UMAP plot displays a joint embedding of RNA and ATACseq data from the astrocytoma sample, with cells grouped by cell type (Fig. 8). Each color represents a different cell type, as indicated in the legend. Cell types include Oligodendrocytes (blue and purple) form distinct clusters, indicating subpopulations or different cell states within the oligodendrocyte lineage (Fig. 8). Neurons (pink and black) cluster separately, reflecting neuronal diversity. Proliferating Cells (possibly cancerous)/astrocytoma (green) are clustered independently, and Macrophages (red) are clustered separately, representing immune cells in the tumor microenvironment (Fig. 8). The distinct clustering patterns show that RNA and ATAC features jointly highlight similar cell types and structures within the data. This visualization highlights the cellular diversity within the astrocytoma

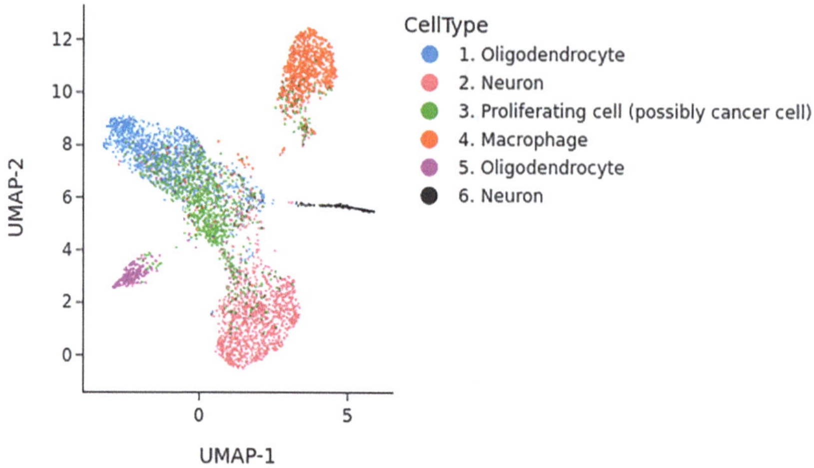

Fig. 8 The UMAP plot presents a joint embedding of RNA and ATAC data from the astrocytoma sample, with cells grouped by type. Each color, as shown in the legend, represents a specific cell type. Oligodendrocytes (blue and purple) form distinct clusters, suggesting subpopulations or varying cell states within the lineage. Neurons (pink and black) cluster separately, reflecting neuronal diversity. Astrocytomas / proliferating cells, potentially cancerous (green), form an independent csssslustler, while macrophages (red) are distinctly clustered, representing immune cells within the tumor microenvironment

sample, showing both normal cell types and potentially tumor-associated populations (Fig. 8). The integration of RNA and ATAC data provides a comprehensive view offering a cohesive understanding of cellular composition and regulatory landscapes within the tumor.

Acknowledgments

This work has been supported by the start-up funds to CG from the University of Nebraska Medical Center. The authors would like to thank the Bioinformatics and Systems Biology Core (BSBC) at UNMC for providing the computational infrastructure. BSBC is partly supported by multiple NIH awards (5P20GM103427, 5P30CA03672).

References

1. Huang WE, Zhou J (2012) When single cell technology meets omics, the new toolbox of analytical biotechnology is emerging. Curr Opin Biotechnol 23(1):1. https://doi.org/10.1016/j.copbio.2011.12.020
2. Canzler S, Schor J, Busch W, Schubert K, Rolle-Kampczyk U, Seitz H et al (2020) Prospects and challenges of multi-omics data integration in toxicology. Arch Toxicol 94(2):371–388. https://doi.org/10.1007/s00204-020-02656-y
3. Koutsogiannouli E, Papavassiliou AG, Papanikolaou NA (2013) Complexity in cancer biology: is systems biology the answer? Cancer Med 2(2):164–177. https://doi.org/10.1002/cam4.62
4. Lai AY, Wade PA (2011) Cancer biology and NuRD: a multifaceted chromatin remodelling

complex. Nat Rev Cancer 11(8):588–596. https://doi.org/10.1038/nrc3091
5. Heo YJ, Hwa C, Lee GH, Park JM, An JY (2021) Integrative multi-omics approaches in cancer research: from biological networks to clinical subtypes. Mol Cell 44(7):433–443. https://doi.org/10.14348/molcells.2021.0042
6. Menyhart O, Gyorffy B (2021) Multi-omics approaches in cancer research with applications in tumor subtyping, prognosis, and diagnosis. Comput Struct Biotechnol J 19:949–960. https://doi.org/10.1016/j.csbj.2021.01.009
7. Lu M, Zhan X (2018) The crucial role of multiomic approach in cancer research and clinically relevant outcomes. EPMA J 9(1):77–102. https://doi.org/10.1007/s13167-018-0128-8
8. Benesch MGK, Cherkassky L, Nurkin SJ (2024) Multi-omic analysis: a possible platform toward personalized and adaptable cancer treatment. Ann Surg Oncol 31(8):4831–4833. https://doi.org/10.1245/s10434-024-15449-9
9. Allaway KC, Gabitto MI, Wapinski O, Saldi G, Wang CY, Bandler RC et al (2021) Genetic and epigenetic coordination of cortical interneuron development. Nature 597(7878):693–697. https://doi.org/10.1038/s41586-021-03933-1
10. Swanson E, Lord C, Reading J, Heubeck AT, Genge PC, Thomson Z et al (2021) Simultaneous trimodal single-cell measurement of transcripts, epitopes, and chromatin accessibility using TEA-seq. Elife 10, https://doi.org/10.7554/eLife.63632
11. Mimitou EP, Lareau CA, Chen KY, Zorzetto-Fernandes AL, Hao Y, Takeshima Y et al (2021) Scalable, multimodal profiling of chromatin accessibility, gene expression and protein levels in single cells. Nat Biotechnol 39(10):1246–1258. https://doi.org/10.1038/s41587-021-00927-2
12. Stoeckius M, Hafemeister C, Stephenson W, Houck-Loomis B, Chattopadhyay PK, Swerdlow H et al (2017) Simultaneous epitope and transcriptome measurement in single cells. Nat Methods 14(9):865–868. https://doi.org/10.1038/nmeth.4380
13. Peterson VM, Zhang KX, Kumar N, Wong J, Li L, Wilson DC et al (2017) Multiplexed quantification of proteins and transcripts in single cells. Nat Biotechnol 35(10):936–939. https://doi.org/10.1038/nbt.3973
14. Cao J, Cusanovich DA, Ramani V, Aghamirzaie D, Pliner HA, Hill AJ et al (2018) Joint profiling of chromatin accessibility and gene expressisson in thousands of single cells. Science 361(6409):1380–1385. https://doi.org/10.1126/science.aau0730
15. Chen S, Lake BB, Zhang K (2019) High-throughput sequencing of the transcriptome and chromatin accessibility in the same cell. Nat Biotechnol 37(12):1452–1457. https://doi.org/10.1038/s41587-019-0290-0
16. Ma S, Zhang B, LaFave LM, Earl AS, Chiang Z, Hu Y et al (2020) Chromatin potential identified by shared single-cell profiling of RNA and chromatin. Cell 183(4):1103–1116. e1120. https://doi.org/10.1016/j.cell.2020.09.056
17. Butler A, Hoffman P, Smibert P, Papalexi E, Satija R (2018) Integrating single-cell transcriptomic data across different conditions, technologies, and species. Nat Biotechnol 36(5):411–420. https://doi.org/10.1038/nbt.4096
18. Zhang K, Zemke NR, Armand EJ, Ren B (2024) A fast, scalable and versatile tool for analysis of single-cell omics data. Nat Methods 21(2):217–227. https://doi.org/10.1038/s41592-023-02139-9
19. GPTCelltype (2024) https://github.com/Winnie09/GPTCelltype

Part V

Discovery and Analysis of Regulatory RNAs in Cancer

Chapter 12

A Comprehensive Pipeline for Long Non-Coding RNA Discovery and Characterization in Cancer

Chittibabu Guda, Sankarasubramanian Jagadesan and Avinash Veerappa

Abstract

Long non-coding RNAs (lncRNAs) represent a significant yet underexplored component of the human genome, with emerging evidence linking them to various biological processes and diseases. Despite over 205,000 lncRNAs documented in existing databases, a substantial number remain unidentified, highlighting the complexity of the human genome. LncRNAs engage in intricate regulatory mechanisms that influence gene expression, chromatin dynamics, and cellular metabolism, particularly in the context of cancer. While substantial progress has been made in cataloging lncRNAs and elucidating their functions, the unknown and uncharacterized information is more than the known. A functional framework to effectively profile lncRNAs from the plethora of RNA-seq datasets of cancer remains inadequate, particularly regarding best practices to discover novel and known lncRNAs. Therefore, a comprehensive approach is needed to address these gaps and to harness the biological and pathological potential of lncRNAs in cancers. Here, we present a comprehensive framework that incorporates proven and reliable tools for rigorous quality control, adapter trimming, alignment, post-processing, differential analysis, interaction prediction, and visualization of lncRNAs from RNA-seq data. We employed this framework to examine the dysregulation of lncRNAs derived from human breast tumors. This dataset includes 29 pairs of luminal A subtype tumors and matched normal breast tissues from the same patients. The framework identified 4390 lncRNAs that were dysregulated with log2 fold change (log2fc) thresholds of ≤ -1 and $\geq +1$, with a *p*adj of ≤ 0.05. The log2FoldChange ranged from −7.75 to 5.76, indicating significant variability in differential expression. LncRNA interaction analysis revealed one of the significantly upregulated lncRNAs ENSG00000256513 (log2fc 5.47 with *p*adj of 3.76E−15) showing base-pairing contact with ENST00000582008 (LINC00667/lncOCMRL1). We propose that the upregulated lncRNA ENSG00000265613 may enhance malignancy by stabilizing the RNA target ENSG00000582008, particularly given its established role in oncogenesis. This interaction appears crucial for regulating stability, translation, or splicing, potentially influencing oncogenic and tumor-suppressive processes. This lncRNA data analysis framework effectively facilitated the identification of critical lncRNA dysregulations and potential oncogenic contacts, revealing the role of ENSG00000265613 in enhancing malignancy through its association with ENSG00000582008 in luminal A breast cancer.

Keywords Long intergenic non-coding RNA, lncRNA, Gene expression, RNA-seq, Ccancer, Breast cancer, Luminal-A, ENSG00000265613, LINC00667, lncOCMRL1

Archana Dhasarathy and Motoki Takaku (eds.), *Cancer Epigenetics*, Methods in Molecular Biology, vol. 3005,
https://doi.org/10.1007/978-1-0716-5072-1_12,

Abbreviations

lncRNA	Long Non-Coding RNA
ncRNA	Non-Coding RNA
miRNA	MicroRNA
circRNA	Circular RNA
piRNA	PIWI-Interacting RNA
RNA-seq	RNA Sequencing
TF	Transcription Factor
UTR	Untranslated Region
eRNA	Enhancer RNA
GRCh38	Genome Reference Consortium Human Build 38
GTF	Gene Transfer Format
GFF	General Feature Format
QC	Quality Control
CPAT	Coding Potential Assessment Tool
DESeq2	Differential Expression Sequencing 2
FEELnc	Flexible Extraction of LncRNAs
FASTQ	A file format for storing sequence data
BAM	Binary Alignment Map
STAR	Spliced Transcripts Alignment to a Reference
GC	Guanine-Cytosine Content
Q20/Q30	Quality Score Metrics (20 or 30 Phred score)
adj-*p*/*p*adj	Adjusted *p*-value
LINC	Long Intergenic Non-Coding RNA
RRI	RNA-RNA Interaction
EMT	Epithelial-Mesenchymal Transition
Enrichr	Enrichment Analysis Resource
WebGestalt	Web-based Gene Set Analysis Toolkit
gProfiler	Gene Function Prediction and Functional Annotation Tool
PCG	Protein-Coding Gene
SRA	Sequence Read Archive
HDD	Hard Disk Drive

SSD	Solid State Drive
m6A	N6-Methyladenosine
***p*-value**	Probability Value
FDR	False Discovery Rate
log2fc	Logarithm (base 2) of Fold Change
HTML	Hypertext Markup Language
JSON	JavaScript Object Notation
KMER	Substring of Length K in Sequencing
Phred	Base Calling Quality Score
Chr	Chromosome
MT	Mitochondrial DNA (chrM)
SRA Toolkit	Sequence Read Archive Toolkit
SRR	Sequence Read Archive Run Accession Number
LncRRIsearch	Long Non-Coding RNA-RNA Interaction Search Tool
RBP	RNA-Binding Protein
STRING	Search Tool for the Retrieval of Interacting Genes/Proteins
sed	Stream Editor for Filtering and Transforming Text
GEO	Gene Expression Omnibus
NTPM	Normalized Transcripts Per Million

1 Introduction

It is estimated that approximately 75% of the human genome is transcribed into RNA, while only 3% is transcribed into protein-coding mRNAs [1]. Long non-coding RNAs (lncRNAs) are ubiquitously transcribed and, similar to mRNAs, exhibit tissue-specific expression [2, 3]. LncRNAs belong to a broad category of non-coding RNAs (ncRNAs) that also includes microRNA (miRNA), circular RNA (circRNA) and PIWI-interacting RNA (piRNA) [4, 5]. LncRNAs are linear in structure and more than 200 nucleotides long and are transcribed from exon, intron, intergenic region or 5′/3′-untranslational regions (UTRs) and fold into complicated secondary structures, which facilitate their interactions with DNA, RNA and proteins [6–8]. LncRNAs can function in cis or trans as scaffolds to regulate gene expression through various mechanisms, including epigenetic modulation of chromatin, modulating protein–protein interactions, modulating transcription factors (TF) to bind to promoters and enhancers, and the related downstream signaling pathways [9–12]. LncRNAs also play a role as miRNA decoys to prevent targeted mRNA degradation [13–15].

LncRNAs are prevalent in the human genome, with over 98% of genome being non-coding, suggesting that a significant portion of lncRNAs remains to be discovered. LncRNAs have been shown to undergo somatic copy-number amplifications in cancer, suggesting their potential involvement in tumor development [16]. Unlike traditional protein-coding genes, lncRNA transcription frequently initiates at enhancer regions rather than promoters, giving rise to enhancer RNAs (eRNAs), which are a subset of lncRNAs [17]. Interestingly, these sequences exhibit stronger selective pressure than neutrally evolving promoter and exon region sequences [18, 19]. GENCODE identified 9277 [18] lncRNA genes producing 14,880 transcripts, while subsequent efforts documented 27,919 lncRNA genes [20] and over 58,000 lncRNAs [21], many of them previously unannotated. Additional studies have expanded this catalog by thousands more, including novel intergenic lncRNAs linked to demographic factors and clinical outcomes. These findings complement existing lncRNA annotation databases, which collectively document over 205,000 lncRNAs across more than 50 tissues and cell lines [21].

IncRNAs are known to regulate gene expression and have been implicated in the pathogenesis of various complex human diseases, including cancers [22, 23]. Dysregulation of lncRNAs has been implicated in tumorigenesis, metastasis, and prognosis across various malignancies, including breast and prostate cancers [24, 25]. Furthermore, lncRNAs have been associated with a wide range of non-malignant conditions, such as coronary artery diseases, autoimmune disorders, and neurological disorders [22, 26]. For this reason, lncRNAs have shown significant potential as diagnostic, prognostic, and therapeutic targets.

Interestingly, lncRNAs have emerged as critical regulators of cancer mechanism through multiple mechanisms. LncRNAs function as epigenetic regulators, modulating gene expression patterns in cancer cells [27, 28]. Additionally, lncRNAs demonstrate significant influence over cellular metabolism in cancer, affecting key metabolic pathways that support tumor growth, metastasis and survival [29, 30]. Through their interaction with RNA-binding proteins, lncRNAs have been found to play a crucial role in regulating genomic instability in cancer cells [31, 32]. They are also involved in alternative splicing mechanisms, contributing to cancer progression by generating protein variants that promote malignancies [33–35]. Furthermore, lncRNAs participate in complex regulatory networks through miRNA sponging, where they sequester and inhibit microRNA function [36–38]. Notably, recent studies have revealed that lncRNAs play a major role in conferring drug resistance in cancer cells, leading to their emergence as promising therapeutic targets in strategies aimed at overcoming cancer therapy resistance [39–41].

Given the importance of lncRNAs in cancer biology, this chapter focuses on describing a robust computational and experimental pipeline for their discovery and functional annotation. In the sections that follow, we will detail the computational requirements and outline the key methods required to implement this pipeline, providing a practical framework for advancing lncRNA research in cancer.

2 Computational Requirements/Materials

For the lncRNA analysis pipeline described here, significant computational resources are required due to the complexity and scale of transcriptome-level analysis, especially when handling large datasets. Below is a detailed breakdown of the computational requirements for efficiently processing and analyzing the data.

2.1 Operating System

- **Recommended:** Linux (Ubuntu 20.04 LTS or AlmaLinux or RedHat or macOS.
- **Minimum:** Windows Subsystem for Linux (WSL) for Windows users.

2.2 Hardware Requirements

2.2.1 Memory (RAM)

- **Recommended:** At least **128 GB of RAM** for analyzing large transcriptomic datasets or processing multiple samples concurrently.
- **Minimum: 64 GB of RAM** for smaller datasets

2.2.2 CPUs (Processing Power)

- **Recommended: 16–32 CPU cores** to parallelize tasks such as genome indexing, alignment, and transcript assembly.
- **Minimum: 8 CPU cores** for smaller-scale analyses.

2.2.3 Disk Storage

- **Recommended: At least 2 TB of SSD storage** for handling large reference genome files, intermediate results, and raw data efficiently.
- **Minimum: 1 TB of HDD storage**

2.3 Software Requirements

2.3.1 Operating System

- Linux (Ubuntu or AlmaLinux or RedHat recommended) or macOS.

2.3.2 Programming Languages

- **R:** Version 4.0 or higher (for DESeq2 and visualization).
- **Python:** Version 3.8 or higher (for fastp, CPAT, and enrichment analyses).

2.3.3 Required Tools and Libraries

- **Quality Control:**
  - FastQC: For evaluating raw read quality.
  - fastp: For trimming and filtering low-quality reads.
- **Alignment and Transcript Assembly:**
  - STAR: For aligning reads to the reference genome and quantifying gene expression.
  - Cufflinks/StringTie: For transcript assembly and quantification.
- **Annotation and Functional Analysis:**
  - gffcompare: For comparing assembled transcripts against reference annotations.
  - FEELnc: For identifying functional interactions between lncRNAs and protein-coding genes.
- **Coding Potential Analysis:**
  - CPAT: For assessing the coding potential of lncRNAs.
- **Differential Expression and Enrichment Analysis:**
  - DESeq2: For identifying differentially expressed lncRNAs.
  - clusterProfiler: For pathway enrichment analysis.
- **Visualization:**
  - ggplot2 and pheatmap: For creating plots such as volcano plots, heatmaps, and histograms.
- **General Tools:**
  - awk and gffread: For genome file manipulation and extracting lncRNA sequences.

2.3.4 GENCODE Annotations

- GRCh38.p14 Long non-coding RNA transcript sequences Fasta file downloaded from https://ftp.ebi.ac.uk/pub/databases/gencode/Gencode_human/release_47/gencode.v47.lncRNA_transcripts.fa.gz
- GTF file for GRCh38.p14 downloaded from the **GENCODE** database at https://ftp.ebi.ac.uk/pub/databases/gencode/Gencode_human/release_47/gencode.v47.long_noncoding_RNAs.gtf.gz

2.3.5 R and Python Packages

- **R Packages:**
  - DESeq2, ggplot2, pheatmap, ggrepel: For statistical analysis and visualization.
  - org.Hs.eg.db: For converting gene symbols and IDs during enrichment analysis.
- **Python Packages:**
  - CPAT: For coding potential assessment.
  - General-purpose packages: pandas, numpy, matplotlib for custom scripting.

3 Methods

The code for the respective programs, R scripts, and best-practice parameters used in this lncRNA analysis is publicly accessible through the GitHub repository at https://github.com/GudaLab/LncRNA-framework. To obtain these resources, navigate to the repository, click on the "Code" dropdown menu, and select the "Download ZIP" option. Once downloaded, extract the contents of the ZIP file into a directory of your choice. These resources provide comprehensive scripts and guidelines to facilitate the replication of the analysis and allow for customization of the pipeline for related studies.

3.1 Creating Environment, Installing Packages and Required Dependencies for lncRNA Analysis

We recommend creating a dedicated *Conda* environment to isolate the tools and their dependencies, preventing conflicts with other software programs and ensuring reproducibility.

```
# Create a conda environment
conda create -n lncRNA_pipeline python=3.9
# Activate the environment
conda activate lncRNA_pipeline
```

The tools installed in this step play critical roles in processing and analyzing lncRNA data from high-throughput transcriptome sequencing. Inside Conda environment, we use conda repository to install some of the required packages.

```
# Conda Install tools
conda install -c bioconda sra-tools fastqc fastp
star samtools stringtie gffcompare subread
```

```
(base) aveerappa@lambda01:/storage5/work/aveerappa/lncrna$ conda create -n lncRNA_pipeline python=3.9
Retrieving notices: ...working... done
Channels:
 - conda-forge
 - defaults
Platform: linux-64
Collecting package metadata (repodata.json): done
Solving environment: done

## Package Plan ##

  environment location: /home/aveerappa/.conda/envs/lncRNA_pipeline

  added / updated specs:
    - python=3.9


The following packages will be downloaded:

    package                    |            build
    ---------------------------|-----------------
    liblzma-5.6.3              |       hb9d3cd8_1         109 KB  conda-forge
    libsqlite-3.47.2           |       hee588c1_0         853 KB  conda-forge
    openssl-3.4.0              |       hb9d3cd8_0         2.8 MB  conda-forge
    python-3.9.21              |h9c0c6dc_1_cpython        22.5 MB  conda-forge
    setuptools-75.6.0          |     pyhff2d567_1         756 KB  conda-forge
    tzdata-2024b               |       hc8b5060_0         119 KB  conda-forge
    wheel-0.45.1               |     pyhd8ed1ab_1          61 KB  conda-forge
    ------------------------------------------------------------
                                           Total:        27.2 MB

The following NEW packages will be INSTALLED:

  _libgcc_mutex      conda-forge/linux-64::_libgcc_mutex-0.1-conda_forge
  _openmp_mutex      conda-forge/linux-64::_openmp_mutex-4.5-2_gnu
  bzip2              conda-forge/linux-64::bzip2-1.0.8-h4bc722e_7
  ca-certificates    conda-forge/linux-64::ca-certificates-2024.8.30-hbcca054_0
  ld_impl_linux-64   conda-forge/linux-64::ld_impl_linux-64-2.43-h712a8e2_2
  libffi             conda-forge/linux-64::libffi-3.4.2-h7f98852_5
  libgcc             conda-forge/linux-64::libgcc-14.2.0-h77fa898_1
  libgcc-ng          conda-forge/linux-64::libgcc-ng-14.2.0-h69a702a_1
  libgomp            conda-forge/linux-64::libgomp-14.2.0-h77fa898_1
  liblzma            conda-forge/linux-64::liblzma-5.6.3-hb9d3cd8_1
  libnsl             conda-forge/linux-64::libnsl-2.0.1-hd590300_0
  libsqlite          conda-forge/linux-64::libsqlite-3.47.2-hee588c1_0
  libuuid            conda-forge/linux-64::libuuid-2.38.1-h0b41bf4_0
  libxcrypt          conda-forge/linux-64::libxcrypt-4.4.36-hd590300_1
  libzlib            conda-forge/linux-64::libzlib-1.3.1-hb9d3cd8_2
  ncurses            conda-forge/linux-64::ncurses-6.5-he02047a_1
  openssl            conda-forge/linux-64::openssl-3.4.0-hb9d3cd8_0
  pip                conda-forge/noarch::pip-24.3.1-pyh8b19718_0
  python             conda-forge/linux-64::python-3.9.21-h9c0c6dc_1_cpython
  readline           conda-forge/linux-64::readline-8.2-h8228510_1
  setuptools         conda-forge/noarch::setuptools-75.6.0-pyhff2d567_1
  tk                 conda-forge/linux-64::tk-8.6.13-noxft_h4845f30_101
  tzdata             conda-forge/noarch::tzdata-2024b-hc8b5060_0
  wheel              conda-forge/noarch::wheel-0.45.1-pyhd8ed1ab_1


Proceed ([y]/n)? y


Downloading and Extracting Packages:
Preparing transaction: done
Verifying transaction: done
Executing transaction: done
#
# To activate this environment, use
#
#     $ conda activate lncRNA_pipeline
#
# To deactivate an active environment, use
#
#     $ conda deactivate

(base) aveerappa@lambda01:/storage5/work/aveerappa/lncrna$ conda activate lncRNA_pipeline
```

Use pip to install CPAT and igraph packages within the same environment.

```
# Pip Install tools
pip install CPAT==3.0.5 python-igraph
```

```
(lncRNA_pipeline) aveerappa@lambda01:/storage5/work/aveerappa/lncrna$ conda activate lncRNA_pipeline
(lncRNA_pipeline) aveerappa@lambda01:/storage5/work/aveerappa/lncrna$ conda install -c bioconda sra-tools fastqc fastp star samtools stringtie gffcompare subread
Channels:
 - bioconda
 - conda-forge
 - defaults
Platform: linux-64
Collecting package metadata (repodata.json): done
Solving environment: done

## Package Plan ##

  environment location: /home/aveerappa/.conda/envs/lncRNA_pipeline

  added / updated specs:
    - fastp
    - fastqc
    - gffcompare
    - samtools
    - sra-tools
    - star
    - stringtie
    - subread


The following packages will be downloaded:

    package                    |            build
    ---------------------------|-----------------
    alsa-lib-1.2.13            |       hb9d3cd8_0         547 KB  conda-forge
    c-ares-1.34.3              |       hb9d3cd8_1         200 KB  conda-forge
    cairo-1.18.2               |       h3394656_1         956 KB  conda-forge
    curl-8.10.1                |       hbbe4b11_0         169 KB  conda-forge
    fastp-0.24.0               |       h125f33a_0         3.0 MB  bioconda
    fastqc-0.12.1              |       hdfd78af_0        11.1 MB  bioconda
    fontconfig-2.15.0          |       h7e30c49_1         259 KB  conda-forge
    gffcompare-0.12.6          |       h9948957_4         177 KB  bioconda
    harfbuzz-10.1.0            |       h0b3b770_0         1.5 MB  conda-forge
    isa-l-2.31.0               |       h4bc722e_2         156 KB  conda-forge
    libexpat-2.6.4             |       h5888daf_0          72 KB  conda-forge
    libglib-2.82.2             |       h2ff4ddf_0         3.7 MB  conda-forge
    liblzma-devel-5.6.3        |       hb9d3cd8_1         368 KB  conda-forge
    libnghttp2-1.64.0          |       h161d5f1_0         632 KB  conda-forge
    libssh2-1.11.1             |       hf672d98_0         297 KB  conda-forge
    libxml2-2.13.5             |       h8d12d68_1         674 KB  conda-forge
    ncbi-vdb-3.1.1             |       h9948957_3        10.7 MB  bioconda
    openjdk-23.0.1             |       h68779a4_1       181.4 MB  conda-forge
    ossuuid-1.6.2              |    hf484d3e_1000          56 KB  conda-forge
    perl-5.22.2.1              |                0        15.0 MB  conda-forge
    perl-threaded-5.32.1       |       hdfd78af_1           5 KB  bioconda
    perl-uri-1.71              |                0          37 KB  bioconda
    perl-xml-libxml-2.0124     |                0         214 KB  bioconda
    perl-xml-namespacesupport-1.11|              0           6 KB  bioconda
    perl-xml-sax-0.99          |                0          24 KB  bioconda
    perl-xml-sax-base-1.08     |                0          13 KB  bioconda
    pixman-0.44.2              |       h29eaf8c_0         372 KB  conda-forge
    samtools-1.21              |       h50ea8bc_0         462 KB  bioconda
    sra-tools-3.1.1            |       h4304569_2        57.9 MB  bioconda
    star-2.7.11b               |       h43eeafb_3         8.7 MB  bioconda
    stringtie-2.2.3            |       h43eeafb_0         360 KB  bioconda
    subread-2.0.8              |       h577a1d6_0        24.1 MB  bioconda
    xorg-libx11-1.8.10         |       h4f16b4b_1         818 KB  conda-forge
    xorg-libxrandr-1.5.4       |       hb9d3cd8_0          29 KB  conda-forge
    xorg-libxrender-0.9.11     |       hb9d3cd8_2          37 KB  conda-forge
    xorg-libxt-1.3.1           |       hb9d3cd8_0         371 KB  conda-forge
    xz-5.6.3                   |       hbcc6ac9_1          23 KB  conda-forge
    xz-gpl-tools-5.6.3         |       hbcc6ac9_1          33 KB  conda-forge
    xz-tools-5.6.3             |       hb9d3cd8_1          88 KB  conda-forge
    ------------------------------------------------------------
                                           Total:       324.4 MB

  perl-xml-sax-base  bioconda/linux-64::perl-xml-sax-base-1.08-0
  pixman             conda-forge/linux-64::pixman-0.44.2-h29eaf8c_0
  pthread-stubs      conda-forge/linux-64::pthread-stubs-0.4-hb9d3cd8_1002
  samtools           bioconda/linux-64::samtools-1.21-h50ea8bc_0
  sra-tools          bioconda/linux-64::sra-tools-3.1.1-h4304569_2
  star               bioconda/linux-64::star-2.7.11b-h43eeafb_3
  stringtie          bioconda/linux-64::stringtie-2.2.3-h43eeafb_0
  subread            bioconda/linux-64::subread-2.0.8-h577a1d6_0
  xorg-libice        conda-forge/linux-64::xorg-libice-1.1.1-hb9d3cd8_1
  xorg-libsm         conda-forge/linux-64::xorg-libsm-1.2.4-he73a12e_1
  xorg-libx11        conda-forge/linux-64::xorg-libx11-1.8.10-h4f16b4b_1
  xorg-libxau        conda-forge/linux-64::xorg-libxau-1.0.11-hb9d3cd8_1
  xorg-libxdmcp      conda-forge/linux-64::xorg-libxdmcp-1.1.5-hb9d3cd8_0
  xorg-libxext       conda-forge/linux-64::xorg-libxext-1.3.6-hb9d3cd8_0
  xorg-libxfixes     conda-forge/linux-64::xorg-libxfixes-6.0.1-hb9d3cd8_0
  xorg-libxi         conda-forge/linux-64::xorg-libxi-1.8.2-hb9d3cd8_0
  xorg-libxrandr     conda-forge/linux-64::xorg-libxrandr-1.5.4-hb9d3cd8_0
  xorg-libxrender    conda-forge/linux-64::xorg-libxrender-0.9.11-hb9d3cd8_2
  xorg-libxt         conda-forge/linux-64::xorg-libxt-1.3.1-hb9d3cd8_0
  xorg-libxtst       conda-forge/linux-64::xorg-libxtst-1.2.5-hb9d3cd8_3
  xz                 conda-forge/linux-64::xz-5.6.3-hbcc6ac9_1
  xz-gpl-tools       conda-forge/linux-64::xz-gpl-tools-5.6.3-hbcc6ac9_1
  xz-tools           conda-forge/linux-64::xz-tools-5.6.3-hb9d3cd8_1
  zstd               conda-forge/linux-64::zstd-1.5.6-ha6fb4c9_0


Proceed ([y]/n)? y


Downloading and Extracting Packages:

Preparing transaction: done
Verifying transaction: done
Executing transaction: done
(lncRNA_pipeline) aveerappa@lambda01:/storage5/work/aveerappa/lncrna$
```

Next, use R to install required packages from the Bioconductor repository.

```
#R Install tools
module load R

# Invoke R
R
```

```
# Install the required R packages
if (!requireNamespace("BiocManager", quietly = TRUE))
install.packages("BiocManager")
BiocManager::install(version = "3.20")

# Install clusterProfiler using BiocManager
BiocManager::install("clusterProfiler")

# Install ggplot2 in R
install.packages("ggplot2")

# Install igraph in R
install.packages("igraph")
```

In the last step, install NCBI's *SRA Toolkit* to download raw sequencing data from NCBI.

```
# Download the SRA Toolkit (replace version and OS
type as needed)
wget    https://ftp-trace.ncbi.nlm.nih.gov/sra/sdk/
3.2.0/sratoolkit.3.2.0-ubuntu64.tar.gz

# Extract the downloaded file
tar -xvzf sratoolkit.3.2.0-ubuntu64.tar.gz

# Step 4: Add the SRA Toolkit to PATH
echo "export PATH=/usr/local/sratoolkit/bin:\$PATH"
>> ~/.bashrc

# Step 5: Reload the shell configuration
source ~/.bashrc

# Step 6: Verify the installation
fastq-dump --version

# Step 7: Configure the toolkit interactively
vdb-config -interactive
```

3.2 Sample Cohort and Raw Data Download

As an example, publicly available bulk poly(A)-selected RNA-seq data from human breast tissue samples will be utilized to illustrate the utility of the lncRNA pipeline in cancer research. Poly(A)-selected RNA-seq primarily captures poly-A transcripts, which includes most protein-coding mRNAs. Since a significant fraction of lncRNAs lacks poly(A) tails, relying solely on poly(A)-enriched RNA-seq data would lead to incomplete lncRNA profiling. Therefore, we recommend using total RNA-seq to profile lncRNAs. However, this framework can be used both on Poly(A)-selected RNA-seq as well as total RNA-seq (with depleted rRNA) datasets. The chosen dataset for this walkthrough includes normal ($n = 29$) and paired tumor samples ($n = 29$) from the Luminal A subtype of breast cancer. Tumor and normal tissue samples were reported to be collected from the same breast of each patient, with a minimum distance of 4 cm between the sampling sites to ensure sample integrity [42]. The RNA-seq data were originally generated using the Illumina HiSeq 4000 platform and made available through the NCBI Gene Expression Omnibus (GEO) under the accession number GSE233242.

In the following code snippet, we will use NCBI's SRA Toolkit [43] to download a fastq file of a sample from the NCBI repository using its unique SRR24709142 identifier.

```
#Downloading fastq from NCBI
fastq-dump SRR24709142 --split-3--gzip
```

```
jsankarasubramani@lambda01:/storage5/work/subu/GSE233242$ fastq-dump SRR24709142 --split-3 --gzip
Read 59280637 spots for SRR24709142
Written 59280637 spots for SRR24709142
```

3.3 Data Preprocessing

3.3.1 Quality Control

Pre-processing is performed with fastqc [44] and fastp [45].

```
#Running fastqc on the paired-end fastq files
fastqc SRR24709142_1.fastq.gz SRR24709142_2.fastq.
gz -o qc_results
```

```
jsankarasubramani@lambda01:/storage5/work/subu/GSE233242$ fastqc SRR24709142_1.fastq.gz SRR24709142_2.fastq.gz -o qc_results
application/gzip
application/gzip
Started analysis of SRR24709142_1.fastq.gz
Approx 5% complete for SRR24709142_1.fastq.gz
Approx 10% complete for SRR24709142_1.fastq.gz
Approx 15% complete for SRR24709142_1.fastq.gz
Approx 20% complete for SRR24709142_1.fastq.gz
Approx 25% complete for SRR24709142_1.fastq.gz
Approx 30% complete for SRR24709142_1.fastq.gz
Approx 35% complete for SRR24709142_1.fastq.gz
Approx 40% complete for SRR24709142_1.fastq.gz
Approx 45% complete for SRR24709142_1.fastq.gz
Approx 50% complete for SRR24709142_1.fastq.gz
Approx 55% complete for SRR24709142_1.fastq.gz
Approx 60% complete for SRR24709142_1.fastq.gz
Approx 65% complete for SRR24709142_1.fastq.gz
Approx 70% complete for SRR24709142_1.fastq.gz
Approx 75% complete for SRR24709142_1.fastq.gz
Approx 80% complete for SRR24709142_1.fastq.gz
Approx 85% complete for SRR24709142_1.fastq.gz
Approx 90% complete for SRR24709142_1.fastq.gz
Approx 95% complete for SRR24709142_1.fastq.gz
Analysis complete for SRR24709142_1.fastq.gz
Started analysis of SRR24709142_2.fastq.gz
Approx 5% complete for SRR24709142_2.fastq.gz
Approx 10% complete for SRR24709142_2.fastq.gz
Approx 15% complete for SRR24709142_2.fastq.gz
Approx 20% complete for SRR24709142_2.fastq.gz
Approx 25% complete for SRR24709142_2.fastq.gz
Approx 30% complete for SRR24709142_2.fastq.gz
Approx 35% complete for SRR24709142_2.fastq.gz
Approx 40% complete for SRR24709142_2.fastq.gz
Approx 45% complete for SRR24709142_2.fastq.gz
Approx 50% complete for SRR24709142_2.fastq.gz
Approx 55% complete for SRR24709142_2.fastq.gz
Approx 60% complete for SRR24709142_2.fastq.gz
Approx 65% complete for SRR24709142_2.fastq.gz
Approx 70% complete for SRR24709142_2.fastq.gz
Approx 75% complete for SRR24709142_2.fastq.gz
Approx 80% complete for SRR24709142_2.fastq.gz
Approx 85% complete for SRR24709142_2.fastq.gz
Approx 90% complete for SRR24709142_2.fastq.gz
Approx 95% complete for SRR24709142_2.fastq.gz
Analysis complete for SRR24709142_2.fastq.gz
```

Similarly, the fastqc should be run on reads of all samples. For this walkthrough, we will describe the outputs of fastqc for the Read 1 of SRR24709142 sample. The output of fastqc provides a graphical understanding of the quality metrices of sequence reads as shown in the Fig. 1. The fastqc report for sample SRR24709142 provides an

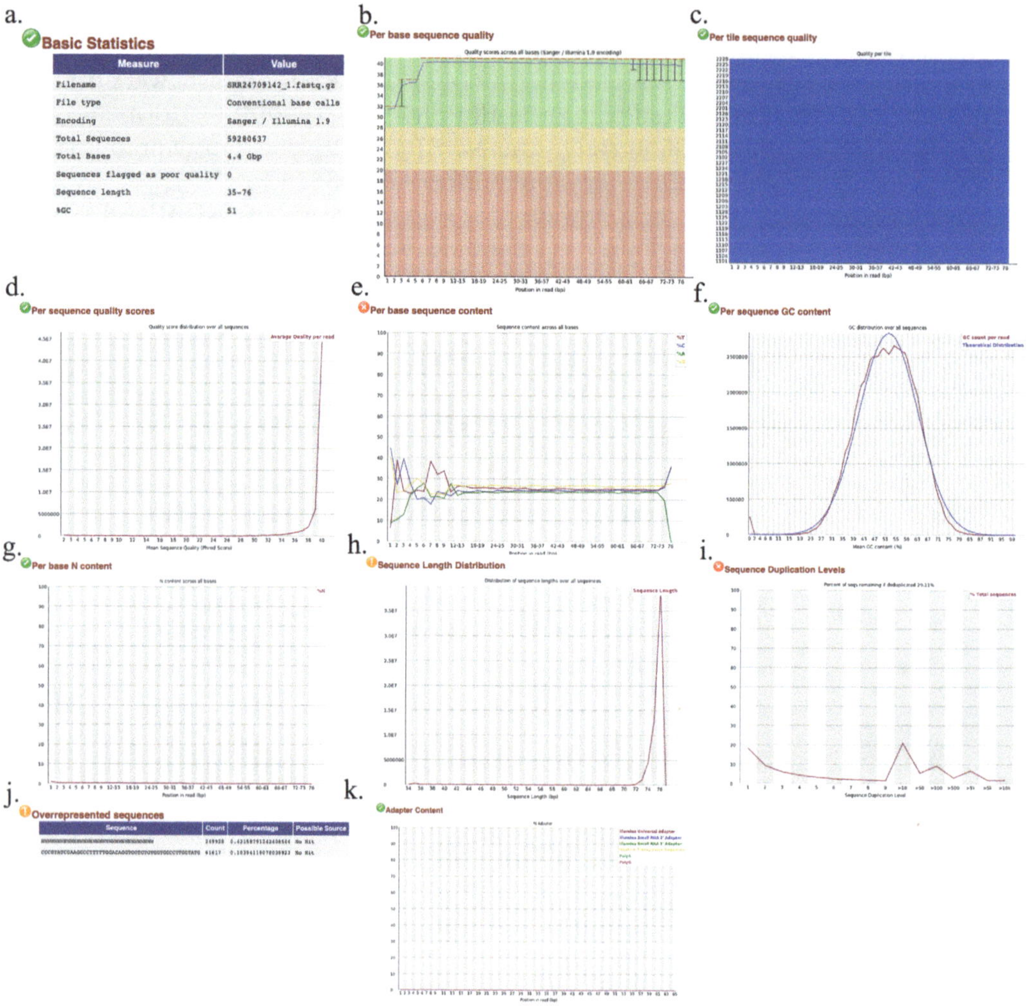

Fig. 1 Quality control metrics from FastQC for sample SRR24709142. This figure summarizes the quality control metrics generated by FastQC for the RNA-seq sample SRR24709142. **a** Overview of sequencing data including filename, encoding type, total sequences, sequence length range (35–76 bp), and GC content (51%). **b** Per Base Sequence Quality shows quality scores (Phred) across all base positions, showing high-quality scores (>Q30) throughout most of the read, with a slight drop at the 3' end. **c** Per Tile Sequence Quality Visualization of quality variations across sequencing tiles, highlighting uniform quality. **d** Per Sequence Quality Scores provides the distribution of average quality scores per read, indicating most reads have high-quality scores clustering near Q40. **e** Per Base Sequence Content displays the proportion of A, T, G, and C bases across all positions, revealing slight imbalances at the initial bases, stabilizing after position 10. **f** Per Sequence GC Content shows GC content distribution across all sequences, showing a normal distribution. **g** Per Base N Content shows the proportion of ambiguous 'N' bases at each position, showing no significant occurrences, confirming high data quality. **h** Shows the Sequence Length Distribution of sequence lengths, clustering predominantly around 76 bp, typical for RNA-seq data. **i** Sequence Duplication rates showing ~29% unique sequences, reflecting moderate duplication levels due to library preparation. **j** Identification of two overrepresented sequences, likely technical artifacts. **k** Absence of detectable adapter contamination across all read positions, confirming clean sequencing data

overview of sequencing quality, indicating approximately 59.3 million reads with a sequence length of 35–76 bp and a GC content of 51%, well within the expected range (Fig. 1a). For poly(A)-selected datasets, 50–60 million paired-end reads are generally sufficient. No sequences were flagged as poor quality, confirming the dataset's integrity. Per base quality scores (Fig. 1b) show most positions exceeding Q30, with slight drops at the 3' end necessitating trimming. Tile quality (Fig. 1c) reveals uniformity across most tiles with minor localized deviations, while average sequence quality scores (Fig. 1d) cluster near Q40, confirming excellent overall quality. Base content analysis (Fig. 1e) identifies imbalances at early positions, typical of RNA-seq datasets, which stabilize after position 10 and can be mitigated through trimming. GC content (Fig. 1f) aligns with theoretical expectations, confirming the absence of contamination or GC bias. No significant ambiguous 'N' bases were observed (Fig. 1g), and sequence lengths primarily clustered at 76 bp (Fig. 1h), with moderate duplication levels of 29% (Fig. 1i) reflecting acceptable PCR bias. Overrepresented sequences (Fig. 1j), including technical artifacts, and adapter content (Fig. 1k), showed minimal contamination, ensuring reliable downstream analysis. Further, issues such as 3' end quality drops, early base imbalances, and moderate duplication levels are resolvable with trimming, confirming the dataset's high quality for alignment, transcript assembly, and lncRNA analysis.

3.3.2 Trimming and Filtering

Since the Per Base Sequence Quality metric from the fastqc report showed a decline in quality at the 3' ends of reads, trimming low-quality bases is essential to ensure accurate alignment and downstream analyses. To address this, employ the fastp tool for quality control and read trimming. Fastp also includes comprehensive quality profiling, both before and after filtering, providing metrics such as quality curves, base content, KMER, Q20/Q30 scores, GC ratio, duplication, and adapter content. The tool filters out poor-quality reads, including those with low quality, excessive 'N' bases, or short lengths. It trims low-quality bases from the 5' and 3' ends using a sliding window strategy. Fastp generates comprehensive output in a JSON format for further interpretation and visualizes quality control results in a single HTML report.

```
#Trimming and Filtering using fastp
fastp -i SRR24709142_1.fastq.gz -I SRR24709142_2.
fastq.gz -o SRR24709142_1_clean.fastq.gz -O
SRR24709142_2_clean.fastq.gz -h SRR24709142_report.
html --trim_poly_g --cut_mean_quality 20 --length_
required 50
```

```
jsankarasubramani@lambda01:/storage5/work/subu/GSE233242$ fastp -i SRR24709142_1.fastq.gz -I SRR24709142_2.fastq.gz -o SRR24709142_1_clean.fastq.gz -O SRR24709142_2_clean.fastq.gz -h SR
R24709142_report.html --trim_poly_g --cut_mean_quality 20 --length_required 50
WARNING: you specified the options for cutting by quality, but forogt to enable any of cut_front/cut_tail/cut_right. This will have no effect.
Read1 before filtering:
total reads: 59280637
total bases: 4463770624
Q20 bases: 4427939910(99.1973%)
Q30 bases: 4370837699(97.9181%)

Read2 before filtering:
total reads: 59280637
total bases: 4464520230
Q20 bases: 4411841063(98.82%)
Q30 bases: 4342999251(97.2781%)

Read1 after filtering:
total reads: 58678074
total bases: 4429415959
Q20 bases: 4403765770(99.4209%)
Q30 bases: 4348219439(98.1669%)

Read2 after filtering:
total reads: 58678074
total bases: 4428507512
Q20 bases: 4395199369(99.2479%)
Q30 bases: 4332144111(97.824%)

Filtering result:
reads passed filter: 117356148
reads failed due to low quality: 1038158
reads failed due to too many N: 36286
reads failed due to too short: 130682
reads with adapter trimmed: 126904
bases trimmed due to adapters: 2353125

Duplication rate: 31.2441%

Insert size peak (evaluated by paired-end reads): 118

JSON report: fastp.json
HTML report: SRR24709142_report.html

fastp -i SRR24709142_1.fastq.gz -I SRR24709142_2.fastq.gz -o SRR24709142_1_clean.fastq.gz -O SRR24709142_2_clean.fastq.gz -h SRR24709142_report.html --trim_poly_g --cut_mean_quality 20
--length_required 50
fastp v0.23.2, time used: 307 seconds
```

The above code outputs an html file showing the below metrices. The fastp report shown in Fig. 2 for Read 1 of sample SRR24709142 indicates high-quality paired-end sequencing data with a mean read length of 75 bp and an insert size peak of 118 bp (Fig. 2a). Before filtering, the dataset contained 118.56 million reads with 8.93 giga-base-pairs (Gbps), achieving 99.0% Q20 and 97.6% Q30 scores, a GC content of 51.5%, and a duplication rate of 31.24%. Post-filtering, 117.35 million reads (8.86 Gbp) were retained, with improved quality metrics (99.3% Q20 and 98.0% Q30) and stable GC content (51.6%). Only 1.02% of reads were removed due to low quality (0.87%), excessive 'N' bases (0.03%), or short lengths (0.11%), demonstrating minimal data loss (Fig. 2a). Adapter contamination was negligible, accounting for only 0.026% of reads (Fig. 2b). Insert Size Estimation plot in Fig. 2c reveals a peak insert size around 118 bp, consistent with the expected fragment length distribution for RNA-seq libraries. The Per Base Quality (Fig. 2d) and KMER Content (Fig. 2f) analyses revealed minor quality drops and overrepresented sequences at early positions, addressed effectively through trimming. The Per Base Sequence Content (Fig. 2e) showed early nucleotide imbalances typical of RNA-seq data, which stabilized after position 10.

Thus, fastp preprocessing improved data integrity by removing low-quality bases, adapters, and artifacts, retaining 98.98% of data and optimizing it for downstream tasks. With these preprocessing steps, the dataset is now optimized for downstream tasks.

3.4 STAR Index Generation

After pre-processing the reads using fastqc and fastp, the next step involves preparing the reference genome for alignment. This is achieved by first generating index files using the STAR [46] aligner prior to aligning the FASTQ reads to the reference genome. STAR stands for Spliced Transcripts Alignment to a Reference and is available at https://github.com/alexdobin/STAR. The generation of index files is a critical step that optimizes and accelerates the alignment process during downstream analyses. Indexing creates a structured,

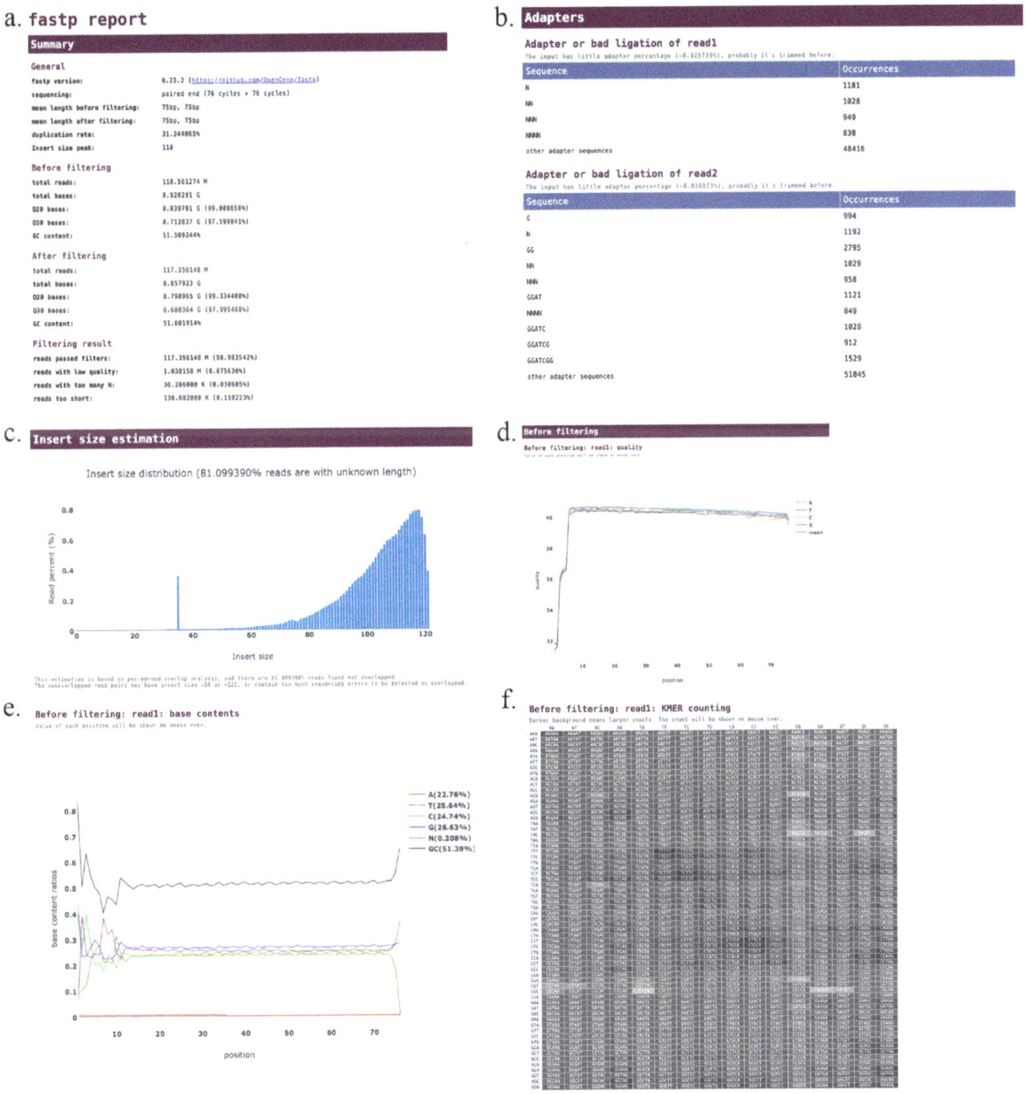

Fig. 2 Quality control and filtering metrics from Fastp for Sample SRR24709142. **a** Summarizes pre- and post-filtering statistics, including total reads, GC content, and filtering results. **b** Highlights adapter or bad ligation sequences for both reads, indicating minimal contamination. **c** Displays the insert size distribution with a peak at 118 bp. **d** Shows the base quality scores before filtering, with consistently high Phred scores across all positions. **e** Depicts the base content distribution before filtering, with slight imbalances at the start of reads. **f** Visualizes KMER content for Read 1 before filtering, identifying overrepresented sequences

searchable database of the reference genome, allowing the alignment tool to rapidly access specific regions without the need to sequentially scan the entire sequence. We recommend using GRCh38.p14 reference genome and the lncRNA GENCODE GTF annotation file for compatible transcriptome assembly. Below code line is used to generate index files for the reference genome.

```
#Indexing the reference genome using STAR
STAR --runThreadN 50 --runMode genomeGenerate --
genomeDir
human_genome/STAR_index --genomeFastaFiles
human_genome/GRCh38.p14.genome.fa --sjdbGTFfilehu-
man_genome/gencode.v47.long_noncoding_RNAs.gtf
```

```
jsankarasubramani@lambda01:/storage5/work/subu/GSE210909$ STAR --runThreadN 50 --runMode genomeGenerate --genomeDir human_genome/STAR_index --genomeFastaFiles hu
_genome/GRCh38.p14.genome.fa --sjdbGTFfile human_genome/gencode.v47.long_noncoding_RNAs.gtf
        STAR --runThreadN 50 --runMode genomeGenerate --genomeDir human_genome/STAR_index --genomeFastaFiles human_genome/GRCh38.p14.genome.fa --sjdbGTFfile huma
enome/gencode.v47.long_noncoding_RNAs.gtf
        STAR version: 2.7.10b   compiled: 2022-11-01T09:53:26-04:00 :/home/dobin/data/STAR/STARcode/STAR.master/source
Dec 05 22:34:12 ..... started STAR run
Dec 05 22:34:29 ... starting to generate Genome files
Dec 05 22:35:13 ..... processing annotations GTF
Dec 05 22:35:24 ... starting to sort Suffix Array. This may take a long time...
Dec 05 22:35:39 ... sorting Suffix Array chunks and saving them to disk...
Dec 05 22:54:09 ... loading chunks from disk, packing SA...
Dec 05 23:08:45 ... finished generating suffix array
Dec 05 23:08:45 ... generating Suffix Array index
Dec 05 23:14:38 ... completed Suffix Array index
Dec 05 23:14:39 ..... inserting junctions into the genome indices
Dec 05 23:16:11 ... writing Genome to disk ...
Dec 05 23:20:29 ... writing Suffix Array to disk ...
Dec 05 23:40:48 ... writing SAindex to disk
Dec 05 23:40:55 ..... finished successfully
```

3.5 Alignment to Reference Genome

In the next step, perform alignment of fastq reads against the reference genome using STAR aligner. STAR performs alignment using sequential maximum mappable seed search in uncompressed suffix arrays followed by seed clustering and stitching procedure.

The below code command performs the core alignment of cleaned RNA-seq (post-trimming by fastp) reads to the reference genome and produces a BAM (Binary Alignment Map) file. The quantMode parameters (--quantMode TranscriptomeSAM GeneCounts) ensure that both gene-level expression counts and transcriptome alignments are generated, facilitating subsequent quantification and differential expression analysis of lncRNAs . Additionally, using multiple threads (--runThreadN 50) and on-the-fly file decompression (--readFilesCommand zcat) enhances computational efficiency, which is crucial when processing large-scale RNA-seq datasets.

```
#Alignment to Reference Genome using STAR
STAR --runThreadN 50 --genomeDir human_genome/STAR_
index --readFilesIn SRR24709142_1_clean.fastq.gz
SRR24709142_2_clean.fastq.gz --readFilesCommand
zcat --outSAMtype BAM SortedByCoordinate --outFile-
NamePrefix
SRR24709142_ --quantMode TranscriptomeSAM GeneCounts
--outSAMstrandField intronMotif
```

```
jsankarasubramani@lambda01:/storage5/work/subu/GSE233242$ STAR --runThreadN 50 --genomeDir human_genome/STAR_index --readFilesIn SRR24709142_1_clea
n.fastq.gz SRR24709142_2_clean.fastq.gz --readFilesCommand zcat --outSAMtype BAM SortedByCoordinate --outFileNamePrefix SRR24709142_ --quantMode Tr
anscriptomeSAM GeneCounts --outSAMstrandField intronMotif
        STAR --runThreadN 50 --genomeDir human_genome/STAR_index --readFilesIn SRR24709142_1_clean.fastq.gz SRR24709142_2_clean.fastq.gz --readFile
sCommand zcat --outSAMtype BAM SortedByCoordinate --outFileNamePrefix SRR24709142_ --quantMode TranscriptomeSAM GeneCounts --outSAMstrandField intr
onMotif
        STAR version: 2.7.10b   compiled: 2022-11-01T09:53:26-04:00 :/home/dobin/data/STAR/STARcode/STAR.master/source
Jan 15 09:34:38 ..... started STAR run
Jan 15 09:34:40 ..... loading genome
Jan 15 09:34:54 ..... started mapping
Jan 15 09:38:15 ..... finished mapping
Jan 15 09:38:21 ..... started sorting BAM
Jan 15 09:39:43 ..... finished successfully
```

3.6 Transcript Assembly

Aligning FASTQ reads to the reference genome produces an aligned BAM file, which contains the mapping information of sequencing reads. However, this alignment alone does not assemble or quantify full-length transcripts. To address this, we can either use Cufflinks [47] or StringTie [48] programs. Code blocks for both tools are provided below and users can make use of any one program depending on the availability of the tool in their setup. For this walkthrough, we will use Cufflinks. Cufflinks can be accessed at http://cole-trapnell-lab.github.io/cufflinks/ whereas, StringTie is available at https://ccb.jhu.edu/software/stringtie/.

Cufflinks command assembles transcripts and estimates their abundance using the aligned RNA-seq reads provided in a BAM file. This step builds on the alignment data generated by STAR. Cufflinks reconstructs transcript structures, including novel isoforms or previously unannotated lncRNAs. The inclusion of the GENCODE annotation file (-g) ensures consistency with known lncRNA annotations while enabling the discovery of novel transcripts. The results generated by Cufflinks can be used to generate differential expression findings, explore alternative splicing events, or identify new lncRNA candidates.

```
# Transcript Assembly using cufflinks
cufflinks -g human_genome/gencode.v47.long_noncoding_
RNAs.gtf -o
SRR24709142_cufflinks_out  -p  8  SRR24709142_Aligned.
sortedByCoord.out.bam
```

```
jsankarasubramani@lambda01:/storage5/work/subu/GSE233242$ cufflinks -g human_genome/gencode.v47.long_noncoding_RNAs.gtf -o SRR24709142_cufflinks_ou
t -p 8 SRR24709142 Aligned.sortedByCoord.out.bam
```

Given that RNA sequencing generates millions of short reads, StringTie uses a network flow algorithm to efficiently assemble these reads into transcripts. This approach enables the generation of more complete and accurate reconstructions of genes while providing precise estimates of expression levels. The output GTF file (SRR24709142_transcripts.gtf) contains the assembled transcripts and their abundances, serving as input for further downstream analyses. Additionally, the -B option prepares outputs compatible with Ballgown, enabling visualization and statistical analysis of transcript expression data.

```
#Transcript Assembly using stringtie
stringtie SRR24709142_Aligned.sortedByCoord.out.bam -G
human_genome/gencode.v47.long_noncoding_RNAs.gtf -o
SRR24709142_transcripts.gtf -p 8 -e -B
```

```
jsankarasubramani@lambda01:/storage5/work/subu/GSE233242$ stringtie SRR24709142_Aligned.sortedByCoord.out.bam -G human_genome/gencode.v47.long_noncoding_RNAs.gtf -o SRR24709142_transcri
pts.gtf -p 8 -e -B
```

3.7 Merging Transcripts

In the next step, manually create a mergelist.txt file using the vi editor, which lists the paths to the GTF files generated for each sample by Cufflinks. This file is required for the subsequent merging of transcript assemblies across samples, providing input paths to all individual GTF files.

```
#Manually create a mergelist.txt file
vi mergelist.txt
Press insert button to edit the file.
```

```
#Type the following lines
SRR24709142_cufflinks_out/transcripts.gtf
SRR24709143_cufflinks_out/transcripts.gtf
SRR24709144_cufflinks_out/transcripts.gtf
SRR24709145_cufflinks_out/transcripts.gtf
SRR24709146_cufflinks_out/transcripts.gtf
SRR24709147_cufflinks_out/transcripts.gtf
SRR24709148_cufflinks_out/transcripts.gtf
SRR24709149_cufflinks_out/transcripts.gtf
SRR24709150_cufflinks_out/transcripts.gtf
SRR24709151_cufflinks_out/transcripts.gtf
SRR24709152_cufflinks_out/transcripts.gtf
SRR24709153_cufflinks_out/transcripts.gtf
SRR24709154_cufflinks_out/transcripts.gtf
SRR24709155_cufflinks_out/transcripts.gtf
SRR24709156_cufflinks_out/transcripts.gtf
SRR24709157_cufflinks_out/transcripts.gtf
SRR24709158_cufflinks_out/transcripts.gtf
SRR24709159_cufflinks_out/transcripts.gtf
SRR24709160_cufflinks_out/transcripts.gtf
SRR24709161_cufflinks_out/transcripts.gtf
SRR24709162_cufflinks_out/transcripts.gtf
SRR24709163_cufflinks_out/transcripts.gtf
SRR24709164_cufflinks_out/transcripts.gtf
SRR24709165_cufflinks_out/transcripts.gtf
SRR24709166_cufflinks_out/transcripts.gtf
SRR24709167_cufflinks_out/transcripts.gtf
SRR24709172_cufflinks_out/transcripts.gtf
SRR24709173_cufflinks_out/transcripts.gtf
SRR24709174_cufflinks_out/transcripts.gtf
SRR24709175_cufflinks_out/transcripts.gtf
SRR24709182_cufflinks_out/transcripts.gtf
SRR24709183_cufflinks_out/transcripts.gtf
SRR24709194_cufflinks_out/transcripts.gtf
SRR24709195_cufflinks_out/transcripts.gtf
SRR24709196_cufflinks_out/transcripts.gtf
SRR24709197_cufflinks_out/transcripts.gtf
SRR24709198_cufflinks_out/transcripts.gtf
SRR24709199_cufflinks_out/transcripts.gtf
SRR24709204_cufflinks_out/transcripts.gtf
SRR24709205_cufflinks_out/transcripts.gtf
SRR24709206_cufflinks_out/transcripts.gtf
SRR24709207_cufflinks_out/transcripts.gtf
SRR24709208_cufflinks_out/transcripts.gtf
SRR24709209_cufflinks_out/transcripts.gtf
SRR24709210_cufflinks_out/transcripts.gtf
SRR24709211_cufflinks_out/transcripts.gtf
SRR24709212_cufflinks_out/transcripts.gtf
SRR24709213_cufflinks_out/transcripts.gtf
SRR24709214_cufflinks_out/transcripts.gtf
SRR24709215_cufflinks_out/transcripts.gtf
SRR24709218_cufflinks_out/transcripts.gtf
SRR24709219_cufflinks_out/transcripts.gtf
SRR24709220_cufflinks_out/transcripts.gtf
SRR24709221_cufflinks_out/transcripts.gtf
```

```
SRR24709222_cufflinks_out/transcripts.gtf
SRR24709223_cufflinks_out/transcripts.gtf
SRR24709224_cufflinks_out/transcripts.gtf
SRR24709225_cufflinks_out/transcripts.gtf
press the escape button and type :wq to save the file

# Merge GTFs
cat $(<mergelist.txt)>merged.gtf
```

This step ensures that all transcript assemblies from individual samples are included in the merging process, which is critical for generating a unified transcriptome annotation. Unlike using automated commands to generate the list, this manual method allows precise control over the file content, enabling users to verify and include specific samples.

3.8 Clearing Duplicates

In the next step, remove duplicate entries from the merged GTF file, ensuring that each transcript or feature is uniquely represented in the final annotation. Duplicate entries may arise during the merging process, especially when similar transcripts are assembled across multiple samples.

```
#After gtf merge, clear duplicates
awk '!seen[$1, $4, $5]++' merged.gtf > merged_with-
out_duplicates.gtf
```

```
jsankarasubramani@lambda01:/storage5/work/subu/GSE233242$ awk '!seen[$1, $4, $5]++' merged.gtf > merged_without_duplicates.gtf
```

This step ensures that the final merged transcript annotation is free from redundancies, providing a clean and accurate representation of the transcriptome. Duplicates in the GTF file can interfere with downstream analyses, such as quantification, functional annotation, or visualization, by artificially inflating transcript counts or introducing ambiguities. By retaining only unique entries based on chromosome and genomic coordinates, this step maintains the integrity of the merged annotation file. In the next step, remove the duplicate transcript identifier entries using the below code blockpython Remove_Duplicate_transcript_ID.py merged_without_duplicates.gtf output_merged_without_duplicates.gtf

3.9 Removing the Mitochondrial Genome

In the next command, use awk to remove the mitochondrial chromosome (chrM, also referred to as "MT") from the genome FASTA file. The resulting file excludes sequences associated with the mitochondrial genome, which is often omitted in RNA-seq analyses due to its unique characteristics and potential to introduce alignment biases.

```
#Remove chrM from the reference genome
awk '/^>/{p=($0!=">chrM MT")} p' human_genome/
GRCh38.p14.genome.fa >
human_genome/GRCh38.p14.genome_no_chrM.fa
```

Removing chrM from the reference genome is a common preprocessing step in RNA-seq pipelines, particularly for analyses focused on lncRNAs derived from nuclear transcripts, as it reduces noise from mitochondrial sequences.

3.10 Extracting lncRNAs

In the next step, use gffread [48] to extract sequences of lncRNA from the genome based on transcript annotations provided in a GTF file. The output is a FASTA file containing only the lncRNA sequences, filtered to exclude mitochondrial sequences since the input genome file does not include chrM.

```
#extract lncRNA sequences from genome fasta
gffread -w filtered_lncRNAs.fa -g
human_genome/GRCh38.p14.genome_no_chrM.fa
cuffmerge_out/merged_without_duplicates.gtf
```

This step extracts the actual nucleotide sequences of lncRNAs from the reference genome based on the unified annotation file, allowing for further sequence-based analyses, such as functional annotation, secondary structure prediction, or target interaction studies. By excluding mitochondrial sequences through the use of the GRCh38.p14.genome_no_chrM.fa file, the extracted dataset focuses exclusively on nuclear lncRNAs .

3.11 Coding Potential Assessment

The next command uses CPAT [49] (Coding Potential Assessment Tool) to evaluate the coding potential of lncRNA sequences. Although we used a GTF file containing only lncRNAs, we added this filtering step as an additional safeguard to ensure the GTF does not contain any transcripts with protein-coding potential, since some lncRNAs have been shown to code for proteins [50,51]. This step is particularly useful when starting with a mixed transcript GTF, as it effectively identifies and removes protein-coding transcripts. The pre-trained models "Human_Hexamer.tsv" and "Human_logitModel.RData" can be downloaded from the link https://github.com/liguowang/cpat/tree/master/prebuilt_models.

```
#Assessing Coding Potential (Download the pre-built
models from here https://github.com/liguowang/cpat/
tree/master/prebuilt_models)
cpat.py -g filtered_lncRNAs.fa -x Human_Hexamer.tsv -d
Human_logitModel.RData -o cpat_output.txt
```

```
jsankarasubramani@lambda01:/storage5/work/subu/GSE233242$ cpat.py -g filtered_lncRNAs.fa -x Human_Hexamer.tsv -d Human_logitModel.RData -o cpat_out
put.txt
```

The output file (cpat_output.txt) provides a detailed classification of each sequence, which can be used to filter out coding transcripts and retain only high-confidence lncRNA candidates for downstream analyses, such as differential expression, functional annotation, or target prediction.

3.12 Cross-Referencing with GTF File to Update Annotations

In the next step, the output file from the above step (cpat_output.txt) is filtered to retain only transcripts with low coding potential scores (i.e., non-coding). A typical threshold for non-coding classification is applied (coding potential score <0.38 for human lncRNAs).

```
#Cross-referencing with GTF file (filtered_gencode.
gtf) to update annotations
awk '$4 < 0.38' cpat_output.txt > high_confidence_ln-
cRNAs.txt
```

```
jsankarasubramani@lambda01:/storage5/work/subu/GSE233242$ awk '   < 0.38' cpat_output.txt > high_confidence_lncRNAs.txt
```

This command filters the 4th column (coding potential score) and extracts only those classified as non-coding. Below snapshot shows the first few rows of the filtered lncRNAs that show low coding potential scores of <0.38.

ENST00000832833.1	97	0	0	0	0.00039596
ENST00000832834.1	20	0	0	0	0.00039763
ENST00000831747.1	80	6	0.3299	-2.901853	2.98E-11
ENST00000831637.1	265	0	0	0	0.00039233
ENST00000831738.1	95	0	0	0	0.000396
ENST00000831177.1	141	0	0	0	0.000395
ENST00000831323.1	65	0	0	0	0.00039665
ENST00000831737.1	36	0	0	0	0.00039728
ENST00000831517.1	18	0	0	0	0.00039768
ENST00000831505.1	21	0	0	0	0.00039761
ENST00000831470.1	29	0	0	0	0.00039744
ENST00000831204.1	30	0	0	0	0.00039741
ENST00000831333.1	31	0	0	0	0.00039739
ENST00000831602.1	32	0	0	0	0.00039737
ENST00000831340.1	33	0	0	0	0.00039735
ENST00000831272.1	36	0	0	0	0.00039728
ENST00000831312.1	38	0	0	0	0.00039724
ENST00000831165.1	40	0	0	0	0.0003972
ENST00000831170.1	44	0	0	0	0.00039711
ENST00000831550.1	45	0	0	0	0.00039709
ENST00000831218.1	48	0	0	0	0.00039702
ENST00000831217.1	56	0	0	0	0.00039685
ENST00000831529.1	57	0	0	0	0.00039683
ENST00000831414.1	70	0	0	0	0.00039654
ENST00000831523.1	35	0	0	0	0.00039731
ENST00000831522.1	56	0	0	0	0.00039685
ENST00000831521.1	145	0	0	0	0.00039492
CUFF.3694.1	626	0	0	0	0.00038464
CUFF.15339.1	559	0	0	0	0.00038606
CUFF.38798.1	657	0	0	0	0.00038399
ENST00000831099.1	21	0	0	0	0.00039761

In the next step, the high-confidence lncRNAs "high_confidence_lncRNAs.txt" are cross-referenced with GTF files (merged_without_duplicates.gtf) to update annotations, ensuring they align with transcript IDs of interest. The high-confidence lncRNAs identified in "high_confidence_lncRNAs.txt" are used as input for functional annotation (FEELnc_filter.pl) to classify lncRNAs into various functional categories (intergenic or antisense).

```
#Extract transcript IDs from the filtered CPAT output
awk '{print $1}' high_confidence_lncRNAs.txt >
```

```
high_confidence_transcripts.txt
#Cross-reference the filtered transcripts with the
GTF file
grep -Ff high_confidence_transcripts.txt merged_with-
out_duplicates.gtf > updated_filtered_gencode.gtf
#Verify the updated GTF file (optional)
head updated_filtered_gencode.gtf
```

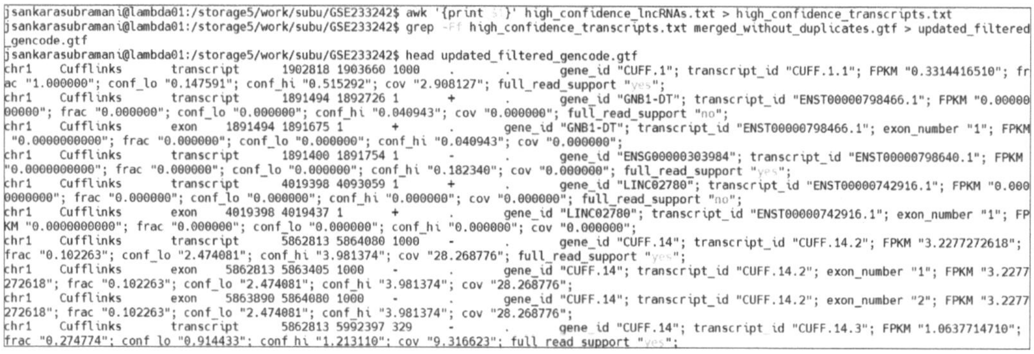

```
jsankarasubramani@lambda01:/storage5/work/subu/GSE233242$ awk '{print $1}' high_confidence_lncRNAs.txt > high_confidence_transcripts.txt
jsankarasubramani@lambda01:/storage5/work/subu/GSE233242$ grep -Ff high_confidence_transcripts.txt merged_without_duplicates.gtf > updated_filtered_gencode.gtf
jsankarasubramani@lambda01:/storage5/work/subu/GSE233242$ head updated_filtered_gencode.gtf
chr1	Cufflinks	transcript	1902818 1903660 1000	.	.	gene_id "CUFF.1"; transcript_id "CUFF.1.1"; FPKM "0.3314416510"; frac "1.000000"; conf_lo "0.147591"; conf_hi "0.515292"; cov "2.908127"; full_read_support "yes";
chr1	Cufflinks	transcript	1891494 1892726 1	+	.	gene_id "GNB1-DT"; transcript_id "ENST00000798466.1"; FPKM "0.0000000000"; frac "0.000000"; conf_lo "0.000000"; conf_hi "0.040943"; cov "0.000000"; full_read_support "no";
chr1	Cufflinks	exon	1891494 1891675 1	+	.	gene_id "GNB1-DT"; transcript_id "ENST00000798466.1"; exon_number "1"; FPKM "0.0000000000"; frac "0.000000"; conf_lo "0.000000"; conf_hi "0.040943"; cov "0.000000";
chr1	Cufflinks	transcript	1891400 1891754 1	-	.	gene_id "ENSG00000303984"; transcript_id "ENST00000798640.1"; FPKM "0.0000000000"; frac "0.000000"; conf_lo "0.000000"; conf_hi "0.182340"; cov "0.000000"; full_read_support "yes";
chr1	Cufflinks	transcript	4019398 4093059 1	+	.	gene_id "LINC02780"; transcript_id "ENST00000742916.1"; FPKM "0.0000000000"; frac "0.000000"; conf_lo "0.000000"; conf_hi "0.000000"; cov "0.000000"; full_read_support "no";
chr1	Cufflinks	exon	4019398 4019437 1	+	.	gene_id "LINC02780"; transcript_id "ENST00000742916.1"; exon_number "1"; FPKM "0.0000000000"; frac "0.000000"; conf_lo "0.000000"; conf_hi "0.000000"; cov "0.000000";
chr1	Cufflinks	transcript	5862813 5864080 1000	-	.	gene_id "CUFF.14"; transcript_id "CUFF.14.2"; FPKM "3.2277272618"; frac "0.102263"; conf_lo "2.474081"; conf_hi "3.981374"; cov "28.268776"; full_read_support "yes";
chr1	Cufflinks	exon	5862813 5863405 1000	-	.	gene_id "CUFF.14"; transcript_id "CUFF.14.2"; exon_number "1"; FPKM "3.2277272618"; frac "0.102263"; conf_lo "2.474081"; conf_hi "3.981374"; cov "28.268776";
chr1	Cufflinks	exon	5863890 5864080 1000	-	.	gene_id "CUFF.14"; transcript_id "CUFF.14.2"; exon_number "2"; FPKM "3.2277272618"; frac "0.102263"; conf_lo "2.474081"; conf_hi "3.981374"; cov "28.268776";
chr1	Cufflinks	transcript	5862813 5992397 329	-	.	gene_id "CUFF.14"; transcript_id "CUFF.14.3"; FPKM "1.0637714710"; frac "0.274774"; conf_lo "0.914433"; conf_hi "1.213110"; cov "9.316623"; full_read_support "yes";
```

3.13 Quantification of Transcript Abundance

In the next step, we use featureCounts [52] for quantifying transcript abundance from aligned RNA-seq reads. The output provides raw count data for each transcript, which is essential for downstream analyses such as differential expression. This step bridges the alignment and expression analysis phases of the pipeline.

```
featureCounts -a cuffmerge_out/merged.gtf -o
SRR24709142_counts.txt -T 8 -p -B -C
SRR24709142_Aligned.sortedByCoord.out.bam
```

```
jsankarasubramani@lambda01:/storage5/work/subu/GSE233242$ featureCounts -a merged_without_duplicates.gtf -o SRR24709142_counts.txt -T 8 -p -B -C SRR24709142_Aligned.sortedByCoord.out.bam

       SUBREAD
  v2.0.1

//========================== featureCounts setting ===========================\\
||                                                                            ||
||             Input files : 1 BAM file                                       ||
||                           o SRR24709142_Aligned.sortedByCoord.out.bam      ||
||                                                                            ||
||             Output file : SRR24709142_counts.txt                           ||
||                 Summary : SRR24709142_counts.txt.summary                   ||
||              Annotation : merged_without_duplicates.gtf (GTF)              ||
||      Dir for temp files : ./                                               ||
||                                                                            ||
||                 Threads : 8                                                ||
||                   Level : meta-feature level                               ||
||              Paired-end : yes                                              ||
||      Multimapping reads : not counted                                      ||
|| Multi-overlapping reads : not counted                                      ||
||   Min overlapping bases : 1                                                ||
||                                                                            ||
||          Chimeric reads : not counted                                      ||
||        Both ends mapped : required                                         ||
||                                                                            ||
\\============================================================================//

//================================= Running ==================================\\
||                                                                            ||
|| Load annotation file merged_without_duplicates.gtf ...                     ||
||    Features : 2304463                                                      ||
||    Meta-features : 93038                                                   ||
||    Chromosomes/contigs : 164                                               ||
||                                                                            ||
|| Process BAM file SRR24709142_Aligned.sortedByCoord.out.bam...              ||
||    Paired-end reads are included.                                          ||
||    Total alignments : 74228246                                             ||
||    Successfully assigned alignments : 4149394 (5.6%)                       ||
||    Running time : 2.07 minutes                                             ||
||                                                                            ||
|| Write the final count table.                                               ||
|| Write the read assignment summary.                                         ||
||                                                                            ||
|| Summary of counting results can be found in file "SRR24709142_counts.txt.  ||
|| summary"                                                                   ||
||                                                                            ||
\\============================================================================//
```

The featureCounts command assigns aligned RNA-seq reads to genomic features using the merged GTF annotation file that was processed in the above steps. The '-a' parameter specifies the input annotation file ('cuffmerge_out/merged.gtf'), which contains transcript definitions, ensuring alignment between transcript quantification and the unified annotation. The output is saved as 'SRR24709142_counts.txt' ('-o'), a file containing raw read counts for each transcript. The process is accelerated by utilizing 8 threads ('-T 8'), enabling efficient parallel computation. The '-p' option ensures accurate counting for paired-end reads by considering both mates of each read pair. Additionally, the '-B' parameter requires that both reads in a pair be properly aligned to a feature, while '-C' filters out read pairs mapped to different chromosomes, enhancing the reliability of the count data. The input BAM file, "SRR24709142_Aligned.sortedByCoord.out.bam", provides the alignment data for quantification, with reads already sorted by genomic coordinates for efficient processing. Together, these parameters ensure precise and computationally efficient quantification of transcript abundance.

In the next step, we use a set of commands to process and consolidate transcript abundance data generated by featureCounts into a combined matrix. Each row represents a transcript, and each column corresponds to a sample.

```
#Extract the 1st and 7th column
for file in *_counts.txt; do
sed '1d' "$file" | awk -v OFS="\t" '{print $1, $7}' | sed
's/_Aligned\.sortedByCoord\.out\.bam//g' >
"processed_$file"
done

#Extract the 1st row
cut -f1 processed_SRR24709142_counts.txt > combined_
file.txt

#Loop through all other files and extract only the
2nd column
for file in processed_*_counts.txt; do
awk '{print $2}' "$file" | paste combined_file.txt
- > temp.txt
mv temp.txt combined_file.txt

done
```

First, all files with the suffix _counts.txt are iteratively processed to clean and extract the relevant information. The header row is removed using sed, and the first and seventh columns, containing transcript IDs and raw counts, are extracted using awk. The transcript identifiers are further cleaned by removing the suffix _Aligned.sortedByCoord.out.bam with sed, and the results are saved as new files prefixed with processed_. Next, the first column (transcript IDs) is extracted from a representative processed file and saved to combined_file.txt, creating a base file for consolidating data. Subsequently, the script iterates through all processed files to extract the second column (containing raw counts for each sample) and appends it to the growing matrix in combined_file.txt. This ensures that each row represents a transcript, and each column corresponds to a sample. By iteratively updating the combined matrix, the final output provides a comprehensive dataset of transcript abundance across all samples, enabling multi-sample comparisons in tools like DESeq2 or edgeR. This step ensures the data is formatted consistently for downstream lncRNA analyses, such as differential expression and clustering.

3.14 Differential Expression Analysis

The next step, performed in R, carries out differential expression analysis (DEA) on the combined transcript count data using DESeq2 [53]. Unlike the previous preprocessing and quantification steps, this analysis is implemented in the R.

The table structure of input 'Luminal_A_vs_control.txt' should be as below;

```
#Differential Expression Analysis
setwd("C:/Users/Path/Downloads")
library(DESeq2)
library(ggplot2)
library(ggrepel)
samples  <-  read.table("Luminal_A_vs_control.txt",
header = TRUE)
count_data <- read.table("combined_file.txt", header
= TRUE, row.names = 1)
samples <- samples[match(colnames(count_data), sam-
ples$sample_name), ]
dds  <-  DESeqDataSetFromMatrix(countData  =  count_
data, colData = samples, design = ~ group)
dds <- DESeq(dds)
res <- results(dds)
res$gene <- rownames(res)
res$diffexpressed <- "NO"
res$diffexpressed[res$log2FoldChange >= 1 & res$padj
<= 0.05] <- "UP"
res$diffexpressed[res$log2FoldChange <= -1 & res$p-
adj <= 0.05] <- "DOWN"
top_genes <- rbind(
res[res$diffexpressed == "UP", ][order(res[res$diff-
expressed == "UP", ]$padj), ][1:25, ],
res[res$diffexpressed == "DOWN", ][order(res[res$dif-
fexpressed == "DOWN", ]$padj), ][1:25, ]
)
write.csv(as.data.frame(res),  file  =  "differential_
expression_results_with_labels.csv")
```

The table structure of input 'combined_file.txt' should be as below;

```
jsankarasubramani@lambda01:/storage5/work/subu/GSE233242$ head Luminal_A_vs_control.txt
sample_name     group
SRR24709142     Luminal_A
SRR24709144     Luminal_A
SRR24709146     Luminal_A
SRR24709148     Luminal_A
SRR24709150     Luminal_A
SRR24709152     Luminal_A
SRR24709154     Luminal_A
SRR24709156     Luminal_A
SRR24709158     Luminal_A
```

```
jsankarasubramani@lambda01:/storage5/work/subu/GSE233242$ head combined_file.txt
Geneid SRR24709142 SRR24709143 SRR24709144 SRR24709145 SRR24709146 SRR24709147 SRR24709148 SRR24709149 SRR24709150 SRR24709151 SRR247091525
RR24709153 SRR24709154 SRR24709155 SRR24709156 SRR24709157 SRR24709158 SRR24709159 SRR24709160 SRR24709161 SRR24709162 SRR24709163 SRR2
4709164 SRR24709165 SRR24709166 SRR24709167 SRR24709172 SRR24709173 SRR24709174 SRR24709175 SRR24709182 SRR24709183 SRR24709194 SRR247091955
RR24709196 SRR24709197 SRR24709198 SRR24709199 SRR24709204 SRR24709205 SRR24709206 SRR24709207 SRR24709208 SRR24709209 SRR24709210 SRR2
4709211 SRR24709212 SRR24709213 SRR24709214 SRR24709215 SRR24709218 SRR24709219 SRR24709220 SRR24709221 SRR24709222 SRR24709223 SRR247092245
RR24709225
ENSG00000306476 60 4 12 13 0 17 1 1 4 11 12 22 1 0 160 23 54 6 7 10 1 1
12 2 3 17 1 3 3 8 7 0 13 1 164 1 2 4 1 3 2 9 43 1 0 1
15 8 0 10 2 18 6 16 1 14 1 2
ENSG00000294910 29 8 8 10 1 18 0 3 16 6 8 9 3 2 100 27 22 12 28 6 7 7
19 2 4 25 1 1 2 2 4 1 13 7 84 0 2 9 9 5 0 16 24 1 0 2
22 6 1 2 0 13 5 12 9 17 3 1
CUFF.5 50 28 15 33 6 48 50 15 14 20 10 25 11 16 69 80 50 21 28 26 22 7 9
4 17 47 6 7 2 6 14 8 55 9 30 4 6 58 10 19 6 25 32 1 2 3 3
4 6 8 17 20 29 6 22 2 19 14 17
CUFF.6 0 0 0 0 0 0 0 0 0 0 2 0 0 0 0 1 0 0 0 0 0 0 0
0 0 0 0 0 0 0 0 0 0 0 0 0 0 0 0 0 0 0 0 0 0 0 0
0 0 0 0 0 2 0 0 0 0 0
GNB1-DT 24 26 6 35 11 43 19 13 9 11 11 6 8 13 89 38 26 50 28 12 22 26 1
0 5 65 57 3 14 2 9 1 49 33 19 43 1 1 50 10 25 19 29 5 1 8 9
21 24 6 19 16 22 24 55 24 58 22 11
ENSG00000233234 0 0 0 0 0 0 0 0 0 0 0 0 0 0 0 0 0 0 0 0 0 0
0 0 0 0 0 0 0 0 0 0 0 0 0 0 0 0 0 0 0 0 0 0 0 0
0 0 0 0 0 0 0 0 0 0 0 0
ENSG00000295089 0 0 0 0 0 0 0 0 0 0 0 0 0 0 0 0 0 0 0 0 0 0
0 0 0 0 0 0 0 0 0 0 0 0 0 0 0 0 0 0 0 0 0 0 0 0
0 0 0 0 0 0 0 0 0 0 0 0
ENSG00000302855 0 0 0 0 0 0 0 0 0 0 0 0 0 0 0 0 0 0 0 0 0 0
0 0 0 0 0 0 0 0 0 0 0 0 0 0 0 0 0 0 0 0 0 0 0 0
0 0 0 0 0 0 0 0 0 0 0 0
ENSG00000300015 0 0 0 0 0 0 0 0 0 0 0 0 0 0 0 0 0 0 0 0 0 0
0 0 0 0 0 0 0 0 0 0 0 0 0 0 0 0 0 0 0 0 0 0 0 0
0 0 0 0 0 0 0 0 0 0 0 0
```

The purpose of this step is to identify lncRNA transcripts that are significantly differentially expressed between the experimental groups (control vs. Luminal_A). We utilize DESeq2 program which adjusts for sequencing depth and variability. The integration of sample metadata and transcript counts (tables shown above) ensures accurate modeling of the experimental design. Additionally, the categorization of transcripts as UP, DOWN, or NO highlights biologically relevant changes in expression, which are crucial for interpreting the role of lncRNAs in cancer.

The next R code block generates a volcano plot using ggplot [54] to visualize the results of differential expression analysis and writes the labeled results to a CSV file. The plot provides a clear representation of the relationship between statistical significance (*p*-value) and the magnitude of expression changes (log2 fold change) while highlighting differentially expressed genes (DEGs).

```
#R code to visualize volcano plot
ggplot(res, aes(x = log2FoldChange, y = -log10(pval-
ue), color = diffexpressed)) +
geom_point(alpha = 0.6) +
geom_text_repel(data = top_genes, aes(label = gene),
size = 3, box.padding = 0.5, point.padding = 0.3,
max.overlaps = 30) +
theme_minimal() +
scale_color_manual(values = c("UP" = "red", "DOWN"
= "blue", "NO" = "gray")) +
```

```
labs(title = "Volcano Plot", x = "log2 Fold Change",
y = "-log10 p-value", color = "Regulation") +
theme(legend.position = "top", plot.title = ele-
ment_text(hjust = 0.5))
```

This visualization step provides a critical summary of the DEA, making it easier to identify and interpret biologically significant lncRNAs . The volcano plot (Fig. 3) highlights key transcripts that are significantly upregulated or downregulated, visually distinguishing them from non-significant genes. By labeling the top 50 most differentially expressed genes (top_genes), it facilitates focused follow-up analyses, such as functional annotation or pathway enrichment (Fig. 3). Exporting results to a labeled CSV file ensures reproducibility and allows for further integration. This step bridges statistical results with biological interpretation, underscoring the role of lncRNAs in gene regulation and disease progression.

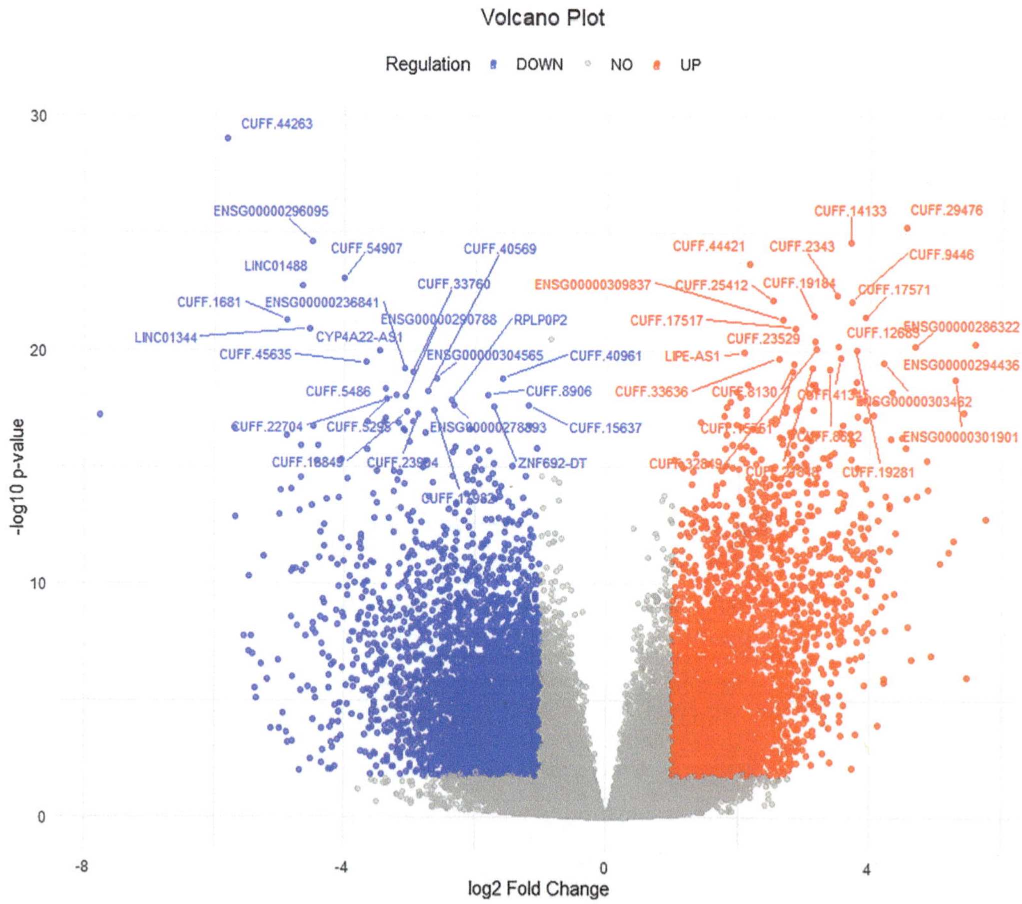

Fig. 3 Volcano Plot of differentially expressed lncRNAs. Figure illustrates the relationship between the log2FC and *p*adj for differentially expressed lncRNAs. Significantly upregulated lncRNAs are shown on the right, while significantly downregulated lncRNAs are on the left, based on a *p*adj ≤0.05 and absolute log2FC ≤−1 and ≥+1. Non-significant lncRNAs are clustered near the center. Highlighted points represent the top 50 differentially expressed lncRNAs

The results from the differential expression analysis include 75,109 transcripts. The log2FoldChange ranged −7.75 to 5.76 indicating significant variability in differential expression. Negative values represent downregulated genes while positive values represent upregulated genes. Genes with *p*adj ≤0.05 and absolute log2FoldChange ≤−1 and ≥+1 are considered significantly differentially expressed. In the output table ("differential_expression_results_with_labels.csv"), genes labeled "UP" or "DOWN" indicate significant differential expression, while "NO" indicates no significant expression change. Table 1 lists the distribution of lncRNAs across different log2 fold change ranges.

3.14.1 Downregulated lncRNAs

- −7: Only one lncRNA is extremely downregulated, showing a log2 fold change of −7.
- **−5 to −5.99**: 16 lncRNAs are moderately downregulated in this range.
- **−4 to −4.99**: 83 lncRNAs show strong downregulation.
- **−3 to −3.99**: 241 lncRNAs are moderately downregulated.
- **−2 to −2.99**: 1083 lncRNAs are slightly downregulated.
- **−1 to −1.99**: The largest cluster of downregulated lncRNAs (4011) falls within this range.

3.14.2 Upregulated lncRNAs

- **5 to 5.99**: Only 8 lncRNAs exhibit extreme upregulation
- **4 to 4.99**: 39 lncRNAs are strongly upregulated.
- **3 to 3.99**: 262 lncRNAs show moderate upregulation.
- **2 to 2.99**: 1062 lncRNAs are mildly upregulated.
- **1 to 1.99**: 3018 lncRNAs exhibit mild upregulation.

In the next step, we will visualize the datapoints from Table 1 in a histogram using matplot [55] (Fig. 4) using the below code.

```
import matplotlib.pyplot as plt

# Data from the table
log2_fold_change_ranges = [
"-7", "-5 to -5.99", "-4 to -4.99", "-3 to -3.99",
"-2 to -2.99", "-1 to -1.99",
"1 to 1.99", "2 to 2.99", "3 to 3.99", "4 to 4.99",
"5 to 5.99"
]
number_of_lncRNAs = [1, 16, 83, 241, 1083, 4011,
3018, 1062, 262, 39, 8]

# Create the histogram
plt.Fig.(figsize=(10, 6))
plt.bar(log2_fold_change_ranges, number_of_lncRNAs,
color='skyblue', edgecolor='black')
plt.xlabel("Log2 Fold Change Range")
plt.ylabel("Number of lncRNAs")
```

Table 1

Summarizes the distribution of lncRNAs across different log2 fold change ranges, reflecting the extent of their upregulation or downregulation

Log2fold change range	Number of lncRNAs
−7	1
−5 to −5.99	16
−4 to −4.99	83
−3 to −3.99	241
−2 to −2.99	1083
−1 to −1.99	4011
5 to 5.99	8
4 to 4.99	39
3 to 3.99	262
2 to 2.99	1062
1 to 1.99	3018

```
plt.title("Distribution of lncRNAs Across Log2 Fold
Change Ranges")
plt.xticks(rotation=45, ha='right')
plt.tight_layout()
plt.show()
```

In the next step, we will utilize the below code block to visualize top 50 up- and down-regulated DEGs in a heatmap between the healthy and luminal A breast cancer subtype.

```
#R code to visualize Heatmap
library(pheatmap)
normalized_counts <- counts(dds, normalized = TRUE)
significant_genes <- res[!is.na(res$padj) & res$padj
<= 0.05, ]
significant_gene_names <- rownames(significant_genes)
heatmap_data <- normalized_counts[significant_gene_
names, ]
heatmap_data_log <- log2(heatmap_data + 0.0001)
samples$group <- factor(samples$group, levels =
unique(samples$group))
ordered_columns <- order(samples$group)
heatmap_data_log <- heatmap_data_log[, ordered_col-
umns]
annotation_col <- data.frame(Cohort = sam-
ples$group[ordered_columns])
rownames(annotation_col) <- colnames(heatmap_data_
log)
unique_cohorts <- levels(samples$group)
```

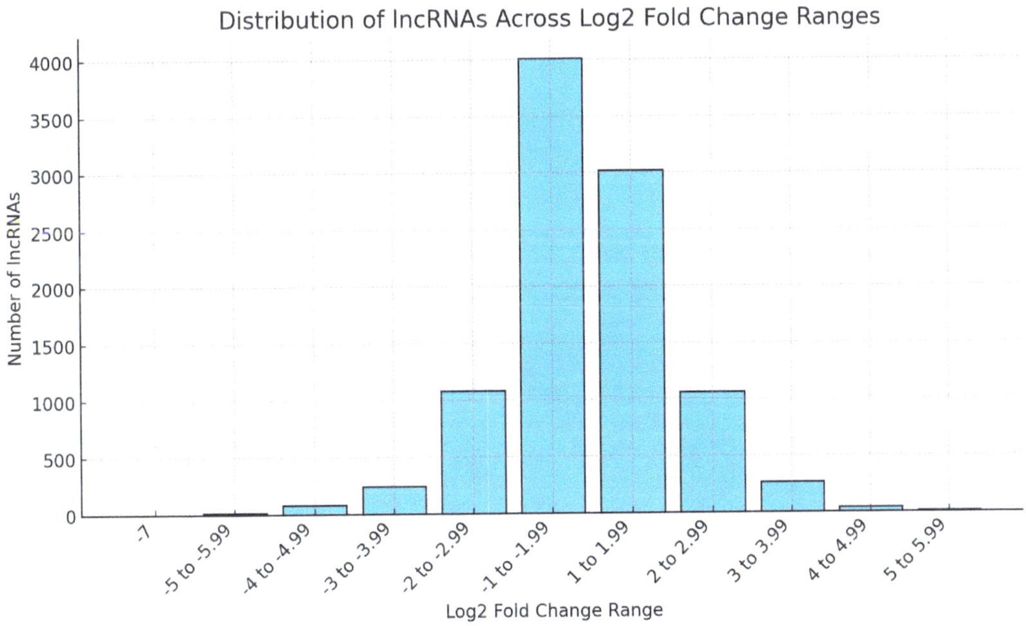

Fig. 4 Distribution of lncRNAs across log2 fold change ranges. Figure shows the distribution of differentially expressed lncRNAs across various log2fc ranges

```
annotation_colors <- list(Cohort = setNames(rain-
bow(length(unique_cohorts)), unique_cohorts))
pheatmap(
heatmap_data_log,
annotation_col = annotation_col,
annotation_colors = annotation_colors,
cluster_rows = TRUE,
cluster_cols = FALSE,
show_rownames = FALSE,
show_colnames = TRUE,
color = colorRampPalette(c("blue", "white", "red"))
(50),
main = "Heatmap of All Significant Genes"
)
```

The heatmap provides an overview of expression patterns for the most significant transcripts, emphasizing clustering and group-specific expression differences (Fig. 5).

Auditing the Table 2 showed LNCDAT (log2fc 5.76), ENSG00000294436 (log2fc 5.59) and ENSG00000256513 (log2fc 5.47) exhibiting strong upregulation with highly significant *p*-values (1.55E−13, 4.76E−21 and 3.76E−15), indicating robust expression. In contrast, lncRNAs ENSG00000238176 (log2fc −7.75), CUFF.44263 (log2fc −5.80) and ENSG00000302504 (log2fc −5.69) showed substantial downregulation, also accompanied by highly significant *p*-values, suggesting significant decreases in their

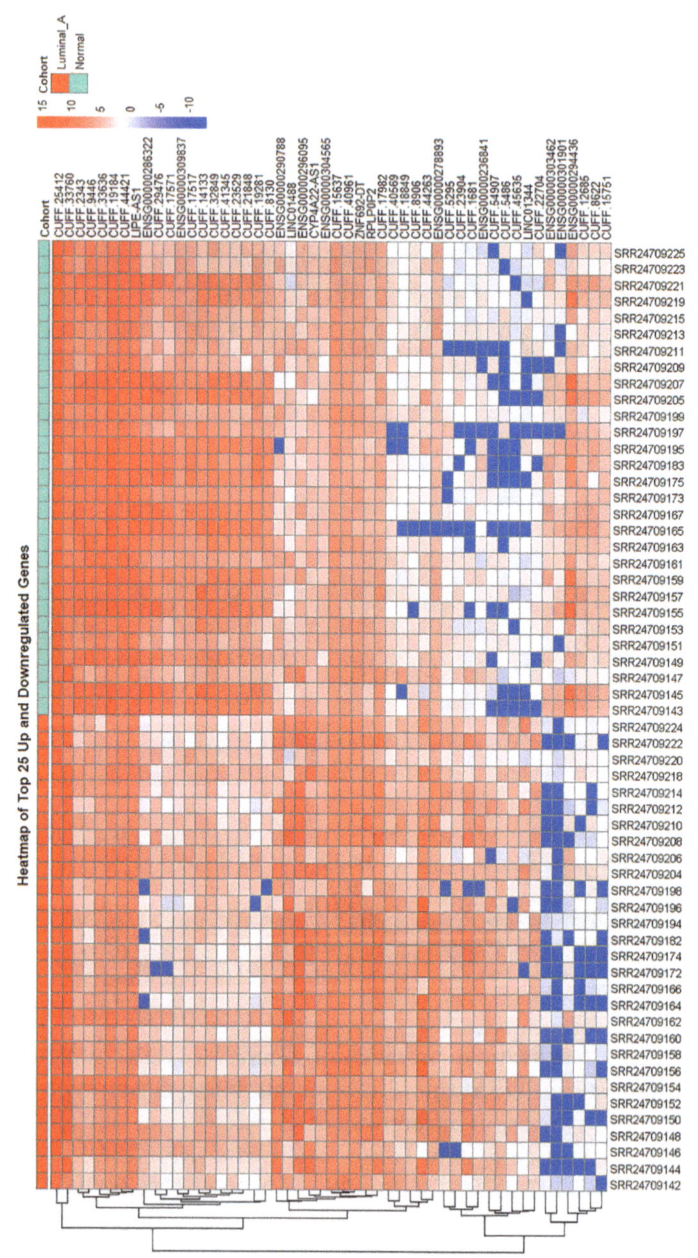

Fig. 5 Heatmap of top 50 differentially expressed lncRNAs (top 25 each). Rows represent lncRNAs, and columns represent individual samples, with color intensity indicating relative expression levels. Clustering patterns reveal distinct expression differences between conditions

Table 2
Summarizes the distribution of top 10 upregulated and downregulated lncRNA transcripts

lncRNAs[a]	baseMean	log2FoldChange	lfcSE	stat	*p*-value	*padj*	Diffexpressed
LNCDAT	*6.72937829*	*5.762237798*	*0.78053174*	*7.38245161*	*1.55E−13*	*2.21E−11*	*UP*
ENSG00000294436	147.660554	5.591506783	0.59393834	9.41428829	4.76E−21	1.23E−17	UP
ENSG00000256513	5.11869771	5.474208397	1.11815049	4.89577068	9.79E−07	1.56E−05	UP
ENSG00000298736	5.28929193	5.421125275	0.62519125	8.67114704	4.28E−18	3.01E−15	UP
ENSG00000301901	6.4668336	5.291749587	0.58604516	9.02959352	1.72E−19	2.23E−16	UP
CUFF.48932	4.06537977	5.26223174	0.74102648	7.10127354	1.24E−12	1.29E−10	UP
ENSG00000299094	32.2887967	5.190175083	0.74775909	6.94097225	3.89E−12	3.43E−10	UP
ENSG00000294455	31.4106502	5.059216335	0.74608122	6.78105307	1.19E−11	9.07E−10	UP
CUFF.36924	4.72803774	4.929572118	0.92861221	5.30853682	1.11E−07	2.41E−06	UP
ENSG00000305882	7.86536912	4.879837579	0.62928738	7.75454543	8.87E−15	1.87E−12	UP
ENSG00000238176	*60.8517681*	*−7.754163106*	*0.8980015*	*−8.6349111*	*5.88E−18*	*3.76E−15*	*DOWN*
CUFF.44263	380.535126	−5.804750372	0.51233042	−11.330091	9.31E−30	4.58E−25	DOWN
ENSG00000302504	5.37390058	−5.695151716	0.67164386	−8.479422	2.26E−17	1.17E−14	DOWN
ENSG00000307583	54.427574	−5.670464544	0.76610886	−7.4016433	1.35E−13	1.96E−11	DOWN
LINC02262	4.90812179	−5.539773291	0.98377937	−5.6311135	1.79E−08	4.92E−07	DOWN
ENSG00000293085	33.0833574	−5.463406599	0.831601	−6.5697451	5.04E−11	3.13E−09	DOWN
ENSG00000287448	71.2856904	−5.457623243	1.01373648	−5.3836706	7.30E−08	1.67E−06	DOWN
ENSG00000303361	7.52621892	−5.432963865	0.96266108	−5.6436933	1.66E−08	4.63E−07	DOWN
CUFF.25089	9.47098957	−5.412450779	1.01533291	−5.3307154	9.78E−08	2.17E−06	DOWN
LINC00052	16.4638903	−5.373683792	1.15020523	−4.6719347	2.98E−06	4.07E−05	DOWN

[a]Ensembl IDs (ENSGxxxxxxxxxxx), Cuff IDs (CUFF.xxxxx), and LINC IDs (LINCxxxxx)

expression levels (Table 2). Overall, these findings reveal a distinct pattern of lncRNA regulation, with several lncRNAs showing pronounced upregulation while others are markedly downregulated.

The list of differentially expressed lncRNAs includes a variety of gene identifiers, such as Ensembl IDs (ENSGxxxxxxxxxxx), Cuff IDs (CUFF.xxxxx), and LINC IDs (LINCxxxxx) (Table 2) (output table differential_expression_results_with_labels.csv). While Ensembl IDs and LINC IDs correspond to known, annotated lncRNA genes, the Cuff IDs represent potentially novel lncRNA transcripts identified in this study. Cufflinks program assigns Cuff IDs to transcripts during assembly when they cannot be matched to existing annotated genes in the GTF. Any transcript with a Cuff ID can be considered a potentially novel transcript.

3.15 Cross-Referencing Differentially Expressed lncRNAs with Genomic Positions

The next step is to understand the annotations such as, "Feature" type, "Start" and "End" coordinates of the differentially expressed lncRNAs to understand the location and sequence information. Further, we extract the nucleotide sequence, translate the DNA sequence to RNA sequence and make use of several programs to model potential lncRNA interactions with either -RNA [56–58], -proteins [59–61]. However, for this walkthrough, we will utilize the webserver 'LncRRIsearch' [56] to identify and model potential interactions with other RNAs, and we will use the 'FEELnc_filter.pl' [62] program to filter candidate lncRNAs to identify those that interact with protein-coding genes.

3.15.1 Auditing Few DE lncRNAs

Auditing the LNCDAT gene which has a log2fold change of 5.76 shows below table structure extracted from the gtf providing brief overview of the coordinates of LNCDAT gene. The sequences from the coordinates can be obtained to model potential lncRNA interactions.

	A	B	C	D	E	F	G
1	Chromosome	Assembler	Feature	Feature_Start	Feature_End	Strand	Annotation
2	chr6	Cufflinks	transcript	166167920	166171172	+	gene_id "LNCDAT"; transcript_id "ENST00000675446.1"; FPKM "0.0000000000"; frac "0.000000"; conf_lo "0.000000"; conf_hi "0.000000"; cov "0.000000"; full_read_support "no";
3	chr6	Cufflinks	exon	166167920	166168150	+	gene_id "LNCDAT"; transcript_id "ENST00000675446.1"; exon_number "1"; FPKM "0.0000000000"; frac "0.000000"; conf_lo "0.000000"; conf_hi "0.000000"; cov "0.000000";
4	chr6	Cufflinks	exon	166168343	166171172	+	gene_id "LNCDAT"; transcript_id "ENST00000675446.1"; exon_number "2"; FPKM "0.0000000000"; frac "0.000000"; conf_lo "0.000000"; conf_hi "0.000000"; cov "0.000000";
5	chr6	Cufflinks	transcript	166168013	166168706	+	gene_id "LNCDAT"; transcript_id "ENST00000675895.1"; FPKM "0.0000000000"; frac "0.000000"; conf_lo "0.000000"; conf_hi "0.000000"; cov "0.000000"; full_read_support "no";
6	chr6	Cufflinks	exon	166168013	166168150	+	gene_id "LNCDAT"; transcript_id "ENST00000675895.1"; exon_number "1"; FPKM "0.0000000000"; frac "0.000000"; conf_lo "0.000000"; conf_hi "0.000000"; cov "0.000000";
7	chr6	Cufflinks	exon	166168277	166168706	+	gene_id "LNCDAT"; transcript_id "ENST00000675895.1"; exon_number "2"; FPKM "0.0000000000"; frac "0.000000"; conf_lo "0.000000"; conf_hi "0.000000"; cov "0.000000";
8	chr6	Cufflinks	transcript	166168014	166171172	+	gene_id "LNCDAT"; transcript_id "ENST00000674611.1"; FPKM "0.0000000000"; frac "0.000000"; conf_lo "0.000000"; conf_hi "0.000000"; cov "0.000000"; full_read_support "no";
9	chr6	Cufflinks	transcript	166168018	166171376	+	gene_id "LNCDAT"; transcript_id "ENST00000676186.1"; FPKM "0.0000000000"; frac "0.000000"; conf_lo "0.000000"; conf_hi "0.000000"; cov "0.000000"; full_read_support "no";
10	chr6	Cufflinks	exon	166168018	166168598	+	gene_id "LNCDAT"; transcript_id "ENST00000676186.1"; exon_number "1"; FPKM "0.0000000000"; frac "0.000000"; conf_lo "0.000000"; conf_hi "0.000000"; cov "0.000000";
11	chr6	Cufflinks	exon	166168690	166168770	+	gene_id "LNCDAT"; transcript_id "ENST00000676186.1"; exon_number "2"; FPKM "0.0000000000"; frac "0.000000"; conf_lo "0.000000"; conf_hi "0.000000"; cov "0.000000";
12	chr6	Cufflinks	exon	166170072	166171376	+	gene_id "LNCDAT"; transcript_id "ENST00000676186.1"; exon_number "3"; FPKM "0.0000000000"; frac "0.000000"; conf_lo "0.000000"; conf_hi "0.000000"; cov "0.000000";
13	chr6	Cufflinks	transcript	166168123	166170979	+	gene_id "LNCDAT"; transcript_id "ENST00000675059.1"; FPKM "0.0000000000"; frac "0.000000"; conf_lo "0.000000"; conf_hi "0.000000"; cov "0.000000"; full_read_support "no";
14	chr6	Cufflinks	exon	166168123	166168150	+	gene_id "LNCDAT"; transcript_id "ENST00000675059.1"; exon_number "1"; FPKM "0.0000000000"; frac "0.000000"; conf_lo "0.000000"; conf_hi "0.000000"; cov "0.000000";
15	chr6	Cufflinks	exon	166168277	166168598	+	gene_id "LNCDAT"; transcript_id "ENST00000675059.1"; exon_number "2"; FPKM "0.0000000000"; frac "0.000000"; conf_lo "0.000000"; conf_hi "0.000000"; cov "0.000000";
16	chr6	Cufflinks	exon	166170072	166170979	+	gene_id "LNCDAT"; transcript_id "ENST00000675059.1"; exon_number "4"; FPKM "0.0000000000"; frac "0.000000"; conf_lo "0.000000"; conf_hi "0.000000"; cov "0.000000";
17	chr6	Cufflinks	transcript	166168144	166171172	+	gene_id "LNCDAT"; transcript_id "ENST00000675788.1"; FPKM "0.0000000000"; frac "0.000000"; conf_lo "0.000000"; conf_hi "0.000000"; cov "0.000000"; full_read_support "no";
18	chr6	Cufflinks	exon	166168144	166168703	+	gene_id "LNCDAT"; transcript_id "ENST00000675788.1"; exon_number "1"; FPKM "0.0000000000"; frac "0.000000"; conf_lo "0.000000"; conf_hi "0.000000"; cov "0.000000";
19	chr6	Cufflinks	exon	166170072	166171172	+	gene_id "LNCDAT"; transcript_id "ENST00000675788.1"; exon_number "2"; FPKM "0.0000000000"; frac "0.000000"; conf_lo "0.000000"; conf_hi "0.000000"; cov "0.000000";
20	chr6	Cufflinks	transcript	166168240	166170980	+	gene_id "LNCDAT"; transcript_id "ENST00000675188.1"; FPKM "0.0000000000"; frac "0.000000"; conf_lo "0.000000"; conf_hi "0.000000"; cov "0.000000"; full_read_support "no";
21	chr6	Cufflinks	exon	166168240	166168770	+	gene_id "LNCDAT"; transcript_id "ENST00000675188.1"; exon_number "1"; FPKM "0.0000000000"; frac "0.000000"; conf_lo "0.000000"; conf_hi "0.000000"; cov "0.000000";
22	chr6	Cufflinks	exon	166170072	166170980	+	gene_id "LNCDAT"; transcript_id "ENST00000675188.1"; exon_number "2"; FPKM "0.0000000000"; frac "0.000000"; conf_lo "0.000000"; conf_hi "0.000000"; cov "0.000000";

For this walkthrough, use ENSG00000256513, a lncRNA found upregulated with a log2 foldchange of 5.47 and an adj *p*-value of 1.56E-05. Upon auditing the gtf file, below is the table structure displaying the 'Feature' type, 'Start' and 'End' coordinates of the upregulated lncRNA ENSG00000256513.

Chromosome	Assembler	Feature	Feature_Start	Feature_End	Strand	Annotation
chr12	Cufflinks	transcript	28821975	28825831	-	gene_id "ENSG00000256513"; transcript_id "ENST00000537327.1"; FPKM "0.0000000000"; frac "0.000000"; conf_lo "0.000000"; conf_hi "0.000000"; cov "0.000000"; full_read_support "no";
chr12	Cufflinks	exon	28821975	28822273	-	gene_id "ENSG00000256513"; transcript_id "ENST00000537327.1"; exon_number "1"; FPKM "0.0000000000"; frac "0.000000"; conf_lo "0.000000"; conf_hi "0.000000"; cov "0.000000";
chr12	Cufflinks	exon	28825698	28825831	-	gene_id "ENSG00000256513"; transcript_id "ENST00000537327.1"; exon_number "2"; FPKM "0.0000000000"; frac "0.000000"; conf_lo "0.000000"; conf_hi "0.000000"; cov

3.16 LncRNA Interaction Prediction

Several methods exist to predict the interaction of lncRNAs with either RNA [56–58], and proteins [59–61]. In this walkthrough, we will demonstrate the use of 'LncRRIsearch' to identify lncRNA/RNA targets, and 'FEELnc_filter.pl' to identify interacting protein targets and perform pathway enrichment analysis.

3.16.1 Using 'LncRRIsearch' to Identify lncRNA or mRNA Targets

In the next step, use 'LncRRIsearch' [56] webserver located at http://rtools.cbrc.jp/LncRRIsearch/ to identify potential targets of ENSG00000256513 lncRNA. In the Query (lncRNA) search bar, enter the chosen lncRNA 'ENSG00000256513', keep the Energy threshold to default at '−16 kcal/mol' and species as 'Human' and click the 'Submit'. Alternatively, provide the 'Target (lncRNA or mRNA)' if any of them are known. Since the lncRNA interactome is still in its infancy, it is best to proceed without entering any targets. Upon submission, 'LncRRIsearch' server will generate the top RNA targets based on lowest sum of energy (kcal/mol) utilized for the probable pairing. Table 3 lists the potential targets for the chosen lncRNA ENSG00000256513.

In the next step, we performed a literature review on the potential RNA targets listed in Table 3 in the context of cancers with special emphasis on 'breast cancer'. We found ENST00000582008 which is also known by the gene symbol as LINC00667 or lncOCMRL1 [63] to be special interest due to its evidentiary roles in several cancers including breast cancer [64, 65]. We therefore performed lncRNA-RNA interaction between the differentially expressed lncRNA ENSG00000256513 and its predicted target ENST00000582008 (LINC00667/lncOCMRL1) (Fig. 6a, b).

The lncRNA interaction analysis using 'LncRRIsearch' server predicted the number of local base-pairing interactions to be 2 indicating the presence of two specific regions where the lncRNA and target RNA bind (Fig. 6c). Further, evidence confirms that expression of ENSG00000256513 and ENST00000582008 transcripts in breast tissue re-enforcing our findings of differentially expressed genes (Fig. 6d). The Sum of Local Base-Pairing Interaction Energies for the two interactions were −36.61 kcal/mol suggesting a strong overall interaction between the two RNAs (Fig. 6e). The Minimum Local Base-Pairing Interaction Energy was found to be −18.79 kcal/mol. This is the strongest interaction observed indicating a highly stable RNA duplex between the ENSG00000256513 lncRNA and ENST00000582008 (LINC00667/lncOCMRL1) (Fig. 6f). The interaction is visualized using dot-bracket notation and a circular plot. The arcs in the circular plot represent complementary base-pairing, which is crucial for understanding the interaction mechanism. The specific coordinates for the lncRNA and target RNA indicate where the binding occurs (Fig. 6f).

Another interaction site utilizes the Binding Energy of −17.82 kcal/mol, slightly weaker than the first interaction (−18.79 kcal/mol) but still highly stable, indicating a significant secondary interaction site (Fig. 6g). Similarly, the secondary interaction

Table 3
Potential targets for the chosen lncrna ENSG00000256513 based on sum of energy utilized for pairing

Number	Sum of energy	Min of energy	Transcript ID	Transcript name	Genomic distance
1	−55.17	−18.71	ENST00000424890	AP001626.2-001	10,000,000,000
2	−37.06	−18.88	ENST00000606125	XXbac-BPGBPG55C20.2-001	10,000,000,000
3	−36.61	−18.79	ENST00000582008	LINC00667-001	10,000,000,000
4	−36.61	−18.79	ENST00000577439	LINC00667-003	10,000,000,000
5	−35.07	−17.72	ENST00000397609	FAM98B-001	10,000,000,000
6	−32.41	−16.26	ENST00000624168	RP11-720D4.2-001	10,000,000,000
7	−29.78	−29.78	ENST00000637938	PLEKHG4B-001	10,000,000,000
8	−25.67	−25.67	ENST00000566945	CTD-2199O4.3-001	10,000,000,000
9	−25.67	−25.67	ENST00000606203	CTD-2199O4.7-001	10,000,000,000
10	−20.61	−20.61	ENST00000554046	RP11-794A8.1-001	10,000,000,000
11	−19.52	−19.52	ENST00000589281	RP11-95O2.5-001	10,000,000,000
12	−19.43	−19.43	ENST00000368599	FAM26E-001	10,000,000,000
13	−19.19	−19.19	ENST00000602761	RP5-884M6.1-004	10,000,000,000
14	−19.03	−19.03	ENST00000438649	AC007246.3-004	10,000,000,000
15	−18.95	−18.95	ENST00000423008	AC091729.9-001	10,000,000,000
16	−18.91	−18.91	ENST00000608382	ZBTB25-001	10,000,000,000
17	−18.88	−18.88	ENST00000426083	AC007246.3-003	10,000,000,000
18	−18.78	−18.78	ENST00000500117	RP11-410D17.2-001	10,000,000,000
19	−18.51	−18.51	ENST00000546784	RP11-454K7.1-001	10,000,000,000
20	−18.36	−18.36	ENST00000380680	UBL3-001	10,000,000,000
21	−18.25	−18.25	ENST00000444042	CYP4A22-AS1-001	10,000,000,000

22	−18.09	−18.09	ENST00000480931	RP11-723O4.2-002	10,000,000,000
23	−18.08	−18.08	ENST00000498297	RP11-723O4.2-001	10,000,000,000
24	−17.89	−17.89	ENST00000369833	GSTM4-002	10,000,000,000
25	−17.69	−17.69	ENST00000560477	RP11-429B14.3-001	10,000,000,000
26	−17.67	−17.67	ENST00000294008	SLX4-001	10,000,000,000
27	−17.53	−17.53	ENST00000228284	SART3-001	79,696,748
28	−17.47	−17.47	ENST00000433688	C11orf95-004	10,000,000,000
29	−17.39	−17.39	ENST00000589723	RP11-567M16.1-001	10,000,000,000
30	−17.35	−17.35	ENST00000426187	AC073133.2-001	10,000,000,000
31	−17.18	−17.18	ENST00000395184	ARHGAP24-001	10,000,000,000
32	−17.16	−17.16	ENST00000338970	RPL14-001	10,000,000,000
33	−17.04	−17.04	ENST00000256379	MED6-001	10,000,000,000
34	−16.87	−16.87	ENST00000318184	OVCH1-001	601,739
35	−16.84	−16.84	ENST00000227880	SLC15A3-001	10,000,000,000
36	−16.82	−16.82	ENST00000275300	SLC22A3-001	10,000,000,000
37	−16.73	−16.73	ENST00000624919	RP4-671O14.6-001	10,000,000,000
38	−16.71	−16.71	ENST00000334976	EMC10-001	10,000,000,000
39	−16.56	−16.56	ENST00000256682	ARF3-001	20,109,891
40	−16.52	−16.52	ENST00000398092	RP11-302B13.5-001	20,077,671
41	−16.47	−16.47	ENST00000373839	PHACTR4-001	10,000,000,000
42	−16.46	−16.46	ENST00000619666	RP11-156K23.3-001	10,000,000,000

(continued)

Table 3
(continued)

Number	Sum of energy	Min of energy	Transcript ID	Transcript name	Genomic distance
43	−16.42	−16.42	ENST00000627551	RP11-573D15.8-018	10,000,000,000
44	−16.38	−16.38	ENST00000370447	ZNF326-004	10,000,000,000
45	−16.3	−16.3	ENST00000405356	NOLC1-201	10,000,000,000
46	−16.29	−16.29	ENST00000329451	LMNTD2-001	10,000,000,000
47	−16.26	−16.26	ENST00000629019	LINC00284-013	10,000,000,000
48	−16.25	−16.25	ENST00000397333	DDX51-001	103,310,762
49	−16.2	−16.2	ENST00000247815	HELB-001	37,476,713
50	−16.14	−16.14	ENST00000608335	RP11-508N22.12-001	10,000,000,000
51	−16.1	−16.1	ENST00000402394	CGREF1-001	10,000,000,000
52	−16.1	−16.1	ENST00000513892	RP11-468H14.2-001	10,000,000,000
53	−16.08	−16.08	ENST00000586954	LINC00662-006	10,000,000,000
54	−16.08	−16.08	ENST00000393841	SLC29A1-201	10,000,000,000
55	−16.07	−16.07	ENST00000570301	RP11-1055B8.3-002	10,000,000,000
56	−16.06	−16.06	ENST00000301280	CHAF1A-001	10,000,000,000
57	−16.06	−16.06	ENST00000612859	B3GALNT2-003	10,000,000,000
58	−16.01	−16.01	ENST00000559756	AC074212.5-001	10,000,000,000

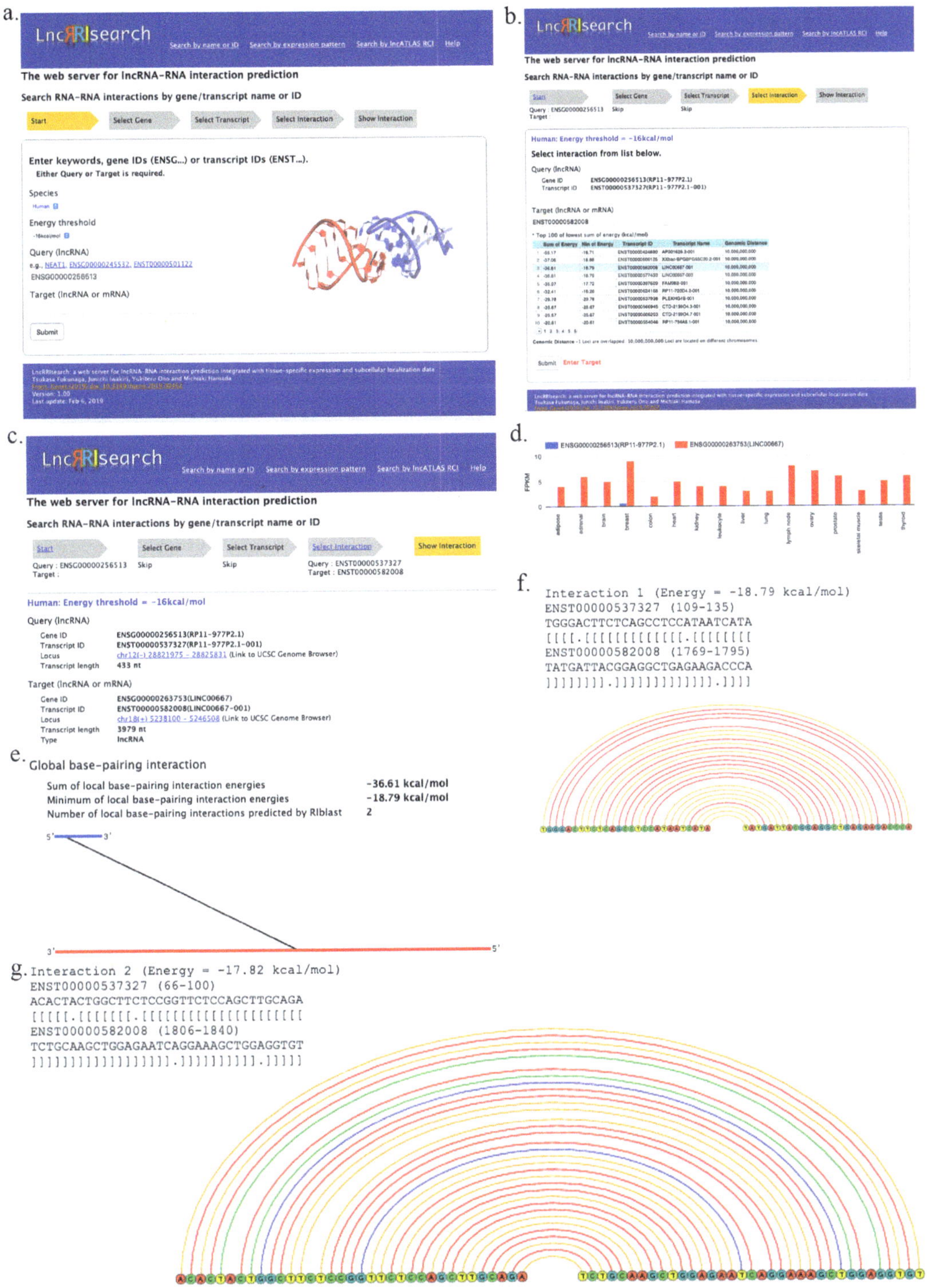

Fig. 6 Predicted interaction of lncRNA ENSG00000256513 with ENST00000582008 using LncRRIsearch. This figure illustrates the RNA-RNA interaction analysis of the lncRNA ENSG00000256513 with its target ENST00000582008 (LINC00667) using the LncRRIsearch tool. **a** Shows the input parameters for interaction prediction, while **b** lists the top potential targets ranked by interaction energy. **c** Highlights the selected

is also visualized through arcs and dot-bracket notation for pinpointing functional regions of the lncRNA . This interaction occurs at a different region compared to Interaction 1, indicating that the lncRNA may bind to multiple regions of the target RNA to enhance its regulatory effects (Fig. 6g). The presence of a secondary binding site emphasizes the potential multi-functional role of ENSG00000265613 in regulating ENSG00000582008.

ENST00000582008 (LINC00667/lncOCMRL1) has been reported to be upregulated in breast cancer tissues and associated with poor prognosis [64, 65]. ENST00000582008 (LINC00667/lncOCMRL1) has been shown to promote breast cancer cell proliferation and migration through a m6A-dependent feedback loop involving KIAA1429 [64]. LncRNAs have been shown to directly interact with target RNAs to stabilize or promote translation [66]. In this study, we believe the interaction shown to occur between the differentially expressed lncRNA ENSG00000256513 and its predicted target ENST00000582008 (LINC00667) may play a role in promoting the oncogenesis by regulating the downstream effectors. Studies have also shown higher ENSG00000582008 (LINC00667/lncOCMRL1) expression to promote the growth and metastasis of tumor cells through induction of epithelial-mesenchymal transition (EMT) via the suppression of RRM2 ubiquitination and thereby promoting the proliferation, invasion, and migration of tumor cells [63]. Since these layers of evidence have pointed at ENSG00000582008 (LINC00667/lncOCMRL1) to act as an oncogene, we believe the lncRNA ENSG00000265613 could promote malignancy by either stabilizing or enhancing the expression of its RNA target ENSG00000582008. Given the significant differential upregulated expression of lncRNA ENSG00000265613 in the current luminal A subtype of breast cancer and the several layers of ENSG00000582008's role in oncogenesis, we believe this pairing interaction seems critical in regulating their stability, translation, or splicing, resulting in driving oncogenic or tumor-suppressive processes in cancer.

3.16.2 Using 'FEELnc' to Identify Protein-Coding Targets (Functional Annotation)

In the next step, we utilize FEELnc (FlExible Extraction of LncRNAs) to filter candidate lncRNAs to identify those that are biologically relevant and likely to interact with protein-coding genes. It also ensures that the lncRNAs meet specific criteria for length, biotype, and proximity to functional genomic regions. This process is essential

Fig. 6 (continued) targetinteraction, with **d** showing the expression profiles of the interacting transcripts across tissues. **e** Summarizes the global base-pairing interaction, including the total energy (−36.61 kcal/mol) and two local interaction sites. **f**, **g** Provide visualizations of the two local interactions, including dot-bracket notations and circular plots indicating strong base-pairing stability suggesting a strong interaction between the two transcripts

for prioritizing lncRNAs and their target genes for experimental validation.

```
#Functional Annotation
FEELnc_filter.pl -i filtered_gencode.gtf -a gencode.
v47.annotation.gtf  -b  transcript_biotype=protein_
coding -l >
candidate_lncRNA.gtf
```

```
jsankarasubramani@lambda01:/storage5/work/subu/GSE233242$ FEELnc_filter.pl    filtered_gencode.gtf    gencode.v47.annotation.gtf    transcript_biotype=protein_coding    > candidate_ln
cRNA.gtf
Possible precedence issue with control flow operator at /storage4/apps/conda_packages/feelnc.0.2/lib/site_perl/5.26.2/Bio/DB/IndexedBase.pm line 805.
Filtered transcripts will be available in file: 'filtered_gencode.feelncfilter.log'
Parsing file 'filtered_gencode.gtf'...
Parse input file:              [                                                            ]
> Filter size (200): 48
> Filter monoexonic (0): 639
> Filter biexonicsize (25): 26
>> Transcripts left after fitler(s): 34145
Parsing file 'gencode.v47.annotation.gtf'...
Parse input file:              [                                                            ]
chr12
chr6
chrXrsect fileA:               [                                                            ]
chr21
chr22sect fileA:               [                                                            ]
Intersect fileA:               [                                                            ]
Intersect fileA:               [                                                            ]
chr1rsect fileA:               [                                                            ]
Intersect fileA:               [                                                            ]
chr4rsect fileA:               [                                                            ]
Intersect fileA:               [                                                            ]
chr14sect fileA:               [                                                            ]
chr19sect fileA:               [                                                            ]
Intersect fileA:               [                                                            ]
Intersect fileA:               [                                                            ]
chr20sect fileA:               [                                                            ]
Intersect fileA:               [                                                            ]
chrYrsect fileA:               [                                                            ]
chr18sect fileA:               [                                                            ]
Intersect fileA:               [                                                            ]
Intersect fileA:               [                                                            ]
chr10sect fileA:               [                                                            ]
chr17sect fileA:               [                                                            ]
Intersect fileA:               [                                                            ]
Intersect fileA:               [                                                            ]
chr13sect fileA:               [                                                            ]
chr15sect fileA:               [                                                            ]
Intersect fileA:               [                                                            ]
Intersect fileA:               [                                                            ]
chr11sect fileA:               [                                                            ]
chr3rsect fileA:               [                                                            ]
Intersect fileA:               [                                                            ]
Intersect fileA:               [                                                            ]
chr7rsect fileA:               [                                                            ]
chr9rsect fileA:               [                                                            ]
Intersect fileA:               [                                                            ]
chr5rsect fileA:               [                                                            ]
Intersect fileA:               [                                                            ]
Intersect fileA:               [                                                            ]
chr16sect fileA:               [                                                            ]
chr8rsect fileA:               [                                                            ]
Intersect fileA:               [                                                            ]
Intersect fileA:               [                                                            ]
chr2rsect fileA:               [                                                            ]
Intersect fileA:               [-                                                           ]
Intersect fileA:               [                                                            ]
Intersect fileA:               [                                                            ]
Intersect fileA:               [
```

The above code refines the list of lncRNAs , ensuring they meet specific criteria for annotation and relevance to protein-coding genes, which is critical for understanding their functional roles. Once you have the output 'candidate_lncRNA.gtf', proceed to the next step 'awk' step.

```
#extract_gene_name
awk  'BEGIN{FS="\t";  OFS="\t"}  {match($9,  /gene_id
"([^"]+)"/, a); if (a[1] != "") printa[1]}' candi-
date_lncRNA.gtf | sort | uniq > lncRNA_target_genes.
txt
```

```
jsankarasubramani@lambda01:/storage5/work/subu/GSE233242$ awk 'BEGIN{FS="\t"; OFS="\t"} {match(   , /gene_id "([^"]+)"/, a); if (a[1] != "") print a[1]}' candidate_lncRNA.gtf | sort |
uniq > lncRNA_target_genes.txt
```

This 'awk' command extracts the gene IDs from the candidate lncRNA GTF file and generates a list of unique target genes. This clean list of protein-coding genes potentially regulated by the candidate lncRNAs , which is crucial for downstream functional analysis and pathway enrichment studies.

Table 4

An overview of various tools used in quality control, trimming, alignment, post-processing, differential analysis, interaction prediction, and visualization along with their parameters/functions used in developing the best practices approach for lncRNA analysis from RNAseq data

Step	Program	Purpose	Parameter	Function
Quality control	FastQC	Perform quality control on raw sequencing reads.	-o	Outputs data
Trimming and filtering	fastp	Trim low-quality bases, remove adapters, and filter short reads.	-i SRR24709142_1.fastq.gz	Input file for read 1
			-I SRR24709142_2.fastq.gz	Input file for read 2
			-o SRR24709142_1_clean.fastq.gz	Output file for trimmed read 1
			-O SRR24709142_2_clean.fastq.gz	Output file for trimmed read 2
			--trim_poly_g	Removes poly-G tails
			--cut_mean_quality 20	Removes low-quality bases below a mean Phred score of 20
			--length_required 50	Ensures trimmed reads are at least 50 bp long
STAR index generation	STAR	Generate genome index files for efficient alignment.	--runThreadN 50	Enables multithreading for faster indexing.
			--runMode genomeGenerate	Specifies genome indexing mode
			--genomeDir	Directory to store index files
			--genomeFastaFiles	Input genome FASTA file.
			--sjdbGTFfile	Annotation file to incorporate splice junctions
			--sjdboverhang	Length of the genomic sequence around splice junctions

Table 4
(continued)

			--runThreadN 50	Utilizes multiple threads for faster alignment.
			--genomeDir	Path to the pre-built genome index
			--readFilesIn	Input files for paired-end reads.
			--readFilesCommand zcat	Unzips the input FASTQ files for processing.
Alignment to reference genome	STAR	Align cleaned reads to the reference genome and generate sorted BAM files.	--outSAMtype BAM SortedByCoordinate	Outputs aligned reads in BAM format, sorted by genomic coordinates.
			--quantMode TranscriptomeSAM GeneCounts	Simultaneously generates transcript-level quantification data.
			--outSAMstrandField intronMotif	Adds strand information for spliced alignments.
			-g human_genome/gencode.v47.long_noncoding_RNAs.gtf	Provides reference annotation for guided assembly.
Transcript assembly	Cufflinks	Assemble transcripts from aligned reads into a transcriptome.	-o SRR24709142_cufflinks_out	Specifies the output directory for assembled transcripts.
			-p 8	Utilizes 8 threads for faster processing.
Merging transcripts	Custom Script	Consolidate transcript GTF files from multiple samples.	echo/vi	Ensures a comprehensive dataset for transcript merging in the next step.
Clearing duplicates	awk	Remove duplicate entries from the merged GTF file.	'!seen[$1, $4, $5]++'	Identifies unique entries based on specific fields.

(continued)

Table 4
(continued)

Step	Program	Purpose	Parameter	Function
Removing mitochondrial genome	awk	Exclude mitochondrial sequences.	'/^>/{p=($0!=">chrM MT")} p'	Filters out mitochondrial chromosome entries.
Extracting lncRNAs	gffread	Extract lncRNA sequences from the genome.	-w filtered_lncRNAs.fa	Specifies the output file for extracted lncRNA sequences.
			-g GRCh38.p14.genome_no_chrM.fa	Input genome file without mitochondrial sequences.
Coding potential assessment	CPAT	Predict coding potential of lncRNAs.	-g filtered_lncRNAs.fa	Input lncRNA sequences.
			-x Human_Hexamer.tsv	Hexamer usage frequency file
			-d Human_logitModel.RData	Logistic regression model for coding prediction.
			-o cpat_output.txt	Output file for coding potential results.
Cross-referencing CPAT output with GTF	awk	Filtering lncRNAs to retain those with low coding potential scores	<0.38	Threshold for non-coding classification is applied
Quantification of transcript abundance	featureCounts	Quantify transcript expression.	-a merged_without_duplicates.gtf	Input GTF file with unique transcripts.
			-o SRR24709142_counts.txt	Output file for expression counts.
			-T 8	Multithreading for faster processing.
			-p	Paired-end reads
			-B	Requires both paired reads to map before counting.
			-C	Discards chimeric fragments.

Differential expression analysis	DESeq2	Identify differentially expressed lncRNAs.	≤ -1 and $\geq +1$ log2fold change	Incorporates statistical modeling to detect significant changes in expression.
LnCRNA-RNA interaction	LncRRIsearch	To predict DE lncRNA-RNA interaction prediction	−16 kcal/mol	Energy threshold
			Differentially expressed query lncRNA	To identify lncRNA or mRNA targets using the energy threshold
Functional annotation	FEELnc_filter.pl	Filter lncRNAs interacting with protein-coding genes.	-i	Input GTF file with unique transcripts.
			-a	Reference annotation for comparison.
			-b	Filters based on biotype
Pathway enrichment analysis	clusterProfiler	Map lncRNA target genes to KEGG pathways.	All defaults	Provides functional insights into lncRNA regulatory roles.

3.17 Pathway Enrichment Analysis

The next step performs pathway enrichment analysis to identify the biological pathways associated with the lncRNA-targeted protein-coding genes identified in the previous step. By mapping the target genes (from the FEELnc step) to known pathways, this step helps infer the functional roles of lncRNAs in cellular processes, particularly in the context of cancer.

```
#Pathway Enrichment Analysis
library(clusterProfiler)
library(org.Hs.eg.db)
target_genes <- read.table("lncRNA_target_genes.
txt", header = FALSE, stringsAsFactors = FALSE)
gene_symbols <- as.character(target_genes$V1)
gene_vector_entrez <- bitr(gene_symbols, fromType =
"ENSEMBL", toType = "ENTREZID", OrgDb = org.Hs.eg.
db)
enrich_res <- enrichKEGG(gene = gene_vector_en-
trez$ENTREZID, organism = "hsa")
write.csv(as.data.frame(enrich_res@result), file =
"pathway_enrichment_results_ENSEMBL_ID.csv")
```

```
> target_genes <- read.table("lncRNA_target_genes.txt", header=FALSE)
> library(org.Hs.eg.db)
> gene_symbols <- as.character(target_genes$V1)
> gene_vector_entrez <- bitr(gene_symbols, fromType = "ENSEMBL", toType = "ENTREZID", OrgDb = org.Hs.eg.db)
'select()' returned 1:many mapping between keys and columns
Warning message:
In bitr(gene_symbols, fromType = "ENSEMBL", toType = "ENTREZID",  :
  96.01% of input gene IDs are fail to map...
> enrich_res <- enrichKEGG(gene = gene_vector_entrez$ENTREZID, organism = "hsa")
> write.csv(as.data.frame(enrich_res), file="pathway_enrichment_results_ENSEMBL_ID.csv")
```

Auditing the 'pathway_enrichment_results_ENSEMBL_ID.csv' file shows the below information. Only one pathway, 'Spinocerebellar ataxia' was found enriched in this analysis. Though pathway enrichment analysis is an essential and critical component of any expression analysis, however, in the context of lncRNA analysis, given its lack of in-depth annotations in the pathways makes it difficult to make any conclusions and users should proceed with caution.

Several enrichment analysis webservers are available for analyzing the genes extracted in the previous step, including g:Profiler (https://biit.cs.ut.ee/gprofiler/gost), WebGestalt (https://www.webgestalt.org) and Enrichr (https://maayanlab.cloud/Enrichr/enrich).

pathway_id	category	subcategory	ID	Description	BgRatio	RichFactor	FoldEnrichment	zScore	pvalue	p.adjust	qvalue	geneID	Count
hsa05017	Human Diseases	Neurodegenerative disease	hsa05017	Spinocerebellar ataxia	144/8538	0.00694444	59.29166667	7.634897947	0.01686578	0.016865777	NA	6315	1

Table 4 presents a list of tools used in this framework, along with their respective parameters and stated purposes, highlighting best practices for analyzing lncRNA dysregulation in RNA-seq datasets. Thus, this lncRNA framework effectively facilitates the identification of critical lncRNA dysregulations and potential oncogenic interactions in the current cohort of luminal A breast cancer. We believe that the robustness of this framework can be applied to detect lncRNA dysregulations in various cancers and other diseases.

4 Notes

4.1 Data Preprocessing

The 'fastqc' tool is employed to evaluate the quality of raw sequencing reads, a critical step in ensuring data integrity prior to downstream analysis in the lncRNA pipeline, whereas fastp is used to trim adapters, filters low-quality reads, and generate QC reports. In the code, 'fastqc' is used to analyze paired-end FASTQ files ('SRR24709142_1.fastq.gz' and SRR24709142_2.fastq.gz') and generate quality reports saved in the specified output directory ('./qc_results/'). This process identifies potential issues such as low-quality bases, adapter contamination, and sequencing biases. The results, including detailed HTML and summary reports, guide corrective measures like trimming or filtering to enhance data reliability. Performing this initial quality control step ensures the accuracy and robustness of subsequent analyses, including alignment and transcript assembly.

4.2 Transcript Assembly

The StringTie command provided in the code block plays a critical role in the pipeline by quantifying transcript expression levels while maintaining consistency with the known lncRNA annotations provided by GENCODE.

4.3 Coding Potential Assessment

The coding potential assessment is a critical step in distinguishing true lncRNAs from potential protein-coding lncRNA transcripts. This ensures that the pipeline focuses exclusively on functional lncRNAs, avoiding false positives caused by coding transcripts misclassified as lncRNAs during assembly or annotation. The use of pre-trained human-specific models, such as the hexamer composition and logistic regression model, enhances the accuracy of predictions. By leveraging pre-trained models, CPAT identifies whether a given transcript is likely to encode a protein or function as a non-coding RNA. This step is crucial for validating the non-coding nature of lncRNA candidates in the pipeline.

Acknowledgments

This work has been supported by the start-up funds to CG from the University of Nebraska Medical Center. The authors would like to thank the Bioinformatics and Systems Biology Core (BSBC) at UNMC for providing the computational infrastructure. BSBC is partly supported by multiple NIH awards (5P20GM103427, 5P30CA03672).

References

1. Ling H, Fabbri M, Calin GA (2013) MicroRNAs and other non-coding RNAs as targets for anticancer drug development. Nat Rev Drug Discov 12(11):847–865. https://doi.org/10.1038/nrd4140
2. Ernst C, Morton CC (2013) Identification and function of long non-coding RNA. Front Cell Neurosci 7:168. https://doi.org/10.3389/fncel.2013.00168
3. Ward M, McEwan C, Mills JD, Janitz M (2015) Conservation and tissue-specific transcription patterns of long noncoding RNAs. J Hum Transcr 1(1):2–9. https://doi.org/10.3109/23324015.2015.1077591
4. Burgos M, Hurtado A, Jimenez R, Barrionuevo FJ (2021) Non-coding RNAs: lncRNAs, miRNAs, and piRNAs in sexual development. Sex Dev 15(5–6):335–350. https://doi.org/10.1159/000519237
5. Chen X, Fan S, Song E (2016) Noncoding RNAs: new players in cancers. Adv Exp Med Biol 927:1–47. https://doi.org/10.1007/978-981-10-1498-7_1
6. Dhanoa JK, Sethi RS, Verma R, Arora JS, Mukhopadhyay CS (2018) Long non-coding RNA: its evolutionary relics and biological implications in mammals: a review. J Anim Sci Technol 60:25. https://doi.org/10.1186/s40781-018-0183-7
7. Zampetaki A, Albrecht A, Steinhofel K (2018) Long non-coding RNA structure and function: is there a link? Front Physiol 9:1201. https://doi.org/10.3389/fphys.2018.01201
8. Graf J, Kretz M (2020) From structure to function: route to understanding lncRNA mechanism. BioEssays : news and reviews in molecular, cellular and developmental biology 42(12):e2000027. https://doi.org/10.1002/bies.202000027
9. Roberts TC, Morris KV, Weinberg MS (2014) Perspectives on the mechanism of transcriptional regulation by long non-coding RNAs. Epigenetics 9(1):13–20. https://doi.org/10.4161/epi.26700
10. Mercer TR, Mattick JS (2013) Structure and function of long noncoding RNAs in epigenetic regulation. Nat Struct Mol Biol 20(3):300–307. https://doi.org/10.1038/nsmb.2480
11. Sun W, Yang Y, Xu C, Guo J (2017) Regulatory mechanisms of long noncoding RNAs on gene expression in cancers. Cancer Genet 216–217:105–110. https://doi.org/10.1016/j.cancergen.2017.06.003
12. Rinn JL, Chang HY (2012) Genome regulation by long noncoding RNAs. Annu Rev Biochem 81:145–166. https://doi.org/10.1146/annurev-biochem-051410-092902
13. Chen LL (2016) Linking long noncoding RNA localization and function. Trends Biochem Sci 41(9):761–772. https://doi.org/10.1016/j.tibs.2016.07.003
14. Yoon JH, Abdelmohsen K, Gorospe M (2014) Functional interactions among microRNAs and long noncoding RNAs. Semin Cell Dev Biol 34:9–14. https://doi.org/10.1016/j.semcdb.2014.05.015
15. Kugel JF, Goodrich JA (2013) The regulation of mammalian mRNA transcription by lncRNAs: recent discoveries and current concepts. Epigenomics 5(1):95–102. https://doi.org/10.2217/epi.12.69
16. Akrami R, Jacobsen A, Hoell J, Schultz N, Sander C, Larsson E (2013) Comprehensive analysis of long non-coding RNAs in ovarian cancer reveals global patterns and targeted DNA amplification. PLoS one 8(11):e80306. https://doi.org/10.1371/journal.pone.0080306
17. Kim TK, Hemberg M, Gray JM (2015) Enhancer RNAs: a class of long noncoding RNAs synthesized at enhancers. Cold Spring Harb Perspect Biol 7(1):a018622. https://doi.org/10.1101/cshperspect.a018622
18. Derrien T, Johnson R, Bussotti G, Tanzer A, Djebali S, Tilgner H, Guernec G, Martin D, Merkel A, Knowles DG, Lagarde J, Veeravalli L, Ruan X, Ruan Y, Lassmann T, Carninci P, Brown JB, Lipovich L, Gonzalez JM, Thomas M, Davis CA, Shiekhattar R, Gingeras TR, Hubbard TJ, Notredame C, Harrow J, Guigo R (2012) The GENCODE v7 catalog of human long noncoding RNAs: analysis of their gene structure, evolution, and expression. Genome Res 22(9):1775–1789. https://doi.org/10.1101/gr.132159.111
19. Pegueroles C, Gabaldon T (2016) Secondary structure impacts patterns of selection in human lncRNAs. BMC Biol 14:60. https://doi.org/10.1186/s12915-016-0283-0
20. Hon CC, Ramilowski JA, Harshbarger J, Bertin N, Rackham OJ, Gough J, Denisenko E, Schmeier S, Poulsen TM, Severin J, Lizio M, Kawaji H, Kasukawa T, Itoh M, Burroughs AM, Noma S, Djebali S, Alam T, Medvedeva YA, Testa AC, Lipovich L, Yip CW, Abugessaisa I, Mendez M, Hasegawa A, Tang D, Lassmann T, Heutink P, Babina M, Wells CA, Kojima S, Nakamura Y, Suzuki H, Daub CO, de Hoon MJ, Arner E, Hayashizaki Y, Carninci P, Forrest AR (2017) An atlas of human long non-coding RNAs with accurate 5' ends. Nature 543(7644):199–204. https://doi.org/10.1038/nature21374

21. Xu J, Bai J, Zhang X, Lv Y, Gong Y, Liu L, Zhao H, Yu F, Ping Y, Zhang G, Lan Y, Xiao Y, Li X (2017) A comprehensive overview of lncRNA annotation resources. Brief Bioinform 18(2):236–249. https://doi.org/10.1093/bib/bbw015
22. Li J, Xuan Z, Liu C (2013) Long non-coding RNAs and complex human diseases. Int J Mol Sci 14(9):18790–18808. https://doi.org/10.3390/ijms140918790
23. Rafiee A, Riazi-Rad F, Havaskary M, Nuri F (2018) Long noncoding RNAs: regulation, function and cancer. Biotechnol Genet Eng Rev 34(2):153–180. https://doi.org/10.1080/02648725.2018.1471566
24. Kumar S, Prajapati KS, Singh AK, Kushwaha PP, Shuaib M, Gupta S (2021) Long non-coding RNA regulating androgen receptor signaling in breast and prostate cancer. Cancer Lett 504:15–22. https://doi.org/10.1016/j.canlet.2020.11.039
25. Zhu YS, Zhu J (2022) Molecular and cellular functions of long non-coding RNAs in prostate and breast cancer. Adv Clin Chem 106:91–179. https://doi.org/10.1016/bs.acc.2021.09.005
26. Plewka P, Raczynska KD (2022) Long intergenic noncoding RNAs affect biological pathways underlying autoimmune and neurodegenerative disorders. Mol Neurobiol 59(9):5785–5808. https://doi.org/10.1007/s12035-022-02941-0
27. Tachiwana H, Saitoh N (2021) Nuclear long non-coding RNAs as epigenetic regulators in cancer. Curr Med Chem 28(25):5098–5109. https://doi.org/10.2174/0929867328666210215114506
28. Li Y, Li W, Hoffman AR, Cui J, Hu JF (2021) The nucleus/mitochondria-shuttling LncRNAs function as new epigenetic regulators of mitophagy in cancer. Front Cell Dev Biol 9:699621. https://doi.org/10.3389/fcell.2021.699621
29. Lin YH (2020) Crosstalk of lncRNA and cellular metabolism and their regulatory mechanism in cancer. Int J Mol Sci 21(8). doi:https://doi.org/10.3390/ijms21082947
30. Xu Y, Qiu M, Shen M, Dong S, Ye G, Shi X, Sun M (2021) The emerging regulatory roles of long non-coding RNAs implicated in cancer metabolism. Molecular therapy : the journal of the American Society of Gene Therapy 29(7):2209–2218. https://doi.org/10.1016/j.ymthe.2021.03.017
31. Yang K, Liang X, Wen K (2022) Long non-coding RNAs interact with RNA-binding proteins to regulate genomic instability in cancer cells (Review). Oncol Rep 48(4), https://doi.org/10.3892/or.2022.8390
32. Huang H, Li L, Wen K (2021) Interactions between long non-coding RNAs and RNA-binding proteins in cancer (Review). Oncol Rep 46(6), https://doi.org/10.3892/or.2021.8207
33. Ouyang J, Zhong Y, Zhang Y, Yang L, Wu P, Hou X, Xiong F, Li X, Zhang S, Gong Z, He Y, Tang Y, Zhang W, Xiang B, Zhou M, Ma J, Li Y, Li G, Zeng Z, Guo C, Xiong W (2022) Long non-coding RNAs are involved in alternative splicing and promote cancer progression. Br J Cancer 126(8):1113–1124. https://doi.org/10.1038/s41416-021-01600-w
34. Kansara S, Sawant P, Kaur T, Garg M, Pandey AK (2024) LncRNA-mediated orchestrations of alternative splicing in the landscape of breast cancer. Biochim Biophys Acta Gene Regul Mech 1867(2):195017. https://doi.org/10.1016/j.bbagrm.2024.195017
35. Liu Y, Liu X, Lin C, Jia X, Zhu H, Song J, Zhang Y (2021) Noncoding RNAs regulate alternative splicing in cancer. J Exp Clin Cancer Res 40(1):11. https://doi.org/10.1186/s13046-020-01798-2
36. Paraskevopoulou MD, Hatzigeorgiou AG (2016) Analyzing MiRNA-LncRNA interactions. Methods Mol Biol 1402:271–286. https://doi.org/10.1007/978-1-4939-3378-5_21
37. Du Z, Sun T, Hacisuleyman E, Fei T, Wang X, Brown M, Rinn JL, Lee MG, Chen Y, Kantoff PW, Liu XS (2016) Integrative analyses reveal a long noncoding RNA-mediated sponge regulatory network in prostate cancer. Nat Commun 7:10982. https://doi.org/10.1038/ncomms10982
38. Ballantyne MD, McDonald RA, Baker AH (2016) lncRNA/MicroRNA interactions in the vasculature. Clin Pharmacol Ther 99(5):494–501. https://doi.org/10.1002/cpt.355
39. Zhao W, Shan B, He D, Cheng Y, Li B, Zhang C, Duan C (2019) Recent progress in characterizing long noncoding RNAs in cancer drug resistance. J Cancer 10(26):6693–6702. https://doi.org/10.7150/jca.30877
40. Jiang W, Xia J, Xie S, Zou R, Pan S, Wang ZW, Assaraf YG, Zhu X (2020) Long non-coding RNAs as a determinant of cancer drug resistance: Towards the overcoming of chemoresistance via modulation of lncRNAs. Drug Resist Updat 50:100683. https://doi.org/10.1016/j.drup.2020.100683
41. Pan JJ, Xie XJ, Li X, Chen W (2015) Long non-coding RNAs and drug resistance. Asian Pac J Cancer Prev 16(18):8067–8073. https://doi.org/10.7314/apjcp.2015.16.18.8067
42. Li SY, Hammarlund JA, Wu G, Lian JW, Howell SJ, Clarke RB, Adamson AD, Goncalves CF, Hogenesch JB, Anafi RC, Meng QJ (2024) Tumor circadian clock strength influences metastatic potential and predicts patient prognosis in luminal A breast cancer. Proc Natl Acad

Sci USA 121(7):e2311854121. https://doi.org/10.1073/pnas.2311854121
43. SRA N (2025) The SRA toolkit. The NCBI SRA (Sequence Read Archive), SRA Toolkit Release 3.2.0 edn. NCBI, https://github.com/ncbi/sra-tools
44. Andrews S (2010) FastQC: a quality control tool for high throughput sequence data. Available online at: http://www.bioinformaticsbabrahamacuk/projects/fastqc. doi:https://doi.org/10.1093/bioinformatics/btx373
45. Chen S, Zhou Y, Chen Y, Gu J (2018) fastp: an ultra-fast all-in-one FASTQ preprocessor. Bioinformatics 34(17):i884–i890. https://doi.org/10.1093/bioinformatics/bty560
46. Dobin A, Davis CA, Schlesinger F, Drenkow J, Zaleski C, Jha S, Batut P, Chaisson M, Gingeras TR (2013) STAR: ultrafast universal RNA-seq aligner. Bioinformatics 29(1):15–21. https://doi.org/10.1093/bioinformatics/bts635
47. Trapnell C, Williams BA, Pertea G, Mortazavi A, Kwan G, van Baren MJ, Salzberg SL, Wold BJ, Pachter L (2010) Transcript assembly and quantification by RNA-Seq reveals unannotated transcripts and isoform switching during cell differentiation. Nat Biotechnol 28(5):511–515. https://doi.org/10.1038/nbt.1621
48. Pertea M, Pertea GM, Antonescu CM, Chang TC, Mendell JT, Salzberg SL (2015) StringTie enables improved reconstruction of a transcriptome from RNA-seq reads. Nat Biotechnol 33(3):290–295. https://doi.org/10.1038/nbt.3122
49. Wang L, Park HJ, Dasari S, Wang S, Kocher JP, Li W (2013) CPAT: Coding-Potential Assessment Tool using an alignment-free logistic regression model. Nucleic acids Res 41(6):e74. https://doi.org/10.1093/nar/gkt006
50. Yi Q, Feng J, Lan W, Shi H, Sun W, Sun W (2024) CircRNA and lncRNA-encoded peptide in diseases, an update review. Mol Cancer 23(1):214. https://doi.org/10.1186/s12943-024-02131-7
51. Ruiz-Orera J, Messeguer X, Subirana JA, Alba MM (2014) Long non-coding RNAs as a source of new peptides. eLife 3:e03523. https://doi.org/10.7554/eLife.03523
52. Liao Y, Smyth GK, Shi W (2014) featureCounts: an efficient general purpose program for assigning sequence reads to genomic features. Bioinformatics 30(7):923–930. https://doi.org/10.1093/bioinformatics/btt656
53. Love MI, Huber W, Anders S (2014) Moderated estimation of fold change and dispersion for RNA-seq data with DESeq2. Genome Biol 15(12):550. https://doi.org/10.1186/s13059-014-0550-8
54. Wickham H (2016) ggplot2: elegant graphics for data analysis. Springer-Verlag, New York
55. Okura Y (2023) [[Fundamentals] 4. Visualization in Python programming: how to use Matplotlib and seaborn]. Nihon Hoshasen Gijutsu Gakkai Zasshi 79(7):723–731. https://doi.org/10.6009/jjrt.2023-2228
56. Fukunaga T, Iwakiri J, Ono Y, Hamada M (2019) LncRRIsearch: a web server for lncRNA-RNA interaction prediction integrated with tissue-specific expression and subcellular localization data. Front Genet 10:462. https://doi.org/10.3389/fgene.2019.00462
57. Das G, Das T, Parida S, Ghosh Z (2024) LncRTPred: predicting RNA-RNA mode of interaction mediated by lncRNA. IUBMB life 76(1):53–68. https://doi.org/10.1002/iub.2778
58. Rehmsmeier M, Steffen P, Hochsmann M, Giegerich R (2004) Fast and effective prediction of microRNA/target duplexes. RNA 10(10):1507–1517.https://doi.org/10.1261/rna.5248604
59. Tian X, Shen L, Wang Z, Zhou L, Peng L (2021) A novel lncRNA-protein interaction prediction method based on deep forest with cascade forest structure. Sci Rep 11(1):18881. https://doi.org/10.1038/s41598-021-98277-1
60. Lv G, Xia Y, Qi Z, Zhao Z, Tang L, Chen C, Yang S, Wang Q, Gu L (2023) LncRNA-protein interaction prediction with reweighted feature selection. BMC Bioinforma 24(1):410. https://doi.org/10.1186/s12859-023-05536-1
61. Zhao G, Li P, Qiao X, Han X, Liu ZP (2021) Predicting lncRNA-protein interactions by heterogenous network embedding. Front Genet 12:814073. https://doi.org/10.3389/fgene.2021.814073
62. Wucher V, Legeai F, Hedan B, Rizk G, Lagoutte L, Leeb T, Jagannathan V, Cadieu E, David A, Lohi H, Cirera S, Fredholm M, Botherel N, Leegwater PAJ, Le Beguec C, Fieten H, Johnson J, Alfoldi J, Andre C, Lindblad-Toh K, Hitte C, Derrien T (2017) FEELnc: a tool for long non-coding RNA annotation and its application to the dog transcriptome. Nucleic Acids Res 45(8):e57. https://doi.org/10.1093/nar/gkw1306
63. Lu N, Jiang Q, Xu T, Gao Q, Wang Y, Huang Z, Huang Z, Xu X (2024) LncOCMRL1 promotes oral squamous cell carcinoma growth and metastasis via the RRM2/EMT pathway. J Exp Clin Cancer Res 43(1):267. https://doi.org/10.1186/s13046-024-03190-w
64. Ren S, Zhang Y, Yang X, Li X, Zheng Y, Liu Y, Zhang X (2022) N6-methyladenine-induced LINC00667 promoted breast cancer progression through m6A/KIAA1429 positive feedback loop. Bioengineered 13(5):13462–13473. https://doi.org/10.1080/21655979.2022.2077893s

65. Qiu X, Dong J, Zhao Z, Li J, Cai X (2019) LncRNA LINC00668 promotes the progression of breast cancer by inhibiting apoptosis and accelerating cell cycle. Onco Targets Ther 12:5615–5625. https://doi.org/10.2147/OTT.S188933

66. Yoon JH, Abdelmohsen K, Gorospe M (2013) Posttranscriptional gene regulation by long noncoding RNA. J Mol Biol 425(19):3723–3730. https://doi.org/10.1016/j.jmb.2012.11.024

Chapter 13

Small RNA Analysis in Cancer Cells

Victoria Silva, Srikrishna Kandooru, Damien Parrello, Brooke Froelich, Mike Hill, Yagna P. R. Jarajapu and Sergei Nechaev

Abstract

Small RNAs (sRNAs), broadly defined as RNA molecules shorter than ~200 nucleotides in size, include miscellaneous products of transcription and post-transcriptional processing that are variably abundant in cancer cells. While some types of sRNAs, such as microRNA (miRNA), are relatively well-characterized, the biological roles of many others remain unknown. Regardless of functionality, sRNA transcriptomes may reflect the identity of cell types, tissues, and cancers. While deep sequencing remains the best approach for sRNA profiling, there are distinct technical challenges for sRNA extraction, library preparation, sequencing and data analysis. Here, we describe a workflow for the preparation of sRNAs from cancer cells, sRNA libraries for sequencing, and basic bioinformatic analyses of sRNA sequencing data. We describe the preparation and analysis of standard, miRNA-enriched libraries that contain 5'-monophosphorylated RNAs and 3′-hydroxyl ends, as well as a custom protocol for the analysis of sRNAs containing 5′-cap modifications. The chapter should facilitate the analysis of different sRNA classes in cancer cells.

Keywords Small RNA, RNA-seq, Genomics, Cancer, Epigenetics

1 Introduction

Eukaryotic cells produce an abundance of ribonucleic acid molecules shorter than ~200 nucleotides (nt) in size. These so-called small RNAs (sRNAs) may be direct intermediates of transcription or products of post-transcriptional processing of larger RNA molecules [1–4] and have been implicated in numerous biological processes including cancers [5, 6]. While some sRNA classes, such as microRNA (miRNA), have relatively well-established molecular functions, most have no known roles. Regardless of the functionality of individual RNAs, sRNA transcriptomes likely reflect processes related to cell type identity as well as cancers [7–9]. Deep sequencing has been the main strategy for sRNA profiling, and many sRNAs are known only because of it [10–13]. However, apart from miRNAs, sRNA transcriptome profiling is not nearly as well-established as the conventional analyses of gene expression based on mRNA.

Archana Dhasarathy and Motoki Takaku (eds.), *Cancer Epigenetics*, Methods in Molecular Biology, vol. 3005, https://doi.org/10.1007/978-1-0716-5072-1_13,

The preparation of sRNA sequencing libraries aims to enclose the molecules of interest between universal adapters to prime cDNA synthesis and PCR amplification (Fig. 1). The first step is to attach the 3′-adapter, usually with the single-stranded-specific T4 RNA ligase 1. The next step is to attach the 5′-end adapter, which can be accomplished with one of several strategies. The most common and perhaps most versatile strategy involves ligation of the RNA to a universal adapter using the T4 RNA ligase 1 [14]. This strategy can be combined with enzymatic treatments and incorporate additional purification or size-selection steps for custom protocols to prepare specific sRNA populations. Another strategy for sRNA capture is RNA self-circularization [15]. All sRNA library procedures have one important technical distinction. Unlike other libraries, such as those for DNA- or regular mRNA-sequencing, sRNA libraries are stranded. Distinct sequences of sRNA adapters allow one to identify the 5′- and 3′-ends of sRNA molecules.

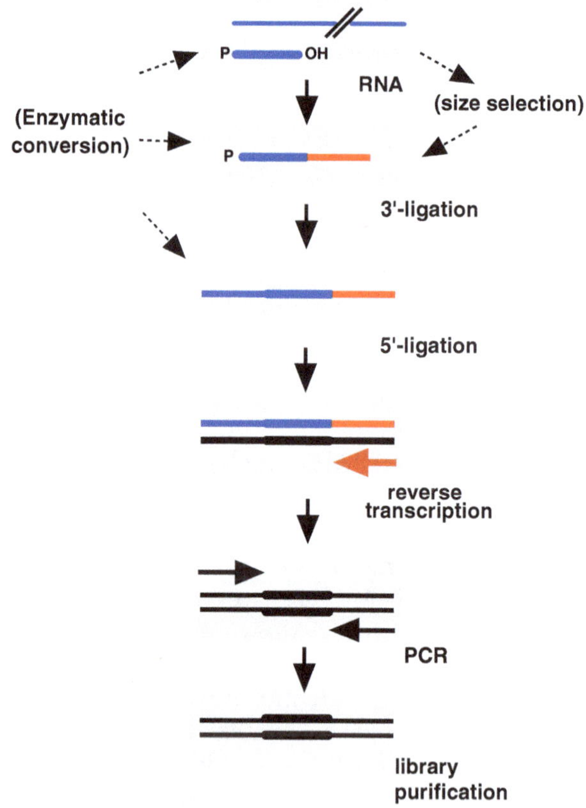

Fig. 1 A scheme of sRNA library preparation. Shown are the main steps of sRNA capture using the most commonly used double ligation strategy. Individual steps are highlighted: target RNA is shown in blue, single-stranded RNA adapters in thin lines in red and blue, and DNA in black. Optional steps are shown in parentheses and dashed arrows

Only RNA molecules that contain the 5′-monophosphate (5′-P) and 3′-hydroxyl (3′-OH) groups can be ligated to an adapter with T4 RNA ligase. Therefore, preparing sRNA libraries from unaltered cellular RNA enriches for molecules that contain 5′-P and 3′-OH ends, such as miRNA. Profiling monophosphorylated sRNA, including miRNA, has been the main application of sRNA sequencing. An extension of this procedure is to select molecules with modified ends. To do so, RNA is treated with modifying enzymes to convert the ends of the sRNAs of interest into 5′-P and 3′-OH. This strategy has been used to capture sRNAs that contain the 5′-cap modification to map transcription start sites at and outside of genes [16, 17], or study RNA polymerase pausing [18–20]. Another application involves selecting for molecules with normally modified 3′-ends to further explore the world of sRNA [11]. These and other modifications can be used in conjunction with the common sRNA library preparation procedure. Here, we outline purification of sRNAs, including isolation, preparation of libraries, sequencing considerations, and basic bioinformatic data analysis for monophosphorylated and 5′-capped sRNAs. Whenever possible, we refer the user to commercial kits and emphasize our observations and changes to the protocols that we have made based on experience.

2 Materials

2.1 Cell Culture Reagents

1. Growth media appropriate for cells of interest (e.g., DMEM/F-12 with 10% fetal bovine serum for MCF-7 breast adenocarcinoma cells).
2. Cell culture incubator at 5% CO_2; 0.25% Trypsin-EDTA; Phosphate-Buffered Saline (PBS; calcium-magnesium free, sterile).
3. Any additional supplements/antibiotics as needed for cells of interest: We do not normally use antibiotics in mammalian cell culture.
4. A refrigerated centrifuge supporting 15 or 50 mL centrifuge tubes.

2.2 RNA Extraction

1. Trizol or Trizol LS reagents (*see* **Note 1**).
2. Chloroform (*see* **Note 2**).
3. Ethanol (100%/96% and 70% (v/v)).
4. Nuclease-free water as well as co-precipitant carrier (for example, GlycoBlue co-precipitant, Ambion) for ethanol precipitation.
5. A refrigerated microcentrifuge.

6. Nuclease-free plastics. For all handling of RNA, libraries, and organic solutions, filter pipette tips are strongly recommended to avoid material carryover.

2.3 RNA Modifying Enzymes

Preparation of libraries for sRNAs containing 5′-P and 3′-OH ends, such as miRNA, does not need additional enzymes beyond those included in the sRNA library preparation kit. The following enzymes are used to enrich for sRNAs harboring 5′-end cap modifications: RNA 5′-polyphosphatase, Terminator 5′-monophosphate-depended RNA exonuclease (both available from Biosearch Inc. *(Lucigen)), a decapping enzyme (for example, RppH from New England Biolabs, NEB), Thermopol 10× buffer for RppH (NEB, sold separately), Calf Intestinal Alkaline Phosphatase (CIP), T4 Polynucleotide Kinase (PNK).

2.4 Library Preparation Reagents

1. An Illumina-compatible library kit for small RNA library preparation that is based on dual ligation and supports multiplexing (indexing), such as the NEBNext Small RNA Library Prep kit. Other kits, including those based on reverse transcriptase template switching or RNA self-circularization, may be used based on one's preference (*see* **Notes 3 and 4**).
2. Magnetic cleanup beads: solid-phase reversible immobilization (SPRI) beads (such as AMPure XP Beads, Beckman), if not included with the library preparation kit.
3. A magnetic rack for 200 μL and 1.7 mL tubes.

2.5 Size Selection Supplies (see Notes 5 and 6)

1. Precast 6% Tris Borate EDTA (TBE) gels such as Novex (Thermo).
2. TBE buffer.
3. Gel Breaker tubes (IST engineering or 0.5 mL Eppendorf tubes pierced at the bottom 3–4 times with an 18-gauge needle).
4. Centrifuge spin filters (such as cellulose acetate 0.22 μm centrifuge tube filters from Agilent or Corning) for elution from PAGE gels.
5. Elution buffer (Tris-EDTA + 300 mM NaCl).
6. A UV-transilluminator supporting ethidium bromide staining detection for manual gel selection.
7. a UV-transparent gel tray.
8. Metal razor blades, such as contractors' single-edge razor blades.
9. A gel imaging system for documentation.

2.6 Bioinformatics Data Analysis

1. UNIX workstations with 64 GB or more RAM, with at least 16 available processor cores.
2. Software needed for data analysis. For our sRNA data analysis, the following software should be installed on the workstations: quality control (FastQC [21] or MultiQC [22]); read

trimming (Trimmomatic [23] or Cutadapt [24]), aligning (STAR [25] or bowtie [26], BEDTools (bedtools genomecov) [27], deepTools (Bamcoverage, MultiBamSummary, MultiBigWigSummary [28]), peak calling (MACS2 [29] and/or HOMER software [30]), featureCounts [31], mirDeep2 [32]. In addition to data analysis, access to at least 1–2 TB of frequently backed up storage space is desirable (*see* **Note 7**).

3. A genome browser accessible on a personal computer, such as the Integrative Genomics Viewer (IGV) browser [33] and/or an account on the UCSC genome browser [34].

3 Methods

3.1 RNA Extraction

Before starting, pre-chill a microcentrifuge. Cells can be processed fresh or, alternatively, flash-frozen pellets can be stored at −80°C after removing media. Cells can be processed directly without washing. Collect fresh cells as a pellet and set on ice. If starting with frozen cells, do not defrost them and immediately proceed to RNA extraction.

1. Add Trizol and mix by vortexing or pipetting with a 1000 μL micropipette, taking care not to aspirate any material into the pipette tip filter (*see* **Note 8**). If vortexing, take care not to spill any phenol: keep the tube secured with your index finger while vortexing and change gloves if any leaks occur. Make sure to lyse cells fully so that the Trizol suspension is particulate-free and clear. This may take up to several minutes.
2. Follow the manufacturer's protocol for Trizol lysis, incubation, centrifugation and chloroform-induced phase separation steps of the RNA extraction protocol. Briefly, incubate the resuspended sample for 5 min on ice, add chloroform (1/5th of the regular Trizol volume used or 0.2 mL per 0.75 mL of Trizol LS), mix vigorously by shaking or briefly vortexing immediately after adding chloroform to each sample. After shaking, incubate for 2 min at room temperature and spin at maximum speed for 10 min in a refrigerated microcentrifuge.
3. Transfer the clear (aqueous) phase into a new 1.7 mL Eppendorf tube. To estimate the volume, use a 200 μL pipette set at 100 μL, transferring the aqueous phase in 100 μL steps to measure the volume. When transferring the aqueous phase into a new tube after centrifugation, avoid taking any of the white interphase that contains protein and DNA, or the pink phenolic phase.

4. To the aqueous phase, add 0.6 volumes of chloroform, vortex for 10 s, incubate at room temperature for 2 min and spin in the microcentrifuge for 2 min (*see* **Note 9**). Transfer the aqueous phase into a new tube, making sure to avoid taking any traces of chloroform with the aqueous phase (*see* **Note 10**). Note the approximate volume of the water phase.
5. Add 1 μL of GlycoBlue co-precipitant (or glycogen) to the water phase, close the tube and mix briefly by flicking, then add 3 volumes of 96–100% ethanol. After inverting several times to mix fully, incubate samples for at least 30 min at room temperature (*see* **Note 11**).

STOPPING POINT. Reactions can be stored at −20°C overnight.

6. Spin the tubes in a refrigerated microcentrifuge at maximum speed for 15 min. After the spin, a small but visible blue (if using GlycoBlue, preferred) or white (if using glycogen) pellet should form on the far side of the tube (*see* **Note 12**).
7. Withdraw ethanol by pipetting, add ~500 μL 70% ethanol and spin samples for 2 min at max speed in a refrigerated microcentrifuge (*see* **Note 13**).
8. Take out 70% ethanol by pipetting, keeping an eye on the pellet. Leave the last 10–50 μL of ethanol to avoid accidentally taking in the pellet.
9. Close the lid and spin the empty tubes for 5 s at maximum speed. Place tubes on a rack at room temperature with lids closed. Using a small (10 or 20 μL) pipette tip, take out all ethanol without disturbing the pellet. No traces of ethanol should be visible inside the tube at this point. Leave samples at room temperature with lids open. If 1–2 samples are processed, let the pellets dry at room temperature for 2–3 min before resuspending. If more than 1–2 samples are processed, there is no need to wait: after processing the last sample, the first one should be ready to resuspend (*see* **Note 14**).
10. Resuspend each pellet in a desired amount of nuclease-free water. Use 30–50 μL for RNA prepared from cell pellets or, if the entire prep is to be used for library preparation, in the maximum volume allowed by the kit (6 μL for the NEB Small RNA kit, for example) (*see* **Note 15**). Close the tube lid and place the sample on ice while resuspending the rest of the samples. Measure the concentration and assess purity and store resuspended RNA at −80°C until use (*see* **Notes 16 and 17**).
11. For preparation of RNA from exosomes, *see* **Notes 18 and 19**.

3.2 Measuring RNA

Measurements take 1–2 μL of an RNA sample.

1. If high yields of RNA are expected, usually >100 ng, such as when preparing RNA from whole cells, measure RNA using the Nanodrop or equivalent microvolume spectrophotometer, or a fluorescence-based method such as Qubit or equivalent.
2. If low yields of RNA are expected, such as after gel size selection or cap selection, do not measure RNA and instead use the entire sample for library preparation.

3.3 Enzymatic Selection

Treatment with RNA-modifying enzymes is used for the preparation of libraries that capture RNA enriched for ends other than 5′-P and 3′-OH. Below we give an example for the selection of RNAs containing 5′-cap and 3′-OH ends (*see* **Note 20**).

1. In a 1.7 mL Eppendorf tube, treat RNA with 1 μL RNA polyphosphatase in 20 μl of 1x polyphosphatase reaction buffer for 30 min at 37°C.
2. Add 300 μL Trizol reagent and run RNA purification as described above (see Sect. 3.1). Resuspend in 17 μL H_2O.
3. Add 2 μL 10× Buffer A and 1 μL Terminator exonuclease. Incubate for 60 min at 30°C.
4. Add 300 μL Trizol reagent and run RNA purification as described above (see Sect. 3.1). Resuspend in 45 μL H_2O.
5. Add 5 μL Thermopol buffer and 1 μL RppH enzyme, and incubate for 60 min at 30°C.
6. Add 300 μL Trizol reagent and run RNA purification as described above (see Sect. 3.1). Resuspend in nuclease-free water. The volume should be the maximum allowed by the next step in library preparation (6 μL for NEB Small RNA library kit, for example).
7. Transfer the entire sample into a 0.2 mL PCR tube and begin library preparation.

STOPPING POINT. The protocol can be stopped at an ethanol precipitation step (steps 2, 4, or 6 in the current section). After adding 70% ethanol, samples can be stored at –20°C for at least several days.

3.4 Library Preparation

To prepare small RNA libraries, follow the chosen kit protocol up to the PCR step (*see* **Notes 21 and 22**). Some additional considerations are helpful:

1. For easier handling, use 200 μL PCR tube strips rather than individual tubes and do not use more than 6 samples per 8-tube strip.
2. Use the lowest concentrations (highest dilution) of adapters allowed by the kit. For example, if the dilution option is specified for low amounts of starting RNA, use the option.

3. If real-time monitoring of amplification is used, such as for low or unknown starting amounts of RNA, tube caps should be switched to flat-top optical caps before qPCR.

3.5 Library PCR

Apart from amplification, the PCR step introduces indexes (also referred to as barcodes), which are 6–10 nt sequence identifiers usually present in PCR primers to uniquely tag each library. This enables pooling and sequencing of multiple libraries in the same flow cell (*see* **Notes 23 and 24**).

1. If sequencing a small number of samples, such as on a personal sequencer, maintaining index diversity helps avoid ambiguities in sample calling. Choose the index combinations per manufacturer's recommendations for the number of samples. If planning to use a large-scale sequencer, such as at a commercial or academic sequencing facility, consult with the facility about indexing prior to library preparation.
2. Use the material after reverse transcription to set up PCR with appropriate index primers using reagents from the kit.
3. If following the kit's recommended starting RNA amounts, such as, for example, for miRNA libraries prepared from known amounts of RNA, set up the PCR according to the kit recommendations.
4. If working with limiting or unknown amounts of RNA or using a custom protocol such as 5′-cap RNA selection or nuclear run-on assay, monitor library PCR in real time to avoid overamplification. To do so, add fluorescent dye (5% (v/v) of water-based EvaGreen or recommended dilution of SYBR Green dye) into the reaction. The extra 5% volume does not impact PCR even for PCR instruments with a 50 μL maximum supported reaction volume. Use the cycling conditions suggested by the kit manufacturer, except set the number of PCR cycles to 40. Monitor amplification after each PCR extension step (*see* **Note 25**). Stop before the reaction has reached ~50% of the maximum value (*see* **Notes 26 and 27**).
5. Analyze libraries by polyacrylamide TBE gel electrophoresis (Fig. 2) or capillary electrophoresis. PAGE can be readily combined with size selection. Capillary electrophoresis (such as the Bioanalyzer or Tapestation) is more expensive, but allows one to check the size distribution, expending only a small portion (< 10%) of the library material. Using a gel allows one to combine analysis with size selection by loading the entire library on a gel.

3.6 Post-PCR Cleanup and Size Selection

Small RNA library preps may contain empty adapter dimers with no insert. Empty adapter amplicons (~125 bp for Illumina adapters) are close in size to amplicons with inserts containing ~18 nt miRNAs (145–150 nt), making size selection beneficial.

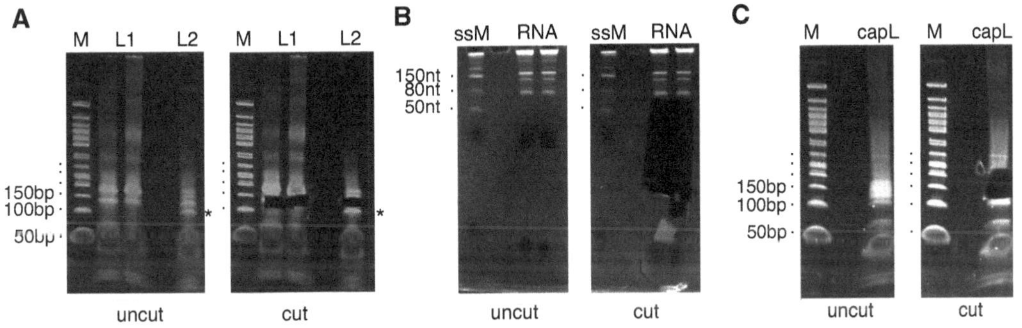

Fig. 2 Gel analysis and size selection of sRNA samples. (**A**) Separation of sRNA libraries prepared from whole cells using the NEB sRNA kit using total RNA as input without additional modifications. Libraries after PCR were resolved on a 6% TBE vertical PAGE alongside the 50-bp ladder and gel-stained with ethidium bromide. Library 2 (L2) but not library 1 (L1) contains empty ~125 bp dimers indicated with stars on the right. The target area containing the ~145 bp band containing mature miRNAs was cut off from the gel (right). Gel selection of sRNA libraries is not necessary, but is beneficial for in-depth mature miRNA profiling. (**B**) Separation of total RNA prepared with Trizol on a denaturing 15% urea-TBE polyacrylamide gel alongside the single-strand RNA ladder (left) and size selection of RNA ~20–70 nt (right). Size selection is beneficial for the preparation of libraries from 5′-capped sRNA. Care is taken to avoid the ~75 bp visible band. (**C**) The capped RNA library prepared from the RNA in (**B**) was resolved in a 6% PAGE. Care is taken to avoid the ~125 bp band corresponding to empty linker dimers

1. For libraries prepared with commercial kits using recommended amounts of total RNA from cells or nuclei, we have noted little to no adapter contamination. For these libraries, follow manufacturer-recommended steps to remove primers, usually using SPRI magnetic bead purification. Elute the library in the recommended volume of water or kit's elution buffer (usually 15–30 μL).
2. For libraries prepared with home-made reagents or commercial kit libraries using low amounts of material, perform gel size selection after PCR. Load each library onto 1–2 lanes of a 6% TBE gel (we use ready gels from Novex, Thermo). It is possible to load the entire 50 μL PCR mixed with 10 μL DNA loading buffer (such as 6× gel loading buffer, NEB) on two adjacent lanes of the gel. If concentrating, ethanol-precipitate the PCR reaction and resuspend in ~15 μL water plus 6× gel loading dye (NEB) (*see* **Notes 28 and 29**). Load a 100 bp or 50 bp dsDNA size ladder on the same gel.
3. Run the gel in 0.5–1× nuclease-free TBE at 180–200 V until the faster dye travels ~3/4 of the gel's height. Stop the run, disassemble the gel cassette and stain in TBE-ethidium bromide (~10 μL ethidium bromide per 50 mL TBE) with gentle shaking for 2 min. We use the lids of 1000 μL pipette tip boxes for staining these gels.

4. Transfer the gel onto a UV-transparent tray, such as a tray from agarose gel electrophoresis. Take a picture of the gel, including the marker, before cutting (*see* **Note 30**).
5. Cut out the desired portion of the gel under a transilluminator using a razor blade (*see* **Note 31**). If any of the target bands remain in the gel, cut them again and add to the existing sample. Take a picture of the gel before and after cutting (Fig. 2).
6. Transfer the gel slurry from a sample into a gel breaker tube inserted into a 2-mL Eppendorf tube, taking care not to drop the gel piece. Carry gel slices on the razor blade to the tube over the gel tray for easier retrieval if dropped. If the gel breaker tube cap does not close, push the gel pieces into the gel breaker tube with a pipette tip held by hand.
7. Close the gel breaker tube cap and collect the crushed gel material at the bottom of the 2 mL tube by spinning at maximum speed in a microcentrifuge for 2 min. Make sure no sample remains inside the gel breaker tube. If any sample remains, repeat the spin or extract the remaining pieces with a clean pipette tip.
8. Discard the gel-breaker tube and elute DNA with 400 μL of Gel Elution Buffer (TE + 300 mM NaCl) with gentle shaking at room temperature for 2 h. After incubation, transfer the slurry into a cellulose acetate membrane spin column (*see* **Note 32**).
9. To the eluate collected in the tube included with the spin column, add 30 μL of 5 M NaCl and 3 volumes (1200 μL) of ethanol, mix by inverting several times, incubate at room temperature for 10+ min, then recover DNA as done above for RNA. Resuspend the library in 15–20 μL TE or water.

STOPPING POINT. Libraries can be stored at −20°C for several days or at −80°C indefinitely.

3.7 Library Quality Control

1. Measure libraries by UV-absorbance or fluorescence. A typical yield of post-PCR library DNA is 200–1000 ng. If measuring UV absorbance, such as using Nanodrop, obtain the full spectrum, not just 260/280 readings (*see* **Note 33 and** Fig. 3).
2. Run 1 μL of the purified library on capillary electrophoresis, such as Bioanalyzer or Tapestation, to assess DNA size range. If capillary electrophoresis is not available, run ~100–300 ng of library on an agarose or polyacrylamide gel stained with a dye such as Ethidium Bromide, alongside a dsDNA size marker.

3.8 Sequencing

1. Prior to submitting libraries to a sequencing facility, determine whether the facility requires each library separately or could also accept combined libraries as a pool. (*see* **Note 34**).

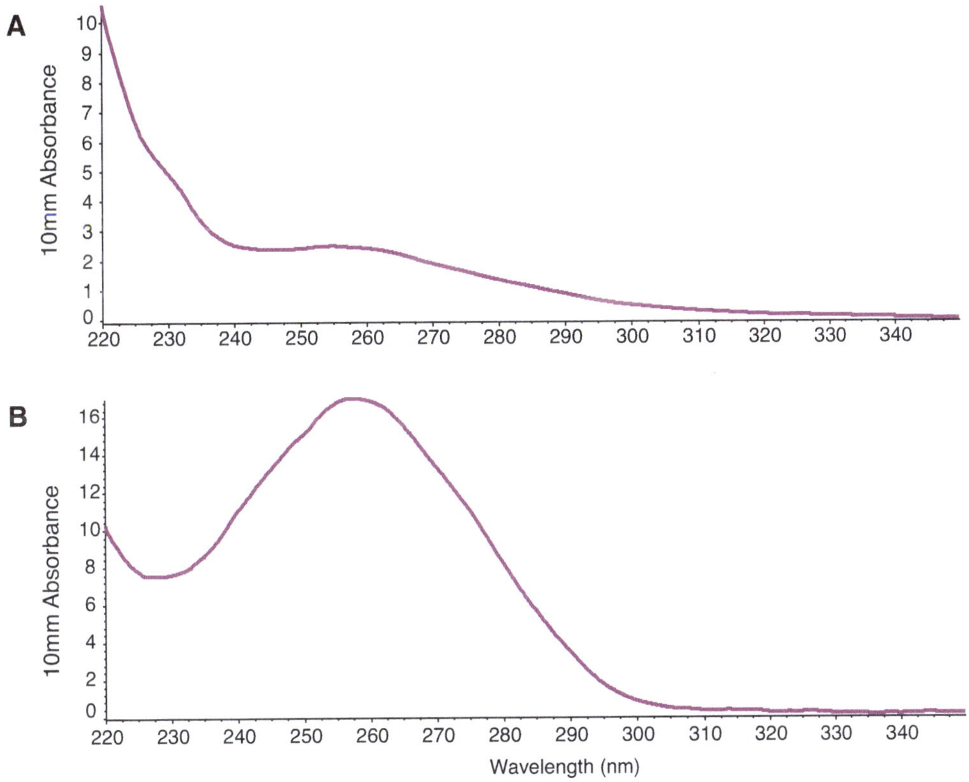

Fig. 3 DNA absorbance spectra after polyacrylamide gel purification. UV–VIS absorbance spectra of nucleic acids obtained with a nano-spectrophotometer (Nanodrop). (**A**) PAGE-purified sRNA library is shown after purification from a gel, ethanol precipitation and resuspension in water. (**B**) Pure nucleic acid (RNA) is shown for comparison. The library was used for sequencing without additional purification. The high absorbance area around 220–240 nm is reproducible for PAGE purification and does not affect quantitation or sequencing performance

2. Send the library to a sequencing facility along with the indexing information. For a remote facility, send libraries on dry ice using a common carrier (*see* **Notes 35 and 36**).

4 Bioinformatics

The overall scheme of sRNA data analysis includes technical quality assessment, adapter trimming, alignment to the genome and/or to known RNA species, and visualization (Fig. 4). Additional steps depend on the question at hand and may include high-resolution signal enrichment mapping, RNA quantitation, genomic feature enrichment, etc. These in-depth secondary analyses are not covered here in detail and should be co-developed with a bioinformatician.

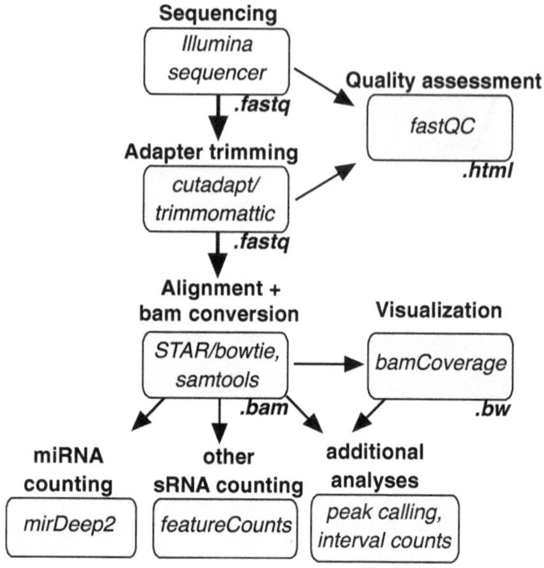

Fig. 4 A scheme of sRNA data analysis. An overall scheme of steps involved in a typical analysis of sRNA data. The tools we use commonly are listed inside the corresponding rectangle. The file type extension is shown underneath

4.1 Data Quality Control

1. For technical quality assessment, use fastQC or multiQC, such as in the following line.

   For Read 1:

```
fastqc -t 10 -o output_dir file_R1.fastq.gz
```

   For Read 2:

```
fastqc -t 10 -o output_dir file_R2.fastq.gz
```

   - The "-t" option specifies the number of CPU threads to use. Adjust the number of (10 in this case) based on your system's resources.
   - The -o option specifies the output directory to save FastQC reports to.
   - Replace file_R1.fastq.gz and file_R2.fastq.gz with your file name. You can use UNIX wildcards or loop through each file to process.

2. **Trim adapters from fastq files**. We use cutadapt or trimmomatic, although other tools can be used based on preference (*see* **Note 37**). To trim reads, collect the sequence information for adapters, 3′-end for R1 read trimming and (if desired) 5′-end for R2 file trimming. This information is available as part of the small RNA kit. An example command for cutadapt for NEB small RNA adapters is:

```
cutadapt -j 10 --minimum-length=10 -a ADAPTER_
SEQUENCE \
-o file.cutadapt.fastq.gz file.fastq.gz
```

- The "-j" option specifies the number of CPU threads to use (10 in this case).
- --minimum-length=10: This parameter discards any reads that are shorter than 10 bases after trimming.
- -a ADAPTER_SEQUENCE: This option specifies the actual adapter sequence to be trimmed from the reads. For NEB Small RNA kits, the argument to trim the R1 read is -a AGATCGGAAGAGCACACGTCTGAACTCCAGTCAC
- `-o file_R1.fastq.gz`: This specifies the output file where the trimmed reads will be saved.

Additional options are available with cutadapt and other tools. Refer to the software manuals and examples for more information.

3. **Run FastQC on trimmed reads.** After trimming, use FastQC to verify proper adapter trimming. The command below will run FastQC on all fastq files containing trimmed reads. We usually discard unpaired files generated by Trimmomatic:

```
fastqc -t 10 -o output_dir file.cutadapt.fastq.gz
```

4. Verify proper adapter trimming by examining FastQC output for the same samples before and after trimming. Proper trimming should leave no adapters, and the sequence length distribution should change to reflect the actual size of RNAs in the library (Fig. 5) (*see* **Notes 38 and 39**).

4.2 Alignment

Many sequence aligners are suitable for small RNA alignment. We show an example of STAR. Prior to aligning, a reference genome index must be available (*see* **Note 40**).

1. **Create an index for the genome (if one does not exist).**

```
STAR --runThreadN 10 --runMode genomeGenerate \
--genomeDir /path/to/STAR.index/ --genomeFasta-
Files /path/to/genome.fa \
--sjdbOverhang 50
```

The genome.fa file is a whole genome sequence with one line per chromosome. Whole genome sequences can be downloaded from many public sources. This is a stable link from the UCSC Genome Browser for the human hg38 genome: https://hgdownload.soe.ucsc.edu/goldenPath/hg38/bigZips/hg38.fa.gz. Unzip the file using the gunzip hg38.fa.gz command after downloading.

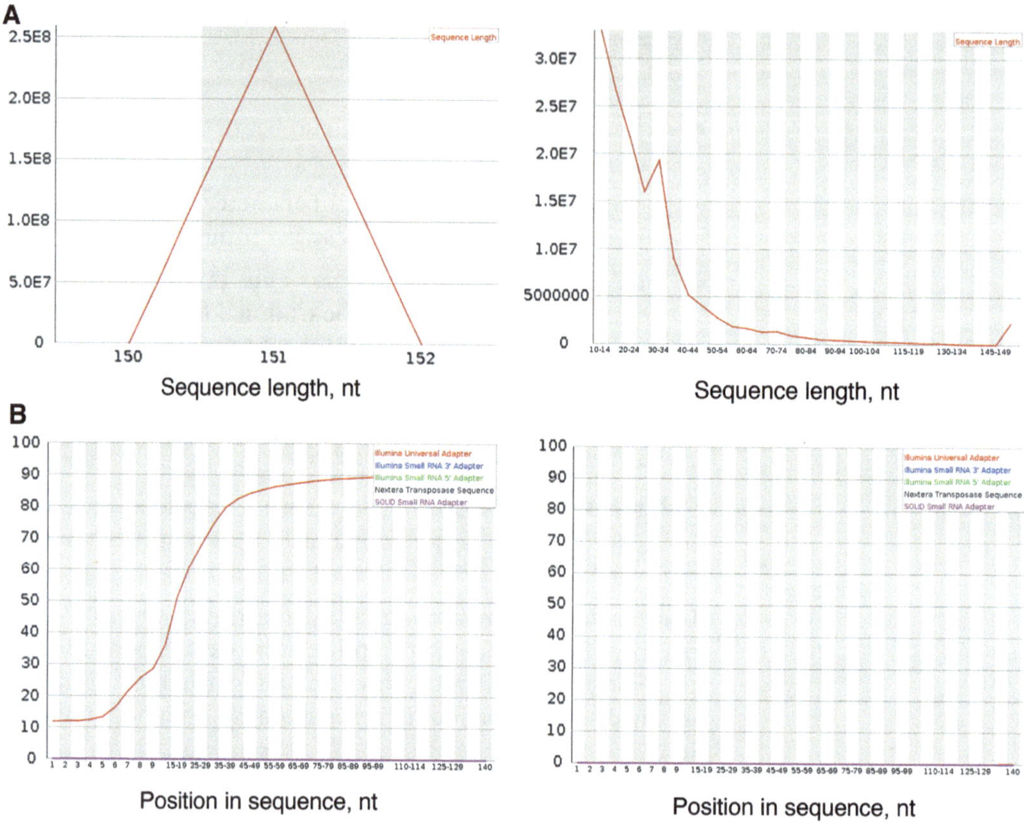

Fig. 5 Adapter trimming of sRNA data. Examples of fastQC output to assess adapter trimming of a sequenced small RNA library before and after trimming. The library was prepared from commercial (MCF-7) control exosomes (Systembio) using RNA extracted with the Norgen exosome RNA extraction kit. (**A**) The size distribution before trimming is uniform, as sequenced (left), and shows an abundance of short inserts (right). (**B**) Adapter content graph before trimming (left) and after trimming (right) shows successful removal of the sequencing adapter

2. **Align to the reference genome.**
   Start with the following command (*see* **Note 41**):

```
STAR --runThreadN 10 --genomeDir /path/to/your/
STAR.index/ \
--outFileNamePrefix file.cutadapt. --outFilterMis-
matchNoverLmax 0.05 \
--outFilterMatchNmin 10 --outFilterScoreMinOver-
Lread 0 \
--outFilterMatchNminOverLread 0 --alignIntronMax
1 \
--readFilesCommand zcat --readFilesIn file.cut-
adapt.fastq.gz
```

- --runThreadN 10: The number of threads (10) to use.
- --genomeDir /path/to/your/STAR.index/: The path to the folder where the STAR index for the appropriate genome is located.

- --outFileNamePrefix file.cutadapt.: The prefix for the output files. This is a quirk of STAR, as other aligners usually require the full name for output files.
- --outFilterMismatchNoverLmax 0.05: The maximum ratio of mismatches to read length, in this case 5% mismatches relative to the read length. For a typical miRNA, approximately one mismatch would be allowed.
- --alignIntronMax 1: This limits the maximum intron size allowed for the alignment to 1 base. This effectively disables spliced alignment. This option is often used when aligning reads from small RNA or sRNA sequencing, which generally do not contain introns.
- --readFilesCommand zcat: Tells STAR to use the zcat command to decompress the gz compressed input fastq.gz file.
- --readFilesIn file.cutadapt.fastq.gz: This specifies the input file

The following additional parameters are technical, and we do not change them:

- --outFilterMatchNmin 10: The minimum number of bases that must match for a read to be considered valid.
- --outFilterScoreMinOverLread 0: The value of 0 disables filtering by the minimum alignment score.
- --outFilterMatchNminOverLread 0: The value of 0 disables the filtering by the proportion of matched bases over the read length, meaning that even short matches relative to the read length will be allowed.

3. **Convert STAR output to bam format.**

```
samtools view -h -S -b -q 10 file.cutadapt.Aligned.
out.sam > \ file.cutadapt.Aligned.bam
rm file.cutadapt.Aligned.out.sam
```

samtools view parameters:

- -h: This option includes the header from the input SAM file (file.cutadapt.Aligned.out.sam) in the output BAM file. The header contains important information about the reference genome, read groups, and other metadata.
- -S: This flag tells Samtools that the input file is in SAM format. This flag is optional in more recent versions of Samtools, as it can usually infer the input format based on the file extension (.sam), but it ensures compatibility with older versions of Samtools.
- -b: This option specifies that the output should be in BAM format. BAM is the binary, compressed version of a SAM file, which is more efficient in terms of storage and processing speed.

- q 10: This flag filters the reads based on their mapping quality score (MAPQ). Specifically, it retains only the reads with a MAPQ score of 10 or higher, meaning that reads with lower confidence (those that align to multiple places or have poor quality) will be discarded.
- file.cutadapt.Aligned.out.sam: The input SAM file that contains the aligned reads after the initial alignment step (in this case, aligned reads from the file.cutadapt.Aligned.out.sam file).
- >: Redirects the output to a specified file. In this case, it directs the output BAM file to file.cutadapt.Aligned.q10.bam.
- file.cutadapt.Aligned.q10.bam: The output BAM file, which will contain only the reads with a mapping quality score of 10 or higher from the original SAM file. The result is a filtered BAM file in a compressed binary format.

The rm command deletes the sam file.

4. **Sort and index the bam file.**
   Run the following commands:

```
samtools sort file.Aligned.bam > file.sorted.bam
samtools index file.sorted.bam
rm file.Aligned.bam
```

   The sorted and indexed bam file along with its index .bai file are suitable for downstream applications. The original file.cutadapt.Aligned.bam file can be deleted.

4.3 Counting miRNA

For quantifying miRNAs, we use miRDeep2 tools, specifically quantifier.pl.

1. **Create bowtie index:**

```
bowtie-build /path/to/genome.fa /path/to/index/
genome
```

bowtie-build parameters.

- `/path/to/genome.fa` – path to the genome.fa file that is downloaded from a public repository (see above).
- `/path/to/index/genome` – path to and prefix of the genome index files. Several files will be generated, such as /path/to/index/genome.1.ebwt, etc. Only the prefix part (without .1.ebwt) should be specified.

2. **Run mapper.pl**

```
mapper.pl file.fastq -e -h -m -p /path/to/index/
genome -s file.fa \
-t file.arf
```

Mapper.pl parameters (*see* **Note 42**).

- file.fastq: unzipped fastq file. Use the trimmed fastq file.
- -e: specifies that input is in fastq format.
- -h: parse the input into fasta format.
- -m: collapse reads.
- -p: path to bowtie genome index, including prefix.
- -s output processed reads to a file (file.fa).
- -t output read mappings to a file (file.arf). This file is required for other tools in this software package.

3. **Run quantifier.pl**

```
quantifier.pl -p ./hairpin.fa -m ./mature.fa \
-r ./file.fa -t hsa
```

Quantifier.pl parameters:

- -p ./hairpin.fa: precursor miRNA sequences downloaded from the database. Download the latest revision from miRbase.org. Run the following command from the working directory: wget https://www.mirbase.org/download/hairpin.fa
- -m ./mature.fa: mature miRNA sequences downloaded from the database. Download the latest revision from miRbase.org. Run the following command from the working directory: wget https://www.mirbase.org/download/mature.fa
- -r ./file.fa: the user data fasta file generated by mapper.pl.
- -t hsa: selects miRNA from a certain species, in this case, human (hsa).

Quantifier.pl outputs counts and precursor information for each miRNA in a csv file that can be used in downstream applications, including comparisons between samples.

4.4 Counting Other Small RNA

This counting strategy is different from that of miRNA, as it relies only on the overlap with features in the annotation file. Tools such as featureCounts [31] or HTseq [35] can be used to quantify overlaps with annotated features. The following example uses featureCounts:

```
featureCounts -a small_RNAs.gtf -g gene_id -T 14 -O
-s 1 -M -o file_small_RNA_counts.txt file.sorted.bam
```

featureCounts parameters:

- -a small_RNAs.gtf: the path to the file containing annotations. In this example, the file small_RNAs.gtf is in the current directory (*see* **Note 43**).

- -g gene_id: the attribute field by which to group. The argument (in this case, gene_id) should match exactly those available in the GTF file (*see* **Note 44**).
- -T 14: the number of CPU threads.
- -O: this flag assigns reads that overlap with more than one annotated feature to each of those features.
- -s 1: specifies that the data are strand-specific, and the argument 1 indicates only the correct strand matches are counted.
- -o file_small_RNA_counts.txt: output file name with counts.
- file.sorted.bam: input bam file that is sorted and indexed. The index .bai file with the same name prefix should be in the same directory as the bam file.

The last two columns of the output file will contain the size of a feature (such as annotated RNA) and the read count overlapping with this feature. These numbers can be used for downstream applications.

4.5 Visualizing data

Visualizing in a genome browser helps assess the data, present examples of loci, and visually validate outcomes such as peak calling or differential abundance analyses.

1. **Generate .bw files.**

   We use bamcoverage. Small RNA sequencing retains the original RNA strand information. To accommodate strand coverage, run bamcoverage separately for each strand to generate forward and reverse strand .bw files. Use the following example commands:

```
bamCoverage -p 10 --filterRNAstrand forward --Off-
set 1 -bs 1 \
--bam file.sorted.bam -o file.forward.5prime.bw

bamCoverage -p 10 --filterRNAstrand reverse --Off-
set 1 -bs 1 \
--bam file.sorted.bam -o file.reverse.5prime.bw
```

   Bamcoverage parameters:

   - -p 10: the number of CPU threads. Adjust according to your computer.
   - --filterRNAstrand: only consider reads mapping to one of the strands (forward or reverse).
   - --Offset 1: this option returns only the first nucleotide of the read. This effectively defines the position of the original 5′-end of RNA (*see* **Note 45**).

2. **Visualize .bw files.**

   a. **Genome browsers**. The local (IGV) or cloud (UCSC) genome browsers can be used. The .bw files can be dropped directly onto the IGV browser window (Fig. 6). To show

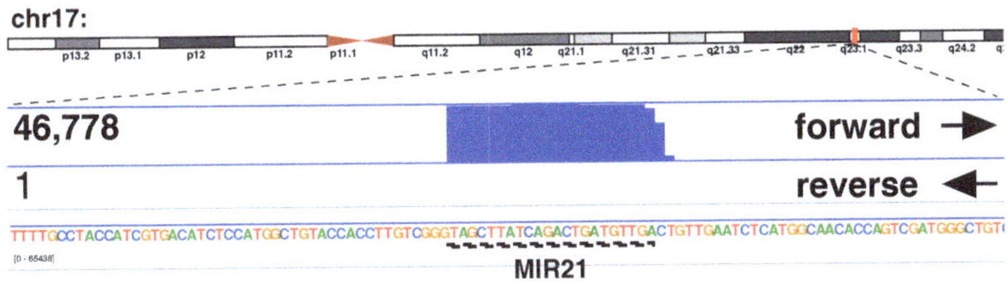

Fig. 6 Visualizing sRNA library data. A browser shot, made with IGV, was saved as a .svg file and edited with vector illustration software, showing a human hg38 genome region of chromosome 17 surrounding the miRNA21 locus. The graph shows read coverage with bin size 1 that was auto scaled independently for each strand. Numbers on the left show the relative signal for each strand, with the reverse strand having no sequence reads in this area. The mature miR21 sequence known to be in the forward strand in hg38 is underlined with a dashed line

.bw files on the UCSC browser, the files should be hosted on an ftp server. Alternatively, it is possible to upload bedgraph files, which can be generated with bamCoverage using *-of bedgraph* option.

4.6 Identifying regions with enriched signal

For uniquely aligned RNAs such as promoter-associated RNA, it may be useful to identify regions of their enrichment across the genome, which are commonly referred to as peaks. Identifying peaks in an unbiased way could reveal regulatory elements and/or transcription start sites. While peak calling is a common step in sequencing data analysis, small RNA (sRNA) data are inherently noisy, so that their distribution is not bell-shaped, but instead focused at individual nucleotides. Therefore, peak finders are generally not intended for identifying small RNA abundance. For miRNAs, we identify individual RNAs as described above (see Sect. 4.4).

While the exact strategy to define signal enrichment depends on a specific project, one possible way is to identify nucleotide positions (RNAs) containing more than a given threshold number (such as N>5 reads at the same position, depending on sequencing depth). For genome-associated RNAs, for example, for PRO-seq or for TSS-associated RNAs, we try peak callers such as macs2 or quantify the signal around annotated features. Check peak calling visually to verify expected results. For nascent RNA (de novo identification), a web-based tool that identifies regulatory regions (based on .bw files separated by strand) is available [36]. Regardless of the strategy, outputs of any peak finders should be checked against browser shots visually to verify peak calling is as expected.

5 Notes

1. RNA preparation based on guanidinium thiocyanate and phenol, such as Trizol or Trizol LS, gives the highest recovery of small RNAs. If RNA is not limiting, a column-based small RNA kit that recovers small RNAs, such as the Qiagen miRNeasy kit, may be used.
2. Many column-based kits may also use guanidinium thiocyanate and phenol-based reagents. If phenol/chloroform use is not acceptable, such as in an educational setting, an organic-free small RNA extraction kit, such as the Norgen microRNA (miRNA) purification kit, may be considered, although we have not tested these phenol-free kits extensively.
3. Preparation of libraries from sRNAs with naturally occurring 5′-P and 3′-OH ends, such as miRNA, does not need additional enzymes beyond those in the sRNA library preparation kit.
4. If using custom sRNA protocols such as for affinity-labeled RNA [37, 38], library preparation components, including T4 ssRNA ligase 1, linkers, reverse transcriptase and PCR reagents, can be ordered individually. Custom protocols may require purification or size selection after each ligation step.
5. For the preparation of miRNA libraries using the amounts of starting material that are specified in manufacturers' kits, such as >100 ng of total RNA, size selection is usually not necessary. Size selection is essential when low inputs or enzyme-selected material is used, particularly for the preparation of 5′-capped libraries.
6. A confounding problem for sRNA libraries is the formation of linker dimers that do not contain inserts, but reduce the sequencing yield for the library, and possibly interfere with libraries co-sequenced in the same flow cell. While commercial sRNA kits employ strategies to minimize empty dimers, lower amounts of input material result in more dimers, necessitating size selection.
7. To maintain workstations, some familiarity with UNIX administration is required, so for smaller groups and those new to bioinformatics, shared remote-access workstation(s) with preinstalled software and centrally managed, backed-up storage are the best option. For large-scale analysis, access to an institutionally maintained high-performance computing cluster is strongly advised. Even with improving cloud-based data analysis platforms such as Galaxy [39], we find that it is still beneficial to do some data analysis locally. We recommend that every user gain some familiarity with bioinformatics to

improve communication with bioinformaticians and to handle publicly available data more readily.

8. For smaller sample amounts (~10^5 cells), use regular Trizol, 300–500 μL per sample. For larger cell amounts, use Trizol LS, which allows one to run the entire preparation in a lesser volume to fit in one test tube, a factor when simultaneously handling multiple samples. Up to ~500 μL of Trizol LS (and ~1 mL of regular Trizol) usually allows the sample to be ethanol-precipitated in one Eppendorf 1.7 mL tube. Do not use 2 mL Eppendorf tubes to precipitate nucleic acids because pellets are less visible due to the shape of the tube. Use two 1.7 mL tubes instead if necessary.
9. This is a change from the manufacturer's protocol—it does not include chloroform extraction. We find that this step helps remove traces of water-dissolved phenol that impedes enzymatic reactions.
10. This step is important because chloroform often enters the pipette tip when transferring the last ~100 μL of the water phase. To minimize chloroform carryover while maximizing sample recovery, take the last ~50 μL of aqueous phase with a 200 μL pipette set at 100 μL, intaking air after 50 μL of liquid to the full 100 μL. Next, aspirate by pipetting the air (not the liquid) back and forth several times: ~1 μL of chloroform may then trickle down the pipette tip. Get rid of this chloroform by touching the wall of the original tube with the tip. Repeat this process several times until no chloroform is seen separating from the water phase inside the 200 μL pipette tip, then deposit the remaining water phase within the pipette tip into the rest of the sample, *without* touching the walls of the new tube with the tip.
11. The maximum volume to process in one 1.7 mL Eppendorf tube is 400 μL of aqueous phase. If a larger volume of water phase is obtained, split the aqueous phase into two tubes equally, add 1 μL of GlycoBlue into each and ethanol precipitate the content in each tube. Samples can be combined during the resuspension step. The additional GlycoBlue has no detrimental effect on subsequent steps.
12. Always spin samples in a centrifuge with the same side of the tube facing outward to see the pellets better. It may take some practice to see pellets at first. Always keep the pellet in sight while pipetting out the ethanol, as pellets can dislodge, especially with 70% ethanol. If the pellet is taken into a tip, carefully pipette it out, leaving some ethanol with it in the tube, then spin the tube for at least a minute at maximum speed. It helps to visualize pellets by placing a source of light in front and covering the bench with white surface, such as bench paper.

13. Sometimes the pellets are not visible because they are spread out over the surface of the tube. Always note the side of the tube on which the pellet is expected. In our experience, spinning with 70% ethanol often compacts the pellet, so it becomes visible.

14. This step eliminates the need for a DNA concentrator (Speedvac) or long air-drying on the bench, which helps avoid over-drying and cross-contamination between RNA and libraries.

15. Whole cell RNA pellets prepared from larger amounts of cells may be visible as a glassy pellet. Take care to dissolve the entire visible pellet without taking any pellet inside the pipette tip because RNA pellet tends to stick to the inside of pipette tips. After the pellet is dissolved, pipette the solution up and down the side of the tube on the same side above the pellet up to ~0.5 mL level in the tube. This collects any material that may have settled in the tube outside of the visible pellet. If no pellet is visible, collect material along the side of the tube as above. If using GlycoBlue, the properly resuspended sample should be visibly blue when viewed against a white surface.

16. For column-based RNA purification, RNA eluted from the column, usually in 30–50 μL, is ready to use. The UV–Vis spectrum of this RNA may have an extra density toward shorter wavelengths: this does not affect subsequent steps negatively. To concentrate, RNA may be precipitated with ethanol and resuspended in a desired volume of nuclease-free water as described above.

17. If using a Qiagen kit or a kit with similarly shaped columns, the spin columns are shaped such that 1–2 μL of the wash buffer containing ethanol remains at the rim of the column and might be carried over into the sample during spin elution. To mitigate this, after spinning the washed empty column for 1 min at max speed per manufacturer's protocol, collect any remaining liquid around the rim of the column (without touching the white membrane) with the smallest pipette (10 μL tip) prior to eluting.

18. Exosomes are small (~30–1000 nm) lipid particles that are secreted by at least some cells and are considered a means of cell-to-cell communication [40, 41]. The RNA content in exosomes is variable, and the available amounts of exosomes is typically low, so that typical exosome preps done from 1 to 10 mL of cell-free serum contain on the order of nanograms of RNA, compared to micrograms for a typical RNA prep from cells. These low amounts are technically limiting for sRNA library preparation. Furthermore, exosome quantitation is not well standardized and is commonly reported as numbers

of particles or amounts of total protein across studies. Given the uncertainties and variability, sequencing of sRNA from exosomes remains challenging. For preparation of RNA from exosomes, try the protocols that maximize RNA recovery and tolerate the lowest amounts of starting material. In addition, always include negative controls, including those for reagents used for library preparation.

19. Low expected RNA yield from exosomes requires additional negative controls to verify the suitability of reagents for preparation, including commercial kits. For example, two commercial kits gave a background signal from their buffers, albeit at lower levels than exosomes (Fig. 7).

20. Further adjustments of enzymatic steps can be used to recover RNA species with desired combinations of 5′- and 3′-ends (for example, see refs. [4, 11]).

21. If a custom library protocol is used instead of a kit, perform ligation 1, ligation 2, and RT-PCR steps (we use SSRT III reverse transcriptase enzyme) as recommended by the enzyme manufacturers. Adapters should be HPLC-purified RNA oligos or DNA-RNA-chimeras with at least 3 bases proximal to the ligation side being RNA bases. The 3′-adapter can be ordered monophosphorylated or pre-adenylated at the 5′-end and should be blocked at the 3′-end. The use of the pre-adenylated 3′-adapter allows one to ligate the 3′-end of

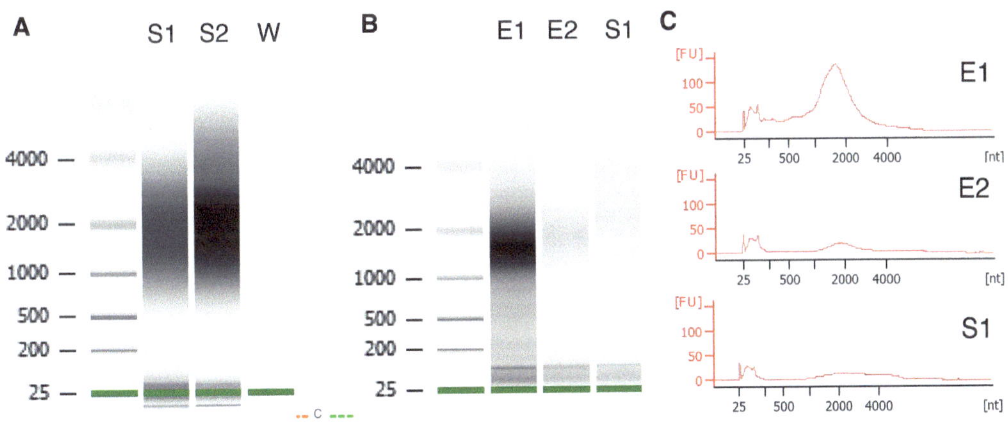

Fig. 7 Testing organic-free kits for RNA exosome preparation. (**A**) Bioanalyzer Pico traces of nuclease-free water used for mock-purification of RNA using the indicated exosome RNA commercial purification kits: Norgen Exosomal RNA purification kit (S1) and SBI Exosome RNA purification kit (S2). Samples were ethanol-precipitated in the presence of GlycoBlue and resuspended in 6ul of water. Water ethanol-precipitated with GlycoBlue is also shown as a negative control (W). (**B**) RNA purified from two preps of patient-derived exosomes (E1 and E2) is resolved alongside the same mock control S1 as in (**A**). Signal is scaled to the highest-signal sample and are therefore comparable. This is not a default Bioanalyzer setting. (**C**) The same Bioanalyzer samples are shown as trace graphs scaled to the highest-signal sample

RNA without ATP by using the T4 RNA ligase 2 truncated (without ATP) [42] to mitigate empty adapter dimer contamination. The 5′-end adapter should be blocked at the 5′-end and contain the 3′-OH group. Detailed sequences of Illumina-compatible adapters are provided elsewhere (for example, [37]).

22. The T4 RNA ligase 1 is sensitive to the identity of the proximal 2–3 bases around the ligation site [43]. Therefore, the use of different kits that may have slightly different linkers may lead to representation biases. For example, Fig. 8 shows differences between NEB (double ligation) and Zymo (self-circularization) kits used to prepare libraries from the same RNA.

23. Small RNA library kits include PCR reagents, so no additional reagents are needed for miRNA libraries. For homemade libraries, we have successfully used high-fidelity PCR enzymes such as Phusion DNA polymerase (Thermo, NEB), or an equivalent SsoAdvanced enzyme (Bio-Rad), with the manufacturers' reaction buffers. For dNTPs, mix equal

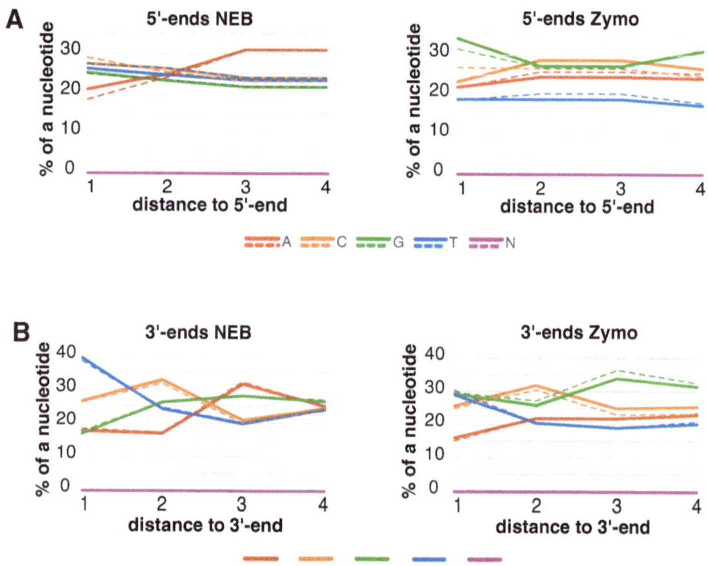

Fig. 8 Nucleotide bias of sRNA data. Comparison of nucleotide content adjacent to adapters was made by preparing sRNA I libraries from the same control RNA sample provided in the Zymo Small RNA Library Prep kit, prepared using the NEB Small RNA library kit (NEBNext Small RNA Library Prep kit, NEB) and the Zymo small RNA library kit (Zymo-Seq miRNA library kit, Zymo). The numbers on the *x*-axes denote the distance from the adapter, i.e., the nucleotide immediately next to the adapter is marked as 1. (**A**) Comparison of 5′-end sides and (**B**) comparison of 3′-end sides of an insert. *y*-axes represent the percentage of reads with an indicated nucleotide in this position. Solid and dashed lines indicate two independent preparations of libraries from the same RNA to show reproducibility

volumes of individual 100 mM stocks and use the resulting solution containing 25 mM each dNTP at 1:125 in PCR. For monitoring amplification of libraries by real-time PCR, add double-stranded DNA-specific intercalating fluorescent dye [such as SYBR Green or Eva Green Dye (20× in water)].

24. The selection of indices is an important step. While indexing is commonly done by core facilities that prepare libraries, small RNA libraries are more likely to be made in the lab, and therefore, indexing is likely to be done by the user. Initially, indexes were available only on one adapter for any type of library. Dual indexing increases the number of possible combinations and thus allows for combining more samples and may reduce sample mis-identification. However, dual indexing for small RNA libraries is not yet widely available [44].

25. Libraries should not take more than 15 cycles for abundant miRNA and 21–22 cycles for low concentration 5'-capped RNAs. To establish this value, we calibrate the instrument by running a positive control reaction with the same volume and amount of dye (usually 50 μL), confirming the calibration approximately every 6 months.

26. When interrupting a reaction, all tubes should be withdrawn promptly before the next cycle's denaturation step has begun. After withdrawing the sample, incubate it at 72°C for 2 min in a separate PCR machine or heating block with a heated lid that holds 200 mL tubes. If only some samples have reached the threshold, take out all samples, separate the desired sample(s), and place the rest back into the PCR machine to resume cycling. Reactions can be kept on ice until all are completed. Pausing mid-run has little to no effect on amplification of the remaining samples.

27. Interrupting (soft pausing without starting the next step) is not usually supported and may not be physically possible in a given PCR instrument. If real-time monitoring and interrupting of PCR is not feasible, it is possible to amplify a portion of the library in a qPCR to completion and then back calculate the number of cycles for the rest of the reaction [37].

28. Avoid loading dyes containing Bromophenol Blue because it overlaps and obscures the sRNA library size range on 6% TBE gels. We use the UV-transparent 6×Gel Loading Dye, Purple (NEB) instead of Bromophenol Blue-containing dyes.

29. Because libraries contain unique indexes, cross-contamination between libraries is not a concern, and libraries can be loaded onto the same gel on adjacent lanes. Do not combine libraries prior to gel loading, as this will not allow one to identify failed libraries.

30. To avoid damaging the material, take care to minimize the time the samples are under UV light: turn off the UV light immediately after picture taking. For manual cutting, make quick marks under UV light, then turn off the UV light and finish excising the marked gel area. It helps to slice the gel pieces into thinner strips that fit into a gel breaker tube, which may be done on the bench without the UV light.
31. We use contractors' single-edge stainless steel razor blades purchased from a hardware store. Because libraries are uniquely indexed, using the same razor blade for all libraries is okay in such a case.
32. Decant the slurry into the spin column, taking care not to spill any. If any material remains in the tube, scoop it into the spin column with a 200 μL pipette tip. To separate the gel from the eluate, spin the column at 2000 × *g* for 2 min. After the spin, close to 400 μL of slurry-free liquid should be collected in the tube. If a lesser volume is seen, the gel may have blocked the membrane. If this happens, gently stir the slurry in the spin column with a clean pipette tip and repeat the spin. Do not spin at high speed as this compresses the gel and blocks the membrane.
33. UV-VIS spectrophotometry gives adequate measurements only for purified DNA, such as after SPRI beads or gel purification, not for raw or ethanol-precipitated PCR reactions. The UV absorption spectrum of DNA extracted from a polyacrylamide gel may be different (Fig. 3). This difference is expected and does not affect quantitation or sequencing.
34. Illumina instruments are currently the practical option for sRNA sequencing. Small RNA libraries can be sequenced on low-throughput or large-scale sequencers, either at an academic or commercial facility. Each sequencing facility has its own way of quantifying and pooling libraries. Usually, fluorescence-based methods work most consistently, and capillary electrophoresis reliably gives size distribution, so we usually have the Core facility perform both steps prior to sequencing. While digital droplet PCR (ddPCR) is excellent for quantitation of libraries, we find it expensive for routine use.
35. We usually send half of each library and retain the rest as backup. This amount is abundantly sufficient for sequencing and re-sequencing if necessary. The actual amounts required for sequencing are minimal. A typical 10 ng/μL library with the average fragment size of 150 nt is at 100 nM. For comparison, the minimum amount to be submitted to one sequencing facility is 15 μL of 4 nM library (or each library in the pool). We aim to submit more than the minimum

requirements, such as 20 μL of 10 nM library or each library in the pool.

36. While any short-read sequencing instrument is capable of sRNA sequencing, we find that a high-throughput Illumina NovaSeq series instrument is the most cost-effective because the run is shared with many other samples from multiple users. Furthermore, a large-scale instrument allows the sequencing facility to adjust the sequencing depth of individual samples and bill the user flexibly, per million reads. We have had success both with commercial and academic facilities, wherein sRNA samples are sequenced together with other samples of different types. While the size of sRNA library DNA (insert plus linker) is smaller than the most frequently used Illumina sequencing length (150 nt), we find that sequencing sRNA libraries at 150 nt does not affect the quality of sequencing.
37. Unlike other types of libraries, we rarely use paired-end format for sRNAs and instead only process either R1 or R2 data files (usually R1) in single-end mode.
38. Some small RNA libraries may contain linkers with extra nucleotides and/or unique molecular identifiers as part of the sequence. These sequence features need to be handled separately. Clipping of the first or last several nucleotides can be done with trimmomatic or cutadapt.
39. It may be helpful to verify how reads map to the expected features (such as the target genome) by manually pasting 2–5 individual read sequences into the blat tool [45] that is accessible, for example, through UCSC Genome Browser. This shows which part of the read matches the genome and whether the match is as expected. Manual examination is especially advisable the first time a new protocol is used.
40. One can reuse the genome index for different projects and share it among users in a commonly accessible location. Care should be taken to make sure the index is appropriate for your project.
41. Alignment metrics, including percent unique alignments, should be recorded and reported for each sample.
42. The trimmed fastq file must be unzipped for this step because the bowtie aligner, which is used specifically by this tool, is an older tool that can only use uncompressed input files. Uncompressing can be done using the gunzip file.fastq.gz command. Compress the fastq file after running the script to save space by running the gzip file.fastq command.
43. The files are usually obtained from the database. For example, an exhaustive list of annotations for the human hg38 genome is downloaded with the command:

```
wget https://ftp.ensembl.org/pub/release-111/gtf/
homo_sapiens/Homo_sapiens.GRCh38.111.gtf.gz
```

After downloading, uncompress the file:

```
gunzip Homo_sapiens.GRCh38.111.gtf.gz
```

The file may need to be filtered to extract specific classes of genes, such as snoRNA, and formatted to avoid errors. Consult a bioinformatician for details.

44. To group by a more human-readable gene_name attribute, the GTF from the database may give error messages and would need to be formatted so that every lane includes gene_name and gene_id attributes. Consult a bioinformatician for details.

45. Another useful option is --Offset -1, which returns the last nucleotide of the read. If the RNA size is shorter than the sequencing read length, this defines the 3′-position of the original RNA. The current example uses R1 reads, so that the first nucleotide corresponds to the 5′-end of the original RNA. If paired-end sequencing is available, it is possible to use R2 reads, in which the first nucleotide extracted with --Offset 1 option defines the 3′-end of the original RNAs.

Acknowledgments

This work was supported by NSF award 1750379 (S.N.) and NIH R01AG056881 (Y.P.R.J.). We thank the UND Genomics core for assistance with sequencing and analysis of the small RNA samples. The Genomics core is supported by the NIH awards U54GM128729 and P20GM104360.

References

1. Lau NC, Lim LP, Weinstein EG et al (2001) An abundant class of tiny RNAs with probable regulatory roles in *Caenorhabditis elegans*. Science 294(5543):858–862. https://doi.org/10.1126/science.1065062
2. Affymetrix, ENCODE Transcriptome Project, Cold Spring Harbor Laboratory, ENCODE Transcriptome Project (2009) Post-transcriptional processing generates a diversity of 5’-modified long and short RNAs. Nature 457(7232):1028–1032. https://doi.org/10.1038/nature07759
3. Taft RJ, Glazov EA, Cloonan N et al (2009) Tiny RNAs associated with transcription start sites in animals. Nat Genet 41(5):572–578. https://doi.org/10.1038/ng.312
4. Nechaev S, Fargo DC, dos Santos G et al (2010) Global analysis of short RNAs reveals widespread promoter-proximal stalling and arrest of Pol II in Drosophila. Science 327(5963):335–338. https://doi.org/10.1126/science.1181421
5. Zhang Z, Zhang J, Diao L et al (2021) Small non-coding RNAs in human cancer: function, clinical utility, and characterization. Oncogene 40(9):1570–1577. https://doi.org/10.1038/s41388-020-01630-3
6. Kim T, Croce CM (2023) MicroRNA: trends in clinical trials of cancer diagnosis and therapy strategies. Exp Mol Med 55(7):1314–1321. https://doi.org/10.1038/s12276-023-01050-9

7. Lu J, Getz G, Miska EA et al (2005) MicroRNA expression profiles classify human cancers. Nature 435(7043):834–838. https://doi.org/10.1038/nature03702
8. Ivey KN, Srivastava D (2010) MicroRNAs as regulators of differentiation and cell fate decisions. Cell Stem Cell 7(1):36–41. https://doi.org/10.1016/j.stem.2010.06.012
9. Volinia S, Calin GA, Liu C-G et al (2006) A microRNA expression signature of human solid tumors defines cancer gene targets. Proc Natl Acad Sci USA 103(7):2257–2261. https://doi.org/10.1073/pnas.0510565103
10. Landgraf P, Rusu M, Sheridan R et al (2007) A mammalian microRNA expression atlas based on small RNA library sequencing. Cell 129(7):1401–1414. https://doi.org/10.1016/j.cell.2007.04.040
11. Lai H, Feng N, Zhai Q (2023) Discovery of the major 15–30 nt mammalian small RNAs, their biogenesis and function. Nat Commun 14(1), 5796. https://doi.org/10.1038/s41467-023-41554-6
12. Metcalf GAD (2024) MicroRNAs: circulating biomarkers for the early detection of imperceptible cancers via biosensor and machine-learning advances. Oncogene 43(28):2135–2142.https://doi.org/10.1038/s41388-024-03076-3
13. Liu S, Huang J, Zhou J et al (2024) NAP-seq reveals multiple classes of structured noncoding RNAs with regulatory functions. Nat Commun 15(1), 2425. https://doi.org/10.1038/s41467-024-46596-y
14. Michael MZ (2006) Cloning microRNAs from mammalian tissues. Methods Mol Biol 342189–342207. https://doi.org/10.1385/1-59745-123-1:189
15. Kwon YS (2011) Small RNA library preparation for next-generation sequencing by single ligation, extension and circularization technology. Biotechnol Lett. https://doi.org/10.1007/s10529-011-0611-y
16. Duttke SH, Chang MW, Heinz S et al (2019) Identification and dynamic quantification of regulatory elements using total RNA. Genome Res 29(11):1836–1846. https://doi.org/10.1101/gr.253492.119
17. Policastro RA, Zentner GE (2021) Global approaches for profiling transcription initiation. Cell Rep Methods 1(5):100081. https://doi.org/10.1016/j.crmeth.2021.100081
18. Henriques T, Scruggs BS, Inouye MO et al (2018) Widespread transcriptional pausing and elongation control at enhancers. Genes Dev 32(1):26–41. https://doi.org/10.1101/gad.309351.117
19. Scheidegger A, Dunn CJ, Samarakkody A et al (2019) Genome-wide RNA pol II initiation and pausing in neural progenitors of the rat. BMC Genomics 20(1), 477. https://doi.org/10.1186/s12864-019-5829-4
20. Day DS, Zhang B, Stevens SM et al (2016) Comprehensive analysis of promoter-proximal RNA polymerase II pausing across mammalian cell types. Genome Biol 17(1), 120. https://doi.org/10.1186/s13059-016-0984-2
21. Andrews S (2010) FastQC: a quality control tool for high throughput sequence data. Available online at: http://www.bioinformatics.babraham.ac.uk/projects/fastqc
22. Ewels P, Magnusson M, Lundin S et al (2016) MultiQC: summarize analysis results for multiple tools and samples in a single report. Bioinformatics 32(19):3047–3048. https://doi.org/10.1093/bioinformatics/btw354
23. Bolger AM, Lohse M, Usadel B (2014) Trimmomatic: a flexible trimmer for Illumina sequencedata.Bioinformatics30(15):2114–2120. https://doi.org/10.1093/bioinformatics/btu170
24. Martin M (2011) Cutadapt removes adapter sequences from high-throughput sequencing reads. EMBnet J 17(1). https://doi.org/10.14806/ej.17.1.200
25. Dobin A, Davis CA, Schlesinger F et al (2013) STAR: ultrafast universal RNA-seq aligner. Bioinformatics 29(1):15–21. https://doi.org/10.1093/bioinformatics/bts635
26. Langmead B, Trapnell C, Pop M et al (2009) Ultrafast and memory-efficient alignment of short DNA sequences to the human genome. Genome Biol 10(3), R25. https://doi.org/10.1186/gb-2009-10-3-r25
27. Quinlan AR, Hall IM (2010) BEDTools: a flexible suite of utilities for comparing genomic features. Bioinformatics 26(6):841–842. https://doi.org/10.1093/bioinformatics/btq033
28. Ramírez F, Ryan DP, Grüning B et al (2016) deepTools2: a next generation web server for deep-sequencing data analysis. Nucleic Acids Res 44(W1):W160–5. https://doi.org/10.1093/nar/gkw257
29. Zhang Y, Liu T, Meyer CA et al (2008) Model-based analysis of ChIP-Seq (MACS. Genome Biol 9(9), R137. https://doi.org/10.1186/gb-2008-9-9-r137
30. Heinz S, Benner C, Spann N et al (2010) Simple combinations of lineage-determining transcription factors prime cis-regulatory elements required for macrophage and B cell identities. Mol Cell 38(4):576–589. https://doi.org/10.1016/j.molcel.2010.05.004
31. Liao Y, Smyth GK, Shi W (2014) featureCounts: an efficient general purpose program for assigning sequence reads to genomic features. Bioinformatics 30(7):923–930. https://doi.org/10.1093/bioinformatics/btt656

32. Friedländer MR, Mackowiak SD, Li N et al (2012) miRDeep2 accurately identifies known and hundreds of novel microRNA genes in seven animal clades. Nucleic Acids Res 40(1):37–52. https://doi.org/10.1093/nar/gkr688
33. Robinson JT, Thorvaldsdóttir H, Winckler W et al (2011) Integrative genomics viewer. Nat Biotechnol 29(1):24–26. https://doi.org/10.1038/nbt.1754
34. Raney BJ, Barber GP, Benet-Pagès A et al (2024) The UCSC Genome Browser database: 2024 update. Nucleic Acids Res 52(D1): D1082–D1088. https://doi.org/10.1093/nar/gkad987
35. Anders S, Pyl PT, Huber W (2015) HTSeq—a Python framework to work with high-throughput sequencing data. Bioinformatics 31(2):166–169. https://doi.org/10.1093/bioinformatics/btu638
36. Danko CG, Hyland SL, Core LJ et al (2015) Identification of active transcriptional regulatory elements from GRO-seq data. Nat Methods 12(5):433–438. https://doi.org/10.1038/nmeth.3329
37. Mahat DB, Kwak H, Booth GT et al (2016) Base-pair-resolution genome-wide mapping of active RNA polymerases using precision nuclear run-on (PRO-seq). Nat Protoc 11(8):1455–1476. https://doi.org/10.1038/nprot.2016.086
38. Tome JM, Tippens ND, Lis JT (2018) Single-molecule nascent RNA sequencing identifies regulatory domain architecture at promoters and enhancers. Nat Genet 50(11):1533–1541. https://doi.org/10.1038/s41588-018-0234-5
39. Community G (2024) The Galaxy platform for accessible, reproducible, and collaborative data analyses: 2024 update. Nucleic Acids Res 52(W1):W83–W94. https://doi.org/10.1093/nar/gkae410
40. Valadi H, Ekström K, Bossios A et al (2007) Exosome-mediated transfer of mRNAs and microRNAs is a novel mechanism of genetic exchange between cells. Nat Cell Biol 9(6):654–659. https://doi.org/10.1038/ncb1596
41. Vestad B, Llorente A, Neurauter A et al (2017) Size and concentration analyses of extracellular vesicles by nanoparticle tracking analysis: a variation study. J Extracell Vesicles 6(1):1344087. https://doi.org/10.1080/20013078.2017.1344087
42. Ho CK, Wang LK, Lima CD et al (2004) Structure and mechanism of RNA ligase. Structure 12(2):327–339. https://doi.org/10.1016/j.str.2004.01.011
43. Jayaprakash AD, Jabado O, Brown BD et al (2011) Identification and remediation of biases in the activity of RNA ligases in small-RNA deep sequencing. Nucleic Acids Res 39(21):e141. https://doi.org/10.1093/nar/gkr693
44. Persson H, Søkilde R, Pirona AC et al (2017) Preparation of highly multiplexed small RNA sequencing libraries. Biotechniques 63(2):57–64. https://doi.org/10.2144/000114574
45. Kent WJ (2002) BLAT–the BLAST-like alignment tool. Genome Res 12(4):656–664. https://doi.org/10.1101/gr.229202

Chapter 14

Isolation of RNA and MicroRNAs from Formalin-Fixed, Paraffin-Embedded Tissues (FFPE), and Cultured Cells for Subsequent Applications

Sofia Marotta, Stella Mrmić, Lucia Mundo, Tommaso Salamone, Eugenio Santacroce, Pankaj Trivedi and Eleni Anastasiadou

Abstract

MicroRNAs (miRNAs) are small, non-coding RNAs that play critical roles in post-transcriptional gene regulation and are involved in various biological processes, including cancer progression. Efficient extraction of miRNAs from cultured cells and tissue samples is crucial for downstream applications, including quantitative real-time PCR (qRT-PCR), next-generation sequencing (NGS), and microarray analysis. This chapter provides reliable protocols for extracting high-quality RNA and microRNAs from formalin-fixed paraffin-embedded (FFPE) tissues and cultured cells. It compares manual and automated methods to guide optimal kit selection. The protocols support reproducible results from challenging samples in research and clinical settings.

Keywords RNA extraction, miRNA isolation, RNA&miRNA integrity, FFPE tissues, Cancer cell lines, Bioanalyzer, TapeStation, Qubit fluorometer, RNA stabilization, RNase contamination prevention

1 Introduction

MicroRNAs (miRNAs) are small, non-coding RNA molecules, about 18–25 nucleotides in length, that play a pivotal role in post-transcriptional gene regulation [1]. Their discovery has revealed a previously unrecognized layer of gene expression control, influencing various cellular processes, including differentiation, proliferation, apoptosis, and immune responses [2]. In cancer research, miRNAs have emerged as essential biomarkers for diagnosis, prognosis, and therapeutic interventions, functioning as tumor suppressors or oncogenes [3, 4].

Victor Ambros and Gary Ruvkun's groundbreaking work in the early 1990s led to the identification of the first miRNA, lin-4, and its target, the lin-14 3′ untranslated region (UTR), in *Caenorhabditis*

These authors contributed equally: Sofia Marotta, Stella Mrmić and Lucia Mundo.

Archana Dhasarathy and Motoki Takaku (eds.), *Cancer Epigenetics*, Methods in Molecular Biology, vol. 3005, https://doi.org/10.1007/978-1-0716-5072-1_14,

elegans [5, 6]. Ruvkun's subsequent discovery of let-7, a conserved miRNA across species, underscored the fundamental role of miRNAs in gene regulation. In recognition of their seminal contributions, Ambros and Ruvkun were jointly awarded the 2024 Nobel Prize in Physiology or Medicine for their discovery of microRNA and its role in post-transcriptional gene regulation [7].

The extraction and analysis of miRNAs from biological samples are essential for understanding their functions and implications in diseases, particularly cancer [8–12]. However, isolating miRNAs presents unique challenges due to their small size, low abundance, and susceptibility to degradation. Efficient extraction methods are crucial for separating miRNAs from other RNA species while preserving their integrity for downstream applications such as quantitative real-time PCR (qRT-PCR), microarray analysis, and next-generation sequencing (NGS) [13, 14].

Isolating miRNAs from cultured cells is relatively straightforward, while extraction from solid tissues requires additional considerations, including effective homogenization, RNA stabilization, and removal of protein and genomic DNA contaminants [15, 16]. Various techniques have been developed to optimize miRNA isolation, including phenol-chloroform extraction (e.g., TRIzol), silica column-based kits, and magnetic bead-based methods [17]. The choice of method depends on factors such as sample type, RNA purity requirements, and intended downstream applications.

This chapter presents optimized protocols for extracting miRNA from both cultured cells and solid tissues, with a focus on ensuring high-quality RNA yields, preventing degradation, overcoming tissue-specific processing challenges, and selecting the most suitable methods based on experimental needs. Additionally, it provides troubleshooting strategies to address common issues such as low yield, RNA degradation, and contamination, thereby helping researchers obtain reliable and reproducible results.

2 Materials

The following reagents and equipment are required to extract high-quality RNA-containing miRNAs from cell lines and paraffin-embedded tissues. Here, we describe a protocol for RNA extraction from human-derived cancer cell lines (i.e., B-cell lymphoma and ovarian cancer cell lines) using the Norgen Total RNA Purification Plus Kit and two FFPE-specific extraction kits: Maxwell CSC RNA FFPE and Qiagen miRNeasy FFPE. However, other commercially available kits can provide comparable results, depending on the application.

2.1 Cell Culture Buffers/Solutions

1. Complete medium RPMI1640, 10% FBS, 1%, 100 U/mL penicillin, 100 μg/mL streptomycin, and 29.2 mg/mL of L-glutamine.
2. Phosphate-buffered saline (PBS).
3. Trypsin-EDTA solution.

2.2 RNA Extraction Reagents from Human-derived Cancer Cell Lines (Primary or Established)

1. TRIzol™ reagent or other phenol-based extraction reagents.
2. Column-based RNA extraction kit (e.g., Qiagen miRNeasy, Zymo Research, Norgen).
3. gDNA removal column.
4. DNase/RNase-free water.
5. Chloroform (for TRIzol extraction).
6. 100% ethanol (for RNA precipitation) and 70% ethanol (to clean up impurities from RNA (e.g., organic solvents and salt).
7. Isopropanol (for RNA precipitation).
8. β-mercaptoethanol (for RNA stability, optional).

2.2.1 Tissue Processing Buffers (Maxwell® CSC RNA FFPE)

1. Lysis Buffer (from extraction kit, if using a column-based method).
2. Proteinase K (optional for difficult-to-dissolve tissues).
3. Mineral Oil (prevents evaporation during incubation).
4. Blue Dye (optional, tracking reagent).
5. $MnCl_2$, 0.09 M (for enzymatic steps, e.g., crosslink reversal).
6. DNase Buffer (supports DNase I activity).
7. DNase I (lyophilized) (removes genomic DNA).
8. Disposable Extraction Cartridges (used in automated extraction).
9. Plungers (for cartridge processing).
10. Elution Tubes (0.5 mL) (RNA collection).
11. Nuclease-free water (for reagent prep and RNA elution).

2.2.2 RNA Extraction Reagents for FFPE Tissue (Qiagen miRNeasy FFPE Kit)

1. RNeasy MinElute Spin Columns.
2. Buffer RNA binding buffer concentrate (RBC).
3. Lysis Buffer Proteinase K digestion (PKD).
4. Proteinase K.
5. RNase-Free DNase I (lyophilized).
6. 100% ethanol.
7. RNase-Free Water (for use with RNase-Free DNase I).
8. DNase Booster Buffer.
9. Buffer RNA purification ethanol-based (RPE).
10. RNase-Free Water.

2.3 Materials Required for RNA Quality Control

1. Total RNA sample (extracted).
2. Agarose.
3. 10× TBE or TAE buffer.

4. DEPC-treated water or RNase-free water.
5. RNA loading dye (with or without formaldehyde, *see* "**Notes**").
6. Ethidium bromide (EtBr) or SYBR Safe (for staining).
7. RNA ladder (e.g., RiboRuler, Invitrogen).

2.4 Equipment for Cell Lysis

1. Microcentrifuge (capable of 12,000–14,000 × *g*).
2. Pipettes and filter tips (RNase-free).

2.5 Equipment for Paraffined Tissue Lysis

1. Microcentrifuge (capable of 12,000–14,000 × *g*).
2. Heat blocks set at 56°C and at 80°C.
3. Maxwell® CSC Instrument (automated system for magnetic bead-based nucleic acid purification).
4. Pipettors and pipette tips for sample preprocessing and transfer into prefilled reagent cartridges (RNase-free and aerosol-resistant).
5. 1.5–2.0 mL tubes for incubation of samples (e.g., microtubes, 1.5 mL).
6. Razor blades (optional; used to scrape tissue off slide-mounted sections).

2.6 For RNA Quality and Quantity Assessment

1. Nanodrop spectrophotometer (for RNA concentration measurement).
2. Bioanalyzer or TapeStation or Qubit™ 4 Fluorometer (for RNA integrity assessment and quantification).
3. Agarose gel electrophoresis system (for RNA quality check).

3 Methods

3.1 RNA Extraction from Cell Lines

For adherent cells:

1. One day before, half a million cells (i.e., 0.5×10^6 ovarian cancer cell line, SKOV3) are plated in T25 flasks. The cells should be in a log phase of growth at the time of collection.
2. Remove the culture medium from the cells.
3. Gently wash the cells with phosphate-buffered saline (PBS) to eliminate residual medium.
4. Add an appropriate volume of trypsin-EDTA solution to detach the cells.
5. Incubate at 37°C until cells detach (usually 2–5 min).
6. Neutralize trypsin by adding a complete growth medium (trypsin/complete medium ratio: 1:2)

7. Transfer the cell suspension to a centrifuge tube.
8. Centrifuge at 1500 rpm × for 5 min to pellet the cells.
9. Aspirate the supernatant carefully.

For suspension cells:

1. Transfer one million cells (i.e., B-cell lymphoma) to a centrifuge tube.
2. Centrifuge at 1500 rpm × for 5 min to pellet the cells.
3. Aspirate the supernatant carefully.

3.1.1 Cell Lysis

1. Resuspend pellet of one million cells in 350 μL of RNA lysis buffer.
2. Optional: Add 3.5 μL of β-mercaptoethanol to the lysis buffer before use to enhance RNA stability.
3. Vortex the mixture for 15 s to ensure complete lysis.

3.1.2 RNA Binding

1. Add 200 μL of 96–100% ethanol to the lysate.
2. Vortex for 10 s to mix thoroughly.
3. Transfer up to 600 μL of the mixture to an RNA purification column, which is placed in a collection tube.
4. Centrifuge at 14,000 × *g* for 1 min.
5. Discard the flow-through and reassemble the column with the collection tube in place.
6. Repeat this step with any remaining lysate.

3.1.3 Column Washing

Wash 1:

1. Apply 400 μL of Wash Solution A to the column.
2. Centrifuge at 14,000 × *g* for 1 min.
3. Discard the flow-through and reassemble the column with the collection tube in place.

Wash 2:

1. Apply 400 μL of Wash Solution B to the column.
2. Centrifuge at 14,000 × *g* for 1 min.
3. Discard the flow-through and reassemble the column with the collection tube in place.

Wash 3:
Repeat Wash 2.

3.1.4 RNA Elution

1. Transfer the purification column to a new 1.7 mL elution tube
2. Add 50 μL of Elution Solution directly to the column's membrane.
3. Centrifuge at 200 × *g* for 2 min, followed by 14,000 × *g* for 1 min to elute the RNA.

3.1.5 RNA Storage

1. Immediately place the eluted RNA on ice for short-term storage or at −20°C.
2. For long-term storage, store the RNA at −80°C.

3.2 RNA Extraction from Tissues

3.2.1 Tissue Digestion (Maxwell® CSC RNA FFPE)

1. Cut 3-μm sequential sections from formalin-fixed paraffin-embedded (FFPE) tissue using a microtome and place them into a 1.5 mL autoclaved nuclease-free microcentrifuge tube.
2. Add 300 μL of mineral oil and vortex for 10 s.
3. Heat at 80°C for 2 min, then let the samples cool to room temperature.
4. Prepare a master mix: 224 μL Lysis Buffer, 25 μL Proteinase K, and 1 μL Blue Dye per sample ($n + 1$).
5. Add 250 μL of the master mix to each sample tube and vortex for 5 s.
6. Centrifuge at 10,000 × *g* for 20 s. If a pellet is present in the lower blue layer, gently resuspend. Leave both layers in the tube.
7. Incubate at 56°C for 15 min.
8. Incubate at 80°C for 1 h.
9. Cool samples at room temperature for 15 min. Prepare the DNase cocktail during this time.
10. Prepare the DNase cocktail by mixing 26 μL of $MnCl_2$ (0.09 M), 14 μL of DNase Buffer, and 10 μL of DNase I (in that order), pipetting after each addition. If the buffer contains precipitate, heat to 56°C for 2 min, then vortex the mixture.
11. Add 50 μL of DNase cocktail to the aqueous (blue) phase of each sample and mix by pipetting 10 times.
12. Incubate samples at room temperature (15–30°C) for 15 min. Meanwhile, prepare Maxwell® cartridges.
13. Centrifuge the samples at full speed for 5 min in a microcentrifuge.
14. Transfer the aqueous (blue) phase to well #1 of a Maxwell® CSC RNA FFPE Cartridge.

3.2.2 Maxwell® CSC RNA FFPE Cartridge Preparation

1. Wear gloves. Set up cartridges, plungers, and elution tubes in the deck tray(s) outside the instrument.
   - Place each cartridge in the tray with well #1 facing the elution tubes.
   - Press down on each cartridge to snap it into position.
   - Make sure cartridges are fully seated in the deck tray.
   - Remove all sealing tape completely.
2. Insert one plunger into well #8 of each cartridge.
3. Place one elution tube into the elution tube position for each cartridge (*see* **Note 4.1**).
4. Add 50 μL of Nuclease-Free Water to the bottom of each elution tube.
   - Elution tubes must remain open during the RNA purification process.
   - Use only the Nuclease-Free Water provided with the kit (*see* **Note 4.1**).

3.2.3 Instrument Run (Maxwell® CSC RNA FFPE Method)

1. Turn on the Maxwell® Instrument and Tablet PC. Launch the software by double-clicking the desktop icon.
2. On the Home screen, select Start.
3. Scan the kit barcode to select the RNA FFPE method automatically.
4. In the "Cartridge Setup" screen, select the cartridge positions to be used, enter any tracking information if required, and touch Proceed.
5. Insert the deck tray into the instrument with the elution tubes facing the back. Press down on the front edge to ensure proper seating.
6. Confirm all setup steps, then touch Start to close the door and begin the extraction run.
7. If the run is interrupted, check for plungers on the plunger bar and perform Clean Up if necessary.
8. Upon completion, the instrument will display a message indicating that the method has ended.
9. Follow the on-screen instructions to open the instrument door. Confirm that plungers are fully inserted in well #8 of each cartridge.
   If plungers remain attached to the plunger bar, refer to the instrument manual and perform the Clean Up procedure.
10. Immediately cap and remove the elution tubes containing RNA to prevent evaporation. Remove the deck tray by

holding it at the sides (*see* **Note 4.1**). For RNA storage, refer to **Note 4.1**, point 7.

11. Discard used cartridges, plungers, and elution tubes as hazardous waste, following institutional biosafety protocols (*see* **Note 4.1**).

3.3 Tissue digestion with QIAGEN miRNeasy FFPE Kit

1. Cut five sections 5 μm thick (Fig. 1A).
2. Add 240 μL Buffer PKD and mix by vortexing (Fig. 1B).
3. Incubate at 56°C for 3 min and then add 10 μL proteinase K to the liquid phase (Fig. 1C).
4. After overnight incubation at 56°C (Fig. 1D), incubate at 80°C for 20 min.
5. Centrifuge at full speed for 5 min.
6. Use a tip to break the ring of a paraffin (Fig. 1E).
7. Transfer the liquid phase in a new 2 mL collection tube (Fig. 1F).

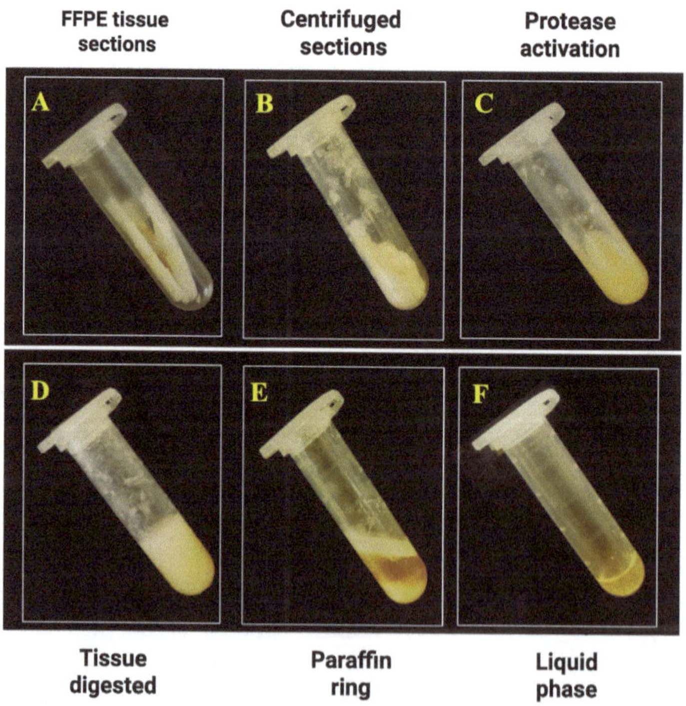

Fig. 1 Deparaffinization workflow of FFPE tissue. Sections of FFPE tissue (**A**) were centrifuged (**B**) and incubated with lysis buffer and protease (**C**). Tissue digested (**D**) was centrifuged to separate the paraffin: a ring of paraffin (first white layer); tissue lysate (second colored aqueous layer); tissue debris (third layer) (**E**) from the liquid phase (**F**)

8. Add DNase Booster Buffer equivalent to a tenth of the total sample volume (approximately 25 μL) and 10 μL DNase I stock solution.
9. Mix by inverting the tube.
10. Centrifuge briefly (30 s) to collect residual liquid from the sides of the tube.

RNA binding:

1. Add 500 μL Buffer RBC to adjust binding conditions, and mix the lysate thoroughly.
2. Add 1750 μL ethanol (100%) to the sample, and mix well by pipetting.
3. Transfer 700 μL of the sample, including any precipitate that may have formed, to an RNeasy.
   MinElute spin column placed in a 2 mL collection tube.
4. Centrifuge for 15 s at ≥8000 × (≥10,000 rpm). Discard the flow-through.
5. Repeat step 3 until the entire sample has passed through the RNeasy MinElute spin column.

Column washing

Wash 1:

1. Add 500 μL Buffer RPE to the RNeasy MinElute spin column.
2. Centrifuge for 15 s at ≥8000 x *g* (≥10,000 rpm). Discard the flow-through.

Wash 2:

1. Add 500 μL Buffer RPE to the RNeasy MinElute spin column.
2. Centrifuge for 2 min at ≥8000 x *g* (≥10,000 rpm) to wash the spin column membrane and discard the collection tube with the low-through.

RNA elution:

1. Place the RNeasy MinElute spin column in a new 2 mL collection tube.
2. Open the lid of the spin column, place the spin columns into the centrifuge with at least one empty position between each column, and orient the lids so that they face in the opposite direction of the rotor's rotation. Centrifuge at full speed for 5 min. Discard the collection tube with the flow-through.
3. Dry the spin column membrane, leaving the lid of the spin column membrane open for 5 min.

4. Place the RNeasy MinElute spin column in a new 1.5 mL collection tube and add 14–30 μL of RNase-free water directly to the spin column membrane. Centrifuge for 1 min at full speed to elute the RNA.

3.4 RNA Quality Control and Preparation

Before proceeding with RNA integrity assessment, it is important to select the most appropriate quality control method based on sample type, available equipment, and required equipment (Figs. 2 and 3).

3.4.1 Prepare 1% Agarose Gel to Check the Quality of the Extracted RNA

1. Weigh 1 g of agarose and add it to 100 mL of 1× TBE or TAE buffer.
2. Microwave until agarose is fully dissolved (~1–2 min).

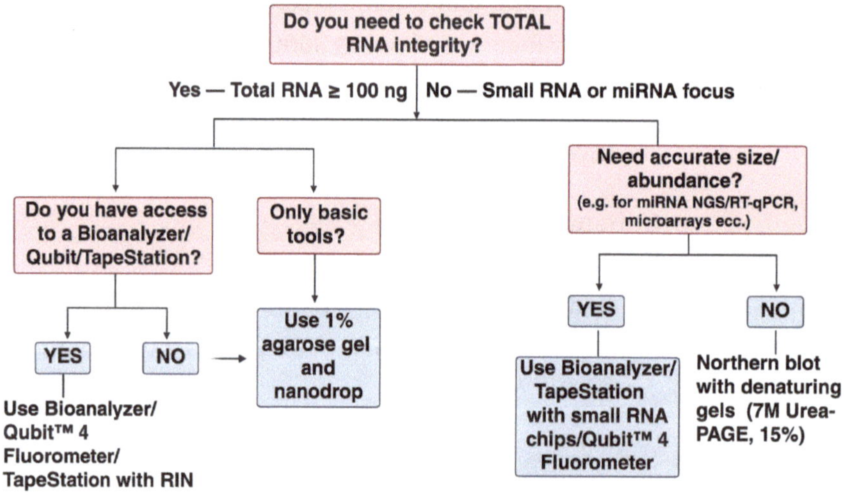

Fig. 2 RNA/miRNA quality control decision tree. Depending on the sample type and available resources, select the most appropriate RNA or miRNA quality control (QC) method

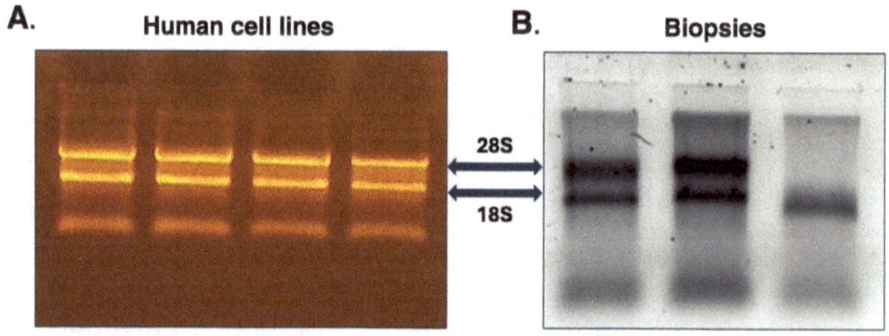

Fig. 3 RNA quality control in 1% agarose gels. **A** One microgram of total RNA from four B-cell lymphoma cell lines displays distinct 28S and 18S rRNA bands, indicating high integrity. **B** RNA from three tumor biopsies: the first two are intact, while the third is degraded. miRNAs are not visible due to their small size. The degraded sample was unsuitable for downstream experiments. The third faint band likely represents degraded RNA fragments, but it did not affect the overall usability of the RNA for downstream applications

3. Cool to ~60°C, then add EtBr (0.5 μg/mL) or SYBR Safe.
4. Pour into the casting tray with a comb and let it solidify (20–30 min).

3.4.2 Prepare RNA Samples

1. Mix 1–2 μg of RNA with RNA loading dye in a 1:1 ratio.
2. For denaturing conditions, use a loading dye based on formamide and heat it to 65–70°C for 5 min, then chill it quickly on ice.
3. For non-denaturing gels, which are sufficient for a rough integrity check, you may skip the heating step.
4. Place the gel into the electrophoresis tank and cover it with 1× TBE/TAE buffer.
5. Load RNA ladder (1–2 μL) in one well.
6. Load RNA samples (~10–15 μL) in separate wells.
7. Run at 80–100 V for ~20–30 min or until the dye front is ~3/4 of the way down the gel.
8. Visualize the RNA with a UV transilluminator or gel documentation system.

3.4.3 Interpretation of Agarose Gel Results

1. For eukaryotic RNA, two distinct sharp bands are shown in Fig. 3A:
   28S ribosomal RNA (rRNA) (upper, brighter band),
   18S rRNA (lower, approximately twofold less intense band),
   A 28S:18S ratio of approximately 2:1 indicates high-quality, intact RNA.

If the RNA exhibits a smear or lacks distinct bands, it is either partially or fully degraded (Fig. 3B). A low-molecular-weight smear at the bottom may suggest the presence of RNA species.

4 Notes

4.1 Comparison of RNA Extraction Kits for Tissue Samples

In our laboratory, we employed two distinct protocols for RNA extraction from tissue specimens: the Maxwell® CSC RNA FFPE Kit and the Qiagen miRNeasy FFPE Kit. Both are designed to recover total RNA, including small RNAs such as microRNAs, from formalin-fixed paraffin-embedded (FFPE) tissues; however, they differ in workflow, automation, and sample handling.

The Maxwell® CSC RNA FFPE Kit uses a magnetic bead-based purification system integrated into an automated instrument, providing high consistency, reducing hands-on time, and minimizing contamination risk. This method is particularly advantageous in high-throughput or clinical environments where reproducibility and standardization are crucial. It features enzymatic digestion, DNase treatment, and automated elution steps, rendering it user-friendly and efficient. It is suggested to use only the elution tubes provided in the Maxwell® CSC RNA FFPE Kit. Other tubes may be incompatible and negatively affect RNA purification. Any spills of specimens or reagents on the deck tray should be cleaned with a detergent–water solution, followed by a bactericidal spray or wipe, and then rinsed with water. Avoid using bleach on any instrument parts. It is also important not to expose RNA samples to UV light before removing the elution tubes from the Maxwell® CSC instrument. Most importantly, all kit components are single-use only and must not be reused.

In contrast, the Qiagen miRNeasy FFPE Kit is a column-based manual method that is well-suited for smaller sample batches and laboratories without access to automation. It provides robust recovery of both large and small RNA species, including miRNAs. Compared to a magnetic bead-based purification system integrated into an automated instrument, it enables higher RNA yields and flexible handling. However, it is more labor-intensive and requires careful manual execution of each step to ensure consistent results (Fig. 1).

While both kits produce RNA of adequate quality for downstream applications, such as qRT-PCR and high-throughput miRNA profiling, the decision between them largely depends on sample throughput, available equipment, and workflow preference. However, to achieve a high-quality and high-quantity RNA extraction, it is essential to start with less paraffin surrounding the tissue.

4.2 Assessment of the RNA Quality

To evaluate whether RNA has degraded during extraction from cell lines, particularly RNA isolated from paraffin-embedded tissues, it is advisable to run the sample on a 1% agarose gel in Tris-borate-EDTA (TBE) buffer using SafeView Classic, a safer alternative to ethidium bromide (EtBr) (Fig. 3A). This nucleic acid stain provides a non-toxic option for visualizing RNA integrity. Before starting the work area, the equipment, such as gel trays, combs, and pipettes, should be wiped down with RNaseZap or sodium hypochlorite (bleach, 2.5–10%) to eliminate RNase contamination. Wearing gloves at all times is mandatory. It is suggested to switch to fresh gloves during the procedure. Detailed information about the materials required for RNA quality control is provided in Sect. 2.3, "Materials Required for RNA Quality Control." Additional best practice guidelines can be found in Sect. 3.4.1–3.4.3, covering the preparation of a 1% agarose gel (see Sect. 3.4.1), RNA sample loading and electrophoresis (see Sect. 3.4.2), and the interpretation of agarose gel results (see Sect. 3.4.3).

4.3 Can the Integrity of Total RNA be Reliably Used to Infer the Preservation of MiRNAs in a Sample?

If the total RNA appears intact, displaying clear 28S and 18S rRNA bands with slight smearing, it indicates minimal RNase activity during the extraction process. This suggests that miRNAs are likely preserved as well, since RNases would have degraded both total RNA and miRNAs.

MiRNAs are approximately 22 nucleotides long, making them too small to be visible on standard 1% agarose gels and more susceptible to loss during extraction, especially if the protocol or column is not optimized for RNAs under 200 nucleotides. Even when rRNA appears intact, lost miRNAs can still occur due to incomplete lysis, poor recovery during phenol-chloroform extraction, the use of kits not designed for small RNA isolation or selective degradation of small RNAs in partially compromised samples. Therefore, if total RNA is intact, miRNAs are likely to be preserved, but not guaranteed. Below is a checklist for miRNA-specific extraction and quality control:

1. **Sample collection and handling**
   - Use RNase-free tubes, pipette tips, and gloves.
   - Snap-freeze tissue or cells in liquid nitrogen immediately.
   - Store samples at –80°C or in RNA stabilization reagent (e.g., RNAlater).
2. **Extraction method**
   - Use a miRNA-compatible extraction kit (e.g., Qiagen miRNeasy, Norgen Total RNA Purification Plus, or TRIzol with small RNA protocol).
   - Follow the protocol for small RNA enrichment, if available.
   - Include an exogenous spike-in control (e.g., cel-miR-39) to monitor recovery.
3. **DNase treatment (optional but recommended)**
   - Treat with RNase-free DNase I to remove genomic DNA contamination.
   - Perform purification after DNase treatment using a silica column or ethanol precipitation.
4. **RNA quantification**
   - To achieve higher yields of RNA, it is advisable to reduce the volume of the Elution buffer from 50 μL to 30 μL (see Sect. 3. Materials, point 2, RNA elution).
   - Measure concentration using a NanoDrop or Qubit, with the RNA High Sensitivity (HS) Assay preferred for improved sensitivity (e.g., Qubit 4 Fluorometer).
   - Check the 260/280 ratio (~2.0 ideal) and 260/230 (>1.8 ideal) for purity (Table 1).
5. **miRNA integrity check**
   - Use Bioanalyzer (Agilent Small RNA chip) or TapeStation to verify miRNA fraction (Table 2).

Table 1
Techniques/instruments for RNA quality testing

Technique/instrument	MiRNA-specific?	What it tells you
Agarose gel electrophoresis	No	rRNA band integrity (28S/18S), degradation smears
UV spectrophotometry (NanoDrop)	No	Purity (260/280, 260/230), but not integrity
Qubit fluorometer (RNA Assay)	No	Accurate quantification, no integrity check
Agilent Bioanalyzer (RNA 6000 Nano/Pico)	No (but informative)	RIN (RNA Integrity Number); shows rRNA peak degradation
Agilent TapeStation	No	RINe (RNA Integrity Estimate); similar to Bioanalyzar

Assessment of total RNA, especially ribosomal RNA (rRNA) integrity, can indirectly suggest miRNA integrity

Table 2
miRNA-specific techniques/instruments to test the quality and quantity

Technique/instrument	miRNA-specific?	What it tells you
Agilent Bioanalyzer (small RNA chip)	Yes	Distribution of small RNAs (10–150 nt), miRNA size peak (18–24 nt)
Agilent TapeStation (small RNA assay)	Yes	Similar output as Bioanalyzer; less sample prep time
qRT-PCR (TaqMan or SYBR)	Yes	miRNA presence and relative quantity (target-specific)
Northern blot (with miRNA probe)	Yes	Size and expression; very specific but laborious
Next-generation sequencing (NGS)	Yes	Profiles the entire miRNA population; includes detection and integrity
Small RNA-specific denaturing PAGE	Yes	High-resolution separation of miRNA and other small RNAs
Droplet digital PCR (ddPCR)	Yes	Absolute quantification of specific miRNAs
Capillary electrophoresis (e.g., Fragment Analyzer)	Yes	High-throughput profiling of miRNAs and small RNAs

- Run denaturing polyacrylamide gel electrophoresis (PAGE) or small RNA-specific gels if needed, 15% acrylamide (19:1 acrylamide:bisacrylamide, 8 M urea, 1× TBE or TBE-Urea buffer).

6. **Confirmation of miRNA presence**
   - Conduct qRT-PCR for a known endogenous miRNA (e.g., miR-16, snRNU6).
   - Compare Ct values among samples to ensure consistency.
   - Include controls without templates and RT to ensure specificity.
7. **Storage**
   - Aliquot total RNA to avoid freeze-thaw cycles.
   - Do not expose RNA samples to UV light before removal.
   - Store RNA at –30°C to –10°C for short-term use or at a temperature below –60°C for long-term storage.

5 Conclusions

High-quality RNA extraction is a critical prerequisite for reliable high-throughput transcriptomic analyses, including RT-qPCR, microarray-based miRNA profiling, and next-generation sequencing (NGS). In this chapter, we present detailed protocols for extracting RNA suitable for miRNA analysis from human cell lines and tissues. We outline two complementary methods for isolating RNA from solid tumor biopsies, including paraffin-embedded tissues. These methods address the specific challenges posed by various biological materials, ensuring optimal RNA yield and integrity across diverse sample types. Since miRNAs are particularly vulnerable to degradation and can often be lost during suboptimal extraction, the protocols and quality control strategies described here are vital for preserving the small RNA fraction. Thorough RNA integrity assessment and method selection tailored to the sample type are essential for achieving accurate and reproducible insights into gene regulation in both experimental and clinical contexts.

Acknowledgments

Pankaj Trivedi is supported by grants from LILT, Italy, Associazione Italiana per la Ricerca sul Cancro (AIRC) and the Italian Ministry of University and Research (PRIN2022). The RM123188E8E9A26E and RM1241905BB85097 Sapienza grants support Eleni Anastasiadou.

References

1. Anastasiadou E, Jacob LS, Slack FJ (2018) Non-coding RNA networks in cancer. Nat Rev Cancer 18(1):5–18. https://doi.org/10.1038/nrc.2017.99
2. Han J, Mendell JT (2023) MicroRNA turnover: a tale of tailing, trimming, and targets. Trends Biochem Sci 48(1):26–39. https://doi.org/10.1016/j.tibs.2022.06.005
3. Anastasiadou E, Faggioni A, Trivedi P, Slack FJ (2018) The nefarious nexus of noncoding RNAs in cancer. Int J Mol Sci 19 (7). https://doi.org/10.3390/ijms19072072
4. Rosato P et al (2012) Differential regulation of miR-21 and miR-146a by Epstein-Barr virus-encoded EBNA2. Leukemia 26(11):2343–2352. https://doi.org/10.1038/leu.2012.108
5. Lee RC, Feinbaum RL, Ambros V (1993) The *C. elegans* heterochronic gene lin-4 encodes small RNAs with antisense complementarity to lin-14. Cell 75(5):843–854. https://doi.org/10.1016/0092-8674(93)90529-y
6. Wightman B, Ha I, Ruvkun G (1993) Posttranscriptional regulation of the heterochronic gene lin-14 by lin-4 mediates temporal pattern formation in *C. elegans*. Cell 75(5):855–862. https://doi.org/10.1016/0092-8674(93)90530-4
7. Calin GA et al (2024) The 2024 Nobel Prize in physiology or medicine: microRNA takes center stage. Noncoding RNA 10(6). https://doi.org/10.3390/ncrna10060062
8. Izzotti A et al (2021) Identification by microRNA analysis of environmental risk factors bearing pathogenic relevance in non-smoker lung cancer. J Pers Med 11(7). https://doi.org/10.3390/jpm11070666
9. Leopizzi M et al (2024) Epstein-Barr virus-encoded EBNA2 downregulates ICOSL by inducing miR-24 in B-cell lymphoma. Blood 143(5):429–443. https://doi.org/10.1182/blood.2023021346
10. Anastasiadou E et al (2019) Epstein-Barr virus-encoded EBNA2 alters immune checkpoint PD-L1 expression by downregulating miR-34a in B-cell lymphomas. Leukemia 33(1):132–147. https://doi.org/10.1038/s41375-018-0178-x
11. Ambrosio MR et al (2017) MicroRNAs sequencing unveils distinct molecular subgroups of plasmablastic lymphoma. Oncotarget 8(64):107356–107373. https://doi.org/10.18632/oncotarget.22219
12. Ambrosio MR et al (2014) The Epstein Barr-encoded BART-6-3p microRNA affects regulation of cell growth and immuno response in Burkitt lymphoma. Infect Agent Cancer 9, 12. https://doi.org/10.1186/1750-9378-9-12
13. Mundo L et al (2017) Unveiling another missing piece in EBV-driven lymphomagenesis: EBV-encoded MicroRNAs expression in EBER-negative Burkitt lymphoma cases. Front Microbiol 8:229. https://doi.org/10.3389/fmicb.2017.00229
14. Luzzi A et al (2014) HIV-1 Tat induces DNMT over-expression through microRNA dysregulation in HIV-related non Hodgkin lymphomas. Infect Agent Cancer 9, 41. https://doi.org/10.1186/1750-9378-9-41
15. Anastasiadou E et al (2021) MiR-200c-3p contrasts PD-L1 induction by combinatorial therapies and slows proliferation of epithelial ovarian cancer through downregulation of β-catenin and c-Myc. Cells 10(3). https://doi.org/10.3390/cells10030519
16. Anastasiadou E et al (2021) MiR-200c-3p maintains stemness and proliferative potential in adipose-derived stem cells by counteracting senescence mechanisms. PLoS One 16(9):e0257070. https://doi.org/10.1371/journal.pone.0257070
17. Anastasiadou E et al (2021) Calcineurin gamma catalytic subunit PPP3CC inhibition by miR-200c-3p affects apoptosis in epithelial ovarian cancer. Genes 12(9). https://doi.org/10.3390/genes12091400

Part VI

Functional Genomics Tools for Epigenetic and Therapeutic Discovery

Chapter 15

Dissecting Epigenetic Drug Response Mechanisms Using CRISPR Knockout Screens

Xinyue Zhou and Rui Lu

Abstract

Epigenetic drugs are widely applied in cancer therapy due to their ability to modify gene expression without altering the DNA sequence. Despite their therapeutic potential, drug resistance frequently occurs, posing a significant challenge in cancer treatment. CRISPR screens have emerged as a powerful tool to address this issue by leveraging the precision of CRISPR-Cas9 gene editing to enable the systematic interrogation of genes. This approach involves the simultaneous targeting of thousands of genes to elucidate their roles in various biological processes, disease mechanisms, and drug responses, providing valuable insights into gene function and potential therapeutic targets. In cancer therapy, CRISPR screens provide a deeper understanding of cancer progression by enabling the identification of essential genes and facilitating the discovery of novel therapeutic targets when combined with epigenetic drugs. Here, we present an overview of the CRISPR screen methodology, which involves introducing guide RNAs targeting specific genes into cells, followed by phenotype selection and analysis.

Keywords CRISPR-Cas9, Epigenetic drug response, Library design, sgRNA, Pooled screen, Phenotype selection, Library QC

1 Introduction

The development of CRISPR-Cas9 technology has revolutionized the field of molecular biology, offering precise and efficient genome editing capabilities. Initially discovered as a bacterial immune system, CRISPR-Cas9 was repurposed for genome editing in eukaryotic cells. Typically, a single guide RNA (sgRNA) is employed that guides the Cas9 enzyme to target a specific gene for disruption or modulation [1–3]. Among its myriad applications, CRISPR screens that contain thousands of sgRNAs targeting different genes stand out as a powerful tool for systematically probing gene functions across the entire genome, allowing for the simultaneous perturbation of multiple genes within a population of cells.

CRISPR screens have shown great promise in improving cancer cell treatment via offering a new understanding of cancer genetics

Archana Dhasarathy and Motoki Takaku (eds.), *Cancer Epigenetics*, Methods in Molecular Biology, vol. 3005,
https://doi.org/10.1007/978-1-0716-5072-1_15,

and cancer biology [4, 5]. This includes the unbiased classification of essential genes for cancer cells, identification of vulnerable and resistant targets for therapeutic treatment, and discovery of novel epigenetic regulators and genetic interactions [6–9]. Various cell types and model systems can be utilized for CRISPR screens, including immortalized cell lines, primary cells, and model organisms such as mice or zebrafish [1, 10, 11]. The choice of cell type depends on the research question and the specific phenotype under investigation. The principal driver of CRISPR screens lies in the selective pressure applied to cells following the introduction of CRISPR libraries by exposing cells to specific conditions or stimuli and then analyzing the resulting phenotypic changes. By interrogating thousands of genes in parallel, CRISPR screens provide a comprehensive understanding of cellular processes and pathways, and have profound implications for drug discovery, personalized medicine, and understanding the genetic basis of diseases [12, 13]. This chapter systematically introduces the detailed process of genome-wide CRISPR screens, including design and generation of sgRNAs library plasmids, sgRNA library QC, lentiviral plasmid transfection, lentiviral transduction into target cells and cell selection, drug treatment, screen library construction, and next-generation sequencing (see Fig. 1).

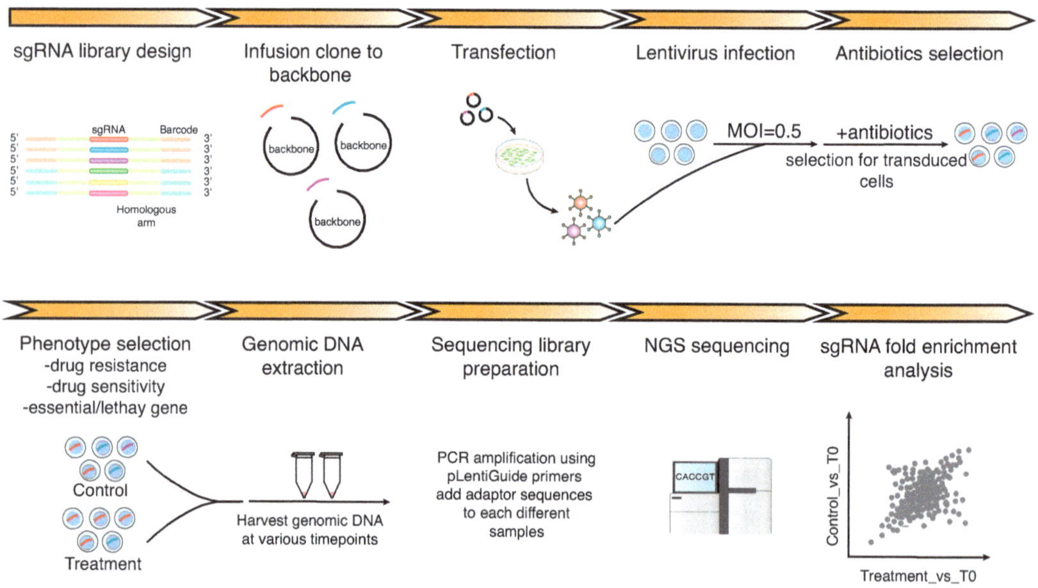

Fig. 1 Schematic flow of pooled CRISPR screen

2 Materials

2.1 Reagents

1. Q5 High Fidelity PCR (New England Biolabs, includes buffer, DNA polymerase and dNTP).
2. Digestion mixture (contains restriction enzyme BsmBI-v2 (New England Biolabs) and 10× Fast Digesting Buffer (Thermo Scientific)).
3. Gibson Assembly Master Mix (New England Biolabs).
4. NEB stable 10-beta *E. coli* cells (New England Biolabs).
5. LB Ampicillin Agar plates and LB broth culture medium.
6. HEK 293FT (Invitrogen).
7. HEK 293FT Culture medium (DMEM + 20% FBS + 1% P/S).
8. Lentiviral production medium (DMEM with NanoFect (Alstem)).
9. Polybrene (EMD Millipore).
10. G-418 Sulfate Geneticin (Neomycin) (GOLDBIO).
11. Gel DNA Recovery Kit (Zymo Research).
12. Plasmid Midiprep kit (Zymo Research).
13. Quick-DNA Miniprep Plus Kit (Zymo Research).

2.2 Oligo, Plasmids, and Primers

1. Oligonucleotide pools (see Sect. 3.1).
2. LRG2.1-Neo (Addgene).
3. pMD2G (Addgene).
4. pSPAX2 (Addgene).
5. Barcode-Specific Forward Primer (see Table 1).
6. Barcode-Specific Reverse Primer (see Table 1).
7. HR Forward Primer: TTCTTGGCTTTATATATCTTGTG-GAAAGGACGAAACACCG.
8. HR Reverse Primer: ATTTAAACTTGCTATGCTGTTTC-CAGCATAGCTCTTAAAC.
9. Lenti-Guide Forward Primer: TCGTCGGCAGCGTCA-GATGTGTATAAGAGACAGAATGGACTATCATATGCT-TACCGTAACTTGAAAGTATTTCG.
10. Lenti-Guide Reverse Primer: GTCTCGTGGGCTCGGAGA TGTGTATAAGAGACAGCTTTAGTTTGTATGTCTGTT GCTATTATGTCTACTATTCTTTCCC.

Table 1
Barcode list

Barcode name	ID	Sequence	Tm	Ta
Barcode1	Barcode1_F	AGGCACTTGCTCGTACGACG	70	71
Barcode1	Barcode1_R	ATGTGGGCCCGGCACCTTAA	73	71
Barcode2	Barcode2_F	GTGTAACCCGTAGGGCACCT	70	69
Barcode2	Barcode2_R	GTCGAGAGCAGTCCTTCGAC	68	69
Barcode3	Barcode3_F	CCGCCCAGAAGCGTCTATGA	70	71
Barcode3	Barcode3_R	CATAACCTCCGGGCATGGGT	70	71
Barcode4	Barcode4_F	GCCTATGCGACAGAGACCGT	70	71
Barcode4	Barcode4_R	GCGGGACTAGGAAGCGATGT	70	71
Barcode5	Barcode5_F	CCGGCCCCATTAGTACAGCA	70	70
Barcode5	Barcode5_R	GGCGATACACGAGTCGACCA	69	70
Barcode6	Barcode6_F	GCGGCTGCAGACATTACACC	69	70
Barcode6	Barcode6_R	GGCGTCAAGTAGCCCAGGAT	70	70

11. Nextera Universal Forward Primer: (see Table 2).
12. Nextera reverse primer: (see Table 2).

2.3 Equipment

1. Thermal Cycler.
2. Bacterial shaker (37°C, 200 rpm).
3. Gel electrophoresis apparatus.
4. Gel imager.
5. Digital dry bath (55°C), Mini incubator (37°C), Water bath (42°C).
6. Centrifuge tubes (1.7, 15, and 50 mL), PCR tubes.
7. Pipette.
8. Centrifuges.
9. 0.45 μm PES filter.
10. Tissue culture plate (10-cm^2 dish, 15-cm^2 dish, and 24-well plate).
11. NanoDrop.
12. Qubit 4 Fluorometer.

Table 2
Nextera index sequences

Index	Sequences
Ad1_noMX	AATGATACGGCGACCACCGAGATCTACACTCGTCGGCAGCGTCAGATGTG
Ad2.1_TAAGGCGA	CAAGCAGAAGACGGCATACGAGATTCGCCTTAGTCTCGTGGGCTCGGAGATGT
Ad2.2_CGTACTAG	CAAGCAGAAGACGGCATACGAGATCTAGTACGGTCTCGTGGGCTCGGAGATGT
Ad2.3_AGGCAGAA	CAAGCAGAAGACGGCATACGAGATTTCTGCCTGTCTCGTGGGCTCGGAGATGT
Ad2.4_TCCTGAGC	CAAGCAGAAGACGGCATACGAGATGCTCAGGAGTCTCGTGGGCTCGGAGATGT
Ad2.5_GGACTCCT	CAAGCAGAAGACGGCATACGAGATAGGAGTCCGTCTCGTGGGCTCGGAGATGT
Ad2.6_TAGGCATG	CAAGCAGAAGACGGCATACGAGATCATGCCTAGTCTCGTGGGCTCGGAGATGT
Ad2.7_CTCTCTAC	CAAGCAGAAGACGGCATACGAGATGTAGAGAGGTCTCGTGGGCTCGGAGATGT
Ad2.8_CAGAGAGG	CAAGCAGAAGACGGCATACGAGATCCTCTCTGGTCTCGTGGGCTCGGAGATGT
Ad2.9_GCTACGCT	CAAGCAGAAGACGGCATACGAGATAGCGTAGCGTCTCGTGGGCTCGGAGATGT
Ad2.10_CGAGGCTG	CAAGCAGAAGACGGCATACGAGATCAGCCTCGGTCTCGTGGGCTCGGAGATGT
Ad2.11_AAGAGGCA	CAAGCAGAAGACGGCATACGAGATTGCCTCTTGTCTCGTGGGCTCGGAGATGT
Ad2.12_GTAGAGGA	CAAGCAGAAGACGGCATACGAGATTCCTCTACGTCTCGTGGGCTCGGAGATGT
Ad2.13_GTCGTGAT	CAAGCAGAAGACGGCATACGAGATATCACGACGTCTCGTGGGCTCGGAGATGT
Ad2.14_ACCACTGT	CAAGCAGAAGACGGCATACGAGATACAGTGGTGTCTCGTGGGCTCGGAGATGT
Ad2.15_TGGATCTG	CAAGCAGAAGACGGCATACGAGATCAGATCCAGTCTCGTGGGCTCGGAGATGT
Ad2.16_CCGTTTGT	CAAGCAGAAGACGGCATACGAGATACAAACGGGTCTCGTGGGCTCGGAGATGT
Ad2.17_TGCTGGGT	CAAGCAGAAGACGGCATACGAGATACCCAGCAGTCTCGTGGGCTCGGAGATGT
Ad2.18_GAGGGGTT	CAAGCAGAAGACGGCATACGAGATAACCCCTCGTCTCGTGGGCTCGGAGATGT

(continued)

Table 2
Continued

Index	Sequences
Ad2.19_AGGTTGGG	CAAGCAGAAGACGGCATACGAGATCCCAACCTGTCTCGTGGGCTCGGAGATGT
Ad2.20_GTGTGGTG	CAAGCAGAAGACGGCATACGAGATCACCACACGTCTCGTGGGCTCGGAGATGT
Ad2.21_TGGGTTTC	CAAGCAGAAGACGGCATACGAGATGAAACCCAGTCTCGTGGGCTCGGAGATGT
Ad2.22_TGGTCACA	CAAGCAGAAGACGGCATACGAGATTGTGACCAGTCTCGTGGGCTCGGAGATGT
Ad2.23_TTGACCCT	CAAGCAGAAGACGGCATACGAGATAGGGTCAAGTCTCGTGGGCTCGGAGATGT
Ad2.24_CCACTCCT	CAAGCAGAAGACGGCATACGAGATAGGAGTGGGTCTCGTGGGCTCGGAGATGT

3 Methods

3.1 Oligo Design and Synthesis

1. Identify the genes of interest and obtain a list of sgRNA sequences through online tools. For the pooled screens, this includes targeting multiple genes through multiple sgRNAs per gene (*see* **Note 1**). Always include negative controls and positive controls for pooled CRISPR screens. Negative controls could be non-targeting sgRNAs or sgRNAs targeting a region with no phenotype of interest.
2. Add the homologous arms (see Fig. 2). The LRG2.1-Neo vector is used as an example in this protocol.
   a. Add the 5′ universal flanking sequence: GTGGAAAGGACGAAACACCG. For sgRNA sequences that do not start with a "G," add an additional "G" to ensure efficient transcription from the U6 promoter.
   b. Add the 3′ universal flanking sequence: GTTTTAGAGCTAGAAATAGC. An example of sgRNA insertion placement is shown below. The homologous region is highlighted in bold, and the desired sgRNA sequence is represented as "NNN".

   5′-ACTTGAAAGTATTTCGATTTCTTGGCT-TTATATATCTT**GTGGAAAGGACGAAACAC-CG**NNNNNNNNNNNNNNNNNNNN**GTTTAAGAGC-TATGCTGGAAACAGC**ATAGCAAGTTTAAATAA-GGCTAGTCC-3′.
3. Design a barcode for multiple sgRNA libraries synthesis (*see* **Note 2**).

   The Custom Array Oligo design would then look like the following:

   5′-**CCGGCCCCATTAGTACAGCA**GTGGAAAG-GACGAAACACCGNNNNNNNNNNNNNNNNNNN-NGTTTAAGAGCTATGCTGGAAACAGC**GGCGATACAC-GAGTCGACCA**-3′.

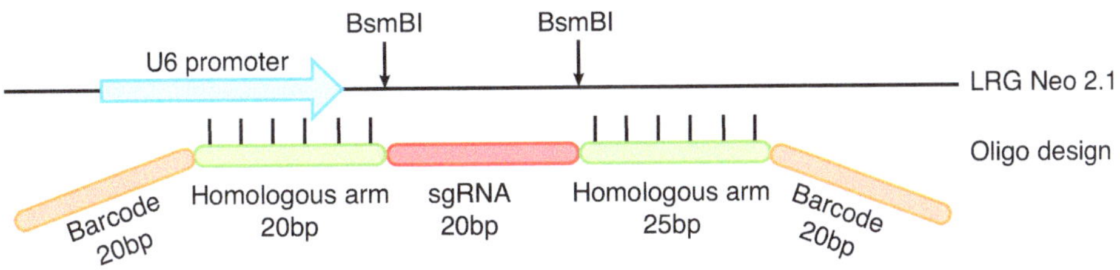

Fig. 2 Schematic plot of oligo design for CRISPR library construction

The barcode sequence is highlighted in bold. You can refer to the following barcode list (Table 1). The barcode used in this example is Barcode 5.

4. Order oligonucleotide pools.

Most single guide RNA (sgRNA) libraries are synthesized as pooled oligonucleotides by commercial providers such as CustomArray Inc. (Bothell, WA). The choice of vendor should be guided by library size, sequence complexity, turnaround time, and synthesis quality, as these factors may vary depending on experimental requirements.

3.2 Library Amplification (see Note 3)

1. Add barcode to synthesized oligonucleotides.

Components	Amount (μL)	Final concentration
Synthesized Oligonucleotides (10 ng/μL)	1 μL	–
Barcode-Specific Forward Primer (10 μM)	2.5 μL	0.5 μM
Barcode-Specific Reverse Primer (10 μM)	2.5 μL	0.5 μM
5× Q5 Reaction Buffer	10 μL	1×
dNTPs (10 mM)	1 μL	200 μM
Q5 HIFI DNA Polymerase	0.5 μL	1.0 U/50 μL PCR
Nuclease-Free Water	To 50 μL	

2. Use the following PCR cycling conditions:

Temperature	Duration	Cycles
98°C	30 s	1
98°C, 72°C	5 s, 15 s	15 (*see* **Note 4**)
72°C	2 min	1
4°C	–	Hold

3. Amplify with homologous primers.

Components	Amount (μL)	Final concentration
Barcode-specific PCR diluted product [1:10] or synthesized oligonucleotides (*see* **Note 5**)	1 μL	–
HR Forward Primer (10 μM)	2.5 μL	0.5 μM
HR Reverse Primer (10 μM)	2.5 μL	0.5 μM
5× Q5 Reaction Buffer	10 μL	1×
dNTPs (10 mM)	1 μL	200 μM
Q5 HIFI DNA Polymerase	0.5 μL	1.0 U/50 μL PCR
Nuclease-Free Water	To 50 μL	

4. Use the following PCR cycling conditions:

Temperature	Duration	Cycles
98°C	30 s	1
98°C, 72°C	5 s, 15 s	15 (*see* **Note 6**)
72°C	2 min	1
4°C	–	Hold

3.3 Digest the LRG2.1-Neo Backbone

1. Backbone digestion.
   Make the following mixture with your vector plasmid:

Components	Amount (μL)
LRG2.1-Neo	3 μg
BsmBI	3 μL
10× Fast Digesting Buffer	5 μL
Nuclease-Free Water	to 50 μL

2. Pipette to mix and incubate at 55°C overnight for BsmBI digestion.
3. Run the digestion product on a 1% agarose gel, making sure to run the gel for enough time to separate the target band from the other bands. Gel purifies the ~8 kb backbone using a gel purification kit as per the manufacturer's instructions.
4. Measure the concentration using NanoDrop. Gel purification products could be stored at −20°C until further use.

3.4 In-Fusion Cloning

1. Gibson Assembly (*see* **Note 7**)

Components	Amount
Digested LRG2.1-Neo backbone (gel purification product)	25 ng
HR-specific PCR product from Sect. 3.2	10 ng
2× Gibson Assembly Master Mix	5 μL
Nuclease-Free Water	to 10 μL

2. Pipette to mix well and incubate the mixture at 55°C for 1 h.

3.5 Heat Shock Transformation

1. Thaw NEB Stable *E. coli* cells on ice for about 10–15 min. Add 5 μL Gibson Assembly product acquired from above into 100 μL NEB Stable *E. coli* cells, gently pipette to mix. Incubate the mixture for 30 min on ice to allow for DNA uptake by the cells. Avoid vortexing during the whole process to prevent shearing of the DNA.
2. Take LB agar plates (containing Ampicillin) into a 37°C incubator for pre-warmup.

3. Heat shock the cell–DNA mixture at 42°C in a water bath for 45 s and then incubate the mixture on ice for 2–5 min to stabilize the cells.
4. Add 2 mL LB broth or NEB Stable 10-Beta medium into the mixture and incubate in a 37°C shaking incubator at 200 rpm for 1 h to allow for recovery and expression of antibiotic resistance protein.
5. After incubation, mix the tubes by gently flicking and inverting them to ensure even distribution of the transformed cells. At this stage, the total volume is 2100 μL.
6. Plate 21 μL (100× dilution) onto one agar plate, and distribute the remaining volume evenly across multiple plates, with ~500 μL volume per plate using a cell spreader.
7. Incubate the plates at 37°C for 24 h to allow transformed colonies to grow (*see* **Note 8**).

3.6 Prepare Library Plasmid

1. The next day, use a cell lifter to collect all the bacteria from the agar plate and combine them into one centrifuge tube. Alternatively, resuspend colonies in 2 mL LB broth and transfer them to a centrifuge tube.
2. Centrifuge the bacterial suspension at 2400 rpm for 15–20 min at room temperature to pellet the cells. Carefully decant the supernatant without disturbing the pellet.
3. Weigh the bacterial pellet to determine the approximate amount. Typically, expect at least 0.2 g of bacterial pellet.
4. Perform the midi-prep procedure as per the manufacturer's instructions. Maintain sterility and handle the reagents with care to avoid contamination.
5. Assess the quality and quantity of the purified plasmid DNA using Nanodrop. Ensure that the concentration and purity of the DNA meet the requirements for downstream applications. Typically, expect at least 500 mg (*see* **Note 9**).
6. Store the purified library plasmid DNA at −20°C or −80°C until further use. Proper storage conditions are essential to maintain the integrity of the DNA for long-term storage.

3.7 Library Plasmid QC

To examine the library plasmid complexity and uniformity, please construct the sequencing library following Sect. 3.11 and proceed to NGS sequencing.

3.8 Lentivirus Production

1. Maintain HEK 293FT cells in a 10-cm round plate with 8 mL of DMEM (w/10% FBS and 1% P/S).
2. Perform transfection when HEK 293FT cells confluency reaches about 70–80%. Prepare mixture A and mixture B using DMEM medium (w/o FBS and P/S) as follows [14].

Mixture A		Mixture B	
DMEM	500 μL	DMEM	500 μL
Nanofect	40 μg	pMD2G	2.5 μg
		pSPAX2	7.5 μg
		Library plasmid	10 μg

Mix by vortexing the tube and incubate at room temperature for at least 5 min.

3. Combine mixture A and mixture B. Mix via vortexing for at least 10 s and incubate at room temperature for another 25 min.
4. Remove all the medium for HEK 293FT cells and add 7 mL of DMEM medium (w/FBS and P/S).
5. Gently add the 1 mL suspension from above dropwise into HEK 293FT cells, without touching the cells or dislodging the cells from the plate. Gently swirl to mix.
6. Collect medium after 48 h incubation, replace with 8 mL DMEM medium (w/10% FBS and 1% P/S). Collect again after 72 h of incubation. Combine the medium harvested at 48 h and 72 h and filter through a 0.45 μm PES filter.
7. Test virus titer accordingly (*see* **Note 10**).

3.9 Lentivirus Transduction

The cells used for CRISPR screen should have a stable expression of CRISPR-cas9, this can be achieved by transducing the cells with CRISPR virus or inserting the sgRNAs library into a CRISPR-Cas9 expressing backbone (e.g., LentiCRISPR-v2). The knockout efficiency of CRISPR-Cas9 needs to be tested before sgRNAs library transduction.

A pilot experiment is needed for testing the transduction rate of the lenti-sgRNAs library using the target cell line and cell number/well to help calculate the amount needed for formal screening.

To ensure that each unique sgRNA has a decent chance of infecting multiple target cells, the cell numbers for transduction at the initial stage are critical [15]. A 1000× coverage rate is used in most of our library screens to limit the risk that a certain sgRNA is excluded from the screen and leads to a false positive result. 1000× coverage represents that 1000 cells are transduced by the median sgRNA. The CRISPR knockout ACER library [16] focused on a subset of epigenetic regulators developed by our lab, which contains ~8000 sgRNAs; then 8000 sgRNAs × 1000 × coverage = 8,000,000 cells must be transduced. However, the coverage rate should be optimized based on different library types and experiment design.

A low multiplicity of infection (MOI), usually between 30 and 50%, should be selected based on the pilot results to encourage both a good transduction rate and the uptake of a sgRNA per cell (*see* **Note 11**).

3.9.1 Pilot Experiment

1. Seed target cells into a 24-well plate with cells and virus. Prepare a mixture of 500 μL volume for transduction as follows:
   Cell suspension (usually use 5×10^5 to 1×10^6 cells/24-well).
   Polybrene (final 8 μg/mL).
   Virus (choose multiple doses, e.g., 50, 100, 150, 300 μL).
   Cell culture medium (optional, only add when the above mixture is less than 500 μL).
2. Spin the plate at 2400 rpm for 90 min at room temperature.
3. Add 500 μL culture medium to a total 1 mL volume, incubate overnight.
4. The next day, spin down cells and replace with fresh cell culture medium.
5. After another 72 h of incubation, collect 300–500 μL cell suspension into a flow tube and check GFP-positive cell percentage using flow cytometry. The GFP-positive cell percentage should be between 30–50%. Supplement the medium with neomycin. Neomycin selection will help achieve a higher GFP-positive cell percentage. A lower dose of neomycin should be used to maintain the cell viability, usually between 400 and 800 μg/mL.
6. Check cell viability and GFP positive cell percentage after 4 days using flow cytometry to decide the virus/cells ratio and neomycin dose. Considering adjusting the virus/cells ratio and neomycin doses to achieve a higher GFP positive percentage and cell viability, so fewer cells are required to maintain for the library screen.

3.9.2 Lentivirus Transduction

1. Calculate the cell number needed for CRISPR library transduction. Take the ACER library for example, there are in total 7580 (~8000) sgRNAs. To aim for 1000× coverage, $8000 \times 1000 = 8{,}000{,}000$ cells are needed for collection post-lentiviral transduction and drug treatment. If two treatment groups and three replicates for each group are included in the experiment design, a total of $8{,}000{,}000 \times 2 \times 3 = 48{,}000{,}000$ cells are needed. Considering the transduction efficiency is likely between 30–50% (based on your pilot experiment result) and the loss of cell viability during neomycin selection and drug treatment, at least 50 million cells should be used for lentiviral transduction.

2. Seed cells into a 24-well plate as tested in Sect. 3.9.1, mix with lentivirus and polybrene. In the case of the ACER library, at least four plates are needed.
3. Spin the plate at 2400 rpm for 90 min at room temperature. Add 500 μL culture medium to a total 1 mL volume per well and incubate overnight.
4. The next day, resuspend all the cells into a 50 mL centrifuge tube, spin down and resuspend with cell culture medium. Evenly distribute the cell suspension into multiple 15-cm dishes at an appropriate density. Supplement the culture medium with neomycin at the optimized dose from the pilot experiment.
5. Check cell viability and GFP-positive cells percentage after another 72 h of incubation using flow cytometry, mark this day as T0/day 4.

3.10 Drug Treatment

1. Harvest T0 samples. Collect the cells for T0 according to the GFP-positive cells percentage. Take the ACER library for example, if there are 50% GFP+ measured at T0/day 4, then 8 million/50% = 16 million cells need to be harvested to fulfill 1000× coverage.
2. Split T0/day 4 cells into different treatment groups and add vehicle or drug treatment. Replicate wells should be included. Passage cells and replace with fresh medium and drug every 3 days. At each passage, maintain culturing with at least 1000 × coverage cell numbers for each sample and discard the rest of the cells (*see* **Note 12**).
3. In the case of the ACER screen, cells were harvested for genomic DNA extraction at T7/day 11 and T14/day 18 after vehicle/drug treatment. Before each harvest, 300–500 μ L cell suspension per sample was collected for flow cytometry to check the cell viability and GFP+ percentage to determine the cell number for genomic DNA. At least 8,000,000 GFP+ cells per sample must be harvested for genomic DNA to meet the 1000× coverage. An earlier (T7/day 11) and a later (T14/ day 18) timepoint is collected to help visualize both the sensitive and resistance hits (*see* **Note 13**). Wash cells with 1× PBS and freeze down the pellet for genomic DNA extraction.

3.11 CRISPR Screen Library Preparation

1. Extract genomic DNA using a genomic DNA extraction kit. Most commercial columns suggest no more than 5 million cells should be loaded, so multiple columns may be used if you have a large number of cells. Measure the quality and quantity of genomic DNA using Nanodrop. Use genomic DNA immediately for library PCR or store at −20°C until further use.
2. Library Analysis PCR 1 (LaPCR1): Amplify sgRNA library (*see* **Note 14**).

Components	Amount (μL/μg)	Final concentration
Library Genomic DNA	5 μg	50–100 ng/μL
Lenti-Guide Forward Primer (10 μM)	5 μL	0.5 μM
Lenti-Guide Reverse Primer (10 μM)	5 μL	0.5 μM
5× Q5 Reaction Buffer	20 μL	1×
Deoxynucleotide triphosphates (dNTPs; at 10 mM)	2 μL	200 μM
Q5 HIFI DNA Polymerase	27 μL	1.0 U/50 μL PCR
Nuclease-Free Water/ddH2O	To 100 μL	

3. Use the following PCR cycling conditions:

Temperature	Duration	Cycles
98°C	30 s	1
98°C, 72°C	10 s, 60 s	15
72°C	5 min	1
4°C	–	Hold

4. Pool all the LaPCR1 products from the same sample together and vortex to mix well.
5. Library Analysis PCR 2 (LaPCR2): Add an index to each sample for multiplexing (Nextera primers see Table 2, *see* **Note 15**).

Components	Amount (μL)	Final concentration
LaPCR1 Product [Diluted 1:5]	1 μL	–
Nextera Universal Forward Primer (2 μM)	1 μL	0.2 μM
Nextera Reverse Primer (2 μ M)	1 μL	0.2 μM
5× Q5 Reaction Buffer	2 μL	1×
Deoxynucleotide triphosphates (dNTPs; at 10 mM)	1 μL	200 μM
Q5 HIFI DNA Polymerase	0.1 μL	1.0 U/50 μL PCR
Nuclease Free Water/ddH2O	To 10 μL	

6. PCR cycling conditions:

Temperature	Duration	Cycles
98°C	30 s	1
98°C, 65°C, 72°C	10 s, 30 s, 30 s	15 (*see* **Note 16**)
72°C	5 min	1
4°C	–	Hold

7. Gel purifies the ~500 bp amplification product using a gel purification kit as per manufacturer's instructions.
8. Measure the DNA concentration using Qubit.
9. Mix the DNA samples and send them out for NGS sequencing.

3.12 Data Analysis

Samples are mixed, and deep sequencing is performed on the NovaSeq PE150 platform. At least 3 million mapped reads (300x coverage for the ACER library) must be acquired for each sample to ensure the quality of the data. After sequencing, sgRNA counts were extracted by MAGeCK count with default parameters [17]. Analysis of gene enrichment and depletion was conducted by using the MAGeCK test in paired mode for two replicates as previously described [18] (*see* **Note 17**).

4 Notes

1. There are lots of choices of online tools for obtaining sgRNA sequences, including CRISPROR, Broad GPP, etc. It is important to pick sgRNAs with high cleavage efficiency and on-target specificity based on the prediction scores that are provided by the bioinformatic tools. Usually, at least 3–10 sgRNAs per gene are recommended for pooled CRISPR screening to ensure a reliable result [1].
2. If multiple sgRNA libraries need to be constructed together, a barcode system [15] can be introduced to distinguish those different sgRNA libraries (if only one library is used, then skip this step and directly go to step 4).
3. The following steps 1 and 2 are only for oligo pools with a barcode sequence. The individual oligo pool can be separated based on PCR using different barcode-specific primers. Skip these steps and directly go to step 3 if no barcode sequence is included.
4. Optimization of cycle numbers is required. Perform multiple cycles (e.g., 10, 15, and 20 cycles) and run 2% agarose gel to optimize the PCR condition.

5. If no barcode sequence is included in the oligo pools, use 10 ng/μL synthesized oligonucleotides directly for PCR at this step; if a barcode sequence is introduced, then use diluted PCR product acquired from step 2 for this step's PCR; however, the dilution factor should be optimized to successfully get the PCR product.
6. Optimization of cycle numbers is required. Perform multiple cycles (e.g., 10, 15, and 20 cycles) and run 2% agarose gel to optimize the PCR condition.
7. Gibson assembly requires fewer manipulations of the sgRNA oligos compared with the conventional enzyme-based methods, which may result in potential bias in the CRISPR library.
8. To help better understand the colony amount needed for each different CRISPR library and other important factors that are mentioned in the rest of the protocol, the ACER library (Addgene) that we designed will be used as an example. There are 7580 sgRNAs in the ACER library, and we would expect 1000× coverage for the ACER library. From previous testing, we normally get approx. 50,000 colonies from 500 μL mixtures growing on one agar plate. To meet 1000× coverage, we would need approx. 8 million colonies, which is about 16 agar plates (~7580 sgRNAs × 1000/50,000 colonies per plate) with 500 μL mixtures per plate.
   The plate number and the volume of sample to be plated can always be adjusted based on the expected coverage, transformation efficiency, and experiment design.
   An alternative electroporation method could be used if the heat shock transformation does not meet the required coverage.
9. Try to get enough library plasmid to be sufficient for all designed experiments and avoid multiple amplifications of the library plasmid. Each amplification will potentially compromise the screen quality.
10. Use fresh viruses for infection if possible. Viruses could be stored at 4°C for at most 1 week and stored at −80°C for long-term storage. Avoid freezing and thawing cycles.
11. If there is low transfection efficiency, optimize transfection reagents and conditions, such as transfection reagent-to-DNA ratio, cell density and viability.
12. The cellular proliferation rate within the group subjected to drug treatment is anticipated to be lower compared to cells treated with the vehicle alone. It is advisable to optimize the drug dosages prior to commencing the CRISPR screen to ensure the manifestation of a robust phenotype while

simultaneously preserving cell viability. Optimal dosages should aim to induce mortality in approximately 20–30% of the cells.

13. Multiple time points can be set to harvest the screen cells for genomic DNA extraction, depending on the specific hypothesis or research aims. The goal is to see significant changes between drug treatment groups and vehicle treatment groups, at least 8-–10 population doubling time is needed before cell harvesting due to differences in cell proliferation caused by gene knockout. For the drug resistance screen, a later time point is suggested to better visualize the sgRNA enrichment. Vice versa, for the drug sensitivity screen, an earlier time point is better to see the sgRNA dropout.

14. Take ACER library as example, to achieve 1000× coverage, at least 8,000,000 cells which is approximately 52.8 μg (the amount of genomic DNA acquired from 1 million cells is typically around 6.6 μg) must be used for library PCR. In this case, multiple tubes per sample are needed for LaPCR1. Optimize the genomic DNA amount used in PCR 1 accordingly. Besides, optimization of cycle numbers is required. Perform multiple cycles (e.g., 10, 15, and 20 cycles) and run 2% agarose gel to optimize the PCR condition. Fewer PCR cycles should be employed to decrease the potential bias during PCR amplification.

15. LaPCR1 fragments are amplified with Nextera Ad1 and Ad2 PCR primers (Table 2). A total of 24 Ad2 indexes allows you to mix at most 24 samples on one pipeline [19].

16. Optimization of the cycle number and dilution factor for the PCR 1 product is required to successfully get the PCR product. Perform multiple cycles (e.g., 10, 15, and 20 cycles) and run 2% agarose gel to optimize the PCR condition. Fewer PCR cycles should be employed to decrease the potential bias during PCR amplification.

17. Insufficient genomic target coverage is a common limitation in pooled CRISPR screens and can be mitigated by increasing sgRNA library complexity, for example by including multiple independent sgRNAs per gene. Viral transduction conditions, including multiplicity of infection (MOI), should be optimized to favor single-sgRNA integration per cell while preserving adequate representation of all library elements throughout the screening process. Prior to screening, the integrity, diversity, and evenness of the sgRNA library should be verified by next-generation sequencing to ensure comprehensive target coverage and to minimize representation bias.

Acknowledgments

We thank the Lu lab members for their input in the experimental design and techniques used in this protocol. This work was supported in part by National Institutes of Health grant R01 CA259480. R.L. is an American Society of Hematology Scholar in Basic Science and a Research Scholar of the American Cancer Society (RSG-22-036-01-DMC).

References

1. Shalem O, Sanjana NE, Zhang F (2015) High-throughput functional genomics using CRISPR-Cas9. Nat Rev Genet 16(5):299–311. https://doi.org/10.1038/nrg3899
2. Jinek M, Chylinski K, Fonfara I, Hauer M, Doudna JA, Charpentier E (2012) A programmable dual-RNA-guided DNA endonuclease in adaptive bacterial immunity. Science 337(6096):816–821. https://doi.org/10.1126/science.1225829
3. Doudna JA, Charpentier E (2014) Genome editing. The new frontier of genome engineering with CRISPR-Cas9. Science 346(6213):1258096. https://doi.org/10.1126/science.1258096
4. Chen M, Mao A, Xu M, Weng Q, Mao J, Ji J (2019) CRISPR-Cas9 for cancer therapy: opportunities and challenges. Cancer Lett 447:48–55. https://doi.org/10.1016/j.canlet.2019.01.017
5. Katti A, Diaz BJ, Caragine CM, Sanjana NE, Dow LE (2022) CRISPR in cancer biology and therapy. Nat Rev Cancer 22(5):259–279. https://doi.org/10.1038/s41568-022-00441-w
6. Tzelepis K, Koike-Yusa H, De Braekeleer E, Li Y, Metzakopian E, Dovey OM, Mupo A, Grinkevich V, Li M, Mazan M, Gozdecka M, Ohnishi S, Cooper J, Patel M, McKerrell T, Chen B, Domingues AF, Gallipoli P, Teichmann S, Ponstingl H, McDermott U, Saez-Rodriguez J, Huntly BJP, Iorio F, Pina C, Vassiliou GS, Yusa K (2016) A CRISPR dropout screen identifies genetic vulnerabilities and therapeutic targets in acute myeloid leukemia. Cell Rep 17(4):1193–1205. https://doi.org/10.1016/j.celrep.2016.09.079
7. Michels BE, Mosa MH, Streibl BI, Zhan T, Menche C, Abou-El-Ardat K, Darvishi T, Członka E, Wagner S, Winter J, Medyouf H, Boutros M, Farin HF (2020) Pooled in vitro and in vivo CRISPR-Cas9 screening identifies tumor suppressors in human colon organoids. Cell Stem Cell 26(5):782–792. https://doi.org/10.1016/j.stem.2020.04.003 e787
8. Zhang H, Zhang Y, Zhou X, Wright S, Hyle J, Zhao L, An J, Zhao X, Shao Y, Xu B, Lee HM, Chen T, Zhou Y, Chen X, Lu R, Li C (2020) Functional interrogation of HOXA9 regulome in MLLr leukemia via reporter-based CRISPR/Cas9 screen. eLife 9. https://doi.org/10.7554/eLife.57858
9. Han K, Jeng EE, Hess GT, Morgens DW, Li A, Bassik MC (2017) Synergistic drug combinations for cancer identified in a CRISPR screen for pairwise genetic interactions. Nat Biotechnol 35(5):463–474. https://doi.org/10.1038/nbt.3834
10. Doench JG, Fusi N, Sullender M, Hegde M, Vaimberg EW, Donovan KF, Smith I, Tothova Z, Wilen C, Orchard R, Virgin HW, Listgarten J, Root DE (2016) Optimized sgRNA design to maximize activity and minimize off-target effects of CRISPR-Cas9. Nat Biotechnol 34(2):184–191. https://doi.org/10.1038/nbt.3437
11. Wang T, Wei JJ, Sabatini DM, Lander ES (2014) Genetic screens in human cells using the CRISPR-Cas9 system. Science 343(6166):80–84. https://doi.org/10.1126/science.1246981
12. Aguirre AJ, Meyers RM, Weir BA, Vazquez F, Zhang CZ, Ben-David U, Cook A, Ha G, Harrington WF, Doshi MB, Kost-Alimova M, Gill S, Xu H, Ali LD, Jiang G, Pantel S, Lee Y, Goodale A, Cherniack AD, Oh C, Kryukov G, Cowley GS, Garraway LA, Stegmaier K, Roberts CW, Golub TR, Meyerson M, Root DE, Tsherniak A, Hahn WC (2016) Genomic copy number dictates a gene-independent cell response to CRISPR/Cas9 targeting. Cancer Discov 6(8):914–929. https://doi.org/10.1158/2159-8290.Cd-16-0154
13. Wang T, Birsoy K, Hughes NW, Krupczak KM, Post Y, Wei JJ, Lander ES, Sabatini DM (2015) Identification and characterization of essential genes in the human genome. Science 350(6264):1096–1101. https://doi.org/10.1126/science.aac7041
14. Zhou X, Zhang P, Aryal S, Zhang L, Lu R (2023) UTX loss alters therapeutic responses

in KMT2A-rearranged acute myeloid leukemia. Leukemia 37(1):226–230. https://doi.org/10.1038/s41375-022-01741-8
15. Canver MC, Haeussler M, Bauer DE, Orkin SH, Sanjana NE, Shalem O, Yuan GC, Zhang F, Concordet JP, Pinello L (2018) Integrated design, execution, and analysis of arrayed and pooled CRISPR genome-editing experiments. Nat Protoc 13(5):946–986. https://doi.org/10.1038/nprot.2018.005
16. Zhou X, Zhang L, Aryal S, Veasey V, Tajik A, Restelli C, Moreira S, Zhang P, Zhang Y, Hope K.J, Zhou Y, Cheng C, Bhatia R, Lu R (2024) Epigenetic regulation of noncanonical menin targets modulates menin inhibitor response in acute myeloid leukemia. Blood 144, 2018-2032. https://doi.org/10.1182/blood.2023023644
17. Li W, Xu H, Xiao T, Cong L, Love MI, Zhang F, Irizarry RA, Liu JS, Brown M, Liu XS (2014) MAGeCK enables robust identification of essential genes from genome-scale CRISPR/Cas9 knockout screens. Genome Biol 15(12), 554. https://doi.org/10.1186/s13059-014-0554-4
18. Li W, Köster J, Xu H, Chen CH, Xiao T, Liu JS, Brown M, Liu XS (2015) Quality control, modeling, and visualization of CRISPR screens with MAGeCK-VISPR. Genome Biol 16, 281. https://doi.org/10.1186/s13059-015-0843-6
19. Buenrostro JD, Giresi PG, Zaba LC, Chang HY, Greenleaf WJ (2013) Transposition of native chromatin for fast and sensitive epigenomic profiling of open chromatin, DNA-binding proteins and nucleosome position. Nat Methods 10(12):1213–1218. https://doi.org/10.1038/nmeth.2688

Check for updates

Chapter 16

Tools for Cancer Research: CRISPR/Cas-Based Gene Editing and Auxin-Induced Degron Systems

Qian Tang and Jian Liu

Abstract

CRISPR/Cas9 and auxin-degron systems represent two powerful and complementary genetic tools that have revolutionized cancer research. The clustered regularly interspaced short palindromic repeats (CRISPR)-Cas (CRISPR-associated protein)-based gene editing system has been widely used in numerous biological fields, significantly enhancing the capacity of researchers to elucidate the underlying mechanisms of biological phenomena. The two most prevalent applications for CRISPR-mediated gene editing are knockout and knock-in. This protocol covers both single-gRNA (sgRNA) and dual-gRNA directed knockout systems. Furthermore, we describe the procedure for constructing an Auxin-Inducible Degron (AID) knock-in-mediated system to induce the rapid degradation of a target protein in cells, thereby investigating its function.

Keywords Gene editing, CRISPR-Cas9, sgRNA, dual-gRNA, AID system

1 Introduction

Recent technologies such as CRISPR/Cas9 and the Auxin-Degron systems provide researchers with unprecedented control over genes and proteins to investigate cancer biology and develop potential new therapies. CRISPR/Cas9 enables precise genome editing to study gene function and create disease models, while auxin-degron systems allow rapid and reversible protein degradation to examine protein roles. Both tools are increasingly used in almost every lab that studies cancer biology.

CRISPR/Cas system originated as a genome defense mechanism of some prokaryotes to defend against viruses and other exogeneous genetic material [1]. The optimized CRISPR/Cas system for gene editing has two key components: the Cas protein with site-specific DNA cleaving endonuclease activity and the RNA component or guide RNA (sgRNA).

Archana Dhasarathy and Motoki Takaku (eds.), *Cancer Epigenetics*, Methods in Molecular Biology, vol. 3005,
https://doi.org/10.1007/978-1-0716-5072-1_16,

Following the induction of the CRISPR/Cas9 system in a eukaryotic cell line, the gRNA guides Cas9 to cleave the target DNA sequence and induce a double-strand break (DSB) in the DNA. The DSBs can then be repaired by the endogenous DNA repair mechanisms in the cell, primarily through non-homologous end joining (NHEJ), or homology-directed repair (HDR) [2]. NHEJ results in the production of short insertions or deletions (indels), which can disrupt the target gene and thus result in a knockout. In contrast, HDR is an error-free pathway to repair DSB that only happens when homologous DNA fragments of the damaged region exist in normal cells [3]. The CRISPR/Cas9-mediated knock-in procedure employs the HDR mechanism. By introducing an exogenous DNA template, provided by a donor plasmid, a series of gene-editing operations, including insertion, deletion, and mutation, can be performed.

CRISPR/Cas system can be implemented in other scenarios. Rather than direct genome alteration, other means can be adopted to modulate the target gene expression. For instance, the deactivated form of Cas9 protein without nuclease activity termed dead Cas9 (dCas9) can bind to the target region [4]. Transcriptional modulation domains fused to dCas9 can activate or inhibit expression of the target gene. Further, target gene transcription can be regulated through the epigenetic modification mediated by dCas9 conjugated with protein domains with histone modification enzymes [5].

CRISPR screening is a high-throughput technique that leverages CRISPR-Cas systems to simultaneously disrupt or modify a number of genes to explore their functions. This technique involves introduction into cells of both the Cas9 enzyme and a collection of gRNAs termed gRNA library targeting a large number of genes [6]. To conduct screening, the gRNA library is transfected into cells either simultaneously with Cas9 or into cells that have been previously engineered to express Cas9.

The Auxin-Inducible Degron (AID) technology enables rapid, targeted degradation of proteins in cells utilizing CRISPR/Cas9-mediated knock-in procedures. The system is based on Auxins, a family of plant hormones important for plant growth. Small molecules of auxins such as indole-3-acetic acid (IAA) can bind to the transport inhibitor response 1 (TIR1) protein and promote its interaction with Skp1-Cul1-F-box (SCF) complex. This interaction recruits E3 ubiquitin conjugating enzyme to polyubiquitylate the AID sequence of the target protein, marking it for subsequent degradation by the proteasome. By introducing exogenous TIR1 and fusing an AID peptide to the target protein, protein degradation can be induced in non-plant cells [7]. The AID2 system, an optimized version of AID technology, employs a mutant version of TIR1 protein [(TIR1 (F79G)] and a specially designed alternative IAA ligand, 5-phenyl-indole-3-acetic acid (5-Ph-IAA), which allows for more rapid degradation at lower inducer concentration [8]. The AID2 system has been applied to model potential anti-cancer drugs by depleting BRD4 and TOP4A proteins [8]. Further, a novel AID system was modified to degrade TP53 protein, a known tumor suppressor, in specific phases of the cell cycle based on the Cell Cycle Status (ROLECCS system) [9]. Finally, AID technology was used as a tool to regulate gene expression temporally and spatially by reprogramming the chromatin state [10].

Here, we will describe methods to utilize both CRISPR/Cas and the AID2 systems for cancer biology research.

2 Materials

2.1 Plasmids

Plasmids	Company, catalog
LentiCRISPRv2	Addgene, 52961
psPAX.2	Addgene, 12260
pMD.2G	Addgene, 12259
Plenti-H1-hU6-dual-gRNA	From Chen lab
PX459	Addgene, 62988
pAY5	Addgene, 140532
PL552	Addgene, 68407
pCAG-Cre	Addgene, 13775
pMK381-AAVS1-CMV-OsTIR1 (F74G)	Addgene, 140536

2.2 Reagents

Reagents
10× loading buffer
Polyethylenimine Linear (PEI) MW40000
Puromycin Dihydrochloride
Fetal Bovine Serum
Luria-Bertani medium (LB) with Ampicillin 100 μg/mL
Luria-Bertani medium (LB) with Kanamycin 100 μg/mL
Luria-Bertani plate (LB) with Ampicillin 100 μg/mL
Luria-Bertani plate (LB) with Ampicillin 100 μg/mL

2.3 Commercial Kit

Reagents
T4 ligase
BsmBI
FastPure® Gel DNA Extraction Mini Kit
BbsI-HF

ClonExpress II One Step Cloning Kit
Phanta ® Max Super-Fidelity DNA Polymerase
EcoRI-HF
BamHI-HF

2.4 Equipment

Equipment
PCR machine
Bacteria shaker
Bacteria incubator
CO2 cell incubator

3 Methods

3.1 Construction of the sgRNA System

1. Use gRNA design websites such as http://crispor.gi.ucsc.edu/ and http://www.e-crisp.org/ and select the top gRNA generated.
2. Add adapters to two oligos of gRNA based on enzyme digestion sites, conduct gRNA ligation according following conditions.

gRNA-oligo1	1 μL
gRNA-oligo2	1 μL
T4 buffer	1 μL
ddH2O	7 μL
Total	10 μL

   Transfer reactions to the PCR machine programmed at 95°C for 5 min; then turn off the PCR machine, cool for 1 h and dilute 200-fold.

3. Set up Enzyme digestion of the LentiCRISPR vector as below:

LentiCRISPRv2	1 μg
BsmBI	5 μL
10× Cutsmart buffer	5 μL
ddH2O add to	50 μL

   Incubate samples at 55°C for 30–60 min.

4. DNA electrophoresis: Add 5.5 μL 10× loading buffer to the product of Step 3, load samples on a 1% agarose gel, and electrophorese for half an hour.

5. Gel extraction is conducted following the protocol of the Gel DNA Extraction Kit.
6. Ligate gRNA to the vector by mixing together the following:

gRNA dilution product	2 μL
Plasmid from Step 3	40–50 ng
10× T4 buffer	1 μL
T4 ligase	1 μL
ddH2O add to	10 μL

Incubate at RT for 2 h.

7. Plasmid transformation:
   a. Take out competent cells on ice.
   b. Add recombined plasmid (do not exceed 1/10 of total volume) to competent cells, and place on ice for 30 min.
   c. Place in a 42°C water bath for 45 s exactly (use a timer), and place on ice immediately for 2 min.
   d. Gently add 900 μL LB liquid medium (preheated at 37°C).
   e. Recovery: Transfer the bacteria to the shaker (37°C, 220 rpm) for 60 min. Warm up the bacterial plate while waiting.
8. Plate cells:
   a. Centrifuge bacteria to pellet cells for 1 min, 2300 × *g*.
   b. Remove most of the supernatant and retain about 100 μL with the pellet.
   c. Gently resuspend the pellet with the remaining liquid and spread evenly on the plate.
   d. Incubate at 37°C overnight.
9. The next day, pick 2–3 individual bacterial colonies and inoculate them in media for sequencing.
10. Perform the plasmid DNA extraction following the protocol of any standard Plasmid miniprep Kit.
11. Sequence the extracted plasmid DNA to determine clone accuracy.
12. Plasmid transfection:
    Transfect HEK293T cells with the newly constructed Lenti-V2 plasmids along with packaging plasmids (e.g., psPAX.2 and pMD.2G) following the protocol of Polyethylenimine Linear (PEI) MW40000 (*see* **Note 1**). Place the dish at 37°C, 5% CO_2 cell incubator for around 10 h, then add 5 mL of complete medium, and place the dish back in the incubator.

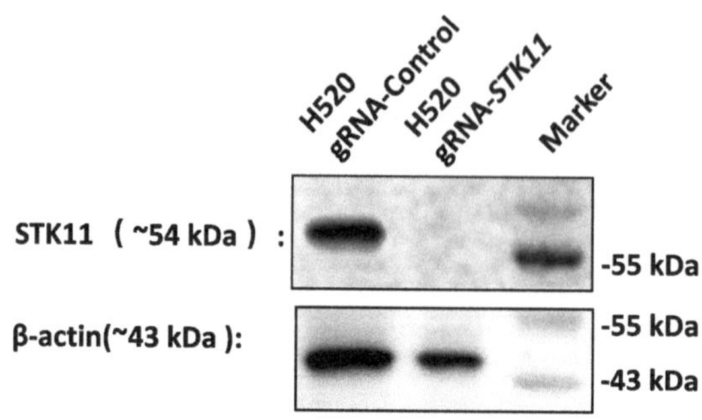

Fig. 1 *STK11* has been knocked out successfully by the LentiCRISPRv2-sgRNA system on the human lung squamous cell line H520, verified by Western blot

13. Virus collection:
    After 48 h of transfection, collect supernatant from the dish, filter with a 0.45 μm pore size filter to collect viruses and discard cells.
14. Virus infection:
    Target cells are transduced with virus for 48–72 h. Infected cells are treated with 2 μg/mL puromycin with 10% FBS in cell culture medium for target gene knockout cells screening.
15. Knockout verification:
    Collect RNA and protein after finishing screening. Perform RT-qPCR and Western Blot to verify target gene knockout efficiency. Here, serine/threonine kinase 11 (*STK11*) was used as the target gene following the above protocol, which was knocked out successfully, verified by Western Blot (see Fig.1).

3.2 Dual-gRNA System Construction

1. As above, select the top gRNAs on the gRNA design website.
2. Digest the vector with the BbsI enzyme as follows:

plenti-H1-hU6-dual-gRNA	5 μg
BbsI-HF (NEB)	1 μL
10× rCutsmart buffer	5 μL
ddH2O add to	50 μL

   Incubate the reaction at 37°C for 1 h.
3. Add loading dye 5.5 μL and load the digest on a 1% DNA gel, and electrophorese samples for half an hour.. Gel extract the linearized vector using FastPure® Gel DNA Extraction Mini Kit. Determine the concentration of the gel-extracted vector DNA using a spectrophotometer such as Nanodrop.
4. Add adapters to the ends of two gRNAs (*see* **Note 2**).

5. The insertion fragment is obtained by PCR using gRNA oligo1 and gRNA oligo2 as the forward primer and the reverse primers, respectively, and plasmid Plenti-H1-hU6-dual-gRNA as the DNA template. (PCR conditions? If the same as above, please indicate which step).
6. Add loading dye 5.5 μL and load the digest on a 1% DNA gel, and electrophorese samples for half an hour. Gel extract the linearized vector using FastPure® Gel DNA Extraction Mini Kit. Determine the concentration of the gel-extracted vector DNA using a spectrophotometer such as Nanodrop.
7. Ligate the insertion fragment to the linear vector as follows using a cloning kit (e.g., ClonExpress II One Step Cloning Kit) (*see* **Note 3**).

Plenti-H1-hU6-dual-gRNA linear vector	150 ng
Insertion fragment	20 ng
5× CE Buffer	4 μL
Exnase	2 μL
ddH2O add to	20 μL

Incubate the ligation reaction at 37 °C for 30 min, and place the ligation reaction on ice to stop the reaction.

8. Transformation plasmids into bacteria as Step 3.1.7. and pick 2–3 bacterial colonies for plasmid prep, followed by sequencing to confirm insertion as Step 3.1.11.
9. Transduce HEK293T cells, collect virus and infect target cells as Step 3.1.12.
10. A quick knockout efficiency check can be performed by PCR. Here, we used PCR to show successful partial knockout of the STK11 gene using the dual-gRNA system (see Fig. 2).

3.3 AID2 System Construction

1. Ligate gRNA to the PX459 plasmid:
   a. Select the top gRNA close to the stop codon of the target gene on the gRNA design website.
   b. Add adapters to two oligos of gRNA based on enzyme digestion sites, and ligate gRNA by mixing together the following in an Eppendorf tube:

gRNA-oligo1	1 μL
gRNA-oligo2	1 μL
T4 buffer	1 μL
ddH2O	7 μL
Total	10 μL

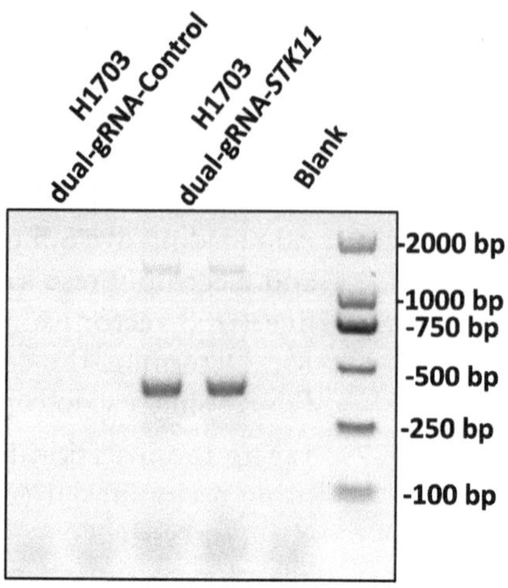

Fig. 2 *STK11* has been knocked out successfully by a dual-gRNA system in the human lung squamous cell line H1703, verified by PCR. Each sample has two replications. Blank means sterilized ddH2O involved to prove the specificity of PCR primers

Place reactions in a PCR machine at 95°C for 5 min, then turn off the PCR machine, cool for 1 h and dilute 200-fold.

c. Perform BbsI enzyme digestion on the PX459 plasmid by mixing together the following reagents in an Eppendorf tube:

PX459	5 μg
BbsI	1 μL
10× Cut buffer	5 μL
ddH2O add to	50 μL

Incubate reaction at 37°C for 60 min.

d. Add 5.5 μL 10× loading dye and load the digest onto a 1% DNA gel. Electrophorese the samples for half an hour (*see* Fig. 3a). Gel extract the linearized vector using FastPure® Gel DNA Extraction Mini Kit. Determine the concentration of the gel-extracted vector DNA using a spectrophotometer such as Nanodrop.

e. Ligate the gRNA to the vector:

gRNA dilution product	2 μL
plasmid from Step c	40–50 ng
10× T4 buffer	1 μL
T4 ligase	1 μL
ddH2O add to	10 μL

Incubate reaction at RT for 2 h.

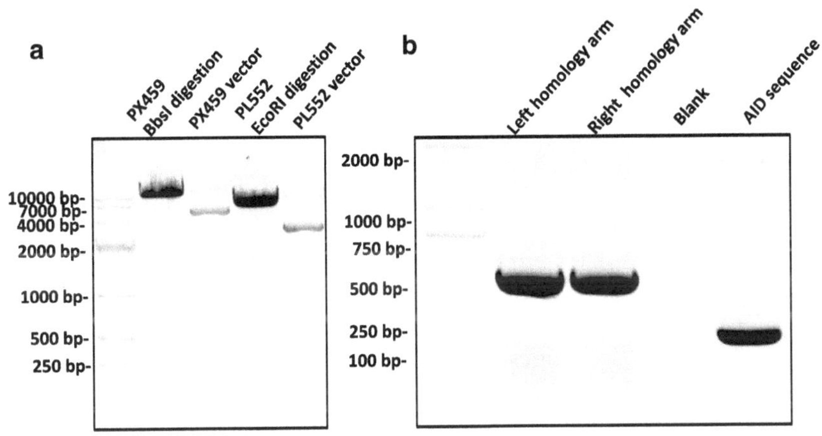

Fig. 3 **a** shows that the PX459 and PL552 vector was linearized by the BbsI and EcoRI enzymes, respectively. **b** shows the left homology arm (500 bp) and right homology arm (500 bp) of the knock-in targeted site, which was obtained by PCR. AID sequence (200 bp) was obtained from the pAY5 plasmid by PCR. The PCR product was extracted from the gel to ligate to the corresponding lineage vector

f. Transform bacteria as above and pick 2–3 individual bacterial colonies for sequencing.

g. Perform plasmid prep.

2. AID2 Donor Plasmid Construction:

a. Preparation of AID2 insertion sequence: Use pAY5 as a template and amplify it by a high-fidelity polymerase under the following conditions (*see* **Note 4**) (see Fig. 3b).

2× Phanta Max Buffer	25 μL
dNTP Mix (10 mM each)	1 μL
Upstream Primer (10 μM)	2 μL
Downstream Primer (10 μM)	2 μL
Super-Fidelity DNA Polymerase	1 μL
DNA Template: pAY5	10 pg–30 ng
ddH2O	up to 50 μL

b. Preparation of homology arms for the AID donor plasmid:

The upstream 500–1000 bp of the target gene stop codon is used as the left homology arm, and the downstream 500–1000bp is used as the right homology arm (*see* **Note 5**). The genome DNA of transfected cells serves as the template for amplification by Phanta® Max Super-Fidelity DNA Polymerase under the above conditions; the amount of template DNA is 50–400 ng (see Fig. 3b).

c. Preparation of linearized vector for AID donor plasmid: Digest the plasmid pL552 with EcoRI-HF as follows:

pL552	4 μg
EcoRI-H	1 μL
10× rCutsmart buffer	5 μL
q.s. with ddH2O to	50 μL

Incubate at 37°C for 1 h.

d. Add 5.5 μL 10× loading dye (how much) and load the digest on a 1% DNA gel, and electrophorese samples for half an hour (see Fig.3a). Gel extract the linearized vector using FastPure® Gel DNA Extraction Mini Kit. Determine the concentration of the gel-extracted vector DNA using a spectrophotometer such as Nanodrop.

e. Ligate the PCR product to the EcoRI-HF digested linearized vector by ClonExpress II One Step Cloning Kit;

Linearized vector	123 ng
Left homology arm	20 ng
AID inserted fragment	8.16 ng
5× CE Buffer	4 μL
Exnase II	2 μL
ddH2O q.s. to	20 μL

Incubate the reaction at 37°C for 30 min, and place the product on ice immediately.

f. Transform bacteria as above and pick 2–3 individual bacterial colonies for sequencing.

g. Perform plasmid prep.

h. Digest the newly constructed PL552 plasmid with the AID fragment and the target gene left arm by BamHI-HF according to the following conditions.

PL552	4 μg
BamHI-HF	1 μL
10× rCutsmart buff	5 μL
ddH2O to	50 μL

Incubate at 37°C for 1 h.

i. Add 5.5 μL 10× loading dye (how much) and load the digest onto a 1% DNA gel. Electrophorese the samples for half an hour (see Fig. 3a). Gel extract the linearized vector using FastPure® Gel DNA Extraction Mini Kit. Determine the concentration of the gel-extracted vector DNA using a spectrophotometer such as Nanodrop.

j. Ligate the right homology arm PCR product to the BamHI-HF digested linearized vector by ClonExpress II One Step Cloning Kit;

Linearised vector	137 ng
Right homology arm	20 ng
5× CE Buffer	4 μL
Exnase	2 μL
ddH2O add to	20 μL

Incubate at 37°C for 30 min, then place the reaction on ice immediately.

k. Transform bacteria as above and pick 2–3 individual bacterial colonies for sequencing.

l. Perform plasmid prep.

m. Transfection: Transfect cells with PX459 containing the gRNA of the target gene and PL552 with the AID fragment and target gene gRNA homology arms (*see***Note 6**).

n. Screening of positive cells: Screen cells by puromycin (*see* **Note 7**). Stop screening when there is no significant cell death.

o. Dilute cells continuously and seed single cells in a 96-well plate as much as possible (*see* **Note 8**). When the density of the monoclonal cells reaches 50–60%, pass cells into a 48-well plate. When cells cover 80% of the 24-well plate, passage cells. Half of the cells are used for genome extraction.

p. Utilize PCR to validate whether the AID fragment sequence has been knocked-in to the genome of cells (*see* **Note 9**).

q. The obtained cells (homozygous for the inserted AID fragment) can be expanded and then transfected with pCAG-Cre, a cre plasmid targeting loxP sites located at the ends of the puromycin resistance gene, and after the excision of the puromycin resistance gene, cells lose their resistance to puromycin, and can be transfected with other plasmids and screened with puromycin afterwards.

r. Similarly, the OsTIR1 insertion sequence can be obtained from pMK381-AAVS1-CMV-OsTIR1 (F74G) by PCR and cloned into the PL552 vector with homology arms of the safe harbor region, which could serve as a donor plasmid to provide OsTIR1 insertion into the incision.

s. Continue to screen by puromycin at the optimal concentration for cells until there is no significant cell death.

t. Utilize PCR and western blot to validate whether the OsTIR1 sequence is inserted and expressed. Select cells with OsTIR1 expression for subsequent experiments.

4 Notes

1. PsPAX2: pMD.2G: pLenti expression vector = 2:1:4
2. To increase the transcription efficiency, a G-C pair is added to the start site of gRNA.
3. To ensure the volume accuracy of adding the sample, the linear vector and insertion fragment can be diluted. The sample amount of each component is supposed to be larger than 1 μL. The highest recombination efficiency is reached at 30 min. A reaction time that is too long or too short will decrease cloning efficiency. The recombinant product can be stored at −20°C for 1 week.
4. It is recommended that the AID sequence be ligated to either the left homology arm or the right homology arm of the stop codon. Take the ligation of the AID sequence and the left homology arm as an example. One primer needs to contain a fragment of the left homology arm and a fragment of the AID sequence. Another primer needs to contain a fragment of the plasmid and a fragment of the AID sequence.
5. Ensure that neither the left nor the right homology arm contains a stop codon.
6. Transfection can be performed by electrotransfection or transfection reagents like Lipo2000 or PEI, etc. Test the transfection efficiency of various transfection methods in different cell lines in advance and select the most effective method.
7. Optimize the screening condition in advance.
8. Cells are cultured in a 96-well plate; ensure that the cell culture medium is replenished and changed often.
9. Homozygous cells of the inserted AID fragment can be used for follow-up experiments.

Acknowledgments

This work was supported by grants to Dr. Jian Liu from the Natural Science Foundation (NSF) of China (General Grant: 82172899 and 82472637), the NSF of Zhejiang Province (Distinguished Young Scholars: LR22H160002).

References

1. Jinek M, Chylinski K, Fonfara I et al (2012) A programmable dual-RNA-guided DNA endonuclease in adaptive bacterial immunity. Science 337(6096):816–821. https://doi.org/10.1126/science.1225829
2. Nambiar TS, Baudrier L, Billon P, Ciccia A (2022) CRISPR-based genome editing through the lens of DNA repair. Mol Cell 82(2):348–388. https://doi.org/10.1016/j.molcel.2021.12.026
3. Chen CC, Feng WR, Lim PX et al (2018) Homology-directed repair and the role of BRCA1, BRCA2, and related proteins in genome integrity and cancer. Annu Rev Canc Biol 2:313–336. https://doi.org/10.1146/annurev-cancerbio-030617-050502
4. Wang CK, Qu YH, Cheng JKW et al (2022) dCas9-based gene editing for cleavage-free genomic knock-in of long sequences. Nat Cell Biol 24(2):268. https://doi.org/10.1038/s41556-021-00836-1
5. Hilton IB, D'Ippolito AM, Vockley CM et al (2015) Epigenome editing by a CRISPR-Cas9-based acetyltransferase activates genes from promoters and enhancers. Nat Biotechnol 33(5):510–U225. https://doi.org/10.1038/nbt.3199
6. Bock C, Datlinger P, Chardon F et al (2022) High-content CRISPR screening. Nat Rev Method Prime 2(1). https://doi.org/10.1038/s43586-022-00098-7
7. Nishimura K, Fukagawa T, Takisawa H et al (2009) An auxin-based degron system for the rapid depletion of proteins in nonplant cells. Nat Methods 6(12):917–U978. https://doi.org/10.1038/nmeth.1401
8. Yesbolatova A, Saito Y, Kitamoto N et al (2020) The auxin-inducible degron 2: technology provides sharp degradation control in yeast, mammalian cells, and mice. Nat Commun 11(1), 5701. https://doi.org/10.1038/s41467-020-19532-z
9. Capece M, Tessari A, Mills J et al (2023) A novel auxin-inducible degron system for rapid, cell cycle-specific targeted proteolysis. Cell Death Differ 30(9):2078–2091. https://doi.org/10.1038/s41418-023-01191-4
10. Kuscu C, Mammadov R, Czikora A et al (2019) Temporal and spatial epigenome editing allows precise gene regulation in mammalian cells. J Mol Biol 431(1). https://doi.org/10.1016/j.jmb.2018.08.001

Chapter 17

The High-Throughput C-Circle Assay for Identifying Modulators of Alternative Lengthening of Telomeres (ALT) Activity

Merrill M. Froney, Sara T. Ledeboer, Michael B. Jarstfer and Samantha G. Pattenden

Abstract

Telomeres are long, repetitive nucleoprotein structures that protect the ends of linear chromosomes during DNA replication. In differentiated cells, telomeres shorten with each cell division, eventually becoming too short to protect chromosome ends from DNA end fusions or recombination. This critical event is known as the "end replication" problem and results in cell senescence. To achieve replicative immortality, tumor cells must evade senescence by maintaining telomere length using one of two telomere maintenance mechanisms: reactivation of telomerase or the alternative lengthening of telomeres (ALT) pathway. The telomerase pathway is normally turned off in differentiated cells, but is reactivated in approximately 85% of cancers. Telomerase uses reverse transcription to elongate telomere ends during DNA replication. The ALT pathway is activated in approximately 15% of cancers. ALT relies on the homologous recombination DNA repair pathway to lengthen telomeres. Activation of ALT is associated with poor prognosis in several cancers, including osteosarcoma and neuroblastoma. ALT is a tumor-specific pathway, which makes it an attractive therapeutic target; however, there are currently no ALT-specific chemical inhibitors. To advance biological understanding of the ALT pathway and to identify potential ALT therapeutics, the high-throughput C-circle assay was developed.

Keywords Drug discovery, High-throughput screening, C-circle assay, Alternative lengthening of telomeres, Cancer

1 Introduction

Telomere length is a key determinant of cell fate. In differentiated cells, telomeres shorten with each cell division, eventually becoming too short to effectively protect chromosome ends, resulting in cell senescence [1, 2]. A key hallmark of cancer is unlimited replicative potential; cancer cells must evade normal senescence by activating a telomere maintenance mechanism (TMM) [3, 4]. TMMs are an attractive target for cancer therapy as they are required for cancer cell

Archana Dhasarathy and Motoki Takaku (eds.), *Cancer Epigenetics*, Methods in Molecular Biology, vol. 3005,
https://doi.org/10.1007/978-1-0716-5072-1_17,

but not normal cell survival [3]. There are two main TMMs activated by cancer cells. The most common TMM is telomerase activation, which is utilized by approximately 85% of cancers [5–8]. This pathway involves the reactivation of telomerase, a reverse-transcription-based pathway utilized for stem cell telomere maintenance during development [8]. Until recently, it was believed that the remaining 15% of cancers utilize the alternative lengthening of telomeres (ALT) pathway; however, that number is likely closer to 40%[9], as tumors can harbor both TMMs. ALT utilizes homologous recombination to lengthen telomeres [10–12] and is most prevalent in tumors of mesenchymal origin [13]. For instance, 65% of osteosarcomas [14], 45% of pediatric glioblastomas [15], and 25% of high-risk neuroblastomas [16] utilize ALT for telomere maintenance. There is a clinical need to understand ALT biology and to develop ALT-specific therapies for the treatment of ALT-positive cancers [17–21].

High-throughput screens enable large compound libraries to be efficiently tested for activity against a given target [22]. Target-based screens are developed when modulation of a specific protein has been identified a priori to affect a pathway of interest. These types of screens are generally conducted in vitro with purified protein but can also be applied to cell models [23]. Phenotypic screens are utilized in cases where a specific cellular target is unknown [24]. Phenotypic screens rely on a biomarker, which is a measurable cellular product that is present only in cells using the biological pathway of interest. Since promising ALT-specific therapeutic targets have not yet been identified, a phenotypic screen based on an ALT-specific biomarker was developed.

Biomarkers that accumulate specifically in ALT positive cells include (a) increased levels of a long RNA called TERRA [25–27]; (b) ALT-associated PML bodies (APBs) [28]; and (c) the accumulation of partially single-stranded extrachromosomal DNA circles (C-circles) [29]. C-circles, a robust biomarker found in all ALT-positive cells, are tightly associated with ALT activity [29]. A single-tube assay was developed for detecting C-circle levels in cells as well as in patient blood serum [30, 31]. This assay, however, was not designed to test hundreds or thousands of samples in parallel. We developed the high-throughput C-circle assay (HT-CCA) to accelerate drug discovery efforts for ALT-specific therapeutics [32].

The C-circle assay is comprised of four steps: (1) cell plating and compound treatment, (2) cell lysis, (3) rolling circle amplification of C-circles, and (4) quantitative polymerase chain reaction (qPCR) to assess relative C-circle levels [30, 32] (Fig. 1). In brief, cells are seeded on 384-well plates, treated with compound, then lysed to release genomic DNA and C-circles. C-circles are specifically amplified over background telomeric DNA by rolling circle amplification with phi29 DNA polymerase. Since C-circles are derived from the C-rich telomere strand ($CCCTAA_n$), they can be selectively amplified over telomeric DNA repeats by omitting deoxycytidine triphosphate (dCTP) from the rolling circle amplification step. Relative C-circle levels are determined

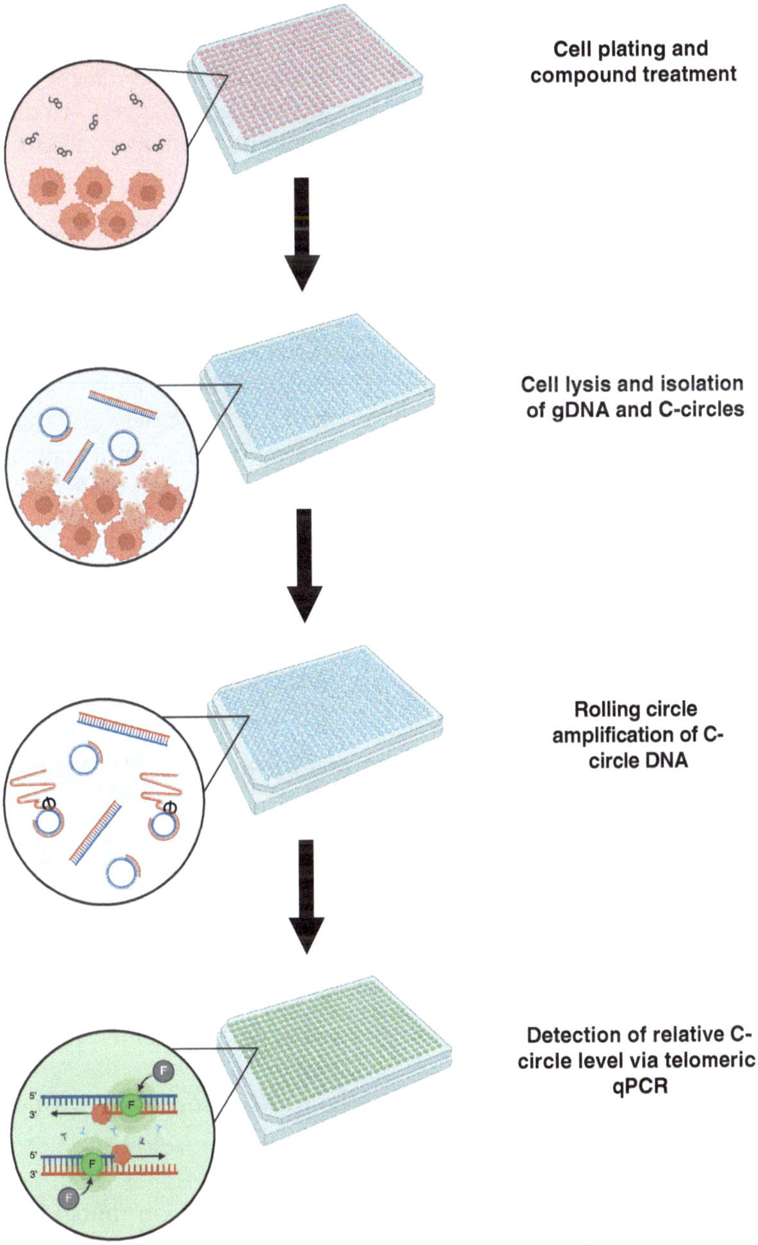

Fig. 1 The high-throughput C-circle assay workflow. Depiction of the main steps of the high-throughput C-circle assay [32]. The assay begins with plating cells followed by treatment with vehicle or compounds. After incubation, cells are lysed to release the C-circle DNA and genomic DNA (gDNA) from the cells. To obtain a robust C-circle signal over background telomeric DNA, C-circles are specifically amplified using rolling circle amplification. Samples are analyzed via relative qPCR to determine the relative C-circle level of samples or to detect hits using the SSMD* method [33]

by qPCR through comparison of a telomeric DNA amplicon to an amplicon from a non-telomeric genomic region. Hit compounds are identified using the SSMD* metric [33] (recommended for compound library screening) or by calculating a relative C-circle level for each sample (recommended for dose titrations/hit validation).

2 Materials

Buffers for each step can be prepared fresh or stored for long-term use. Lysis buffer can be stored at room temperature, while rolling circle amplification (RCA) reaction buffer and resuspended protease are stored at −20°C. If there is a pause between steps in the assay, completed reactions can be stored at 4°C until ready for the following step. This assay can be performed with electronic multichannel pipettes or an automated liquid handler.

2.1 Cell Plating

1. Cell lines: The assay can be optimized for use with any cell line, but we have provided examples that have been optimized for screening [32].
2. 384-well tissue culture-treated plates.
3. McCoy's 5A + 15% fetal bovine serum (FBS) (SaOS-2 cells).
4. Roswell Park Memorial Institute (RPMI) Medium + 10% FBS (SK-N-FI cells).
5. Dulbecco's Modified Eagle Medium (DMEM) + 10% FBS (MG-63 and SH-SY5Y cells).
6. Molecular biology grade dimethyl sulfoxide (DMSO).
7. Compound stocks at 10 mM concentration in 100% DMSO (vehicle).
8. Electronic multichannel (8 or 16 channel) pipette and tips.

2.2 Cell Dissociation, Lysis, and Protease Digestion

1. Lysis buffer [30]: 50 mM KCl (autoclaved), 10 mM Tris-HCl, pH 8.5 (autoclaved), 2 mM $MgCl_2$, 0.5% v/v NP40, 0.5% v/v Tween-20.
2. Serine protease: resuspend 7.5 Anson units (AU) of protease (QIAGEN) in 7 mL of molecular biology grade water (*see* **Note 1**).
3. 1× Phosphate-buffered saline (PBS).
4. V-bottom 384-well plates.
5. Incubators: Biotang Inc. microplate thermoshaker and Eppendorf ThermoMixer C with PCR 384 SmartBlock and ThermoTop.
6. Plate seals: microfilm A plate seals (BioRad), silicone 384-well plate mats, and aluminum plate seals.

Table 1
Rolling circle amplification master mix recipe for one plate

Component	Volume required (μL)
Molecular biology grade water	2580
10× reaction buffer (recipe above)	800
100 mM dATP (*see* **Note 3**)	80
100 mM dGTP (*see* **Note 3**)	80
100 mM dTTP (*see* **Note 3**)	80
100 mM DTT (do not store master mix with DTT—only add fresh. *See* **Note 2**)	80
Total volume	3700

2.3 Rolling Circle Amplification

1. 10× reaction buffer [30]: 330 mM Tris-acetate (pH 7.9), 100 mM Mg-acetate, 660 mM K-acetate, 1% v/v Tween-20, 10 mM dithiothreitol (DTT) (do not add DTT to reaction buffer stocks—only add fresh right before using. *See* **Note 2**).
2. Reaction master mix (volumes required for one 384-well plate) (Table 1).
3. Molecular biology grade water.
4. EquiPhi29 DNA polymerase (Thermofisher Scientific).
5. V-bottom 384-well PCR plates.
6. Eppendorf ThermoMixer C with PCR 384 SmartBlock and ThermoTop.
7. Aluminum plate seals.

2.4 Rolling Circle Amplification Dilution, and Telomeric qPCR

1. Molecular biology grade water.
2. V-bottom 384-well plates.
3. 384-well optically clear PCR plates.
4. Sterile, optically clear plate seals.
5. 2× SYBR green master mix (e.g., BioRad SsoAdvanced SYBR SYBR Green Supermix).
6. Telomere-specific Primers:
   Forward: 5′-CGGTTTGTTTGGGTTTGGGTTTGGGTTT-GGGTTTGGGTT-3′
   Reverse: 5′-GGCTTGCCTTACCCTTACCCTTACCCTTA-CCCTTACCCT-3′.
7. Non-telomeric genomic region primers:
   Forward: 5′-TTCTCCAACTTTGGAAGCCCAGGA-3′
   Reverse: 5′-TGTCTCCTTCTAGGCCCTCACAAT-3′.

8. qPCR system (Thermofisher Scientific QuantStudio 6 Real-Time PCR system and QuantStudio software, or equivalent).
9. Automated cell counter (Countess 3, Thermo Fisher Scientific) or a hemocytometer.

3 Methods

3.1 Cell Plating

1. All cell plating steps should be completed in a certified biological safety cabinet using sterile equipment and consumables.
2. Prepare cell suspensions by dissociating adherent cells from tissue culture plates and counting cells per milliliter (mL) of suspension (*see* **Note 4**).
3. Prepare a cell suspension for vehicle controls for each cell line. SaOS-2, MG-63, SK-N-FI, and SH-SY5Y cells are optimal at 12,500 cells/25 microliter (μL) (5×10^5 cells/mL) per well for 48-h incubations [32] (*see* **Note 5**). To prepare a control suspension, use the following equation:

$$(\textit{starting cell suspension concentration, cells}/\mu L)(x\ \mu L) = (\textit{desired cell suspension concentration, cells}/\mu L)(\textit{desired volume}, \mu L)$$

(*see* **Note 6**). Transfer the desired cell suspension concentration to a fresh conical tube.
4. Centrifuge the cells at 1000 × *g* for 5 min and aspirate the media from the pellet.
5. Resuspend the pellet in the desired volume (5 mL of media if using the example above) of vehicle control media (i.e., if treating with compounds in DMSO that result in a final concentration of 10 μM in 0.1% DMSO, resuspend the pellets in growth media + 0.1% DMSO) (*see* **Note 7**).
6. Use an electronic pipette to transfer 25 μL of vehicle control suspension into the 32 designated control wells (Fig. 2a).
7. Prepare a cell suspension for compound-treated samples for each cell line. Since cell suspensions for treatment wells are plated at a lower volume to accommodate the compound volume, the cell concentration must be calculated for the adjusted volume (*see* **Note 8**).
8. Once transferred, centrifuge the conical tube at 1000 × *g* for 5 min.

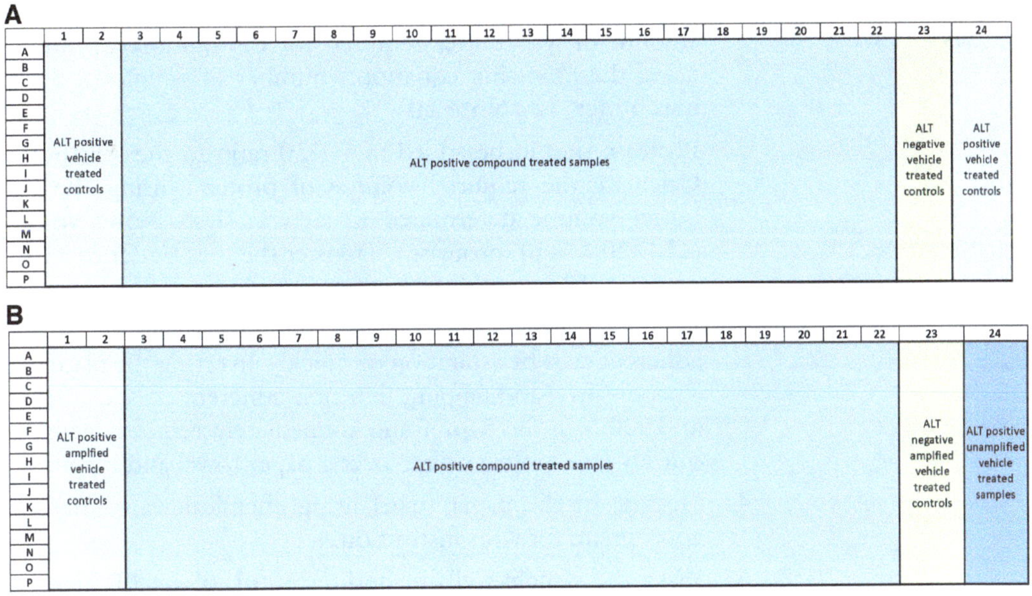

Fig. 2 Example plate layout for cell plating and rolling circle amplification. (**A**) Suggested plate map for cell and compound plating. ALT positive and negative controls should be plated on the edges of the plate, and compound-treated samples (library screening or dose titrations) should be plated in the middle of the plate. (**B**) Suggested plate map for rolling circle amplification. This plate layout matches the initial plating map, so samples can be directly transferred to the same positions in the amplification plate. The last column (column 24) is for unamplified vehicle-treated ALT-positive controls, and these samples should be plated with the "no phi" amplification reaction mix

9. Aspirate media from the pellet and resuspend the pellet in the desired volume of cell growth media (usually 10–15 mL of cell suspension is required for a full 384-well plate).
10. Mix the cell suspension gently by pipetting to prevent cells from settling out of the suspension. Plate samples into a tissue-culture treated 384-well plate using an electronic multichannel pipette. Pipette 22.5 μL of cell suspension into the compound-treated wells (Fig. 2a).
11. Dilute 10 mM in 100% DMSO compound stocks 1:100 using cell growth media. The compounds should now be at 10×10^{-5} M in 1% DMSO.
12. Add 2.5 μL of 10×10^{-5} M compound to the designated treatment well for a final concentration of 10 μM with 0.1% DMSO (*see* **Note 7**).
13. Incubate plates in a humidified tissue culture incubator at 37°C with 5% CO_2 for the desired treatment time. 48 h is recommended, but other incubation times can be accommodated (*see* **Note 9**).

3.2 Cell Dissociation and Lysis

1. Prepare fresh lysis buffer with protease by first calculating the amount of lysis buffer required for the number of samples using the following equation: (number of samples × 50 μL lysis buffer) (*see* **Note 10**).
2. Protease should be added in a 1:20 ratio to the lysis buffer. Calculate the required volume of protease using the lysis buffer volume determined in step 1: (lysis buffer volume (μL)/20 = × μL protease). Transfer the lysis buffer to a fresh conical tube and add the protease, mixing well.
3. In the cell growth plate, remove the media from the wells of adherent cells by aspiration or quickly inverting the plate over a paper towel and tapping. For non-adherent cells, centrifuge at 1000 × *g* for 5 min and immediately remove media by quickly inverting the plate over a paper towel and tapping.
4. Dispose of the paper towel using chemical waste methods appropriate for your institution.
5. Wash the sample wells by adding 20 μL of 1× PBS to each well (do not mix) and remove PBS by inverting the plate over a paper towel and tapping.
6. Add 50 μL of lysis solution to each sample well. If there are any empty wells in the plate, add 50 μL of water to balance the plate for the shaking incubator.
7. Seal the plate using two Microseal A films. Lightly impress a silicone plate mat over the Microfilm A layer so that it is secure, but the seals have not dipped into the sample wells (Fig. 3). Place the cell growth plate cover over the sealed plate and place in the shaking incubator with a second silicone plate mat placed on top.
8. Prepare a balance plate if needed by adding 50 μL of water to all wells of a 384-well plate and sealing in the same manner as the sample plate.
9. Place both plates in a shaking incubator and shake at 56°C at 1000 rpm for 30 min (*see* **Note 11**).
10. Remove plate from the shaking incubator and centrifuge at 1000 × *g* for 30 s. Remove the plate seals and use an electronic multichannel pipette to mix the samples and transfer to a V-bottom 384-well plate.
11. Seal the V-bottom plate with an aluminum seal, top it with a silicone plate mat, and place it in the Thermomixer. Incubate at 56°C for 90 min followed by a 20-min incubation at 70°C to inactivate the protease.
12. Store the lysis plate at 4°C if not using immediately.

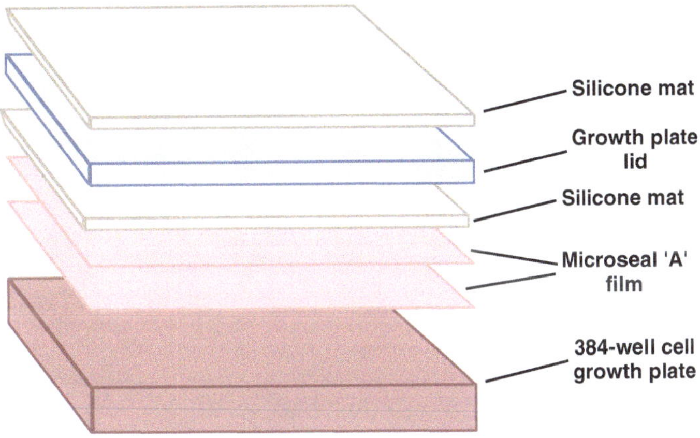

Fig. 3 Plate sealing for heated shaking lysis incubation. Depiction of the sealing layers for the heated shaking lysis step. The heated shaking incubation ensures complete cell dissociation from the plate, but it is imperative that the samples do not jump to other wells. To ensure the plate stays sealed for the duration of the aggressive shaking step, two microseals are used, topped with a silicone mat, then the cell growth plate lid, and finally topped with another silicone mat

3.3 Rolling Circle Amplification

1. Prepare amplification mix for the samples with phi29 DNA polymerase (amplified or "phi" samples) (Table 2). Volumes of each component are dependent on the number of samples (*see* **Note 10**). When preparing the rolling circle amplification master mix, it is critical to add the DTT last, right before use (*see* **Note 2**).
2. Prepare amplification mix for the samples without phi29 DNA polymerase (unamplified or "no phi" samples) (Table 3). Volumes of each component are dependent on the number of samples (*see* **Notes 2 and 10**).
3. Add 4 μL of DNA from the lysis plate to the corresponding wells in a 384-well PCR plate.
4. Add 16 μL of phi or no phi amplification mix to the appropriate wells (Fig. 2b). Centrifuge at 1000 × *g* for 30 s.
5. Incubate plate in the Thermomixer at 45°C (*see* **Note 12**) for the optimal amplification time (amplification time is cell-line dependent, *see* **Note 13**).
6. Incubate the plate at 70°C for 20 min to inactivate the phi29 DNA polymerase.
7. Amplified DNA plates can be sealed with an aluminum plate seal and stored at 4°C if not used immediately.

Table 2
Amplification mix for amplified ("phi") samples

Reaction component	Volume per reaction (multiply this volume by the number of samples) (μL)
Rolling circle amplification master mix	9.25
Molecular biology grade water	6.5
Phi29 DNA polymerase	0.25
Sample DNA	4
Total reaction volume	20

Table 3
Amplification mix for unamplified ("no phi") samples

Reaction component	Volume per reaction (multiply this volume by the number of samples) (μL)
Rolling circle amplification master mix	9.25
Molecular biology grade water	6.75
Sample DNA	4
Total reaction volume	20

3.4 Rolling Circle Amplification Dilution and Telomeric qPCR

1. Prepare qPCR mix for each primer set using the following equations (*see* **Note 14**) (Table 4). Volumes of each component are dependent on the number of samples (*see* **Note 10**).
2. Prepare a dilution plate for the amplified DNA by pipetting 38 μL of molecular biology-grade water into the wells of a V-bottom 384-well plate.
3. Use an electronic pipette to remove 2 μL of DNA from the amplification plate samples and dispense the DNA into the same wells of the dilution plate to bring the well volume to 40 μL. Mix the diluted DNA.
4. Use an electronic pipette to transfer 2 μL of diluted DNA to the appropriate wells in an optically clear 384-well PCR plate (Fig. 2b, one plate per primer set).
5. Add 8 μL of the appropriate qPCR mix to the samples using an electronic pipette and seal the plate. Centrifuge the plate at 1000 × *g* for 1 min.

Table 4
qPCR reaction mix for one primer set

Reaction component	Volume per reaction (multiply this volume by the number of samples) (μL)
2× SYBR green super mix	5.2
10 μM forward primer	0.1
10 μM reverse primer	0.1
Molecular biology grade water	2.6
Diluted amplified DNA	2
Total reaction volume	10

6. Set up the following qPCR reaction in the equipment software and run the experiment:

   Experiment properties: 384-well format, relative analysis (C_T analysis), SYBR green. Set up a plate map in the software to match the samples in the experiment plate. Hold stage (1 cycle): 95°C for 3 min (hold stage duration may vary based on SYBR green manufacturer). PCR stage (40 cycles): 95°C for 1 s followed by 60°C for 20 s. Include a melt curve step after amplification to confirm the presence of a single amplification product.

3.5 Data Analysis for Compound Library Screening: SSMD* Analysis [33]

1. Once C_T values are collected by the qPCR software, calculate a ΔC_T value for each sample:
   - $\Delta C_T = C_{T\,(\text{telomeric DNA})} - C_{T\,(\text{non-telomeric genomic region})}$
   - ΔC_T values represent the relative C-circle level as they are the telomeric DNA expression normalized by background genomic DNA expression.

2. Use all samples (including controls) to calculate an SSMD* score for each sample:

   2.1. Calculate the median ΔC_T of the dataset.
   2.2. Calculate the $|\Delta C_T -$ median $\Delta C_T|$ for each sample.
   2.3. Calculate the median of the $|\Delta C_T -$ median $\Delta C_T|$ values.
   2.4. Calculate the median of absolute deviations (MAD): MAD = 1.4826* median ($|\Delta C_T -$ median $\Delta C_T|$).
   2.5. Calculate the SSMD* value of each sample using the following equation:
   SSMD* = ($\Delta C_{T\,(\text{sample})}$ − median ΔC_T)/($\sqrt{2}$ * MAD).
   2.6. Plot the SSMD* scores as a scatter plot (*see* **Note 15**).
   2.7. Add cutoff lines at $y = +3$MAD and $y = -3$MAD.

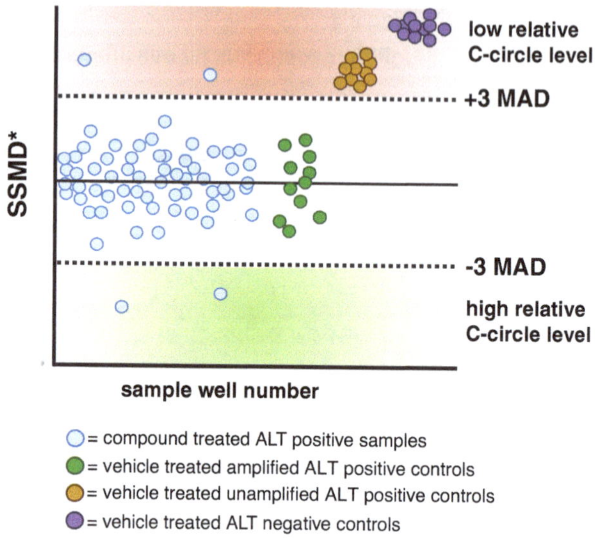

Fig. 4 SSMD* plots reveal screening hits. SSMD* plot of hypothetical data [33]. Green dots represent vehicle-treated amplified ALT-positive samples, which should fall within ±3MAD. Mustard dots represent vehicle-treated unamplified ALT- positive samples, which should fall above +3MAD as they should have detectably lower C-circle levels. Purple dots represent amplified ALT-negative samples, which should fall above +3MAD as they have no C-circles. Blue dots represent treated samples, most of which should fall between ±3MAD. Treated samples that fall above +3MAD have produced detectably lower C-circle levels, and samples that fall below −3MAD produced detectably higher C-circle levels. Treated samples that fall outside of ±3MAD are considered hits and should be investigated further

Untreated and amplified ALT positive control samples should fall within ±3MAD (Fig. 4). ALT negative and unamplified ALT positive control samples should fall above $y = +3$MAD. If controls do not fall as expected on the SSMD* plot, screening results cannot be used as the screening window is not acceptable. If controls are acceptable, treatment samples that fall above $y = +3$MAD or below $y = -3$MAD are hits that should be further investigated for ALT-modulating activity.

3.6 Data Analysis for Compound Dose Titrations: Relative C-Circle Level Analysis

1. Once C_T values are collected, calculate a ΔC_T value for each sample:
   - $\Delta C_T = C_{T\,(\text{telomeric DNA})} - C_{T\,(\text{non-telomeric genomic region})}$.
2. To calculate the relative C-circle level for each sample, linearize the ΔC_T value: relative C-circle level = $2^{-\Delta C_T}$.
3. To visualize the data, graph the data as a bar graph. Use the Mann–Whitney U-test to determine if there is a significant ($p < 0.01$) difference between the amplified ALT positive

controls and the unamplified (no phi) ALT positive controls. If the difference between these two control groups is not significant, then the results are not reliable, and conclusions cannot be drawn. If the difference between amplified and unamplified ALT-positive samples is significant, compare treatment groups and vehicle-treated amplified controls to determine if there is a significant change in C-circle level after compound treatment.

4 Notes

1. After resuspending the protease, aliquot it in small volumes to minimize freeze-thaw cycles, as protease activity may decline with increasing freeze-thaw cycles. For example, if a standard experiment requires 500 μL of protease, freeze it in 500 μL aliquots so the protease only undergoes one freeze/thaw cycle.
2. The rolling circle amplification reaction is critically dependent on DTT. DTT is not stable in solution over time. To promote the most effective amplification, DTT is added to the reaction master mix *directly before the reaction is started*. Reaction buffer and reaction master mixes without DTT can be stored at –20°C until ready for use.
3. The dCTP is omitted to increase the specificity of the rolling circle amplification. dCTP is not required to build the complementary strand of the C-circle, as it does not contain guanine bases. To decrease the background telomere DNA signal and amplify the C-circle signal, dCTP is removed.
4. Cell suspensions should be filtered through a 70 μm cell strainer before counting. Suspensions that are calculated as less than 80% viable should not be used for screening.
5. To optimize a new cell line or optimize the listed cell lines for a different incubation time, cell number titrations should be performed. Start at 100,000 cells/well and perform a two-fold serial dilution to 1562.5 cells/well. Incubate cells for the desired treatment time and then perform the C-circle assay to determine the cell concentration that gives the highest and most significant fold change over unamplified background [32].
6. Cell dilution equation example: if 5 mL of SaOS-2 cell suspension at 5×10^5 cell/mL was needed with a starting concentration of 2×10^6 cell/mL, the following equation would be used: $(2 \times 10^6 \text{ cell/mL})(x \text{ mL}) = (5 \times 10^5 \text{ cell/mL})(5 \text{ mL})$. In this case, $x = 1.25$ mL, so 1.25 mL of the starting suspension would be transferred to a fresh conical tube.

7. Small molecule compounds are often only soluble in 100% DMSO, which is toxic to cells. It is recommended that compound stocks be diluted to keep the final DMSO concentration in the assay at 0.1%. Before screening with any cell line, a DMSO viability curve should be performed on the cell line starting at 25% DMSO and ending at 0.1% DMSO. This ensures that cell viability during screening is not affected by DMSO alone.
8. The volume of cell suspension added to treatment wells is slightly lower to accommodate the extra volume added when compounds are plated. The cell suspension must be adjusted to ensure the proper number of cells per well is plated with the adjusted well volume. For example, SaOS-2 cells are optimal at 12,500 cells per well (total volume is 25 μL). If 2.5 μL of compound is added to the well, that means that the cell suspension needs to be 22.5 μL, and a cell suspension of 12,500 cells/22.5 μL well is required for treatment samples.
9. If incubations are longer than 48 h, it is likely that the media in the wells will begin to evaporate. To avoid losing well volume, store plates in a vented humid secondary container in the incubator or cover with a gas-permeable tissue culture plate seal to avoid media evaporation.
10. It is recommended that when calculating required buffer volumes, the sample number should be the number of samples in the plate plus 10% extra to account for technical errors.
11. If using the shaking incubator model listed in the materials, be sure to preheat the incubator to 56°C for 30 min prior to starting the shaking incubation. This incubator does not reach the displayed temperature until approximately 15 min after it is set. If using a different incubator, it is recommended that the heating block be tested before using it to ensure the temperature is accurate and heating across the block is even.
12. If performing amplification right after lysis, make sure the Thermomixer has fully cooled from the 70°C step of the lysis protocol. Placing the amplification plate on a block that is >45°C will inactivate the phi29 DNA polymerase.
13. Amplification of DNA from any cell line can be optimized. The EquiPhi DNA polymerase used in this protocol has a recommended amplification window of 1–3 h. It is recommended that for all cell lines, amplification reactions should be performed for 1-, 2-, and 3-h incubations. The optimal amplification time is determined as the incubation time that gives the highest and most significant fold-change in relative C-circle level between the phi and no phi samples [32]. For example, SaOS-2 DNA amplifies efficiently after 1 h, and incubations longer than 1 h result in increased signal variability and background amplification. In contrast, SK-N-FI

DNA requires a 3-h amplification to produce the most robust C-circle signal over an unamplified background. Each cell type needs to be optimized prior to screening to ensure robust and reliable amplification.

14. If using a different brand of SYBR mix, the dynamic range for qPCR may vary. Be sure to refer to the user manual before using.
15. When plotting the SSMD*, it is recommended that the control samples be grouped together on the plot (color coding recommended) to keep track of which samples are controls and which are treatment samples.

Acknowledgments

Funding was provided by NIH/NCI R03CA252796 (M.B.J.), a Tier I UNC Lineberger Comprehensive Cancer Center Developmental Award that is supported in part by P30CA016086 Cancer Center Core Support Grant (S.G.P. and M.B.J.), and in part by an Amgen Foundation award to the Office of Undergraduate Research at the University of North Carolina at Chapel Hill (S.T.L. and S.G.P.).

References

1. Griffith JD, Comeau L, Rosenfield S et al (1999) Mammalian telomeres end in a large duplex loop. Cell 97:5032–5514
2. Harley CB, Vaziri H, Counter CM, Allsopp RC (1992) The telomere hypothesis of cellular aging. Exp Gerontol. 27(4):375–382
3. Jafri MA, Ansari SA, Alqahtani MH, Shay JW (2016) Roles of telomeres and telomerase in cancer, and advances in telomerase-targeted therapies. Genome Med 8(1):69
4. Hanahan D, Weinberg RA (2000) The hallmarks of cancer. Cell 100:57–70
5. Masutomi K, Kaneko S, Hayashi N et al (2000) Telomerase activity reconstituted in vitro with purified human telomerase reverse transcriptase and human telomerase RNA component. J Biol Chem 275:22568–22573. https://doi.org/10.1074/jbc.M000622200
6. Bodnar AG, Ouellette M, Frolkis M, Holt SE, Chiu CP, Morin GB, Harley CB, Shay JW, Lichtsteiner S, Wright WE (1998) Extension of life-span by introduction of telomerase into normal human cells. Science 279(5349):349-52
7. Kim NW, Piatyszek MA, Prowse KR, Harley CB, West MD, Ho PL, Coviello GM, Wright WE, Weinrich SL, Shay JW (1994) Specific association of human telomerase activity with immortal cells and cancer. Science 266(5193):2011–5.
8. Greider CW, Blackburn EH (1985) Identification of a specific telomere terminal transferase activity in Tetrahymena extracts. Cell 43(2 Pt 1):405-13. https://doi.org/10.1016/0092-8674(85)90170-9. PMID: 3907856
9. Hakobyan M, Binder H, Arakelyan A (2024) Pan-cancer analysis of telomere maintenance mechanisms. J Biol Chem 300. https://doi.org/10.1016/j.jbc.2024.107392
10. Dunham MA, Neumann AA, Fasching CL, Reddel RR (2000) Telomere maintenance by recombination in human cells. Nat Genet 26(4):447-50
11. Muntoni A, Neumann AA, Hills M, Reddel RR (2009) Telomere elongation involves intra-molecular DNA replication in cells utilizing alternative lengthening of telomeres. Hum Mol Genet 18:1017–1027. https://doi.org/10.1093/hmg/ddn436
12. Draskovic I, Londono Vallejo A (2013) Telomere recombination and alternative telomere lengthening mechanisms. Front Biosci (Landmark Ed) 18(1):1-20
13. Lafferty-Whyte K, Cairney CJ, Will MB et al (2009) A gene expression signature classifying telomerase and ALT immortalization reveals an hTERT regulatory network and

suggests a mesenchymal stem cell origin for ALT. Oncogene 28:3765–3774. https://doi.org/10.1038/onc.2009.238
14. Sanders RP, Drissi R, Billups CA et al (2004) Telomerase expression predicts unfavorable outcome in osteosarcoma. J Clin Oncol 22:3790–3797. https://doi.org/10.1200/JCO.2004.03.043
15. Sturm D, Witt H, Hovestadt V et al (2012) Hotspot mutations in H3F3A and IDH1 define distinct epigenetic and biological subgroups of glioblastoma. Cancer Cell 22:425–437. https://doi.org/10.1016/j.ccr.2012.08.024
16. Dagg RA, Pickett HA, Neumann AA et al (2017) Extensive proliferation of human cancer cells with ever-shorter telomeres. Cell Rep 19:2544–2556. https://doi.org/10.1016/j.celrep.2017.05.087
17. Kent T, Gracias D, Shepherd S, Clynes D (2020) Alternative lengthening of telomeres in pediatric cancer: mechanisms to therapies. Front Oncol 9:1518
18. Sturm D, Pfister SM, Jones DTW (2017) Pediatric gliomas: current concepts on diagnosis, biology, and clinical management. J Clin Oncol 35:2370–2377. https://doi.org/10.1200/JCO.2017
19. Alzoubi H, Minasi S, Gianno F et al (2022) Alternative lengthening of telomeres (ALT) and telomerase reverse transcriptase promoter methylation in recurrent adult and primary pediatric pituitary neuroendocrine tumors. Endocr Pathol. https://doi.org/10.1007/s12022-021-09702-0
20. Schwartzentruber J, Korshunov A, Liu XY et al (2012) Driver mutations in histone H3.3 and chromatin remodelling genes in paediatric glioblastoma. Nature 482:226–231. https://doi.org/10.1038/nature10833
21. Sun H, Chen G, Guo B et al (2023) Potential clinical treatment prospects behind the molecular mechanism of alternative lengthening of telomeres (ALT). J Cancer 14:417–433
22. Strovel J, Sittampalam S, Coussens NP, Hughes M, Inglese J, Kurtz A, Andalibi A, Patton L, Austin C, Baltezor M, Beckloff M, Weingarten M, Weir S (2012/2016) Early Drug Discovery and Development Guidelines: For Academic Researchers, Collaborators, and Start-up Companies. In: Markossian S, Grossman A, Baskir H, Arkin M, Auld D, Austin C, Baell J, Brimacombe K, Chung TDY, Coussens NP, Dahlin JL, Devanarayan V, Foley TL, Glicksman M, Gorshkov K, Grotegut S, Hall MD, Hoare S, Inglese J, Iversen PW, Lal-Nag M, Li Z, Manro JR, McGee J, Norvil A, Pearson M, Riss T, Saradjian P, Sittampalam GS, Tarselli MA, Trask OJ Jr, Weidner JR, Wildey MJ, Wilson K, Xia M, Xu X, (eds) Assay Guidance Manual [Internet]. Bethesda (MD): Eli Lilly & Company and the National Center for Advancing Translational Sciences 2004. PMID: 22553881
23. Croston GE (2017) The utility of target-based discovery. Expert Opin Drug Discov 12:427–429
24. Moffat JG, Vincent F, Lee JA et al (2017) Opportunities and challenges in phenotypic drug discovery: An industry perspective. Nat Rev Drug Discov 16:531–543
25. Lovejoy CA, Li W, Reisenweber S et al (2012) Loss of ATRX, genome instability, and an altered DNA damage response are hallmarks of the alternative lengthening of Telomeres pathway. PLoS Genet 8. https://doi.org/10.1371/journal.pgen.1002772
26. Ng LJ, Cropley JE, Pickett HA et al (2009) Telomerase activity is associated with an increase in DNA methylation at the proximal subtelomere and a reduction in telomeric transcription. Nucleic Acids Res 37:1152–1159. https://doi.org/10.1093/nar/gkn1030
27. Yadav T, Zhang JM, Ouyang J et al (2022) TERRA and RAD51AP1 promote alternative lengthening of telomeres through an R- to D-loop switch. Mol Cell 82:3985–4000.e4. https://doi.org/10.1016/j.molcel.2022.09.026
28. Yeager TR, Neumann AA, Englezou A et al (1999) Telomerase-negative immortalized human cells contain a novel type of promyelocytic leukemia (PML) body. Cancer Res 59:4175–4179
29. Henson JD, Cao Y, Huschtscha LI et al (2009) DNA C-circles are specific and quantifiable markers of alternative-lengthening-of-telomeres activity. Nat Biotechnol 27:1181–1185. https://doi.org/10.1038/nbt.1587
30. Henson JD, Lau LM, Koch S et al (2017) The C-Circle assay for alternative-lengthening-of-telomeres activity. Methods 114:74–84. https://doi.org/10.1016/j.ymeth.2016.08.016
31. Chen YY, Dagg R, Zhang Y et al (2021) The C-circle biomarker is secreted by alternative-lengthening-of-telomeres positive cancer cells inside exosomes and provides a blood-based diagnostic for ALT activity. Cancers 13. https://doi.org/10.3390/cancers13215369
32. Froney MM, Cook CR, Cadiz AM et al (2024) A first-in-class high-throughput screen to discover modulators of the alternative lengthening of telomeres (ALT) pathway. ACS Pharmacol Transl Sci. https://doi.org/10.1021/acsptsci.4c00251
33. Zhang XD (2011) Illustration of SSMD, z score, SSMD*, z* score, and t statistic for hit selection in RNAi high-throughput screens. J Biomol Screen 16:775–785. https://doi.org/10.1177/1087057111405851

Part VII

Epigenetic Markers for Cancer Detection and Diagnosis

Chapter 18

Cell-Free DNA-Based Methods to Map Chromatin

Stephanie Eramo, Abigail Weirich and Srinivas Ramachandran

Abstract

Cell-free DNA (cfDNA) provides a non-invasive means to monitor physiological states and diagnose diseases, with widespread applications in oncology. Our lab utilizes cfDNA as a tool to investigate the state of the tumor and the tumor microenvironment in patients with cancer. This chapter outlines methods for cfDNA extraction from plasma, library preparation for sequencing, and subsequent computational analysis that will enable analysis of cfDNA footprints genome-wide. By examining cfDNA fragment lengths and their genomic origins, we can infer cellular dynamics and identify potential biomarkers for various health conditions, offering valuable insights into the body's internal processes with significant implications for advancing diagnostic and therapeutic approaches.

Keywords cfDNA, Subnucleosome, Chromatin, Epigenomics

1 Introduction

In molecular biology and diagnostics, cell-free DNA (cfDNA) has the potential to make significant advancements that allow us to monitor patient physiology in a less invasive way. When cells die, their genomes become the targets of nucleases. Nucleosomes, transcription factors, and other proteins protect DNA from degradation. These protected DNA fragments are released into the bloodstream [1]. Thus, cfDNA is generated by the action of endogenous nucleases on a chromatinized genome. This means that cfDNA represents a map of chromatin structure that faithfully represents the regulatory state of the genomes of originating cells [2–4] (Fig. 1). The binding of transcription factors and transcription by RNA polymerase II (RNAPII) results in short DNA protections [2]. These states are mappable by the endogenous nucleases that give rise to cfDNA. The binding of transcription factors and the presence of partially disassembled nucleosomes result in smaller footprints in cfDNA (protection is reduced from >147 bp to <100 bp) [2, 5]. When the whole spectrum of cfDNA footprints is captured, we have shown that the enrichment of specific cfDNA fragments that arise from

Archana Dhasarathy and Motoki Takaku (eds.), *Cancer Epigenetics*, Methods in Molecular Biology, vol. 3005,
https://doi.org/10.1007/978-1-0716-5072-1_18,

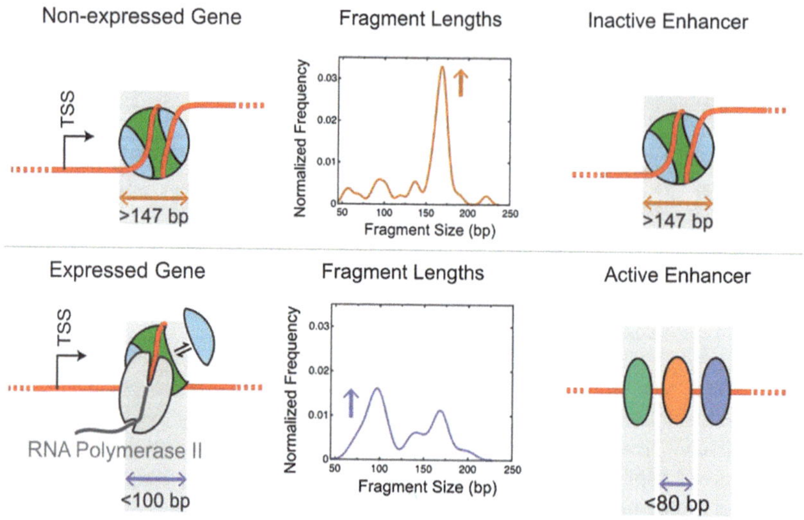

Fig. 1 Schematic of cfDNA chromatin footprints. Chromatin structure at promoters (left) and enhancers (right) reflects the activity of these genetic elements. Inactive promoters and enhancers feature packaging by nucleosomes, leading to long cfDNA fragments (middle, top). Whereas gene or enhancer activation results in short cfDNA fragments generated due to subnucleosomes, subnucleosome or transcription factors (middle, bottom). Abbreviations: TSS, transcription start site

subnucleosomes can reproduce the regulatory landscape of tumor cells [5] and faithfully capture transcriptional activity [2]. This measure, which we call subnucleosomal enrichment (SE), compares the enrichment of cfDNA footprints arising from subnucleosomes to those from nucleosomes at different sites in the genome. Therefore, the locus-specific analysis of cfDNA fragment lengths offers a window into the cell's internal dynamics and the events before the release of cfDNA. This allows us to gain insights into various physiological and pathological processes by analyzing the composition and patterns of cfDNA in the bloodstream, opening doors to new diagnostic and therapeutic possibilities.

Here, we detail a method for extracting cfDNA from plasma, preparing libraries for sequencing, and computational techniques to analyze fragment lengths and locations. The cfDNA extraction protocol was adapted from the UltraPrep method published by Raymond et al. [6], and the library preparation was adapted from the SRSLY method published by Troll et al. [7].

2 Materials

Prepare all solutions using nuclease-free water (NF-H_2O, not DEPC-treated), and use molecular biology grade reagents. Buffer final volumes included in the notes are examples. All buffer volumes will need to be scaled based on how many samples are undergoing cfDNA extraction. For solutions that contain Guanidine Thiocyanate (GITC), add GITC last (*see* **Note 1**). Prepare and store all solutions at room temperature, unless otherwise indicated. Do not pour GITC solutions down the drain. Collect GITC waste in appropriate, clearly labeled containers. Store the waste containers in a designated hazardous waste storage area, ensuring they are sealed and labeled properly.

2.1 Common to Cell-Free DNA Extraction and Single-Stranded DNA Library Preparation

1. Qubit Fluorometer.
2. Tapestation 4150 or 4200.
3. Qubit dsDNA HS Assay Kit.
4. Tapestation High Sensitivity D1000 ScreenTapes.
5. Tapestation High Sensitivity D1000 Sample Buffer & Ladder.

2.2 Cell-Free DNA Extraction

1. Proteinase K Solution, RNA-Grade (20 mg/mL).
2. Silica Superparamagnetic Particles, 2.5% w/v, 0.42 μm.
3. GITC Stock: 8 M GITC (*see* **Note 2**).
4. Digestion Buffer: 50 mM Tris pH 8.0, 5 M GITC, 25 mM EDTA, 25% Tween 20 (*see* **Note 3**). May require storage at 37°C to avoid crystallization.
5. Binding Buffer: 10 mM Tris pH 8.0, 3.5 M GITC, 1 mM EDTA, 2.5% Tween 20, 45% Isopropanol (*see* **Note 4**).
6. Wash Buffer #1: 40 mM Bis-Tris pH 6.0, 3 M GITC, 2 mM EDTA, 5% Tween 20, 30% Isopropanol (*see* **Note 5**).
7. Wash Buffer #2: 50 mM Tris pH 8.0, 0.5 mM EDTA, 80% Ethanol (*see* **Note 6**).
8. TE Buffer: 10 mM Tris pH 8.0, 0.1 mM EDTA (*see* **Note 7**).

2.3 Single-Stranded DNA Library Preparation

1. SRSLY® PicoPlus DNA NGS Library Preparation Base Kit, 96 rxn, Claret Bioscience.
2. i5 uniquely barcoded PCR Primer (dual indexed, barcode sequence denoted by "N" and can be either designed in-house or commercially available barcode sequences can be used, "*" denote phosphorothioate bonds): ATGATACGGCGACCA-CCGAGATCTACACNNNNNNNNACACTCTTTCCCT-ACACGACGCTCTTCCGA*T*C*T.
3. i7 uniquely barcoded PCR Primer (dual indexed, barcode sequence denoted by "N" and can be either designed in-house or commercially available barcode sequences can be used, "*"

denotes phosphorothioate bonds): CAAGCAGAAGACG-GCATACGAGATNNNNNNNNGTGACTGGAGTTCAG-ACGTGTGCTCTTCCGA*T*C*T.

4. AMPure XP Beads, Beckman Coulter (Catalog #A63880, A63881, or A63882).
5. 0.1× TE Buffer: 1 mM Tris pH 8.0, 10 μM EDTA. A 1:10 dilution of TE buffer from Sect. 2.2.

2.4 Sequence Alignment and Analysis

1. Hardware (an example based on our usage): Storage of at least 5 TB for processing 20 samples, RAM 100 GB.
2. Software (versions used are indicated): CentOS Linux version 7, GNU bash version 4.2.46, GNU Awk version 4.0.2, gzip version 1.5, Perl version 5.26.0, Snakemake [8] version 3.13.3, Cutadapt [9] version 1.16, Bowtie 2 [10] version 2.3.2, SAMtools [11] version 1.16.1, BEDTools version 2.31.1 [12].

3 Methods

3.1 Cell-Free DNA Extraction

All steps are performed at room temperature unless otherwise specified. This protocol is presented for the extraction of cfDNA from 2 mL of plasma. For larger volumes, the protocol can be scaled appropriately, or the plasma can be divided into multiple 2 mL aliquots.

1. Spin down plasma samples in a centrifuge at 4°C for 15 min at 1500 to 2000 × *g* to pellet any remaining cells. If a pellet forms, carefully pull off plasma as not to disturb the pellet, and move to a 15 mL conical tube.
2. Bring the plasma up to 2 mL with NF-H_2O (*see* **Note 8**).
3. Add 1.3 mL of digestion buffer and 20 μL of proteinase K solution to the plasma.
4. Incubate this reaction at 56°C for 1 hr in a bead bath.
5. Remove samples from the bead bath. Add 6.6 mL of binding buffer to the digested plasma, followed by the addition of 80 μL of 400 nm silica superparamagnetic particles. Mix by pipetting up and down.
6. Allow the cfDNA to bind to the beads for 10 min.
7. Place the tubes in a magnetic separation rack, and allow the samples to sit for approximately 5 min, or until the slurry becomes clear.
8. Once the beads aggregate into a pellet on the magnet, remove the supernatant, label it as "unbound," and store it at −20°C (*see* **Note 9**).
9. Remove the tubes from the magnetic rack and resuspend the beads with 1 mL of wash buffer #1. At this point, it is best to

move the resuspended beads to 1.5 mL tubes, as it makes the following steps easier.

10. Place the tubes back on the magnet, allow the beads to bind, and discard the supernatant once the solution is clear.
11. Remove the tubes from the magnetic rack and resuspend the beads with 1 mL of wash buffer #2.
12. Place the tubes back on the magnet, allow the beads to bind, and discard the supernatant once the solution is clear.
13. Remove the tubes from the magnetic rack and resuspend the beads with 200 μL of 100% ethanol.
14. Centrifuge briefly to collect any droplets and place all sample tubes onto the magnetic rack.
15. Allow the beads to bind and discard the ethanol once the solution is clear.
16. Spin down any remaining ethanol droplets, allow the beads to bind on the magnet, and discard the ethanol.
17. Leave the tube caps open and dry the beads completely at 37°C for 2 min in the bead bath (*see* **Note 10**).
18. Add 30 μL of 1× TE buffer to each tube and resuspend the beads.
19. Place the tubes back on the magnet, allow the beads to bind, and transfer the supernatant (now the eluate containing cfDNA) into a new, labeled, 0.5 mL Lo-Bind tube.
20. Use a Qubit Fluorometer to check the concentration of cfDNA. Follow instructions from the manufacturer for instrument standardization and for preparing the sample for use with Qubit.
21. **Optional:** Run the samples on the TapeStation 4150 (or 4200) to assess the cfDNA profile prior to library preparation (Fig. 2).
22. Freeze samples at −20°C until library preparation.

3.2 Single-Stranded DNA Library Preparation

Modified from Claret Bio's *SRSLY PicoPlus NGS Library Prep Kit*.

1. Thaw the SRSLY Reaction Master Mix (MM) at room temperature (*see* **Note 11**). This can be left at room temperature through step 8 of the protocol.
2. Thaw the SRSLY NGS Adapters at room temperature and put on ice once thawed (*see* **Note 11**).
3. Thaw cfDNA samples at room temperature.
4. Preheat PCR block (that can hold 0.5 mL tubes) to 98°C with a lid temperature of 105°C.
5. On ice, combine the following in a 0.5 mL PCR tube for each library: 250 pg to 10 ng of input cfDNA (only to a maximum

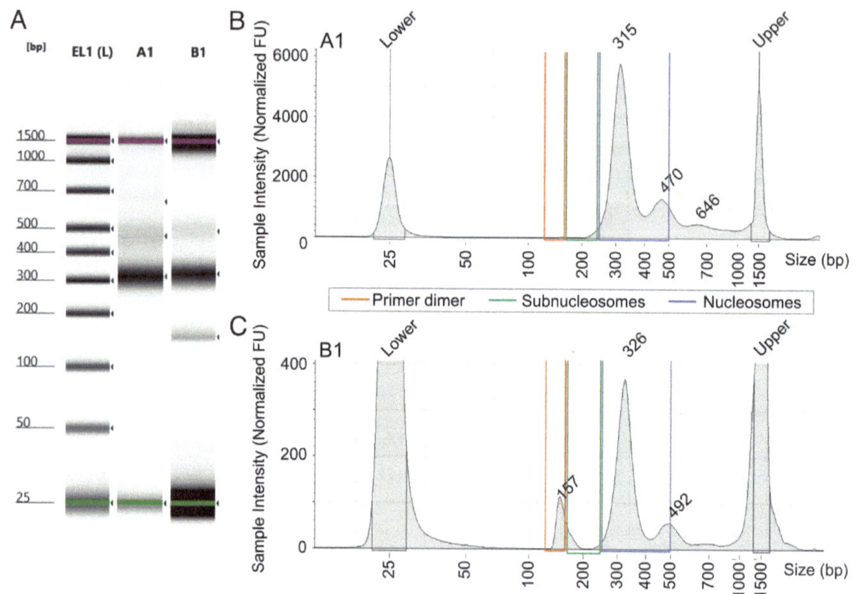

Fig. 2 Expected library fragment peaks from Tapestation 4150 D1000 Tape. (**A**) A successful library with no primer dimer (A1) (*see* **Note 18**) and a library with high primer dimer (B1) are shown next to an "electronic" ladder. B1 is technically not ready for submission and should undergo a second AMPure bead clean at 0.98×. (**B**) and (**C**) Respective plots of intensity vs. fragment length for each library with the primer dimer, subnucleosome, and nucleosome regions defined

volume of 18 μL), the volume of NF-H_2O needed to bring the input cfDNA volume to 18 μL, and 2 μL of the red-capped ssEnhancer from the SRSLY kit. This should make up a total volume of 20 μL.

6. Incubate on ice for 2 min.
7. Heat-shock the tubes at 98°C for 3 min.
8. Immediately return the tubes to ice and incubate for 2 min.
9. Leave tubes on ice and set the 0.5 mL PCR Block to incubate at 37°C with a lid temperature of 45°C.
10. While the PCR Block is cooling, prepare the following master mix (MM1). Multiply the given volumes by the number of libraries being made plus 1 (for overage), as the following volumes only account for one reaction: 2 μL of SRSLY NGS Adapter A, 2 μL of SRSLY NGS Adapter B, and 26 μL of SRSLY Master Mix.
11. Add 30 μL of MM1 to each library tube for a total reaction volume of 50 μL per sample.
12. Vortex each tube thoroughly and spin down in a microcentrifuge.

13. Incubate the tubes at 37°C with a lid temperature of 45°C for 1 hr in the preheated PCR Block.
14. Remove AMPure beads from storage and place at room temperature at least 30 min prior to completion of the sample incubation.
15. After the 1 hr incubation, vortex AMPure Beads until homogeneous.
16. Prepare a mastermix (MM2) of 100% Isopropanol and NF-H_2O. Multiply the given volumes by the number of libraries being made plus 1 (for overage), as the following volumes only account for one reaction: 11.6 μL of 100% Isopropanol and 48.8 μL of NF-H_2O.
17. Add 60 μL of the MM2 to each 50 μL SRSLY ligation reaction.
18. Add 59.4 μL of well-mixed AMPure beads to each tube to reach a final volume of 169.4 μL (*see* **Note 12**).
19. Let the tubes sit at room temperature for 10 min.
20. Place the tubes onto a 0.5 mL magnetic rack for 5 min until the supernatant is clear.
21. Carefully remove and discard the supernatant.
22. Add 200 μL of 80% ethanol to the tubes on the magnetic rack and incubate for 30 sec. Do not move the tubes from the magnet and do not resuspend beads.
23. Remove the ethanol completely without disturbing the bead pellet.
24. Repeat the ethanol wash (steps 22 and 23).
25. Spin the samples down in a microcentrifuge, then place the tubes back on the magnet, and remove all remaining ethanol.
26. Dry the beads for 10 min on the magnet with their lids open.
27. Add 15 μL of NF-H_2O to the beads, resuspend the beads by pipetting up and down, and incubate for 2 min in a non-magnetic rack (*see* **Note 13**).
28. Place the tubes on the magnetic rack for 2 min until the supernatant is clear.
29. Transfer the supernatant (the cfDNA library) to a new, labeled 0.5 mL PCR Tube.

3.3 Library Amplification by PCR

1. Add i5 and i7 uniquely indexed 10 μM primers (*see* **Note 14**) at 2.5 μL each, 5 μL of NF-H_2O, and 25 μL of SRSLY kit provided PCR Master Mix to the purified 15 μL libraries to yield a final reaction volume of 50 μL (*see* **Note 15**).
2. Vortex each tube thoroughly and spin down in a microcentrifuge.

3. Run PCR in 0.5 μL block for ten cycles using the following protocol: **98°C:** 3 min, **98°C:** 10 s, **60°C:** 10 s, **72°C:** 10 s, **72°C:** 1 min, **4°C:** HOLD
4. Proceed to AMPure Bead Purification or store tubes at −20°C until ready to continue.
5. Vortex room temperature AMPure Beads until homogeneous.
6. Add 47.5 μL of AMPure beads to each 50 μL PCR reaction for a final volume of 97.5 μL.
7. Place the tubes onto a 0.5 mL magnetic rack for 5 min until the supernatant is clear.
8. Carefully remove and discard the supernatant.
9. Add 200 μL of 80% ethanol to the tubes on the magnetic rack and incubate for 30 sec. Do not resuspend beads.
10. Remove the ethanol completely without disturbing the bead pellet.
11. Repeat the ethanol wash (steps 9 and 10).
12. Spin the samples down in a microcentrifuge, place the tubes back on the magnet, and remove all remaining ethanol.
13. Dry the beads for 10 min on the magnetic rack with their tops open.
14. Add 30 μL of 0.1× TE buffer to the beads, resuspend by pipetting up and down, and incubate for 2 min in a non-magnetic rack (*see* **Note 13**). If the input DNA was under 6 ng, a second bead clean is required before the final 0.1× TE elution. **In this case:** Add 30 μL of NF-H_2O to the beads, resuspend by pipetting up and down, and incubate for 2 min in a non-magnetic rack (*see* **Note 13**).
15. Place the tubes on the magnetic rack for 2 min until the supernatant is clear.
16. Transfer the supernatant to a new, labeled 1.5 mL Lo-Bind Tube. **In case of second clean**: Transfer NF-H_2O supernatant to a new, labeled 0.5 mL PCR Tube. Repeat 80% ethanol clean as in steps 4 through 13, but add in AMPure Beads at a ratio of 0.98× to the total volume of the library (30 μL). At the end of the clean, follow steps 14 to 16 for 0.1× TE buffer elution.
17. Libraries can now be stored at −20°C until quality analysis has been completed. The samples are ready for sequencing.

3.4 Amplified Library Quality Analysis

1. Use a Qubit Fluorometer to check the concentration of the finished library. Follow instructions from the manufacturer for instrument standardization and for preparing the sample for use with Qubit.

2. Run the finished library on the TapeStation 4150 (or 4200) to check for peaks/bands in the expected areas (Fig. 2). Ideally, there should be minimal primer dimer (*see* **Notes 16 and 17**).
3. For human samples, we typically sequence paired-end (2×151 bp) to a depth of 150 million paired-end reads (~15× coverage of the human genome) on any Illumina sequencer that can sequence TruSeq libraries.

3.5 Processing of sequencing datasets

1. Clone the cfDNA alignment Snakemake pipeline from GitHub (https://github.com/seramo96/cfDNA_pipe) in a high-performance computing cluster with access to multiple cores. The memory and space required for processing ~20 samples are listed under materials. The following files/directories will be part of the repository: config.yaml, code/, data/, rules/ssp_seq_data.snake, results/, Snakefile, and snakemake_run.sh.
2. Ensure all software in Sect. 2.4 is downloaded.
3. In the data directory, create symlinks that point to sequencing fastq files for read 1 and read 2 of the samples that are being analyzed.
4. Edit the config.yaml to contain the sample names and path to the desired alignment genome.

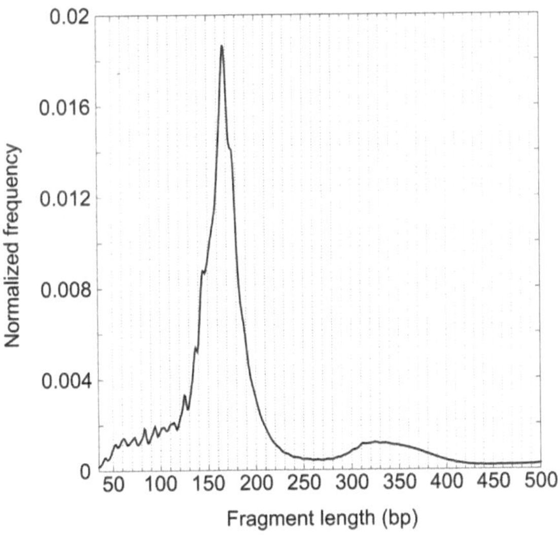

Fig. 3 Representative fragment length distribution obtained after processing of a cfDNA sequencing dataset. After processing raw sequencing reads to remove adapters and then aligning the paired-end reads to the genome, we obtain fragments of various lengths. This plot shows the distribution of the length of cfDNA fragments that map to the human genome from a typical cfDNA sequencing dataset. Fragments 120–200 bp arise from mononucleosomes, 300–400 from dinucleosomes, and <120 bp fragments are subnucleosomal, which consist of both transcription factor footprints and protections from nucleosomal intermediates

5. Run the snakemake_run.sh script after setting up Snakemake to be able to submit jobs in the cluster. Instructions on setting up Snakemake can be found in the Snakemake documentation (https://snakemake.readthedocs.io/en/stable/). Upon successful completion of the pipeline, fragments that map uniquely to the genome will be obtained in a bed file. A typical fragment length distribution obtained after alignment is shown in Fig. 3.

4 Notes

1. 8 M GITC is not soluble at room temperature, so it must be added **last** to all buffers that require it to ensure it does not fall out of solution. It is important to note that this is not the most accurate method of making buffers, as water is added in a specific calculated amount versus the Q.S. (quantity sufficient) method, which takes volume into account. That being said, we have not seen an impact of this method on our results. GITC stock cannot be made in advance, as it requires heat to stay in solution. Plan to make the GITC stock on the same day as the buffers that require it.
2. To make 100 mL of 8 M GITC (molecular weight = 118.16 g/mol), weigh out 94.53 g of GITC. Using a graduated cylinder, measure out 100 mL of water and pour it into a glass beaker containing a stir bar. Mark where the meniscus lies, as this will be important for Q.S. of the final solution. Once marked, dispose of the water, and add the GITC to the beaker. Add just enough water that the solid is submerged, making sure not to exceed the Q.S. marking. Place the beaker on a hot plate and heat the solution to about 100°C. Make sure not to boil the solution. If it starts to boil, lower the temperature. At this point, you should turn on the magnet to stir the solution. Once the GITC is fully dissolved, turn off the magnet. Add water up to the Q.S. line, then restart stirring. Do not remove the beaker from the heat, as the GITC will crystallize very quickly.
3. To make 25 mL of buffer, combine 6.25 mL of 100% Tween 20, 1.25 mL of 1 M Tris pH 8.0, 1.25 mL of 0.5 M EDTA, and 0.625 mL of NF-H_2O into a 50 mL conical tube. After these are all added to the tube, quickly and carefully add 15.625 mL of 8 M GITC.
4. To make 100 mL of buffer, combine 2.50 mL of 100% Tween 20, 1 mL of 1 M Tris pH 8.0, 200 μL of 0.5 M EDTA, 45 mL of 100% isopropanol, and 7.55 mL of NF-H_2O in a 100 mL

or larger glass bottle. After these are all added to the bottle, quickly and carefully add 43.75 mL of 8 M GITC.

5. To make 25 mL of buffer, combine 1.25 mL of 100% Tween 20, 5 mL of 200 mM Bis-Tris pH 6.0, 100 μL of 0.5 M EDTA, 7.5 mL of 100% isopropanol, and 1.775 mL of NF-H_2O into a 50 mL conical tube. After these are all added to the tube, quickly and carefully add 9.375 mL of 8 M GITC.
6. To make 25 mL of buffer, combine 1.25 mL of 1 M Tris pH 8.0, 25 μL of 0.5 M EDTA, 20 mL of 100% ethanol, and 3.725 mL of NF-H_2O into a 50 mL conical tube.
7. To make 15 mL of buffer, combine 150 μL of 1 M Tris pH 8.0, 3 μL of 0.5 M EDTA, and 14.847 mL of NF-H_2O into a 50 mL conical tube.
8. Extractions are generally performed on samples anywhere from 1 to 10 mL. This protocol is directly scalable, so it is possible to do the extractions from plasma volumes greater than 10 mL. A 2 mL extraction is common and is used here as an example.
9. It is important to keep the supernatant as "unbound" because it may contain cfDNA that did not bind to the magnetic particles. If there is little to no yield, extraction can be attempted again on the saved unbound sample. After library preparation and confirmation of quality analysis, these unbound samples can be disposed of.
10. Any remaining ethanol will interfere with library preparation, so be sure all ethanol has evaporated before continuing.
11. For the best performance of the *SRSLY* kit, try to limit freeze-thaw cycles to less than 12. This can be done by aliquoting reagents if necessary. If you use the 96-reaction kit, making libraries in batches of 16 at a time will limit the reagents to a maximum of 8 freeze-thaws, eliminating the need for aliquots.
12. Keep the AMPure beads at room temperature to be used again for the bead clean after PCR (if performing the same day). This eliminates having to wait for them to warm up to room temperature again.
13. Add 3 μL extra of elution reagent to each tube. It will make it easier to remove the full, desired volume of eluent at the end and prevent any bead aspiration.
14. The uniquely indexed i5 and i7 primers can either be purchased from Claret Bio or any company that synthesizes oligonucleotides.

15. To make pipetting the i5 and i7 uniquely indexed 10 μM primers easier, as 2.5 μL is such a small volume, premix the i5 and i7 combos in equal volumes (so they each stay at 10 μM) in labeled tubes and store for further use. Record which primers are in which combo so the indexes can still be used for sequencing/analysis. This way, you can pipette 5 μL of a specific primer combo into a library, making it easier to pipette and lessening human error.
16. If the library concentration is over 2 ng/μL, use a D1000 tape. If it is under 2 ng/μL, use an HSD1000 tape.
17. Ideally, libraries should show bands within the subnucleosomal and nucleosomal regions. A primer dimer is an unintended by-product that occurs when the primers anneal to each other instead of to the target DNA sequence. This can lead to the amplification of unwanted DNA fragments, which can interfere with the efficiency and accuracy of the amplification of cfDNA fragments. Based on sequencing hundreds of libraries, we arrived at a primer dimer cutoff of less than 3% in order to proceed to sequencing. If the primer dimer amount is higher than 3%, one more round of purification with AMPure beads can be performed to reduce the primer dimer amount.
18. The lengths we expect to see in the tapestation (accounting for adapters) are (Fig. 2): Primer dimer at 100–170 bp, subnucleosomal fragments at 170–250 bp, and nucleosomal fragments at 250–550 bp (and potentially higher).

Acknowledgments

This work was supported by the RNA Bioscience Initiative, the University of Colorado School of Medicine, American Cancer Society grant RSG-22-026-01. S.R. is a Pew-Stewart Scholar for Cancer Research, supported by the Pew Charitable Trusts and the Alexander and Margaret Stewart Trust.

References

1. Henikoff S, Church GM (2018) Simultaneous discovery of cell-free DNA and the nucleosome ladder. Genetics 209(1):27–29. https://doi.org/10.1534/genetics.118.300775
2. Ramachandran S, Ahmad K, Henikoff S (2017) Transcription and remodeling produce asymmetrically unwrapped nucleosomal intermediates. Mol Cell 68(6):1038–1053 e1034. https://doi.org/10.1016/j.molcel.2017.11.015
3. Snyder MW, Kircher M, Hill AJ et al (2016) Cell-free DNA comprises an in vivo nucleosome footprint that informs its tissues-of-origin. Cell 164(1-2):57–68. https://doi.org/10.1016/j.cell.2015.11.050
4. Zukowski A, Rao S, Ramachandran S (2020) Phenotypes from cell-free DNA. Open Biol 10(9):200119. https://doi.org/10.1098/rsob.200119
5. Rao S, Han AL, Zukowski A et al (2022) Transcription factor-nucleosome dynamics from plasma cfDNA identifies ER-driven states in breast cancer. Sci Adv

8(34):eabm4358. https://doi.org/10.1126/sciadv.abm4358
6. Raymond CK, Raymond FC, Hill K (2020) UltraPrep is a scalable, cost-effective, bead-based method for purifying cell-free DNA. PLoS One 15(6):e0231854. https://doi.org/10.1371/journal.pone.0231854
7. Troll CJ, Kapp J, Rao V et al (2019) A ligation-based single-stranded library preparation method to analyze cell-free DNA and synthetic oligos. BMC Genomics 20(1):1023. https://doi.org/10.1186/s12864-019-6355-0
8. Molder F, Jablonski KP, Letcher B et al (2021) Sustainable data analysis with Snakemake. F1000Res 10:33. https://doi.org/10.12688/f1000research.29032.2
9. Martin M (2011) Cutadapt removes adapter sequences from high-throughput sequencing reads. 17(1):3. https://doi.org/10.14806/ej.17.1.200
10. Langmead B, Salzberg SL (2012) Fast gapped-read alignment with Bowtie 2. Nat Methods 9(4):357–359. https://doi.org/10.1038/nmeth.1923
11. Li H, Handsaker B, Wysoker A et al (2009) The sequence alignment/map format and SAMtools. Bioinformatics 25(16):2078–2079. https://doi.org/10.1093/bioinformatics/btp352
12. Quinlan AR, Hall IM (2010) BEDTools: a flexible suite of utilities for comparing genomic features. Bioinformatics 26(6):841–842. https://doi.org/10.1093/bioinformatics/btq033

Check for updates

Chapter 19

MicroRNA Extraction from Extracellular Vesicles by Nanowire Devices

Taiga Ajiri, Kunanon Chattrairat and Takao Yasui

Abstract

Extracellular vesicles (EVs) are membrane-bound particles secreted by various cell types; EVs are found abundantly in body fluids, and they play significant roles in intercellular communication. Among the molecules contained in EVs, microRNAs are key regulators of gene expression, and they have been implicated in numerous diseases, including neurodegenerative disorders and cancers. Conventional methods for miRNA analysis, such as ultracentrifugation and immunoaffinity-based techniques, face challenges including the need for large sample volumes and low yield efficiency. To address these issues, our group has developed two types of innovative nanowire devices (inorganic nanowires composed of zinc oxide (ZnO) and organic nanowires composed of cellulose nanofibers) for the efficient capture and analysis of EV-derived miRNAs. In this chapter, we introduce the procedures for fabricating our inorganic and organic nanowire devices, as well as the methods for analyzing miRNAs extracted from EVs using the fabricated devices.

Keywords MicroRNAs, Extracellular vesicles, Nano device, Microfluidic device, Nanowire, Cellulose nano fiber

1 Introduction

Extracellular vesicles (EVs) are membrane vesicles secreted by nearly all types of cells, and they are abundant in body fluids such as blood and urine [1–8]. These EVs are primarily composed of lipids, nucleic acids, proteins, metabolites, and cytoplasmic components derived from the host cells. Once released, EVs are distributed in extracellular spaces, such as body fluids, where they can be taken up by other cells. Therefore, EVs facilitate both short-range and long-range intercellular and intertissue communication. Among the molecules constituting EVs, microRNAs (miRNAs) that bind to mRNA and regulate its transcription play a significant role in intercellular communication. miRNAs derived from EVs have been implicated in the progression of various diseases, and their types and abundances have been reported to function as diagnostic tools. In the field of oncology,

Archana Dhasarathy and Motoki Takaku (eds.), *Cancer Epigenetics*, Methods in Molecular Biology, vol. 3005, https://doi.org/10.1007/978-1-0716-5072-1_19,

EVs have been shown to be involved in multiple processes such as tumor initiation and progression, epithelial-mesenchymal transition (EMT), and suppression of antitumor response, tumor invasion, and angiogenesis, with miRNAs being validated as biomarkers for cancer diagnosis and prognosis prediction [8]. Therefore, analyzing miRNAs in EVs can be leveraged to understand mechanisms of intercellular interactions and for disease diagnosis.

The analysis of miRNAs in EVs typically involves ultracentrifugation or immunoreaction-based separation and enrichment from body fluids or cell culture media, followed by next-generation sequencing (NGS). However, these methods face challenges such as their large sample volume requirement and their low yields [9]. Our group has developed a method for the separation and enrichment of EVs. Our devices employ inorganic or organic nanowires to address these issues.

The inorganic nanowires efficiently capture EVs through electrostatic interactions between the EVs and the surface charges of the nanowires (Fig. 1) [10–16]. The inorganic nanowire device consists of a resin substrate with microchannels and a substrate fabricated with zinc oxide (ZnO) nanowires. This device can efficiently capture EVs from cell culture media and body fluids simply by introducing the liquids into the device. The combination of EV separation and recovery using the inorganic nanowire device and content analysis using NGS has led to the successful detection of tumor-derived molecules, with high recovery performance enabling the detection of trace molecules that could not be identified by conventional methods [10–13, 16].

The organic nanowires capture EVs through pores formed by a cellulose nanofiber network [17]. This device can recover EVs even from minute quantities of body fluids exuded on tissue surfaces. Conventionally, EVs have been recovered from mixed body fluids derived from multiple organs, but this device enables the recovery of independent and locally concentrated EVs with spatial information.

In this chapter, we introduce the procedures for the fabrication of our inorganic and organic nanowire devices and methods for the analysis of miRNAs extracted from EVs using the fabricated devices.

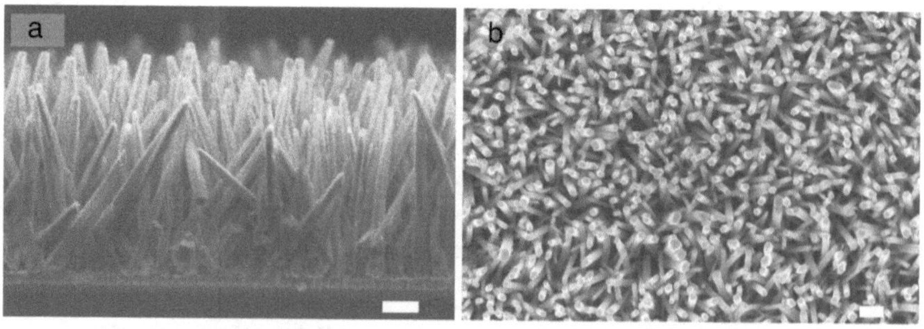

Fig. 1 Field emission scanning electron microscopy images of fabricated nanowires. (**a**) Cross-sectional view. (**b**) Overview. Scale bar: 300 nm

2 Materials

Prepare all solutions using ultrapure water (prepared by purifying deionized water, to attain a sensitivity of 18 MΩ-cm at 25°C) and analytical grade reagents. Prepare 0.22 μm-filtered Phosphate-buffered saline (PBS) (pH 7.4 or pH 7.2) for washing, dilution, etc. Prepare the lysis buffer as 1% Triton X 100 in PBS. The glass substrates were of a grade applicable for lithography technology (e.g., Ra <5 nm). The targets used for sputtering are of the grade used in semiconductor processing. Prepare the RT reaction mix and the PCR reaction mix for RT-PCR reaction and 3′ and 5′ adapters, and the PCR Purification Kit for sequencing of miRNA.

3 Methods

3.1 Fabrication of the Inorganic Nanowire Device

3.1.1 Fabrication of ZnO Nanowire Substrate

1. The glass substrate was cleaned by immersing it in piranha solution, a mixed solution (1:4) of H_2O_2 and H_2SO_4, at 180°C for 2 h. The piranha solution was prepared in a glass container, and heating was done on a hot plate.
2. The substrate was then washed using ultrapure water and dried in flowing nitrogen gas.
3. Photoresist was coated on the substrate using a spin coater (*see* **Note 1**).
4. The coated substrate was baked on a hot plate and then exposed through a microchannel photomask to UV radiation (*see* **Note 2**).
5. The substrate was developed until the pattern appeared.
6. A ZnO layer was deposited on the patterned substrate using a radio-frequency sputter.
7. The ZnO nanowires (NWs) were grown by the hydrothermal method at 95°C for 3 h using 50 mL of growth solution containing 50 mM zinc nitrate hexahydrate ($Zn(NO_3)_2 \cdot 6H_2O$) and 50 mM hexamethylenetetramine (HMTA, $C_6H_{12}N_4$).
8. The patterned NW substrate was cleaned with acetone to remove the remaining photoresist.
9. The patterned substrate was washed with ultrapure water and dried in flowing nitrogen gas.

3.1.2 Fabrication of ZnO Nanowire Device

1. The template substrate, which had a microchannel (width, 2.5 mm; length, 19 mm; and height, 50 μm) with a 12 μm-high herringbone structure, was fabricated by conventional photolithography.

2. The solution to form the polydimethylsiloxane (PDMS) substrate was prepared by mixing PDMS and curing agent in a ratio of 10:1.
3. The mixed solution was poured into the herringbone structure mold and kept in a vacuum chamber for approximately 1 h to eliminate any air bubbles.
4. The filled herringbone structure mold was cured at 80°C for 2 h, with the heating conducted either on a hot plate or in a furnace.
5. After curing, the herringbone-structured PDMS substrate was obtained by cutting and removing it from the mold.
6. The surface of the patterned NW substrate and the herringbone-structured PDMS substrate were subjected to a plasma treatment and then bonded together.
7. This bonded device was heated on a hot plate at 180°C for 3 min to get strong bonding.
8. The herringbone-structured PDMS device was connected to polyether ether ketone (PEEK) tubes for an inlet and outlet (Fig. 2) (*see* **Note 3**).

3.2 Fabrication of the Organic Nanowire Device

1. An aqueous dispersion of cellulose nanofibers (0.2 wt %, 200 mL) was dewatered by suction filtration through a membrane filter (hydrophilic polytetrafluoroethylene (PTFE) membrane, pore diameter of 0.1 μm) (*see* **Note 4**).
2. Tert-butyl alcohol (t-BuOH) (200 mL) was then gently poured onto the nanofibers collected on the filter before the wet sheet of nanofibers was peeled off the filter. This gave a cellulose nanofiber sheet with loosely packed nanostructures.
3. The cellulose nanofiber sheet was then hot-press dried at 110°C for 15 min (1.1 MPa) to completely remove the t-BuOH, resulting in a sheet with densely packed nanostructures (*see* **Note 5**).

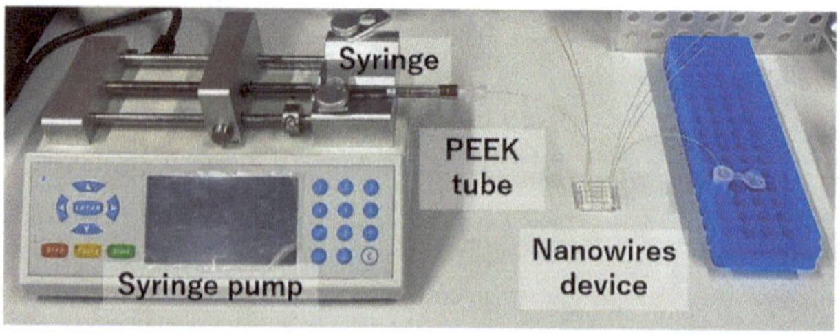

Fig. 2 Example of a liquid flow system using a syringe pump

4. The sheets were cut to the desired size using a cutter or puncher to form the organic nanowire devices (*see* **Note 6**).

3.3 Free miRNA and EV-Derived miRNA Recovery Using the Inorganic Nanowire Device

In the recovery experiment, a syringe pump or other pump can be used to flow liquid to the fabricated device.

1. The samples were introduced and captured with a flow rate of 5 μL/min (*see* **Note 7**).
2. EVs and free miRNAs were captured on the nanowires and the device microchannel was washed using 50 μL PBS (pH 7.4) with a flow rate of 10 μL/min.
3. The thermal step was carried out for the recovery of free miRNAs by supplying 50 μL PBS (pH 7.4) at a flow rate of 5 μL/min while heating the device at 95°C (*see* **Note 8**).
4. The extraction step was carried out using 50 μL lysis buffer at a flow rate of 5 μL/min.
5. The RNA was extracted using the general nucleic acid extraction method (*see* **Note 9**).

3.4 EV-Derived miRNA Recovery Using the Organic Nanowire Device

1. A 10 μL volume of sample was introduced into an organic nanowire device (10 × 10 mm) (*see* **Note 10**).
2. The organic nanowire device was stored and allowed to dry at room temperature for 7 days.
3. The dried organic nanowire device was immersed for 10 s in a Petri dish containing 1 mL of PBS (pH 7.2).
4. The RNA was extracted by spin column chromatography or another separation method (*see* **Note 9**).

3.5 Analysis of Recovered RNA

3.5.1 Quantification of EV-Derived miRNA by Quantitative Reverse Transcription PCR (RT-qPCR) (see Note 11)

1. The RT reaction mix was prepared, and the reverse transcription reaction was run.
2. The PCR reaction mix was prepared, and the real-time PCR was run.

3.5.2 Microarray Analysis of miRNA Expression

1. Purified miRNA was analyzed for miRNA profiling using a microarray (*see* **Note 12**).
2. The results obtained were subject to careful statistical analysis.

3.5.3 Sequencing of miRNA

1. miRNA ligation reactions with 3′ and 5′ adapters were conducted (*see* **Note 13**).

2. The index libraries were added to ligated miRNA solutions, and the mixed miRNA solutions were amplified using the RT-PCR technique (*see* **Note 13**).
3. The PCR products were purified using a commercial PCR Purification Kit (*see* **Note 14**).
4. DNA fragments corresponding to 140–160 bp (the length of small noncoding RNA plus the 3′ and 5′ adapters) were recovered using a gel electrophoresis technique (*see* **Note 15**).
5. After obtaining the DNA fragments, their concentrations were measured (*see* **Note 16**).
6. Single-end reads were analyzed on a DNA sequencer (*see* **Note 17**).

4 Notes

1. OFPR8600LB (Tokyo Ohka Kogyo) or similar photoresist is recommended for use. For OFPR8600LB, the photoresist was coated on the substrate using the spin coater operated at 500 rpm for 10 s and then at 1500 rpm for 120 s.
2. For our fabrication conditions, the length and width of the microchannel were 19 and 2.5 mm, respectively.
3. The sizes of capillaries commonly used for flowing sample solution into microfluidic channels are satisfactory, but we adopted the following sizes: 0.5 mm (outside diameter), 0.26 mm (inside diameter), 10 cm (length).
4. For our conditions and purposes, an ultrasonic homogenizer was used to facilitate the dispersion of the cellulose nanofiber solution.
5. Overnight freeze-drying methods can also be used instead of the hot-press method.
6. For our conditions and purposes, 10 × 10 mm size nanofiber sheets were used for capturing EVs.
7. Any sample volume can be used. Usually, we process volumes ranging from 50 to 1000 μL in a single experiment.
8. For our conditions and purposes, a Peltier temperature controller was used to control the heating of the NW substrate.
9. For our conditions and purposes, the microRNA purification kit (Norgen Biotek Corp.) and its manufacturer's instructions were used.

10. There are no size restrictions, but under our conditions, the nanofiber sheets were cut to the stated size and the experiment was carried out in the stated solution volumes. If the solution accumulated on the sheet in droplet form, it was then evenly spread on the sheet using a micropipette tip.
11. A general quantitative reverse transcription polymerase chain reaction, also called RT-qPCR, can be performed. The RT reaction mix and PCR reaction mix are prepared according to the sequence of interest for quantification. For instance, we have used the following in our previous work: TaqMan® Fast Advanced Master Mix (Thermo Fisher Scientific); and TaqManTM Advanced miRNA Assay (Assay ID, 478490_mir for hsa-miR-200a-3p; Assay ID, 477963_mir for hsa-miR-200b-3p; Assay ID, 002300 for hsa-miR-200c; Assay ID, 000470 for hsa-miR-148a; Assay ID, 477849_mir for hsa-miR-429; Assay ID' 478696_mir for hsa-miR-1299; and Assay ID, 001093 for RNU6B (Thermo Fisher Scientific)).
12. For our conditions and purposes, Toray's 3D-Gene (Toray Industries) human miRNA chips were used for miRNA expression profiling. We analyzed 15 μL of purified miRNA for miRNA profiling using a microarray and the 3D-Gene Human miRNA Oligo chip ver.21 (Toray Industries).
13. For our conditions and purposes, the NEBNext Multiplex Small RNA Library Prep Set for Illumina (New England Biolabs) was used for the preparation of small RNA libraries.
14. For our conditions and purposes, the QIAquick PCR Purification Kit (Qiagen) was used for the PCR products to purify the DNA fragments.
15. For our conditions and purposes, 6% Novex™ TBE gel (Thermo Fisher Scientific) (120 V, 60 min) was used to recover DNA fragments.
16. For our conditions and purposes, a Qubit dsDNA HS Assay Kit (Thermo Fisher Scientific) and a Qubit 2.0 Fluorometer (Thermo Fisher Scientific) were used to measure the concentrations of DNA fragments.
17. For our conditions and purposes, a MiSeq Reagent kit v3 (Illumina) and an Illumina MiSeq (Illumina) were used for the DNA sequencer. For our conditions and purpose, the expression levels of miRNAs between the two groups can be compared (Fig. 3).

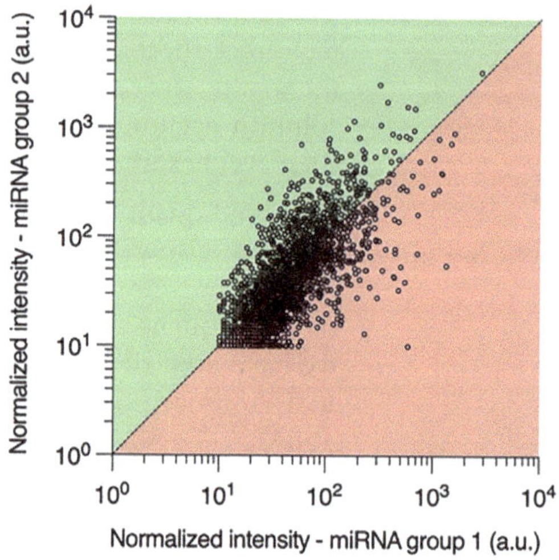

Fig. 3 Scatterplot image of normalized intensities of two miRNA groups. The boundary in each figure represents the same level of miRNA expression for the two groups. Each point corresponds to a different miRNA species

Acknowledgments

This work was supported by the Japan Agency for Medical Research and Development (AMED) Grant No. JP21he2302007, the Moonshot Research and Development Program (Grant Nos. 22zf0127004s0902 and JP22zf0127009) from the AMED, the New Energy and Industrial Technology Development Organization (NEDO) JPNP20004, the Japan Science and Technology Agency (JST) AIP Acceleration Research (JPMJCR23U1), and the JSPS Grant-in-Aid for Scientific Research (A) 24H00792.

References

1. El Andaloussi S, Mäger I, Breakefield XO, Wood MJA (2013) Extracellular vesicles: biology and emerging therapeutic opportunities. Nat Rev Drug Discovery 12(5):347–357. https://doi.org/10.1038/nrd3978
2. Tkach M, Théry C (2016) Communication by extracellular vesicles: where we are and where we need to go. Cell 164(6):1226–1232. https://doi.org/10.1016/j.cell.2016.01.043
3. Kumar MA, Baba SK, Sadida HQ, Marzooqi SA, Jerobin J, Altemani FH, Algehainy N, Alanazi MA, Abou-Samra A-B, Kumar R, Al-Shabeeb Akil AS, Macha MA, Mir R, Bhat AA (2024) Extracellular vesicles as tools and targets in therapy for diseases. Signal Transduct Target Ther 9(1). https://doi.org/10.1038/s41392-024-01735-1
4. Huang X, Yuan T, Tschannen M, Sun Z, Jacob H, Du M, Liang M, Dittmar RL, Liu Y, Liang M, Kohli M, Thibodeau SN, Boardman L, Wang L (2013) Characterization of human plasma-derived exosomal RNAs by deep sequencing. BMC Genomics 14(1):319. https://doi.org/10.1186/1471-2164-14-319
5. Mohr A, Mott J (2015) Overview of microRNA biology. Semin Liver Dis 35(1):003–011. https://doi.org/10.1055/s-0034-1397344
6. O'Brien K, Breyne K, Ughetto S, Laurent LC, Breakefield XO (2020) RNA delivery by extracellular vesicles in mammalian cells and its applications. Nat Rev Mol Cell Biol 21(10):585–606. https://doi.org/10.1038/s41580-020-0251-y

7. Yu Y, Hou K, Ji T, Wang X, Liu Y, Zheng Y, Xu J, Hou Y, Chi G (2021) The role of exosomal microRNAs in central nervous system diseases. Mol Cell Biochem 476(5):2111–2124. https://doi.org/10.1007/s11010-021-04053-0
8. Sun Z, Shi K, Yang S, Liu J, Zhou Q, Wang G, Song J, Li Z, Zhang Z, Yuan W (2018) Effect of exosomal miRNA on cancer biology and clinical applications. Mol Cancer 17(1). https://doi.org/10.1186/s12943-018-0897-7
9. Momen-Heravi F, Balaj L, Alian S, Mantel P-Y, Halleck AE, Trachtenberg AJ, Soria CE, Oquin S, Bonebreak CM, Saracoglu E, Skog J, Kuo WP (2013) Current methods for the isolation of extracellular vesicles. Biol Chem 394(10):1253–1262. https://doi.org/10.1515/hsz-2013-0141
10. Yasui T, Yanagida T, Ito S, Konakade Y, Takeshita D, Naganawa T, Nagashima K, Shimada T, Kaji N, Nakamura Y, Thiodorus IA, He Y, Rahong S, Kanai M, Yukawa H, Ochiya T, Kawai T, Baba Y (2017) Unveiling massive numbers of cancer-related urinary-microRNA candidates via nanowires. Sci Adv 3(12):e1701133. https://doi.org/10.1126/sciadv.1701133
11. Yasui T, Paisrisarn P, Yanagida T, Konakade Y, Nakamura Y, Nagashima K, Musa M, Thiodorus IA, Takahashi H, Naganawa T, Shimada T, Kaji N, Ochiya T, Kawai T, Baba Y (2021) Molecular profiling of extracellular vesicles via charge-based capture using oxide nanowire microfluidics. Biosens Bioelectron 194:113589. https://doi.org/10.1016/j.bios.2021.113589
12. Kitano Y, Aoki K, Ohka F, Yamazaki S, Motomura K, Tanahashi K, Hirano M, Naganawa T, Iida M, Shiraki Y, Nishikawa T, Shimizu H, Yamaguchi J, Maeda S, Suzuki H, Wakabayashi T, Baba Y, Yasui T, Natsume A (2021) Urinary microRNA-based diagnostic model for central nervous system tumors using nanowire scaffolds. ACS Appl Mater Interfaces 13(15):17316–17329. https://doi.org/10.1021/acsami.1c01754
13. Yokoi A, Ukai M, Yasui T, Inokuma Y, Hyeon-Deuk K, Matsuzaki J, Yoshida K, Kitagawa M, Chattrairat K, Iida M, Shimada T, Manabe Y, Chang IY, Asano-Inami E, Koya Y, Nawa A, Nakamura K, Kiyono T, Kato T, Hirakawa A, Yoshioka Y, Ochiya T, Hasegawa T, Baba Y, Yamamoto Y, Kajiyama H (2023) Identifying high-grade serous ovarian carcinoma–specific extracellular vesicles by polyketone-coated nanowires. Sci Adv 9(27). https://doi.org/10.1126/sciadv.ade6958
14. Chattrairat K, Yasui T, Suzuki S, Natsume A, Nagashima K, Iida M, Zhang M, Shimada T, Kato A, Aoki K, Ohka F, Yamazaki S, Yanagida T, Baba Y (2023) All-in-one nanowire assay system for capture and analysis of extracellular vesicles from an ex vivo brain tumor model. ACS Nano 17(3):2235–2244. https://doi.org/10.1021/acsnano.2c08526
15. Sitthisuwannakul K, Sukthai R, Zhu Z, Nagashima K, Chattrairat K, Phanthanawiboon S, Klamchuen A, Rahong S, Baba Y, Yasui T (2024) Urinary dengue NS1 detection on Au-decorated ZnO nanowire platform. Biosens Bioelectron 254:116218. https://doi.org/10.1016/j.bios.2024.116218
16. Chattrairat K, Yokoi A, Zhang M, Iida M, Yoshida K, Kitagawa M, Niwa A, Maeki M, Hasegawa T, Yokoyama T, Tanaka Y, Miyazaki Y, Shinoda W, Tokeshi M, Nagashima K, Yanagida T, Kajiyama H, Baba Y, Yasui T (2024) Discrimination of extracellular miRNA sources for the identification of tumor-related functions based on nanowire thermofluidics. Device 2(6):100363. https://doi.org/10.1016/j.device.2024.100363
17. Yokoi A, Yoshida K, Koga H, Kitagawa M, Nagao Y, Iida M, Kawaguchi S, Zhang M, Nakayama J, Yamamoto Y, Baba Y, Kajiyama H, Yasui T (2023) Spatial exosome analysis using cellulose nanofiber sheets reveals the location heterogeneity of extracellular vesicles. Nat Commun 14(1):6915. https://doi.org/10.1038/s41467-023-42593-9

Index

Archana Dhasarathy and Motoki Takaku (eds.), *Cancer Epigenetics*, Methods in Molecular Biology, vol. 3005, https://doi.org/10.1007/978-1-0716-5072-1

J

L

M

N

P

Q

R

S

T

Kolibri 360 GmbH
Ferdinand-Jühlke-Straße 7
99095 Erfurt, Deutschland
produktsicherheit@kolibri360.de